INDUSTRIAL LOW BACK PAIN

Contemporary Litigation Series

Consulting Editors for the
CONTEMPORARY LITIGATION SERIES

Professor Stephen A. Saltzburg
University of Virginia Law School

Professor Kenneth R. Redden
University of Virginia Law School

INDUSTRIAL LOW BACK PAIN

A Comprehensive Approach

SAM W. WIESEL, M.D.
Professor of Orthopaedic Surgery
The George Washington University
Medical Center
Washington, D.C.

HENRY L. FEFFER, M.D.
Professor of Orthopaedic Surgery
The George Washington University
Medical Center
Washington, D.C.
Director of Medical Research
Health Care Systems, Inc.

RICHARD H. ROTHMAN, M.D., PH.D.
Professor of Orthopaedic Surgery
University of Pennsylvania
Philadelphia, Penna.
Chief of Orthopaedic Section
Pennsylvania Hospital
Philadelphia, Penna.

THE MICHIE COMPANY
Law Publishers
CHARLOTTESVILLE, VIRGINIA

We dedicate this book to Barbara, Daisy, and Marsha,
who make it all worthwhile.

TABLE OF CONTENTS

TABLE OF CONTENTS

TABLE OF CONTENTS

ix

TABLE OF CONTENTS

TABLE OF CONTENTS

TABLE OF CONTENTS

TABLE OF CONTENTS

TABLE OF CONTENTS

TABLE OF CONTENTS

PREFACE

Low back injuries in industry have become a tremendous problem in terms of both health care and financial cost in the United States. The topic is a very emotional one because a large number of the incidents relate to soft tissue trauma which does not always show up objectively. Thus, when a patient's subjective complaint of low back pain is not confirmed by hard findings, a confrontation between the employee and employer is not unusual and the inevitable drawn out claims and legal process can become quite costly and time consuming.

Physicians, lawyers, insurance adjusters, company management, rehabilitation specialists and administrative law judges have to interact so as to assure quality care for the injured worker, while at the same time protecting all parties from abuse. Each of these professionals has a different interest in both the process and the outcome. Unfortunately, there has been no single organized source which brings together all of the needs and concerns of the participants so that the solution can be achieved with appreciation and understanding by all.

The purpose of this text is to create a forum for a group of noted professionals so that each can address his particular area of expertise in such a way as to improve interdisciplinary communication. Thus, the medical section is structured to be comprehensible to lawyers and claims adjusters; whereas, the legal chapter is directed toward the physicians. A major effort has been made to present the consensus of low back thinking in a standardized fashion so that this book can be used as a desk top reference as well as a source book.

Each author has covered his area from a practical point of view based upon the most recent information available. It should prove useful and informative for all professionals who deal with industrial low back pain. The editors are most appreciative of each contribution and are very proud of the final text.

CONTRIBUTORS

William I. Bauer, M.D.
Medical Director
Potomac Electric Power Company
Washington, D.C.

David G. Borenstein, M.D.
Rheumatologist
Associate Professor of Medicine
The George Washington University
Washington, D.C.

Rene Cailliet, M.D.
Professor of Physical Medicine and Rehabilitation
University of California
Director of Physical Medicine and Rehabilitation
Santa Monica Hospital
Los Angeles, Calf.

James L. DeMarce
Associate Director
Division of Coal Mine Workers' Compensation
Office of Workers' Compensation Program
Employment Standards Administration
United States Department of Labor
Washington, D.C.

Henry L. Feffer, M.D.
Professor of Orthopaedic Surgery
The George Washington University
Director of Medical Research
Health Care Systems, Inc.
Washington, D.C.

Gerald Herz
Partner in the Law Firm of:
Friedlander, Misler, Friedlander, Sloan & Herz
Washington, D.C.
Practitioner of workers' compensation law in the District of
Columbia and Maryland since 1964.

CONTRIBUTORS

Leslie D. Michelson
Executive Vice-President
Health Care Systems, Inc.
Washington, D.C.

James L. Mueller
President
Mueller and Zullo, Inc., Alexandria, Va., a consulting firm
managing disability in the workforce. Has had years of research
experience in solving the functional problems which inhibit
disabled persons from entering and returning to employment.

Lorenz K. Y. Ng, M.D.; Gregory A. Grinc, Ph.D.; Janice P. Pazar,
R.N.; and Marianne S. Hallett, M.A., C.R.C.
Director and Staff
Washington Pain Clinic
Washington, D.C.

Kevin M. Quinley
Supervisor and Claims Adjuster
Crawford and Company
Fairfax, Va.

Richard H. Rothman, MD., Ph.D.
Professor of Orthopaedic Surgery
University of Pennsylvania
Chief of Orthopaedic Section
Pennsylvania Hospital
Philadelphia, Penna.

Sam W. Wiesel, M.D.
Professor of Orthopaedic Surgery
The George Washington University
Washington, D.C.

Ronald J. Wisneski, M.D.
Assistant Professor of Orthopaedic Surgery
University of Pennsylvania
Philadelphia, Penna.

CHAPTER 1

SCOPE OF INDUSTRIAL LOW BACK PAIN

William I. Bauer, M.D.

§ 1-1. Prevalence and Incidence.

Low back pain as a health and economic problem unquestionably affects a large proportion of the population

1

of industrialized nations. Since the enormous scope of this problem as an occupational health issue has only recently been recognized, however, no statistics on either the incidence or prevalence of occupational back pain are readily available. To accurately judge the significance of the problem in our society, at least two pieces of data need to be evaluated. The first is the prevalence of the condition; that is, how many individuals at some time in their lives have had attacks? An equally and perhaps even more significant question relates to incidence. How many individuals have this problem in any specified period of time?

The most accurate information has been compiled in Scandinavia, Great Britain, and the United States, and by using these data we are at least able to estimate the frequency of back pain, its potential impact on industry in particular, and on society as a whole. The prevalence of back pain in workers in different Scandinavian industries has been reported by several investigators. In 1954 Hult (see entry 25 in Bibliography at end of this chapter) found that 60% of a sample of Swedish males with different jobs had at some time suffered from back symptoms. In another Swedish general population study (entry 26), lifetime prevalence rates for low back pain in men were found to be between 68% and 70%. In women, the prevalence rate varied between 62% and 81%, with a significant rise as age increased. These and a number of other retrospective studies (entries 13, 23, 24, 56, 59) in Sweden have demonstrated general agreement that between 50% and 80% of adults in that country have at some time suffered back pain.

The incidence rate has also been subject to review. Gyntelberg (entry 22), in a survey in Denmark, found that about 25% of employed males in his study group reported that they had had low back pain during the previous year. In Sweden the one-year incidence rate was found to be similar, with first attacks of low back pain having been found

2

to be highest among the 30-year olds and with a decreasing rate of attack in the older age groups.

In Great Britain, studies by Benn and Wood (entry 6) found that 3.6% of all sickness absent days in 1969-70 were due to back pain. A review of individuals visiting general practitioners showed that about 2% consulted a physician each year because of low back pain; this number increased to 5% between the ages of 50-59. Anderson (entry 2), in another British survey, found that 25% of working men were affected by low back pain each year.

In the United States, Kelsey and Pastides (entry 31) reported that painful back conditions are the most common reason for a decreased work capacity as well as a reduction in leisure time activities of people below the age of 45. Rowe (entry 50) collected data from a group of long term male employees at the Eastman Kodak Company just before their retirement. He was also able to review the medical records and work histories of these men who were involved in jobs requiring all levels of physical activity. Fifty-six percent of them had had severe enough low back pain to require medical care. He also found that the total time lost from work per year from back problems was second only to time lost from upper respiratory tract infections.

These studies and numerous other reports concerning back pain problems in the United States support the conclusion that between 50% and 80% of the adult American working population will suffer from low back pain at some time during their career. It is also estimated that about 25% of them will have low back pain at some time in any given year.

§ 1-2. Annual Rate.

Statistics on the annual rate of low back injuries, that is, the number per hundred workers per year, are useful for comparative purposes in respect to the relative incidence in different working populations. If recurrences are included, about 2% of all employees in the United States have compensable back injuries each year (entries 8, 47). However, as might be expected, the annual rate has been found to vary in different occupational environments. Snook (entry 53) compiled data from the Liberty Mutual Insurance Company which indicate that the annual rates for industrial workers can vary from less than 1% to over 15%, and the incidence among clerical and administrative personnel was found to be considerably less than 1%.

The annual rate of low back pain also varies greatly among different industries. This is reflected in data compiled by the State of California (entry 10). Table I, at end of this chapter, shows a statistical breakdown by industry of the disabling back strain injuries per hundred workers reported under Workers Compensation in California in 1979. To be included in these statistical data, the back injury cases had to involve absences from work for at least one full day or a shift beyond the day of the accident. Using these criteria, it should be noted that the annual rates given in Table 1 are considerably lower than those cited above, apparently for the following reasons: First, only cases requiring employees to lose time were reported; second, the statistics from all types of employees in that particular industry were included and the less vulnerable administrative and clerical employees would tend to suppress the composite annual rate.

Our own experience with employees in an industrial utility setting tends to support the conclusion that from 1% to 2% of industrial employees have job related back problems each year, and that the incidence for administrative and clerical employees is well below 1%. However, the evaluation of incidence data is particularly difficult because of the large number of possible variables. These include the prob-

4

lem of differentiating new injuries, the dilemma relating to the inclusion of work aggravation of non-job related problems, and the pressures to include non-job related back injuries as occurring at work because of a favorable reimbursement potential.

Over the past few years we have kept a close track on the type of complaints which are most often associated with lost time time following job related injuries in our company. We have carefully analyzed each case and have found that between 40% and 45% of all lost work days can be attributed to back pain. This includes new injuries, aggravation of old injuries, and work related aggravation of personal injuries. These figures have helped emphasize the magnitude of the social and economic impact that this problem has on the utility industry as well as on industry in general.

An observation which tends to cast some doubt on the validity of available incidence data relates to the fact that in industries providing a wide range of health and medical services to their employees, irrespective of the causative factor, that is, where medical services are provided whether the injury is work related or personal, the number of work related back injuries has been found to be well below the 1% level. This can be explained either by the fact that under these circumstances fewer legitimate job related accidents are being reported, or else a significant number of back injuries that are reported as occupationally induced when there is a potential monetary advantage are in fact injuries that occurred in a non-occupational setting, but are reported as work related.

At the present time, what does this mean for industry in the United States? Based on estimates derived from the Bureau of Labor Statistics Annual Survey of Injuries and Illness, about 1,000,000 workers suffered back injuries in 1980, accounting for almost one out of every five injuries and illnesses in the workplace.

5

§ 1-3. Economic Impact.

Another perspective can be obtained by reviewing this problem from an economic standpoint, but here too there is a lack of reliable data. Akeson and Murphy (entry 1) estimated that approximately $14 billion was spent in the United States in 1976 on treatment and compensation for low back injuries, and this figure exceeded all other industrial injuries combined. By projecting these data into 1983 on the basis of the consumer price index for all urban medical care, this figure would now be $25.25 billion. In view of the fact that medical costs, technological advancement, and utilization in this area appear to have been greater than the general average for health care because of the widespread use of costly procedures such as CAT scans, myelograms, EMG's, and other costly diagnostic tests; and with the tremendous escalation of compensation costs, these dollar estimates in all likelihood are significantly below the actual current cost.

Another way of looking at costs is by reviewing workers compensation budgets, particularly in companies that are self insured. A review of industrial compensation costs of a number of self insured utility companies shows compensation expenditures in 1983 to be over four times that of 1976. One would expect costs of low back injuries to have increased proportionally and it can also be estimated to be over four times the 1976 expenditure. The total projection from these two methods puts direct costs for back injuries in industry in the range between $25.25 and $56 billion.

§ 1-4. Cost Distribution.

The cost distribution of back injuries has been analyzed by Leavitt, Johnston, and Byer (entry 37). They found that when the disability lasted nine months or less, the average total medical cost of back injuries was no different from that associated with any other injury of the same duration. However as additional time elapsed, back injuries became more expensive than other injuries. Forty-five percent of the total cost related to permanent disability payments, 22% to temporary disability payments, and 33% to medical expenses. An analysis of the distribution of medical costs showed that one-third were physician fees, one-third hospital bills, 7% drugs, 5% appliances, 9% physical therapy, and 12% diagnostic tests.

A high percentage of these costs was found to be concentrated in relatively few cases. Snook (entry 53) reported on data from the Liberty Mutual Insurance Company which showed that 25% of the cases accounted for 90% of the cost. Leavitt, Johnston, and Byer (entry 37) found that in 1971, the total average cost per case was $2,197 but the median total was $404. While these dollar figures appear low by today's standards, the relationship between the figures is very likely still relevant. He found that only 22% of the cases had a total cost above the average. As might be expected, the high cost cases were those involving hospitalization, surgery and litigation.

Attempts have been made to try to predict which types of back accidents will result in prolonged illness and high cost. One might expect that soon after the injury the treating physician should be able to make such a determination, but to date this has not been feasible. Leavitt, Johnston, and Byer (entry 38) screened a large number of low back injury cases with the hope of developing criteria to identify such high cost cases early in the course of the treatment. They found only three variables to be available soon enough to be helpful in predicting those cases which would eventually fall into the high cost category: diagnosis, history of previous back injury, and sex. Factors which were not predictive included the attending physician's prognosis for return to work, employment history, age, and type of accident or

7

injury. They found that injuries occurring in women are more expensive, a fact which has not been corroborated in other studies.

§ 1-5. Factors Affecting Incidence of Back Injuries.

§ 1-5(A). Occupation.

California workers compensation data show that truck drivers have the highest incidence of back injuries (entry 10). Statistics from Wisconsin (entry 63) and Connecticut (entry 30) tend to confirm this, and it has been postulated that the prolonged sitting involved in truck driving is an important contributing factor (entries 29, 43). It is of interest to note, however, that less than 10% of the truck drivers have their initial episode while actually driving the truck; more than half of their low back accidents occur while engaged in the heavy physical work associated with loading and unloading.

§ 1-5(B). Lifting.

In reviewing the general work related factors, Snook, Campanelli, and Hart (entry 55) found that manual handling tasks, that is, lifting, lowering, pushing, pulling, and carrying were implicated as the specific act or movement associated with low back pain in 70% of the injuries. A more detailed look has shown that a significant proportion of the lifting tasks begin lower than 30 inches from the ground, and these are responsible for 75% of the lifting injuries. In tracing the pattern of lifting and placing objects, it was found that approximately half of the chores required lifting objects from floor level.

The amount of lifting that might initiate back pain has also been subject to evaluation. A moderate amount of lifting is apparently tolerated reasonably well but very heavy lifting, especially if it is sudden and unexpected and requires a maximum amount of effort, has been found to be particularly harmful to the back.

§ 1-5(C). Object Weight.

With such a large number and variety of conditions being implicated, it is difficult to sort out any one specific factor as being responsible for low back pain; however, the weight of the object lifted is mentioned more frequently than any other. More than half of the injured workers in a recent study (entry 14) reported lifting objects of at least 60 pounds and 30% reported lifting weights of 100 pounds or more. They all blamed the size of the weight for their problem and three-quarters of them said that they had to do the job without assistance. In addition, almost half of the workers were found to have been handling objects heavier than usual when they were injured.

§ 1-5(D). Object Size.

A related factor appears to be the size or bulkiness of the object lifted. Large boxes and containers are more often implicated as a source of injury than are smaller metal or wooden objects. The actual time involved in moving them appears to be less important since the usual lifting time is rather short; two-thirds of the injured employees reported that they had moved or carried the objects for less than one minute. The number of lifts in a day also does not appear to be a significant contributing factor since 20% of the employees report their injury with the first lift of the day and another 20% were alleged to have made fewer than five lifts before their injury. Over half the employees reported making fewer lifts on the day of the accident than they normally do.

§ 1-5(E). Heavy Physical Work.

A number of investigations have shown an increased incidence of low back symptoms in individuals engaged in occupations which require what is considered to be overall heavy physical exertion. The available data on this topic, however, are conflicting since the correlation of back injuries with heavy work demands has not been a consistent finding. Although this might be explained by the fact that it is difficult to quantitate since there is no standardized definition of heavy physical work, a number of studies have failed to show any increase in back pain, even under the most arduous of conditions. All in all, the relationship between what is usually considered heavy work and low back problems is anything but clear cut.

§ 1-5(F). Bending.

Tasks which require working from a bent over position seem to carry an increased risk of back injury. Bending, along with twisting activities in particular, appears to be responsible for an increased number of incidents although these are difficult to implicate since they are usually associated with lifting.

§ 1-5(G). Vibration.

The role that constant vibration may play in back injuries is still not clear. Frymoyer (entry 19) found vibration to be the main common vocational factor associated with back pain. However as with other factors, it is usually difficult to separate vibration from other possible contributing causes. The same individuals who are subjected to prolonged vibration while driving trucks, taxis, and buses are also involved in prolonged sitting.

§ 1-5(H). General Physical Fitness.

The role of general physical conditioning in employees has also received a considerable amount of attention. Cady,

Bischoff, O'Connell, et al. (entry 9) performed a prospective investigation comparing strength and fitness measurements with the subsequent development of back injuries. They concluded that good physical fitness in fire fighters helped prevent back trouble; the least fit group was ten times more susceptible than the most fit group. Some investigators have found the strength of abdominal and back musculature to be a significant factor in the prevention of low back problems; however, others have seen no such specific association. The difficulty in measuring back and abdominal muscle strength may contribute to this variety in findings. At the present time, it is generally felt that as a contributing factor to back trouble, isolated trunk muscle weakness, while significant, may not be quite as important as poor general physical fitness and muscle strength. Poor general physical fitness is now being recognized in a high percentage of individuals involved in back pain accidents.

§ 1-5(I). Sex.

A number of careful studies have reviewed the sex incidence of employees involved in industrial low back accidents (entries 4, 35). General population reviews have found women to have as many low back complaints as men; however, these findings do not represent what is seen in the workplace. Although women represent 40% of the working population, they develop only 20% of the work related back trouble. There would appear to be several possible explanations for this disparity. Women have traditionally been employed in less physically demanding jobs and in job areas where there are fewer back accidents. Women truck drivers and material handlers are still relatively unusual. Even where job classifications are the same, women have often been in jobs that involve less strenuous activity, those that are considered safer from the accident standpoint.

11

Under these circumstances one would expect fewer back related problems; however, as was pointed out earlier, some studies have found that when women are injured and do suffer back pain, they tend to have longer periods of disability and are responsible for a larger proportion of the more costly injuries. With the current emphasis on placing an increasing number of women into non-traditional jobs, the number of back injuries in women can be expected to increase.

§ 1-5(J). Age.

Age has also been reviewed as a possible relevant factor in back injuries. Although low back problems may occur at any age in the working population, the incidence of injuries has been found to gradually increase from ages 20 to 45, and then remain unchanged or gradually decrease. The average age of employees losing time from work because of back injuries is 36, although this varies somewhat by occupation. As the employees' age increases over 45, the number of new accidents and time lost from work decreases. Possible explanations may be that older employees are more likely to have positions of seniority and therefore be involved in less strenuous activities, or that the older employee may have made personal modifications in the work environment so that he is less likely to be engaged in activities which produce back injuries. In some employment situations, employees over a specified age have been restricted from performing strenuous physical tasks in the belief that they would have fewer back problems; however, there are no statistical data that would confirm that this practice has reduced the number of back injuries.

§ 1-5(K). Psychological Factors.

The emotional factors relating to back pain, particularly in the compensation setting, have lately received increased attention. Psychological testing is ongoing and personality profiles are being seriously evaluated in these individuals. This element of low back pain is felt to be important not only in the diagnosis of the condition but also in its management. These factors will be discussed in depth in Chapter 6; however, it should be mentioned here that it is difficult to determine whether personality and psychopathologic characteristics, which are commonly seen in these patients with low back pain, are primary in the individual or secondary to the somatic condition. To help clear up this controversy, it will be necessary to perform intensive research on this aspect of the epidemiology of low back pain. A prospective study which compares data, such as that obtained from the Minnesota Multi-Phasic Personality Inventory (MMPI) or other appropriate psychologic test on a group of young individuals without low back pain, with a later analysis of these same individuals, some of whom will have developed back pain, would help clarify this question. No large scale studies have yet been reported; however, with the increasingly widespread use of the MMPI in pre-employment evaluation for a number of industries, such as nuclear power generating facilities, it may be possible to obtain such data in the future.

§ 1-5(L). Other Factors.

Frymoyer, Pope, Clements, et al. (entry 20) have recently reviewed a variety of other patient activities in an attempt to delineate potential risk factors. They found low back pain more prevalent in cigarette smokers. Individuals with severe back pain were more likely to be cigarette smokers and had a greater tobacco consumption as measured by both the number of cigarettes smoked per day and the number of years exposed. This correlation had previously been noted by Svensson (entry 57) who speculated that the increased incidence could be explained by an elevated intradiscal pressure in smokers. Other investigators have suggested that coughing and chronic bronchitis rather than smoking is an important etiologic factor in low back pain.

Recreational activities appear to also have some relationship to low back complaints. It has been observed that patients with severe low back pain tend to stop participating in sports between the adolescent and later years, while those with only moderate back pain continue their sports activities largely uninterrupted. There is uncertainty as to whether this behavioral change is the cause or the result of the back pain. It is known, however, that certain sports such as football and gymnastics are associated with up to a fourfold increase in spondylolysis.

Numerous other potential risk factors such as spine geometry, lumbar lordosis, sedentary life style, and obesity have been speculated upon, but in the absence of valid controlled studies, the vague impressions of their possible relation to low back pain must be considered anecdotal.

§ 1-6. Pre-Employment Considerations.

For many years industry has operated on the premise that if employees underwent a comprehensive screening process before being hired, work related accidents and injuries would be reduced. Therefore selection criteria for employment have been proposed to reduce the incidence of low back pain in the workforce. A wide variety of pre-employment practices have been attempted, particularly for positions which may potentially involve heavy physical labor.

§ 1-6(A). Pre-Employment History and Physical Examination.

The most commonly used selection procedure is a pre-employment medical history and physical examination.

14

This type of examination has been performed in a large number of industries for many years. Snook, Campanelli, and Hart (entry 55) analyzed insurance company data from companies which included pre-employment and pre-placement medical histories and physical exams. They could find no significant difference in the incidences of low back injuries in those companies that required examinations, when compared with those that did not. They all had essentially the same rates of back injuries.

Rowe (entry 50), however, in his review of injuries at Kodak, estimated that 10% of workers who will acquire low back disability can be identified on a pre-employment medical examination. He found the best predictor to be a history of back trouble, especially if the back problem occurred without a significant prior injury. He also noted that since the overall incidence of back pain is so high, efforts at prediction might more profitably be aimed at eliminating the individuals who are generally disability prone and therefore likely to lose excessive time with any type of injury, rather than concentrating on the back pain employees alone.

While Rowe's observations are of interest, it has been commonly observed that the person seeking employment is not likely to be candid with the examiner and volunteer a history of back trouble. Therefore, for the screening to be effective in identifying potential back pain problems, the examiner must take great care and follow up on any significant physical or historical finding. Snook, Campanelli, and Hart (entry 55) were unable to evaluate the thoroughness of the pre-employment examination, and therefore the validity of their conclusions is open to question. Before this time-honored and apparently, logical procedure is abandoned, more data are needed to see whether the incidence of back pain can be reduced when the assessments are made under carefully controlled conditions.

§ 1-6(B). Pre-Employment Spine X-Rays.

Pre-employment x-rays have been and continue to be used as a part of the pre-employment examination, particularly in those industries involved in material handling. There have been a number of studies (entries 21, 45, 52), however, that have conclusively shown that x-rays are of little value in identifying potential back problems. The American Occupational Medical Association (AOMA) has given the matter extensive study and has recommended the following guidelines: "Lumbar spine x-ray examinations should not be used as a routine screening procedure for back problems, but rather as a spinal diagnostic procedure available to the physician on appropriate indications for study."

In our review of spine x-rays of employees who have developed back pain, less than one percent of the x-rays revealed any abnormality that would have been detected on a pre-employment x-ray examination and this percent of abnormal films is not significantly different from x-ray findings of employees who have no back complaints. According to the AOMA guidelines, however, individuals with a suspicious history or physical examination would appropriately be selected for x-ray evaluation. The single most significant spinal abnormality which might be found on routine spinal x-rays would be spondylolysis, and even this has only questionable significance in a previously asymptomatic person since the likelihood of increased vulnerability in this group has yet to be substantiated. We would certainly concur with the recommendation that routine pre-employment spinal x-rays not be used since they are not medically justified, are not cost effective, and involve unnecessary radiation.

§ 1-6(C). Ultrasonic Measurements.

A more recently proposed screening procedure has

16

involved the ultrasonic measurement of the spinal canal diameter. MacDonald and his co-workers (entry 40) found, by using this non-invasive, safe procedure, that coal workers with the greatest morbidity from low back pain had narrower spinal canals. They concluded that pre-employment measurement of the spinal canal by ultrasound might identify the workers at greatest risk for back trouble. Feffer (entry 17), however, in an editorial comment following MacDonald's article, noted that these findings have yet to be reproduced by any other investigators, and until the technology can be further refined, the procedure should not be depended upon for general usage.

§ 1-6(D). Physical Ability Testing.

One area which appears to have more promise has been pre-employment strength testing. Various methods of screening physical ability so as to more effectively match job demands with the individual's work capacity have been proposed. Chaffin and his co-workers (entry 12) have demonstrated that specific strength testing using a battery of isometric strength measurements is effective in screening the employees to reduce the likelihood of back injuries. He has shown that a worker's susceptibility to a back injury increases significantly when the lifting requirements of the job approach or exceed the individual's strength capacity as determined by the testing. They also concluded (entry 34) that isometric strength testing was an effective and valid method of reducing back injuries when used in the employee selection process.

Even though pre-employment strength testing is not meant to exclude any one from the work force, there is concern that providing guidelines for job placement will do just that. Rather than being used to assist in determining if the individual is indeed capable of performing the job, there is fear by some that the guidelines will actually be used to reject capable applicants, and the application of this testing has been subjected to very close scrutiny by the legal profession. The possibility for discrimination, particularly against female applicants in the non-traditional workplace, must be carefully considered before any testing program is undertaken. Women are now beginning to compete for physically demanding jobs in increasing numbers and are more frequently found in traditionally male dominated industrial positions. They have less experience in physically demanding activities, however, and given inherent biological differences, the average woman today is not as strong as the average man. The testing program must therefore be non-discriminatory and should be strictly designed to test for the job under consideration.

For a screening program to be effective, there must be a very careful job analysis and the physical requirements of each position must be carefully delineated. Job analysis consists of defining the job and determining just what is required from the employee. This includes identifying the physically demanding tasks which must be performed and determining just what physical prerequisites the employee must have. It has been shown that the occupational demands of one job may not be transferable to another working position even though the tasks involved may be considered similar.

Once the analysis has been completed, a battery of physical performance tests which are linked to the job requirements must be developed. These must then be validated and a predictive rating for the tests developed. Each of these steps may be long, expensive, and time consuming, and only after they have been taken, can such a pre-employment screening procedure be successfully implemented. Even then, many industries have still been reluctant to start such programs because of the potential legal liability. Such concerns have related to risk of injury during the testing process, legal requirements for access to the handicapped, and explosive sex discrimination issues. Thus, even though pre-employment strength testing would appear on the face of it to offer a significant opportunity for industrial cost savings, the development of these programs has been constrained by all of these concerns as well as by high implementation costs.

§ 1-7. Prevention of Occupational Back Injuries.

§ 1-7(A). Training of Workers.

For many years a great deal of time, money, and dedication has gone into teaching employees the correct way to lift. Safety departments have been instructing employees to lift with their legs while keeping their backs straight, and this method has received wide acceptance by organizations such as the National Safety Council. Unfortunately, there have been many studies which cast serious doubt as to the effectiveness of such training. Snook, Campanelli and Hart (entry 55) found just as many back injuries among employees who had been through this type of safety training program as in those employees who had not, and other studies have tended to verify its ineffectiveness. It has also been generally observed that even though employees are instructed in appropriate lifting techniques, and may continue to use these methods for a short period of time, they quickly revert to previous bad habits unless the training is updated frequently. Given the widespread demands for a broad array of safety programs, this constant reinforcement rarely occurs.

§ 1-7(B). Job Design Adjustments.

The redesign of the workplace has been under close

scrutiny of late since there have been a number of studies which have shown that relatively minor design changes can significantly reduce the number of back injuries. In Chapter 8 we will discuss in greater detail the dynamic factors involved as well as the projected outcome; however, it should be pointed out that while many of the changes appear to be simple and inexpensive, their implementation in the workplace, particularly when work patterns have been established for many years, can be quite frustrating to management and labor alike. What appears on the surface to be a minor environmental adjustment may be interpreted by employees, management, or unions as a rule change which requires union approval, job reclassification, change in duty status, or workers' pay adjustment. Thus what, on the face of it, would appear to be a cut and dried issue can very well end up in a complex management negotiation.

Another potential problem relates to the equipment itself. Much of the machinery now in use has been designed for larger, stronger males and is unsuited for smaller women with less strength. Lever arms may be at a height or distance that shorter women cannot reach. Tools may be positioned on high shelves or require strength beyond the capacity of the smaller female. Exhaustion and accidents, as would be expected, are not an unusual outcome. Much of this equipment, however, has been functioning satisfactorily for many years, and there is a natural reluctance to alter or discard it.

§ 1-8. Recovery From Low Back Injuries.

Employees will continue to have accidents that result in low back pain, even with the best of physical conditioning programs, comprehensive pre-employment screening, and meticulous safety precautions. The injured employee's return to work date has been widely used to measure the rate of recovery from low back trauma in the industrial environment, and these data are often quoted in support of the success or failure of various back rehabilitation programs. In spite of a number of qualifying circumstances which should be considered when using these data, they still have been used by a number of investigators as a measure of programmatic success. Bond (entry 8) reported on a study of 226 randomly selected low back cases in Ohio and found that 90% lost less than one week from work. White (entry 60), in an analysis of 12,000 compensation low back cases in Ontario, found that more than 80% were back at work in less than three weeks and 90% in less than six weeks. These and many other studies tend to confirm the impression that most employees with acute back problems return to work in a relatively short time.

There are a number of factors, besides the severity of the accident and the subsequent injury, which affect the return to work date. Individual company policies concerning the availability of appropriate work for employees who are not fully recovered are potentially important limiting factors. The availability of a limited, light, or restricted work environment is frequently mentioned as one of the most significant factors in the encouragement of a rapid return to work. Although there is little published information to support this conclusion, a reduction in time lost due to back injuries has been apparent in work situations where a job modification permits the employee to return to less than full regular duty status. When it is possible to assign the injured employee to a job where he or she is temporarily protected from heavy lifting and repetitive bending after a low back injury, the length of time that employees are out of work can be significantly reduced; if nothing but heavy physical work is available, however, lost time is usually unnecessarily prolonged.

The time away from work in itself may have a significant deleterious effect on the worker's capacity of ever returning to productive employment. McGill (entry 44) has reported that employees of the Weyerhaeuser Company who were off from work for six months had only a 50% possibility of ever returning to productive employment. When the employees were off for over a year, the possibility of returning dropped to 25%, and if they were off over two years, they rarely returned to work. Prolonged absenteeism in itself appears to have a profound adverse psychological effect on these employees. Moreover, Beals and Hickman (entry 5) reported that employees who were off for three or more months with back problems were much more disturbed emotionally than those whose injury was to an extremity or where a group of controls were off work for a similar length of time. The type and extent of these emotional disturbances and their effect on motivation will be discussed in Chapter 6.

§ 1-9. Control and Treatment of the Problem.

Another subject which will be discussed in detail in later chapters, but which deserves limited coverage at this time, is the potential effect of various treatment modalities on the course of the low back pain. One might expect that in a condition as common as low back pain the therapeutic measures available would have been subjected to frequent and careful evaluation, but this appears not to have been the case. Deyo (entry 15) recently reviewed the literature on conservative therapy for low back pain. He found what appeared to be valid studies suggesting improvement in outcome with only three treatment regimes. These were flexion exercises, three oral medications, and one unique traction method. He could find no controlled studies of most of the commonly used treatment modalities such as corsets, bed rest, transcutaneous nerve stimulation, and the various physical therapeutic measures commonly in use. Wiesel (entry 62), however, has recently shown that bed rest, when compared to ambulatory treatment, will decrease the amount of time lost from work by 50%. He also found that the amount of discomfort experienced by the patient decreased 60%.

As noted before, most low back accidents are not serious and the majority of employees are able to return to work in a relatively short period of time regardless of the medical care given, providing it is not by its nature counterproductive. However, we have found that they do better and are able to return to work sooner if an organized, logical approach to the evaluation, diagnosis, and treatment is used. Such an approach is discussed fully in Chapter 3.

The question of malingering or exaggeration of complaints frequently comes up in conjunction with low back pain because subjective symptoms play such an important part in diagnosis. Beals and Hickman (entry 5), in a careful study, concluded that when the subjective complaints were out of proportion to the objective physical findings, there was an unconscious exaggeration of symptoms. They also felt true malingering, which is the conscious effort by the individual to create symptoms which are not present, to be a relatively unusual occurrence. This is a topic which deserves much more careful investigation since in some occupational environments, back pain without confirmatory objective findings is usually construed by the employer to be malingering.

Another topic about which there is much discussion concerns the effect of workers compensation legislation on the low back pain patient and his return to work. When enacted, compensation laws attempted to provide workers with medical care and salary relief for injuries which occurred while on the job. Although each state and the federal government have developed their own unique sets of workers compensation rules and regulations, in most settings the employee who is injured on the job has a free choice of physician or treatment facility and his employer is responsible for payment of the costs involved. This financial obligation, however, is limited to a certain extent by the reasonableness and necessity of care provided. The employee's reimbursement for lost time due to the accident varies widely in different jurisdictions. It is usually a percentage of wages: most often sixty-six and two thirds percent of his current pay with a designated maximum. At the present time the maximum wage reimbursements range from $112 per week in Mississippi to $996 per week in Alaska. In addition, many companies provide a supplemental reimbursement which may increase the injured employee's salary to as much as 100% of his current wage, all of which is tax free. These relatively large monetary payments have been accused of encouraging unnecessary prolongation of complaints by employees.

There is no doubt that much of the unjustified lost time and expense is due to the over-utilization of services by certain health providers. While most health care professionals are dedicated, conscientious individuals, a small minority have seen fit to abuse the workers compensation system. The fee for service reimbursement of these providers has tended to encourage the practice of unnecessarily frequent visits, duplication of diagnostic tests, and the prolonged use of a variety of unproven treatment modalities. This is further complicated by the increasingly common association of ancillary service facilities with primary treating physicians. For example, if the physician caring for a back pain patient has his own x-ray and physical therapy facility, the monetary rewards for over-utilization of these services can be far out of proportion to any possible benefit which might accrue to the patient.

The entire back pain problem is complicated by the lack of generally accepted diagnostic nomenclature. At the present time, individuals with low back pain are designated, for no rhyme or reason, by a variety of diagnoses including: acute low back pain, pinched nerve, acute low back spasm, lumbago, non-neurogenic low back pain, and all sorts of other non-scientific terminology. It is extremely difficult to compare treatment modalities and outcomes when dealing with such imprecise diagnostic nomenclature. The entire subject of a standardized approach will be discussed in some detail in Chapter 3.

A similar problem is found in the description of certain physical findings which mean different things to different examiners and thus make comparative evaluation of treatment response next to impossible. An example is the frequently used straight leg raising test. This is reported by some examiners as positive when the patient complains of low back pain with straight leg raising. Other examiners, however, only consider it positive if pain is referred below the knee. The reporting of many other physical findings suffers from the same degree of imprecision. The consensus of low back experts on these matters will be presented in this book so that hopefully we all can begin to speak the same language.

§ 1-10. Implementation of an Organized Approach.

In 1980, this author undertook a careful review of employees off work due to job related accidents in a public utility company. This analysis revealed that 45% of all time lost from work was caused by back injuries. In an industrial population of 5,250, 3,750 of which were blue collar employees, this amounted to 1,626 days. We also found that the number of back operations performed on these employees seemed unaccountably high as compared to the figures generated in a non-compensation setting. A retrospective review of the medical records of these patients who had undergone surgery revealed that few would satisfy accepted criteria for a quality assurance audit, and the poor results of these operations did much to confirm the fact that these patients were getting anything but optimal medical care. We recognized that, considering the magnitude of the problem, there was a real opportunity to significantly reduce both lost time and costs while at the same time perform a major service for our employees by improving the quality of their medical care.

With this incentive of improved medical care, increased productivity, and reduced medical compensation costs, we initiated a comprehensive back care and follow-up program. We took into consideration many of the factors discussed above and attempted to develop an organized approach to the treatment of employees who had sustained low back injuries. Procedures for pre-employment examination and employee examination at the time of departmental transfer remained unchanged. The diagnostic and treatment protocol which we used is discussed fully in another part of this book.

This program has as its basis the review and examination of each job related back injury by a physician in the company's medical department, as well as by a company orthopaedic surgical consultant, within one week of the injury. A conservative treatment program is implemented during the acute phase of the complaint and instruction is given in preventive back exercise during the recovery phase. All job related accidents must be reported to the company's medical department within twenty-four hours of the time that they occur and the injured party is examined as soon as is practical. The employee involved in such an accident may choose his own personal physician and about two-thirds of them do so, usually an orthopaedic surgeon. The other third or so elect to use the company's in-house professional resources for their entire care.

One morning each week all new back injuries are reviewed at the company's main medical facility by a consulting orthopaedic surgeon. A complete history and physical examination is done on each patient and x-rays are obtained if they are not already available. Depending on the diagnosis and circumstances, a course of treatment is recommended. If the employee has an outside treating physician, his treatment plan is reviewed. If the consulting orthopaedic surgeon agrees with the outside physician's recommendations, this is noted and the patient continues under the care of his private doctor. If the treatment recommended by the outside physician differs significantly from what the consultant feels would be in the best interest of the patient, the outside physician is contacted and treatment differences are discussed. Even though the primary treating physician has the ultimate responsibility and authority under these circumstances, we have found that such personal contact by our medical staff and consulting orthopaedic surgeon has created a steadying influence, particularly in respect to inappropriate intervention. During the acute phase, all cases continue to be followed in the low back clinic, usually on a weekly basis; however, for those patients who require bed rest at home or hospitalization, this supervision is accomplished on a regular basis over the telephone. Long term follow-up may continue weekly or, depending on the severity of the problem and the recommended program, it may be at much less frequent intervals.

When the employee's physical condition has improved to the point that back exercises are indicated, a physical therapist, who is part of the health care team and has been following the employee along with the physicians, instructs the patient in appropriate back exercises. While standardized exercise formats are available, we have found that it is important for the program to be individualized to meet the specific needs of each worker. The distribution of an exercise booklet or sheet without personalized instruction either in a small group setting or on a one to one basis appears to have almost no impact on the patient and will rarely motivate him to comply. We are also firmly convinced that passive physical therapeutic modalities do little to promote recovery and often have a negative psychological effect by creating a dependence.

When, in the opinion of the company medical department and the consulting orthopaedic surgeon, the employee appears able to return to light work, this is then recommended and the physical restrictions and work limitations are clearly outlined. If there is a difference of opinion between the employee's primary treating physician and the company medical department concerning the timing or restrictions for return to work, this difference is discussed with the treating physician over the telephone. Under most circumstances, there is a meeting of the minds and an amicable agreement is made in respect to the decisions under contention.

At this point it is important to have the complete cooperation of company management to be sure that employees actually will be able to return to a job protected by the limitations as recommended by the medical department. Without such cooperation, this type of program would just not be feasible. After the employee returns to work, he continues to be followed closely by the medical department as well as by his primary physician, and the same procedures are used if and when he is fit to return to regular duty.

If conservative management is not successful or if the employee's physician recommends surgery, all relevant medical records, x-rays, CAT scans, myelograms, EMG's, and other diagnostic procedures are reviewed by the company physician and the orthopaedic consultant. If the indications for surgery are there, it is authorized; if they are not and the issue cannot be resolved by direct discussion with the treating physician, a third opinion is obtained from a respected outside independent medical expert. This course, however, has rarely been necessary during the three years that this program has been under way.

In order to ensure that the company's consulting orthopaedic surgeons will have no conflict of interest, they are reimbursed on a monthly basis, do not see these employees as private patients in their offices, and do not perform surgery on them. After its initial implementation, the program has been well accepted by the employees as well as by most of the physicians in the community who have been involved. Situations requiring telephone contact with treating physicians have been infrequent and usually atraumatic. We like to think that even though we occasionally lose a battle, we usually win the war. The doctor may not be willing to back down on a particular issue, but he rarely fails to modify his behavior the next time he has a company employee for a patient. We actually believe that this innovative program in our moderate sized company has made a major impact on the entire metropolitan area orthopaedic community; they seem to take a little longer to get their low back patients into the operating room.

The success of our program has been gratifying. We have reduced the amount of back surgery by over 80% from the pre-program level, and have reduced the total amount of time lost from work due to back injuries from 1,626 days in 1981 to 795 days in 1982, and 304 days in 1983. We have also reduced the average length of time away from work per back injury from 31.3 days in 1981 to 21.5 days in 1982, and 10.5 days in 1983; and the average light duty time from 20.7 days in 1981 to 11.9 days in 1982, and 9.6 days in 1983.

Employee acceptance has been extremely good and those with their own primary care physician usually appreciate having a second opinion concerning diagnosis and treatment. The company's management has been very enthusiastic. Not only are the employees receiving better medical care but there has also been a significant increase in productive time. In addition, compensation, medical and surgical costs have been substantially reduced. There was an initial concern that the close monitoring of back problems might result in an increase in the reporting of other types of accidents. This has not occurred, so in effect there has been a total reduction in time lost from work due to job related accidents.

The success of this low back program has encouraged us to consider similar approaches to other common occupational injuries. Its structured nature, with frequent review of the individual problems, has proved that the general application of good orthopaedic principles in a consistent format helps to bring the problems into perspective. While the impetus for developing this program was to improve employee medical care, a significant secondary gain has been the reduction in time lost from work and the cost savings which have resulted. This systematic approach could be applicable to many medical problems commonly found in an industrial environment.

BIBLIOGRAPHY

1. Akeson, W.H., and Murphy, R.W. Editorial comments: Low back pain. *Clin. Orthop.* 129:2-3, 1977.
2. Anderson, J.A.D. Rheumatism in industry: A review. *Br. J. Ind. Med.* 28:103-121, 1971.
3. Anderson, J.A.D., and Duthie, J.J.R. Rheumatic complaints in dockyard workers. *Ann. Rheum. Dis.* 22:401-409, 1963.
4. Andersson, G.B.J. Low back pain in industry: Epidemiological aspects. *Scan. J. Rehabil. Med.* 11:163-168, 1979.

5. Beals, R.K., and Hickman, N.W. Industrial injuries of the back and extremities. *J. Bone Joint Surg.* (AM) 51:1593-1611, 1972.

6. Benn, R.T., and Wood, P.H.N. Pain in the back. *Rheumatol. Rehabil.* 14:121-128, 1975.

7. Biering-Sorensen, F. A prospective study of low back pain in a general population. Part 1. Occurrence, recurrence and aetiology. *Scand. J. Rehab. Med.* 15:71-79, 1983.

8. Bond, M.B. Low back injuries in industry. *Ind. Med.* 39 (5):28-32, 1970.

9. Cady, L.D., Bischoff, D.P., O'Connell, E.R., Thomas, P.C., and Allan, J.H. Strength and fitness and subsequent back injuries in firefighters. *J. Occ. Med.* 21:269-272, 1979.

10. California, State of. Disabling work injuries under workers compensation involving back strain per 1,000 workers, by Industry, California 1979. San Francisco, Department of Industrial Relations, Division of Labor Statistics and Research, 1980.

11. Chaffin, D.B. Localized muscle fatigue — definition and measurement. *J. Occup. Med.* 15:346-354, 1973.

12. Chaffin, D.B., Herin, G.D., and Keyserling, W.M. Pre-employment strength testing: an updated position. *J. Occup. Med.* 20:403-409, 1978.

13. Delin, O., Hedenrud, B., and Horal, J. Back symptoms in nursing aids in a geriatric hospital. *Scand. J. Rehabil. Med.* 8:47, 1976.

14. Department of Labor Bureau of Labor Statistics. Back injuries associated with lifting. Bulletin 2144, August, 1982.

15. Deyo, R.A. Conservative therapy for low back pain. *JAMA* 250 (8):1057-1062, 1983.

16. Dillane, J.B., Fry, J., and Kalton, G. Acute back syndrome — a study from general practice. *Br. Med. J.* 2:82-84, 1966.

17. Feffer, H.L. Editorial commentary. Ultrasonography of the spine. *J. Occ. Med.* 26:28, 1984.

18. Ferguson, R.J. Low back pain in college football linemen. *J. Bone Joint Surg.* 56A:1300, 1974.

19. Frymoyer, J.W., Clements, J., Wilder, D., Rosen, J., and Pope, M. Epidemiology of low back pain. Part II. Presented at the Seventh Annual Meeting of the International Society for the Study of the Lumbar Spine. New Orleans, Louisiana, May 24-28, 1980.

20. Frymoyer, J.W., Pope, M.H., Clements, J.H., Wilder, D.G., MacPherson, B., and Ashikaga, T. Risk factors in low back pain. *J. Bone Joint Surg.* 65-A(2):213-218, 1983.

21. Gibson, E.S., Martin, R.H., and Terry, C.W. Incidence of low back pain and pre-placement x-ray screening. *J. Occ. Med.* 22(8):515-519, 1980.

22. Gyntelberg, F. One year incidence of low back pain among male residents of Copenhagen aged 40-59. *Dan. Med. Bull.* 21:30-36, 1974.

23. Hirsch, C., Jonsson, B., and Lewin, T. Low back symptoms in a Swedish female population. *Clin. Orthop.* 63:171-176, 1937.

24. Horal, J. The clinical appearance of low back pain disorders in the city of Gothenburg, Sweden. *Acta Orthop. Scand.* (Supp.) 118, 1969.

25. Hult, L. Cervical, dorsal, and lumbar spine syndromes. *Acta Orthop. Scand.* (Supp.) 17, 1-102, 1954.

26. Hult, L. The Munkfors investigation. *Acta Orthop. Scand.* (Supp.) 16:1-76, 1954.

27. Jackson, D.W., Wiltse, L.L., and Cirincion, R.J. Spondylosis in the female gymnast. *Clin. Orthop.* 117:68-73, 1976.

28. Kelsey, J.L. An epidemiological study of acute herniated lumbar intervertebral disc. *Rheumatol. Rehabil.* 14:144-159, 1975.

29. Kelsey, J.L. An epidemiological study of the relationship between occupation and acute herniated lumbar intervertebral disc. *Int. J. Epidemiol.* 4:197-205, 1975.

30. Kelsey, J.L., and Hardy, R.J. Driving of motor vehicles as a risk factor for acute herniated lumbar intervertebral disc. *Am. J. Epidemiol.* 102:63-73, 1975.

31. Kelsey, J.L., Pastides, H., and Bisbee, G.E., Jr. Musculoskeletal disorders: their frequency of occurrence and their impact on the population of the United States. New York: Neale Watson Academic Publications, 1978.

32. Kelsey, J.L., and White, A.A. Epidemiology and impact of low back pain. *Spine* 5(2):133-142, 1980.

33. Kertesz, A., and Kormos, R. Low back pain in workmen in Canada. *Can. Med. Assoc. J.* 115:901-903, 1976.

34. Keyserling, W.M., Herrin, G.D., and Chaffin, D.B. Isometric strength testing as a means of controlling medical incidents on strenuous jobs. *J. Occ. Med.* 22:5, 332-336, 1980.

35. Laubach, L.L. Comparative muscular strength of men and women: a review of the literature. *Aviat. Space and Environ. Med.* 47:534-542, 1976.

36. Lawrence, J.S. Rheumatism in coal miners. Part III. Occupational Factors. *Br. J. Ind. Med.* 12:249-261, 1955.

37. Leavitt, S., Johnston, T., and Beyer, R. The process of recovery: patterns in industrial back injury. Part 1. Cost and other quantitative measures of effort. *Ind. Med.* 40:8, 7-14, 1971.

38. Leavitt, S., Johnston T., and Beyer, R. The process of recovery: patterns in industrial back injury. Part 2. Predicting outcomes from early case data. *Ind. Med.* 40:9, 7-15, 1971.

39. Leavitt, S., Johnston, T., and Beyer, R. The process of recovery: patterns in industrial back injury. Part 4. Mapping the Health Care Process. *Ind. Med.* 41 (2):5-9, 1972.

40. MacDonald, E.B., Porter, R., Hibbert, C., and Hart, J. The relationship between spinal canal diameter and back pain in coal miners: Ultrasonic measurement as a screening test? *J. Occ. Med.* 26:23-28, 1984.

41. Magora, A. Investigation of the relation between low back pain and occupation. Part 1. age, sex, community, education and other factors. *Indus. Med. Surg.* 39:465-471, 1970.

42. Magora, A. Investigation of the relation between low back pain and occupation. Part 2. work history. *Indus. Med. Surg.* 39:504-510, 1970.

43. Magora, A. Investigation of the relation between low back pain and occupation. Part 3. physical requirements: Sitting, standing and weight lifting. *Indus. Med. Surg.* 41:5-9, 1972.

44. McGill, C.M. Industrial back problems, a control program. *J. Occ. Med.* 10(4):174-178, 1968.

45. Montgomery, C.H. Pre-employment back x-ray. *J. Occ. Med.* 18 (7):495-498, 1976.

46. Nachemson, A.L. The lumbar spine: An orthopedic challenge. *Spine* 1:59-71, 1976.

47. National Center for Health Statistics. Acute conditions. Incidence and associated disability. United States — July 1974-June 1975. Series 10 Number 114, Rockville, Maryland, Dept. of HEW, 1977.

48. Partridge, R.E., and Duthie, J.J.R. Rheumatism in dockers and civil servants. A comparison of heavy manual and sedentary workers. *Ann. Rheum. Dis.* 27:559, 1968.

49. Quinet, R.J., and Hadler, N.M. Diagnosis and treatment of backache. *Semin. Arthritis Rheum.* 8:261-287, 1979.

50. Rowe, M.L. Low back disability in industry; updated position. *J. Occup. Med.* 13:476-478, 1971.

51. Rowe, M.L. Low back pain in industry. A position paper. *J. Occup. Med.* 11:161-169, 1969.

52. Rowe, M.L. Are routine spine films on workers in industry cost or risk benefit effective? *J. Occup. Med.* 24 (1):41-43, 1982.

53. Snook, S.H. Low back pain in industry. Proc. workshop Idiopathic Low Back Pain. White & Gordon, C.V. Mosby Co. 1983, pp. 23-38.

54. Snook, S.H., Campanelli, R.A., and Ford, R.J. A study of back injuries at Pratt and Whitney Aircraft. Liberty Mutual Insurance Company Research Center, Hopkinton, Massachusetts, 1980.

55. Snook, S.H., Campanelli, R.A., and Hart, J.W. A study of three preventive approaches to low back injury. *J. Occup. Med.* 20:478-841, 1978.

56. Svensson, H. Low back pain in relation to other diseases and cardiovascular risk factors. Thesis, Gothenburg, Sweden, 1981.

57. Svensson, H., and Andersson, G. Low back pain in 40 to 47 year old men. Work history and environment factors. *Spine* 8:272-276, 1983.

58. Ward, R., Knowelden, J., and Sharrand, W. Low back pain. *Jr. Coll. Gen. Pract.* 15:128-136, 1968.

59. Westrin, C.G. Low back sick-listing. A nosological and medical insurance investigation. *Scand. J. Soc. Med.* (Supp.) 7, 1973.

60. White, A.W. Low back pain in men receiving workman's compensation. *Can. Med. Assoc. J.* 95:50-56, 1966.

61. White, A.M. The compensation back. *Appl. Ther.* 8:871-874, 1966.

62. Wiesel S.W., Cuckler, J.M., Deluca, F., Jones, F., Zeide, M.S., and Rothman, R.H. Acute low-back pain. An objective analysis of conservative therapy. *Spine* 5(4):324-330, 1980.

63. Wisconsin, State of. Workmen's Compensation Data: Back injuries. Statistical Release 3878, Department of Industry, Labor and Human Relations, Workmen's Compensation Division, Bureau of Research and Statistics, Madison 1973.

34

COMPARISON OF DISABLING BACK INJURIES
UNDER WORKERS' COMPENSATION
BY INDUSTRY IN CALIFORNIA, 1979

TABLE I

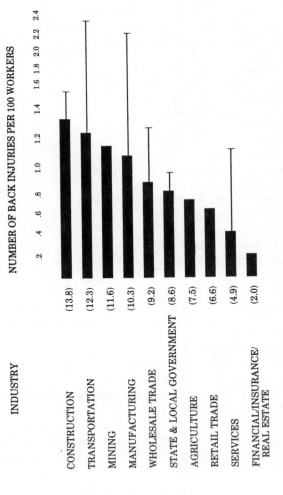

Includes cases involving absence from work for a full day or shift beyond the day of occurrence. The extension lines represent range.

35

CHAPTER 2

ANATOMY AND PATHOPHYSIOLOGY OF LOW BACK PAIN

Richard H. Rothman, M.D., Ph.D.
Ronald J. Wisneski, M.D.

The lumbosacral spine is a complicated structure which enables man to ambulate on his two legs. Because of increased stress and the aging process, this area is susceptible to many problems which translate into pain and disability. It is estimated that 80% of the adult population will at some time in their lives experience low back discomfort of major proportions.

To treat the low back area effectively, one must have a firm knowledge of the normal anatomy and understand the different pathological entities that can cause pain. The purpose of this chapter is to present a working description of the anatomy of the lumbosacral spine and to describe four pathological problems — back strain, disc herniation, spinal stenosis, and spondylolisthesis — that are responsible for the majority of discomfort in the low back. Each of these entities will be defined and the pathophysiology reviewed.

In the succeeding chapters, the above input will be used as baseline information for the development of a standardized diagnostic and treatment protocol which will be presented for use on large patient populations such as exist in an industrial setting.

37

§ 2-1. Normal Anatomy of the Lumbosacral Spine.

In the human, the lumbosacral spine is composed of five lumbar vertebrae, the sacrum and the coccyx. There are also important associated soft tissues such as the intervertebral discs, nerves, and vascular structures. Each will be described.

§ 2-1(A). Lumbar Vertebrae.

The typical lumbar vertebra (Figure 1) consists of a body and a neural arch which encloses the vertebral foramen and through which the spinal cord and cauda equina pass (see entry 23 in Bibliography at end of this chapter). The neural arch has two pedicles on its sides and a lamina for a roof. The spinous process projects dorsally from the lamina in the midline. The transverse processes project laterally from the junction of each pedicle with the lamina. The superior articular processes are found to project upward from the junction of the pedicles to the lamina and by forming true synovial joints with the inferior articular processes of the vertebra above, serve to stabilize the motion segment in respect to both translation and torsion. Thus between any two typical vertebrae, there exists a pair of these articulations to control motion as well as an intervertebral disc to absorb shock.

A deep notch on the lower border of each pedicle matches up with a shallow notch in the top of the pedicle below to form an intervertebral foramen. The foramena are longer in their vertical dimension than in their horizontal one, and their uppermost portion normally serves to accommodate the spinal nerves as they emerge from the spinal canal.

When viewed in cross section, the lumbar canal is seen to be surrounded by the neural arch posteriorly and the posterior surface of the vertebral body anteriorly. The canal itself is triangular in shape with an anterior base. It

38

progressively widens from L1 to the level of the lumbosacral disc. The spinal canal contents in the lumbosacral region include the nerve roots of the cauda equina, which in turn are surrounded by two investing membranes: an outer thick envelope known as the dura mater and an inner thin transparent arachnoid membrane which is in part responsible for the production of the cerebrospinal fluid. It should be noted that in man the spinal cord itself terminates approximately at the level of the inferior margin of the first lumbar vertebra so that pathology below this presents with the typical flaccid paralysis of a lower motor lesion; trauma to the spinal cord, on the other hand, results in a spastic upper motor lesion.

§ 2-1(B). Sacrum and Coccyx.

The sacrum (Figure 2) is a large triangular bone made up of five fused vertebrae and inserted like a wedge between the two pelvic bones. The base of the sacrum articulates with the last lumbar vertebra and an intervertebral disc is interposed between the two. The sacral canal is that portion of the vertebral canal protecting the sacral nerves. These nerves emerge from it through apposing anterior and posterior sacral foramina which are typically situated so as to form communicating canals between the pelvis and the dorsal surface of the sacrum. The coccyx or tail bone is usually composed of four tiny fused vertebrae. It is solid, containing no vertebral canal.

§ 2-1(C). Intervertebral Discs.

The intervertebral discs (Figure 3) together make up approximately 33% of the length of the lumbar spine and are the chief structural link between contiguous vertebrae. They function like universal joints, permitting far greater motion than if the vertebral bodies were in direct contact with each other. Each is made up of a gelatinous nucleus pulposus surrounded by a laminated, fibrous annulus fibrosus and is sandwiched between the cartilagenous end plates of the vertebrae above and below.

The nucleus pulposus (Figure 4) is posterocentrally situated and consists of collagen fibrils enmeshed in a mucoprotein gel. It occupies about 40% of the disc's cross sectional area. It has a high water content — 88% at birth; however, the percentage decreases with age, reflecting both an absolute decrease in available proteoglycans as well as a change in the ratio of different proteoglycans, and this desiccation reduces the functional ability of the nucleus pulposus to withstand stress.

The annulus fibrosus forms the outer boundary of the disc and although it is composed of fibrocartilagenous tissue, its fibrous protein predominates. These fibers are arranged in concentric layers or lamellae and run obliquely from one vertebra to another. Successive layers of the fibers slant in alternate directions so that they cross each other at a variable angle dependent upon the intradiscal pressure. Thus, even though the annulus fibrosus has little in the way of true elastic properties, it can absorb stress by expanding and contracting like a Japanese finger cot. Its peripheral fibers pass over the edge of the cartilagenous end plate to unite with the bone of the vertebral body. The most superficial fibers blend with the anterior and posterior longitudinal ligaments. With age, these fibers deteriorate, become fissured, and lose their capacity to contain the nucleus pulposus. When there is sufficient internal stress, the nuclear material can penetrate the annulus and develop a frank herniation (entry 45).

§ 2-1(D). Ligaments of the Vertebral Column.

The anterior longitudinal ligament is a broad, strong band of fibers extending along the front and sides of the vertebral bodies. Its deepest fibers blend with the intervertebral discs and are firmly bound to each successive vertebral body. The posterior longitudinal ligament, extending along the posterior surface of the vertebral bodies, bounds the front of the spinal canal. In the lumbar region, it becomes narrow as it passes over the vertebral bodies and then expands over the discs. It thus takes on the configuration of a series of hour glasses with the attenuated lateral expansions over the intervertebral discs being most vulnerable to disc herniation.

The ligamenta flava are found posteriorly between adjacent laminae (Figure 5). These yellow ligaments extend from the roots of the articular processes on one side to those on the opposite side, as well as laterally into the intervertebral foramena, forming a portion of the roofs of these apertures. They are attached inferiorly to the superior edges and posterosuperior surfaces of the laminae; superiorly to the inferior and anteroinferior surfaces of the laminae. This unique arrangement, combined with the anterior tilt of the laminae, has the effect of creating an extremely smooth posteroinferior wall which remains smooth and protects the neural elements in spite of what position the spine is bent or twisted into.

§ 2-1(E). Blood Supply of the Lumbar Spine.

There are four lumbar arteries which arise in pairs from the posterior aspect of the aorta in front of the bodies of the first four lumbar vertebrae; and in front of L5, a fifth pair may come off of the middle sacral artery. These vessels (Figure 6) curve posteriorly around the bodies of the vertebrae and give off posterior rami as they pass between the transverse processes. The posterior rami, in turn, furnish the vertebral or spinal branches which supply the bodies of the vertebrae and their ligaments.

The intervertebral discs are without an active blood supply during the adult phase of life. Although up to the age of

41

eight years there are small vessels supplying the discs through the cartilagenous end plates, these are gradually obliterated during the first three decades of life, leaving defects in the cartilage. By the time growth has ceased, the nucleus pulposus and annulus fibrosus no longer have an active blood supply and receive only marginal sustenance from the transfer of tissue fluid across the cartilagenous end plates.

§ 2-1(F). Nerve Supply.

The sinuvertebral nerve is felt to be the major sensory nerve supplying the structures of the lumbar spine. It arises from its corresponding spinal nerve and enters the spinal canal by way of the intervertebral foramen, curving upward around the base of the pedicle and proceeding toward the midline on the posterior longitudinal ligament. It innervates (Figure 7) the posterior longitudinal ligament, the blood vessels of the epidural space, the dura mater and the periosteum. The posterior rami of the posterior primary divisions of the spinal nerve roots, on the other hand, supply sensory fibers to the skin, muscle, fascia, ligaments, and facet joints.

§ 2-2. Pathophysiology of Low Back Pain.

As is the case with any other disease, the intelligent management of a patient with low back pain requires a thorough knowledge of the pathophysiology of the various disorders involved. Appropriate background material will accordingly be presented by first reviewing the epidemiology of back pain, then the pertinent scientific background. This will be followed by a description of the major clinical syndromes involving the low back.

§ 2-2(A). Epidemiologic Factors.

Medical considerations of low back pain are found in the writings of Hippocrates, and indeed are even noted in the Old Testament. Despite decades of clinical and basic science research relating to the causes and treatment of low back pain, the exact etiology of the patient's problem often remains unknown even after a thorough evaluation has been performed by an astute clinician. Dillane (entry 8) in 1966, reported on a large series of patients examined in general practitioner's offices in England and noted that the exact diagnosis could not be made in 79% of the first attacks of low back pain in men and 89% of those in women. In view of these grim figures, White (entry 43), in his introduction to the American Academy of Orthopaedic Surgeons Symposium on Idiopathic Low Back Pain, commented on the significance of low back pain in contemporary western society. He related the following statistics in emphasizing the need for increasing the present state of knowledge of these disorders:

(1) Several studies have indicated that four out of five individuals will have a significant complaint of back pain at one time or another in their lives.

(2) Impairments of the spine are the most frequent cause of time lost from work.

(3) The most consuming problems brought to family general practitioners are "back pain and swelling injury."

(4) Estimates for the cost of treatment in compensation from those suffering from back pain exceed 14 billion dollars per year. This figure does not include loss of productivity estimated at four hours per year per worker.

In terms of disability, epidemiologic studies done in the United States and in Scandinavia have indicated that one half of the adult population at some time is disabled by back pain (entry 39). Disability being defined as the inability to follow an individual's usual gainful employment. Hult found that 35% of light workers and 64% of heavy workers had significant low back pain with or without sciatica (entry 17). Our own studies substantiated the fact that light or sedentary work in no way protects man against back disease.

The average age of onset of significant back pain is 35 years. Of those individuals with chronic back pain, approximately one third will develop sciatica and one half will develop a severe cervical spine disorder at some time in their life. Serial population studies have confirmed this strong statistical association between neck and low back pain secondary to disc degeneration; and would indicate that there is a constitutional disorder of connective tissue operative in these problems, rather than a simple mechanically or traumatically created disorder.

Table I illustrates an approach developed by Keim for the differential diagnosis of back pain. Despite our knowledge of each of these disease entities, a large proportion of low back pain, and particularly that occcurring in American industrial workers, cannot be diagnosed by objective pathologic criteria. Percentages of low back pain without definite diagnoses range from 21% of workers visiting an industrial clinic (entry 34) to 84% of compensable cases of low back pain (entry 22). An equally large portion of low back pain cannot be related to any actual occurrence. Snook (entry 37) has reported that 12% of workers with compensable back injuries reported no obvious relationship between the pain and a specific act or movement. Rowe, in his discussion of low back pain in industry, noted that 65% of workers with low back pain who visited an Industrial Medical Clinic either could think of no circumstances associated with the onset of symptoms or else first became aware of sudden pain while performing usual or customary activities. Rowe further stated that degeneration of an intervertebral disc was the primary cause of symptoms

44

among the men seeking care at the industrial health facility where he performed his study and only 20% of the cases went without an established diagnosis. It is thus apparent that no one is quite sure as to just how many low back cases are truly ideopathic although we know that certain disorders such as degenerative disc disease, facet joint arthrosis, herniated intervertebral discs, fractures, sprains, and spondylolisthesis can be diagnosed as distinct entities.

In most cases where a definite diagnosis is possible, the pain can be attributed to pathology involving the lumbar intervertebral discs and facet joints. Although it seems reasonable to assume that degenerative disc disease is a deteriorative process productive of pain, the changes in the bone and soft tissues of the spine are so ubiquitous that without the presence of symptoms, one hesitates to include them under the heading of a disease process. Lumbar disc degeneration is so common that for incidence data to be valid, great care must be taken to establish specific criteria. If we, as did Vernon Roberts (entry 41), accept the microscopic and gross degenerative changes found in all cadavers as a criterion, then we must accept the fact that degenerative disc disease is universally present in all subjects by middle age.

If we use x-ray evidence of disc degeneration as our criterion, the incidence drops to 83%, as was noted by Kellgren's study wherein he reviewed the lumbar spines of a group of males over the age of 55 (entry 18). Older age groups have a still higher incidence and younger people have a more modest figure. A third criterion of incidence might be pain and disability. Hult, in his Munkfors investigation, noted the incidence of low back pain to be 53% in light workers and 64% in heavy laborers (entry 17).

Much of this is covered in Chapter 4; however, it would be well to keep in mind that some of the rarer causes of back pain should first be excluded before the diagnosis of discogenic disease is assumed. In this respect, MacNab's classification of back pain is simple, concise, and useful. He categorizes the causes as follows: (1) Viscerogenic, (2) Neurogenic, (3) Vasculogenic (4) Psychogenic, and (5) Spondylogenic (entry 24).

§ 2-3. Basic Scientific Knowledge of the Lumbar Spine.

There has been a great deal of basic research involving the lumbar spine taking place over the past decade, and much of the work has been focused on the behavior of healthy and degenerated discs. The research effort has truly been a multidisciplinary one, focusing in on the following relevant factors: biochemical, mechanical, nutritional, immunologic, nociceptive (pain), and psychosocial. A review of our knowledge and understanding in respect to the first five of these will be presented at this time; however, an in-depth analysis of the psychosocial aspects of low back pain will be reserved for Chapter 6.

§ 2-3(A). Biochemical Factors.

From the biochemical point of view, the intervertebral disc has received more attention than any of the other connective tissue structures in the low back. The gross architecture of the disc has already been described. On a microscopic level, it consists of two adjacent hyaline cartilage end plates, an annulus fibrosus composed of sheets of collagen fibers obliquely aligned in layered sheets at angles varying between 40 and 80 degrees (entry 29) (Figure 4), and a nucleus pulposus consisting of a loosely arranged network of collagen fibers and cells in an extracellular matrix. The annular fibers are attached to the end plates as well as for a short distance into the vertebral bodies. The annulus is thicker anteriorly than posteriorly and gradually thins out as it approaches the nucleus pulposus internally.

46

To function efficiently, the disc largely depends on the physical properties of the nucleus pulposus, which in turn, are closely related to its water binding capacity. Unfortunately, the hydration of the disc drops progressively from early life when the water content is 88% to a level of 69% in the eighth decade of life (entry 30). Ultrastructurally, 99% of the tissue mass of the disc is formed by its matrix which contains glycoproteins, proteoglycans, collagen, and other proteins. Most recent research has been based on the hypothesis that structural insufficiency of the disc is guided by alteration in its biochemical composition. This in turn leads to structural change in other components of the motion segment, mainly the facets and their joint capsules (Figure 9). Farfan has appropriately designated this relationship as the "triple joint complex" (entry 13).

Review of the literature in respect to the composition of the normal and degenerated discs shows the bulk of the connective tissue matrix to be collagen, and five of its major subtypes have been identified. Type I, found predominantly in tendons, skin, bones, and ligaments, accounts for about 90% of the collagen in the body. Type II is absent from these tissues but is found in high concentrations in hyaline cartilage. In the human subject, approximately 65% of the collagen in the annulus fibrosus is Type II, with the remainder being Type I. More than 95% of the collagen in the nucleus pulposus is of the Type II variety. It has been shown that the amount and type of collagen varies according to site in the disc, but there is also considerable variation among species as well as according to location in the spine (entry 2). Eyre has commented on the fact that perhaps the most notable feature of the collagen in the disc is the uniquely ordered radial distribution of Types I and II; with Type I being alone around the very outer rim, but then gradually decreasing in content as Type II increases as one progresses inwards toward the nucleus where Type II takes over completely. It has been shown that Type I collagen has a restricted water content and is thus better able to handle tensile stress, while conversely, Type II with its hydroscopic physical properties, is ideally suited to absorb compressive forces.

After collagen, the proteoglycans make up the major component of the extracellular matrix of the disc. These molecules consist of non-collagenous protein cores along with the attached sulfated glucose aminoglycans of chondroitin-4 sulfate, chondroitin-6 sulfate, and keratin sulfate. The proteoglycans are, in turn, attached to long chains of hyaluronic acid by small glycoprotein links. These molecules are found in great abundance within the nucleus pulposus but form only a small proportion of the dry weight of the annulus fibrosus. They are hydrophilic and therefore regulate the fluid content of the nucleus.

Many age related changes have been noted in disc proteoglycans. Various experiments have suggested that the molecules lose their ability to associate with collagen, have a lower molecular weight, have a reduced aggregation potential, and develop an increased keratin sulfate content. It has been suggested that these changes adversely affect the ability of the disc to imbibe water, which in turn decreases its capacity to dissipate energy when loaded.

In addition to the above, the biochemistry of the disc is altered with age, in that there is an accumulation of non-collagenous proteins called betaproteins, and these molecules may represent insoluble proteins resulting from collagen degeneration. Naylor, in his study of human disc material, found them to be present in only 10% of the specimens from subjects under the age of 45; 65% of the discs studied from those older than 45 contained this protein fraction.

§ 2-3(B). Biomechanical Factors.

Much of what is known about the pathomechanics of low back disorders has been learned by studying the mechanical behavior of cadaveric spinal motor segments in a laboratory setting. It seems clear that under flexion, extension, rotation, and shear stress, the load distribution in the motor segment is not confined to the intervertebral disc alone but is shared by the strong anterior and posterior longitudinal ligaments, the facet joints and their capsules, and the other ligamentous structures, such as the ligamentum flavum and the interspinous and supraspinous ligaments, attached to the posterior elements of the spine. In vivo, these same structures, as well as the muscle and fascial attachments intrinsic and extrinsic to the vertebral column, interact to accommodate the load bearing requirements of the spine. With pure compression loads, it has been calculated that the lower lumbar discs may have to bear a load of up to 1,000 kilograms (entry 32), a force which several investigators have calculated will exceed the compressive strength of the vertebral end plates. This excess, however, is probably largely absorbed by other mechanisms such as the hydraulic effect of intra-abdominal pressure produced by contraction of the abdominal wall musculature.

Morris (entry 26) found that when pure compression loads were progressively applied to an intervertebral disc, the vertebral end plate gave way before the nucleus pulposus herniated. The disc itself behaved like a visco-elastic structure and protruded circumferentially. Brown (entry 6) reproduced the results of this experiment and, in addition, noted that with pure compression loading, there is no difference in the failure pattern between normal and degenerated intervertebral discs. These findings may account for the high frequency of Schmorl's nodes in the upper lumbar segments where little shear force is generated during axial compression of the vertebral column.

Nachemson (entry 27) has been able to record in vivo intradiscal pressures by introducing piezo-resistive transducers into the discs of volunteers and data were obtained in various positions, while exercising, and when wearing external supports (Figure 8). This and other similar studies have lent themselves to clinical application in treating patients with low back disorders. It is interesting to note that several investigators have used mathematical models dealing with finite element analysis of loads generated on the lumbar spine and have obtained results which correlate quite well with Nachemson's in vivo measurements.

Farfan (entry 12) has studied the effects of torsional loading on the lumbar spine. His experiments have suggested that abnormal torsional stresses, applied particularly to the lordotic motion segments, may be the mechanism by which radial and circumferential fissures are produced within the annulus fibrosus. Furthermore, these fissures tend to occur in the posterolateral segments of the annulus due to the eccentric geometry of the intervertebral disc. The effects of this torsional stress concentration are also intensified in the presence of asymmetry of the facet joints as well as when there is advanced degeneration of the disc. In both of these situations, the disc tended to fail under fewer degrees of torsion.

When tortional and compressional loads are applied to the lumbar discs, a component of shear loading is also created. The two lowest lumbar discs are affected to a greater degree than those above. It has been shown in vitro that extremely high loads must be applied in a direct horizontal plane to the disc to produce failure in the shear mode. An example of shear failure of the disc can be seen in some cases of high velocity motor vehicle collisions, particularly when the patient has been wearing a seat belt.

Examination of the spinal motion segment under tension reveals that the disc is strongest in the anterior and posterior regions and weakest in the center. It is somewhat paradoxical to note that the tensile strength of the annulus has been found to be greatest in its posterolateral segments. During flexion and extension, tensile stresses develop within the disc as well as in the posterion elements of the motion segment. MacNab (entry 24) has suggested that prolonged tensile stress, such as that which develops when segmental instability is present, leads to a concentration of stress in the outermost region of the annulus. This phenomenon is then radiographically manifested by the formation of traction osteophytes on either side of the intervertebral disc of the unstable motion segment.

Traditionally most of the biomechanical knowledge of the low back deals with the normal and abnormal function of the intervertebral disc. Only recently has there been any real attention devoted to the mechanical function of the posterior elements of the spine. In general, it seems that the anterior elements provide the major support of the lumbar column and absorb various impacts; the posterior structures, in turn, share some of the loads and influence the pattern of motion that is possible. Together, the anterior and posterior elements protect the dural sac which is surrounded by the neural arch. Adams and Hutton (entry 1) have examined cadaveric lumbar spines to determine the mechanical function of the apophyseal joints. They found that these joints resist most of the intervertebral shear force and in lordotic postures, share in resisting intervertebral compression. Furthermore the facet joints serve to prevent excessive motion from damaging the discs. In particular, the posterior annulus is protected in torsion by the facet surfaces and in flexion, by the capsular ligaments of the facet joints.

51

The facet joints, being synovial, can be responsible for symptomatic degenerative changes of aging. According to Anderson (entry 3), the combined degeneration of the discs and apophyseal joints is important not only because of disc bulging and narrowing, osteophyte formation, and direct proliferation of capsular and soft tissue: a combination of factors which can serve to compromise the overall diameter of the spinal canal and lead to a condition known as degenerative spinal stenosis; it also influences motion segment kinematics.

§ 2-3(C). Nutritional Factors.

Urban (entry 38) has recently summarized his experience with investigations into disc nutrition. It seems clear that the intervertebral disc is a relatively avascular structure and beyond 15 to 20 years of age, is devoid of any directly penetrating blood vessels. Diffusion is the main transport mechanism for disc nutrition thereafter. Small uncharged solutes such as glucose and oxygen mainly gain access to the metabolizing cells through the vertebral end plates. The diffusion of negatively charged solutes such as sulfates, which are important for proteoglycan production, occurs mainly through the annulus fibrousus. Since the area available for diffusion of these negatively charged solutes is smaller in the posterior region of the annulus, both collagen and proteoglycan turnover is slower in this critical zone. It has been hypothesized that this nutritional inadequacy in conjunction with the concentration of mechanical stress in the posterolateral annulus together account for the high incidence of intervertebral disc ruptures in this area. Furthermore, it is highly probable that when fissures and cracks do occur in this region, even prior to the stage of rupture, the propensity for healing is liable to be low or inadequate at best.

§ 2-3(D). Immunologic Factors.

In its normal state the intervertebral disc, due to its avascular structure, seems to be an immunologically privileged site. Naylor (entry 28) was the first to suggest that an auto-immune reaction may occur when degenerative changes within the disc expose its chemical constituent parts to the host's normal immune defense mechanism. Bobechko and Hirsch (entry 5) found that the rabbit nucleus pulposus can induce the production of auto-antibodies within the same donor animal. Others have confirmed the occurrence of auto-immune phenomena in the human subject and there is particularly strong evidence that disc sequestration within the spinal canal elicits a strong cell mediated immune response to the autogenous disc material.

One needs to view these findings critically when analyzing the etiology of the acute sciatic pain associated with disc ruptures. It seems clear that mechanical compression on the nerve root or roots which are being irritated by the herniated disc material is an important factor in the production of pain. As far as the inflammation which occurs around the nerve root, however, no one is certain as to whether it is related to an auto-immune phenomenon, ischemic neuropathy from alteration in blood flow patterns, or defects in the neuronal transport mechanism of the nerve root itself. Nevertheless, the fact that inflammation is a key ingredient in the production of root pain from disc herniations has been eloquently demonstrated by the study of Smyth and Wright (entry 36). These investigators passed fine silk threads around lumbar nerve root sheaths at the time of surgery for the removal of herniated discs and the threads were brought out through the skin during wound closure. Postoperatively, when traction was applied to the roots which had not been compressed by the disc herniation, the patient experienced paresthesias alone in the neurotomal distribution of the nerve. When tension was applied to the inflamed root that had previously been compressed by the disc herniation, however, radicular pain was perceived. This study, as well as the favorable results reported in the treatment of sciatica with oral, intradiscal, and epidural steroids, support the concept of inflammation as the critical event in the production of nerve root pain.

§ 2-3(E). Nociceptive Factors.

A nociceptor by definition is an ending of a peripheral nerve fiber which, when activated, is associated with a sensation of pain. As already mentioned in respect to the lumbosacral region, free nerve endings supplied by the sinuvertebral nerves provide sensory innervation to the anterior and posterior longitudinal ligaments, the facet joint capsules, the superficial lamellae of the annulus fibrosus, the dural envelopes, the periosteum of the vertebrae, and the blood vessels. It is known that distension of an intervertebral disc or facet joint by the injection of saline or contrast material can produce pain, presumably by activating the nociceptors in these structures. Howe (entry 16) has suggested that compressional distortion of the dorsal root ganglion following disc herniation may produce scarring which, in turn, could increase the sensitivity of certain nociceptive fibers to mechanical and chemical stimuli, thereby resulting in radicular pain. LaMotte (entry 20), alternatively, has stated that radicular pain could be a referred pain caused by noxious stimulation of nociceptors in deep structures such as facet joints, ligaments, and meninges (Figure 9). It should, however, be emphasized that not all leg pain is sciatica. Stimulation of the outer portion of the disc will cause pain to be referred to other mesenchymal structures from the same embryonic sclerotome. When this type of pain is referred into the buttocks and legs, however, it has a dull, aching quality quite unlike the sharp, lancinating pain of true sciatica.

§ 2-4. Clinical Conditions.

There are many conditions (Table 1, at end of this chapter) which can be present as low back pain in any particular individual. However, the following four most common musculoskeletal problems will be presented in detail: back strain, herniated disc, spinal stenosis, and spondylolisthesis.

§ 2-4(A). Back Strain — Lumbago.

The vast majority of people who have low back discomfort are suffering from a non-radiating type of low back pain called back strain or lumbago. The etiology is not always clear but it probably is a ligamentous or muscular strain secondary to either a specific traumatic episode or the continuous mechanical stress of a postural inadequacy. These may also include patients with a small tear in the annulus fibrosus, and this would account for the frequent prior history of low back pain in patients with a ruptured disc.

These patients' main complaint is back pain and it can be limited to one spot or cover a diffuse area of the lumbosacral spine. At times there may be a referral of pain to the buttocks or posterior thigh since the lower back, buttocks and posterior thigh all originate from the same embryonic tissue or mesoderm. Such referral of pain does not necessarily connote any mechanical compression of the neural elements and should not be called sciatica.

The usual physical findings are limited to local tenderness over the involved area and muscle spasm; however, the attacks will vary in intensity and can conveniently be divided into three categories: mild, moderate, and severe (entry 20). Those placed in the mild group have subjective pain without objective findings and should be able to return to customary activity in less than a week. The moderate group is characterized by a limited range of spinal motion and paravertebral muscle spasm as well as pain, and they should be able to resume full activity in under two weeks. The severe group includes those patients who are tilted forward or to the side. They have trouble ambulating and can take up to three weeks to become functional again.

Since a normal x-ray is to be expected from a patient with a back strain, a radiographic study is usually not necessary on the first visit if the physician feels comfortable with the diagnosis; however, if the response to treatment does not proceed as expected, films should be taken to rule out other more serious problems such as spondylolisthesis or tumor. The prognosis of patients with lumbago is excellent and they will usually recover on target with no lasting impairment. Since the mainstay of successful therapy is bed rest, its interruption for physical therapy at some distant location can be nothing but counterproductive. Although non-steroidal anti-inflammatory medication may be of some help, drugs for the relief of pain and muscle spasm do not seem to alter the course of the attack.

§ 2-4(B). Acute Herniated Disc.

A herniated disc can be defined as the herniation of the nucleus pulposus through the fibers of the annulus fibrosus (entry 33). Most disc ruptures (Figure 10) occur during the third and fourth decade of life while the nucleus pulposus is still gelatinous. The perforations usually arise through a defect just lateral to the posterior midline where the posterior longitudinal ligament is weakest. The two most common levels for disc herniation are L4-5 and L5-S1; pathology at the L2-3 and L3-4 can occur but is relatively uncommon.

Disc herniations at L5-S1 will usually compromise the first sacral nerve root; a lesion at the L4-5 level will most often compress the fifth lumbar root, while a herniation at L3-4 more frequently involves the fourth lumbar root. It should, however, be pointed out that variations in root configuration as well as in the position of the herniation itself can modify these relationships. An L4-5 disc rupture can at times affect the first sacral as well as the fifth lumbar root, and in extreme lateral herniations, the nerve exiting at the same level as the disc will be involved.

Not everyone with a disc herniation has significant discomfort. While a large herniation in a capacious canal may not be clinically apparent since there is no compression of the neural elements, a minor protrusion in a small canal may be crippling since there is not enough room to accommodate both the disc and the nerve root.

Clinically, the patient's major complaint is pain (entry 42). Although there may be a prior history of intermittent episodes of localized low back pain, this is not always the case. The pain not only is present in the back but also radiates down the leg in the distribution of the affected nerve root. It will usually be described as sharp or lancinating, progressing from above downward in the involved leg. Its onset may be insidious or sudden and associated with a tearing or snapping sensation in the spine. Occasionally when the sciatica develops, the back pain may resolve since once the annulus has ruptured, it may no longer be under tension. Finally, the sciatica may vary in intensity; it may be so severe that the patient will be unable to ambulate and he will feel that his back is "locked." On the other hand, the pain may be limited to a dull ache which increases in intensity with ambulation.

On physical examination, there is usually a decreased range of motion in flexion and the patient will tend to drift away from the involved side as he bends. On ambulation, the patient walks with an antalgic gait, holding the involved leg flexed so as to put as little weight as possible on the extremity.

Although neurologic examination may yield objective evidence of nerve root compressions, these findings are often undependable since the involved nerve is often still functional. In addition, such deficit may have little temporal relevance since it may relate to a prior attack at a different level. To be significant, reflex changes, weakness, atrophy, or sensory loss must conform to the rest of the clinical picture.

When the first sacral root is compressed, the patient may have gastroc-soleus weakness and be unable to repeatedly raise up on the toes of that foot. Atrophy of the calf may be apparent and the ankle (Achilles) reflex is often diminished or absent. Sensory loss, if any, is usually confined to the posterior aspect of the calf and lateral side of the foot.

Involvement of the fifth lumbar nerve root can lead to weakness in extension of the great toe and less often to weakness of the everters and dorsiflexors of the foot. An associated sensory deficit can appear over the anterior leg and the dorsomedial aspect of the foot down to the great toe. There are usually no primary reflex changes but on occasion, a diminution in the posterior tibial reflex can be elicited. The absence of this reflex, however, must be asymetrical for it to have any clinical significance.

With compression of the fourth lumbar nerve root, the quadriceps muscle is affected; the patient may note weakness in knee extension and it is often associated with instability. Atrophy of the thigh musculature can be marked. A sensory loss may be apparent over the anteromedial aspect of the thigh and the patellar tendon reflex is usually diminished.

Nerve root sensitivity can be incited by any method which creates tension; however, the straight leg raising test is the one most commonly employed. With the patient supine, one of the examiner's hands is used to stabilize the pelvis while the other one slowly raises the leg by the heel, keeping the knee straight. The test is only considered positive if either pain develops in the leg below the knee or if the patient's radicular symptomatology is reproduced. Back pain alone does not indicate a positive test. As noted in Chapter 5, many variations of this test have been described and all can be useful as long as they are performed and interpreted correctly.

The initial diagnosis of a herniated disc is ordinarily made on the basis of the history and physical examination. Plain x-rays of the lumbosacral spine will rarely add to the diagnosis but should be obtained anyway to help rule out other causes of pain such as infection or tumor. Other tests such as the EMG, CAT scan, and myelogram are confirmatory by nature and can be misinformative when they are used as screening devices.

The treatment for most patients with a herniated disc is non-operative (entry 25) since eighty percent of them will respond to conservative therapy when followed over a period of five years (entry 7). The efficacy of non-operative treatment, however, depends upon a healthy relationship between a capable physician and a well informed patient. If a patient has insight into the rationale for the prescribed treatment and follows instructions, the chances for success are greatly increased.

The most important element in the non-operative treatment of acute disc disease is bed rest (entry 44). This usually can be accomplished at home but the patient must clearly understand that other than for trips to the bathroom, he must stay in bed. Semi-Fowler's position, with the hips and knees comfortably flexed, is ideal since it keeps intradiscal pressure down and reduces nerve root tension. It usually takes at least two weeks of bed rest to obtain relief

59

and this should be followed by gradual mobilization once the pain has substantially eased up. It is unrealistic to expect a patient with a frank disc herniation to return to even sedentary work in less than one month.

Drug therapy is another important part of the treatment and three categories of pharmacological agents are commonly used: anti-inflammatory drugs, analgesics, and muscle relaxants or tranquilizers. In as much as the symptoms of low back pain and sciatica result from an inflammatory reaction as well as mechanical compression, the authors feel that anti-inflammatory medication in the form of two aspirin every four hours should be taken in conjunction with the bed rest. It should be stressed, however, that no medication can take the place of the bed rest. Buffered or enteric coated aspirin can be used by patients with gastrointestinal intolerance. The patient's pain will generally be relieved once the inflammation is brought under control. There may be some numbness or tingling in the involved extremity but this is usually tolerable. Some patients who fail to respond to aspirin may get dramatic relief from a short course of systemic steroids administered in decreasing dosages over a week. In addition, a large array of non-steroidal anti-inflammatories are available for the refractory patient.

Analgesic medication is rarely needed if the patient really stays in bed since the pain is usually adequately controlled by total immobility; however, if the pain is severe enough to require hospitalization, then morphine sulfate is the drug of choice; at home, codeine is recommended.

There is some question whether there actually is a muscle relaxant and all drugs that are so designated probably act as tranquilizers. If one is required, though, methocarbamol and carisoprodol are the ones most frequently used and they can be employed intravenously as well as orally. The use of diazepam (Valium) for this purpose should be discouraged

60

since it actually is a depressant and often will add to the patient's psychological problems.

Eighty percent of those who follow the above regimen will be markedly improved but it requires patience since normally at least six weeks will have evolved before any additional therapy is indicated. Although the non-invasive treatment of a herniated disc can be quite gratifying, it generally takes a significant period of bed rest, and the patient must be aware of the time constraints from the beginning in order to understand the rationale of the measures employed.

§ 2-4(C). Spinal Stenosis.

Spinal stenosis can be defined as a narrowing of the spinal canal (Figure 11) and the mechanical pressure on the neural structures within will depend upon the degree of narrowing (entries 10, 46). Every person's spine, however, becomes narrower with age and one must be familiar with this physiological aging process.

In the first few decades of life, the gross appearance of the spine and its components will remain basically unchanged. The intervertebral discs will maintain their full height, with a thickened, laminated annulus and a tense nucleus. The vertebrae are almost completely ossified except for their apophyseal rings and are square in shape. The facets are well defined, with smooth capsules and normal articular cartilage. The ligamentum flava are only a few millimeters thick, and the space that is available for the neural elements within the canal and the foramina is capacious. Symptoms are unusual even though some developmentally and congenitally narrow canals have much less space available even early in life.

Pronounced changes develop in the lumbar spine between the third and fifth decade of life and the first manifestations of aging show up in the intervertebral discs. In certain individuals, there will be dramatic alterations in a disc and its adjacent structures during this period, with the production of associated symptoms. In the early years the nucleus has a high water content and serves effectively to distribute the stress which develops between adjacent vertebrae. It is at this time in life that nuclear herniations are most likely to occur. As the years pass, the nucleus loses its vigor and the annulus degenerates. The resulting biomechanical insufficiency inevitably results in a transfer of stress posteriorly to facet joints and ligaments which are ill suited to assume compressive, tensile, and shear loads; and capsular strains, hypermobility, and degenerative changes develop. These changes are often manifested radiologically by traction spurs which form anteriorly, one to two millimeters from the disc. The ligamentum flavum is compelled to assume unnatural tensile loads in spite of having become redundant as the total spine length decreases with disc degeneration. The vertebrae themselves also tend to collapse and spread so as to further compromise the space available for the neural elements.

If a disc herniation occurs in a spinal canal that is relatively small, compression of the neural elements will result and the patient will experience symptoms not only of low back pain as mediated through the sinuvertebral nerve supply to the outer margin of the annulus, but also radiating pain in the distribution of the compressed neural elements. In pure terms, this can be thought of as a relative spinal stenosis since the herniated nucleus pulposus is occupying space in a small spinal canal. On the other hand, a similar sized disc herniation in a large spinal canal may cause no symptoms at all because the neural elements have enough room to escape pressure. Thus, symptoms in this age group result from a combination of the disc herniation itself and the size of the canal with which the person was born.

Patients in the fourth decade of life and older can show the hypermobile end-state changes of the aging process. Degeneration both of the facet joints and the intervertebral discs leads to narrowing of the spinal canal. The canal is rimmed by large osteophytes which have developed as a result of the excessive load on the now incompetent intervertebral disc. The facets are hypertrophic and deformed by osteophytic spurs that are encased within the joint capsule. The ligamentum flavum becomes redundant and, in combination with the aforementioned changes, encroaches on the spinal canal and foramena. Although such distortion of the spinal canal occurs to some degree in all active people as they age, not every one suffers significant disability. The symptoms a person will have depend on the original size of the canal; if the spinal canal is small, the changes caused by aging of the disc and facet joints can lead to an absolute stenosis with compression of the neural elements. If, however, the spinal canal is large to begin with, the aging process will only lead to an asymptomatic relative spinal stenosis without neural compression.

In some individuals, the final pathological end-stage of disc degeneration is a fibrous ankylosis between two adjacent vertebrae along with osteophyte formation and a marked narrowing of the disc space. If this is a stable phenomenon, the patient may be relatively free of symptoms or will be aware only of a sense of stiffness in the spine.

As the spine ages, one can also encounter postural alterations with reduction in lordosis as an attempt by the body to decompress the degenerated articular facets by maintaining a flexed rather than an extended posture; however, such postural alterations can lead to chronic muscle tension and become symptomatic. This flexed position also provides more room for the sensitive neural elements which are dynamically compressed in extension.

Although most of the described changes in the motor segment units progress from decade to decade, there is a wide range in the rate of deterioration and it is important to understand that these anatomical alterations do not necessarily dictate symptoms, define disability, or determine prognosis. As the spine ages, these phenomena appear to be tolerated to some degree by all.

For those who do suffer, however, the discomfort can vary from mild annoyance to an inability to walk. The symptom complex is well documented. Patients of either sex, usually not before their fifth decade, will first complain of vague pains, dysesthesias, and paresthesias with ambulation but they will have excellent relief of their symptoms when they are sitting or lying supine. The increased lordotic stance assumed with walking, and particularly walking down grades, is most likely the inciting cause. This symptomatic relationship to posture was verified with the "bicycle test" of Van Gelderan (entry 9); claudication symptoms were not produced while patients leaned forward to bicycle since there was a reduction of lumbar lordosis and a subsequent increase in the central sagital and foraminal dimensions of the canal with this posture. In contrast, vascular claudication symptoms are produced with ambulation up grades because of the increased metabolic demands. The pulses in their feet are also usually either diminished or absent. We are, however, dealing with an age group where varying degrees of both vascular and spinal claudication can coexist, so in cases of uncertainty, arteriography may be indicated.

With maturation of the syndrome, symptoms may even occur at rest. Muscle weakness, atropy, and asymmetrical reflex changes may then appear; however, as long as the symptoms are only aggravated dynamically, neurological changes will only occur after the patient is stressed. The following stress test can be used in an out-patient clinic:

64

After a neurological examination has been performed on the patient, he is asked to walk up and down the corridor until his symptoms occur or he has walked 300 feet. A repeat examination is then done and in many cases the second examination will be positive when the first was negative.

There is an internationally accepted classification of the anatomical state and its clinical syndrome known as lumbar spinal stenosis, and the production of symptoms attributed to these changes can be either localized or generalized in origin. It is important to realize, however, that structural changes in the spinal and foraminal canals that are exaggerated with posture are, as Verbiest noted (entry 40), "conditions, but not absolute determinants of intermittent claudication." The symptoms manifested may vary significantly among patients with similar pathomorphological changes because of the temporal framework in which the neural compression has occurred, the susceptibility of the nerves involved, and the unique functional demands and pain tolerance of each patient.

Plain x-rays are often helpful in visualizing spinal stenosis, particularly degenerative spinal stenosis. One can see the intervertebral disc degeneration, the decreased interpedicular distance, the decreased sagital canal diameter, and the facet degeneration. These findings are a manifestation of the "three joint complex" as described by Kirkaldy-Willis (entry 19) and include the two facet joints as well as the intervertebral disc.

The majority of patients with spinal stenosis, especially the degenerative and combined variety, can be treated non-surgically. Aspirin has been the drug of choice, but the physician must watch for gastric irritation in this older patient population. Finally, a lumbosacral corset is often helpful in reminding the patient to avoid excessive strain. Symptoms are usually intermittent and the individual often needs encouragement in getting through the episode without getting depressed. Non-operative management is preferable as long as the pain is tolerable.

§ 2-5. Spondylolisthesis.

Spondylolisthesis is a spinal condition where all or a part of a vertebra has slipped on another. The word is derived from the Greek "spondylos" meaning vertebra and "olisthesis" meaning to slip. There are five major types of this condition (Table 2, at the end of this chapter) and the etiology of each is different (entry 47). Type II, in which the lesion is in the isthmus or pars interarticularis, has the most clinical importance in persons under the age of 50 and is therefore the one most relevant to a compensation setting. If a defect can be identified, but no slipping has occurred, the condition is termed spondylolysis; if one vertebra has slipped forward on the other (horizontal translation), it is referred to as a spondylolisthesis.

The etiology of the defect in spondylolysis is not clear (entries 4, 31). Although there may be an hereditary component, the lesion is seldom seen in patients under the age of five and is found in five percent of people over the age of seven. The most attractive explanation is that although these children inherit a potential deficiency in the pars, they are not born with any identifiable defect. Between the ages of five and seven, however, they become more active and a stress fracture can develop into a spondylolysis.

Spondylolisthesis has several characteristic features but the forward displacement is easily recognized radiographically on the lateral projection. The degree of slip varies from patient to patient and can range from minimal displacement to complete dislocation of the vertebral body. Increased slipping rarely occurs after the age of 20 unless there has been a severe superimposed injury or surgical

66

intervention without stabilization. The period of most rapid progression coincides with the rapid growth spurt between the ages of nine and fifteen.

The most common clinical manifestation of spondylolisthesis is low back pain. Although the cause of this type of back pain in the adult has been studied extensively, its origin is not clear. There is no clear understanding of how so many patients develop this lesion between the ages of five and seven but still have no back complaints until perhaps age 35 when a sudden twisting or lifting motion will precipitate an acute episode of back and leg pain. Other patients with significant degrees of slipping, however, will go through life with no discomfort.

Although fifty percent of the patients normally cannot associate an injury with the onset of the symptoms, in industry almost all of them will report an associated incident. It is possible to sustain an acute fracture of the pars but it is a very rare occurrence. In any event, a questionable one can be documented by a bone scan within three months of the injury; if the defect is longstanding, it will not light up.

There also frequently is a buildup of a fibrocartilagenous mass at the defect and this can cause pain by irritating the nerve root as it exits. It thus is not unusual in spondylolisthesis to have the patient first complain of back pain, but over time have leg pain develop as the most annoying part of the problem.

Once the symptoms begin, the patient usually has constant low grade back discomfort that is aggravated by activity and relieved by rest. There are some periods during which the pain is more intense than others but unless the picture is complicated by severe leg pain, total incapacitation is rare. The patients are seldom aware of any sensory or motor deficit. At this point it should be reemphasized that in some people, even severe

67

displacement is asymptomatic and gives rise to no disability. It is not uncommon to pick up a previously unrecognized spondylolisthesis on a routine gastrointestinal radiological study of a fifty-year-old patient.

The physical findings of this syndrome are fairly characteristic. In the absence of any radicular pain, the patient exhibits no postural scoliosis, but there is usually an exaggeration of the lumbar curve and a palpable "stepoff" and dimple at the site of the abnormality. Occasionally mild muscle spasm is demonstrable and, in most instances, some local tenderness can be elicited. Although the range of motion is usually complete, some pain can be expected on hyperextension.

Radiographs, particularly the lateral views, confirm the diagnosis (Figure 12). Even the slightest amount of forward slipping of the body of the involved vertebrae is readily discernible, and the oblique views will disclose the actual defect in the pars.

The non-operative treatment of the adult with spondylolisthesis is much the same as that used for backache from other causes. When the symptoms are acute, rest is indicated. If leg pain is a significant problem, then anti-inflammatory medication can be quite beneficial. Exercises should be started once the patient is in a remission and they are usually advised to own a corset for use during occasional strenuous activity.

Once patients experience symptoms secondary to spondylolisthesis, it is unreasonable to expect them to return to heavy work. They should be given some form of light duty involving no heavy lifting or repetitive bending. Surgery, which will be dealt with in Chapter 7, is only indicated to relieve pain. It will not return the employee to heavy work.

BIBLIOGRAPHY

1. Adams, M.A. and Hutton, W.C. The mechanical function of the lumbar apophyseal joints. *Spine.* 8-3:327-330, 1983.
2. Adams, P., Eyre, D.R. and Muir, H. Biochemical aspects of development and aging of human lumbar intervertebral discs. *Rheum. Rehab.* 16:22, 1977.
3. Anderson, B.J.G., Ortengren, R., Nachemson, A. and Elfstrom, G. Lumbar disc pressures and myoelectric back muscle activity during sitting. *Scan. Rehabil. Med.* 6:104-114, 1974.
4. Batts, M., Jr. The etiology of spondylolisthesis. *J. Bone Joint Surg.* 21:879-884, 1939.
5. Bobechko, W.P. and Hirsch, C. Autoimmune response to nucleus pulposus in the rabbit. *J. Bone Joint Surg.,* 47B:574, 1965.
6. Brown, T., Hanson, R. and Yorra, A. Some mechanical tests on the lumbosacral spine with particular reference to the intervertebral discs. *J. Bone Joint Surg.,* 39A:1135, 1957.
7. De Palma, A., and Rothman, R.H. *The Intervertebral Disc.* Philadelphia: W.B. Saunders Co., 1970.
8. Dillane, J.B., Fry, J., Kalton, G. Acute back syndrome — A study from general practice. *Br. Med. J.* 2:82-84, 1966.
9. Dyck, P., and Doyle, J.B. "Bicycle test" of Van Gelderan in diagnosis of intermittent cauda equina compression syndrome. *J. Neurosurg.,* 46:667-670, May 1977.
10. Epstein, B.S., Epstein, J.A., and Jones, M.D. Lumbar spinal stenosis. *Radiologic Clin. N. Amer.,* 15, #2:227-239, August 1977.
11. Eyre, D.R. and Muir, H. Types I and II collagen in intervertebral disc. Interchanging radial distributions in annulus fibrosis. *Biochem. J.,* 157:167, 1976.
12. Farfan, H.F. Mechanical disorders of the Low back. Lea and Febiger, Philadelphia, 1973.
13. Farfan, H.F., Huberdieau, R.M. and Dubow, H.I. Lumbar Intervertebral disc degeneration. *J. Bone Joint Surg.,* 54A:492, 1972.
14. Friis, M.L, Gulliksen, G.C., Rasmussen, P., and Husby, J. Pain and spinal root compression. *Acta. Neurochir.,* 39:241-249, 1977.
15. Hakelius, A., and Hindmarsh, J. The significance of neurological signs and myelography findings in the diagnosis of lumbar root compression. *Acta. Orthop. Scan.,* 43:234-238, 1972.
16. Howe, J.F. A neurophysiologic basis for the radicular pain of nerve root compression in bonica. J.J., Liebeskind, J.C., Albe-Fessard, D.G. (editors). *Advances in Pain Research and Therapy,* Vol. 3, Raven Press, New York, pp. 647-657, 1979.

17. Hult, L. The munkfors investigation. *Acta. Orthop. Scan., Supp.* 16, 1954.
18. Kellgren, J.H. The anatomical source of back pain. *Rheumatol. Rehabil.* 16:3-11, 1977.
19. Kirkaldy-Willis, W.H., and Hill, R.J. A more precise diagnosis for low back pain. *Spine,* 4, #2:102-109, March, April 1979.
20. LaMotte, R.H., Thalhammer, J.G., Torebjork, H.E., et al. Peripheral neural mechanisms of cutaneous hyperalgesia following mild injury by heat. *J. Neurosci. Res.* (In Press, 1982).
21. LaRocca, H. and MacNab, I. Value of pre-employment radiographic assessment of the lumbar spine. *Industrial Med. and Surg.,* Vol. 39:253-358, 1970.
22. Leavitt, S.S., Johnston, T.L., Beyer, R.D. The process of recovery: Patterns in industrial back injury. *Ind. Med. Surg.* 40(B):7-14, 1971.
23. Louis, R. Surgery of the spine, Springer-Verlag, New York, pp. 34-42, 1983.
24. MacNab, I. The traction spur. An indicator of segmental instability. *J. Bone Joint Surg.,* 57A:663-669, 1971.
25. Marshall, L. L. Conservative management of low back pain: A review of 700 cases. *Med. J. Australia,* 1:266-267, 1967.
26. Morris, J.M., Lucas, D.B. and Bresler, B. Role of the trunk in stability of the spine. *J. Bone Joint Surg.,* 43A:327-351, 1961.
27. Nachemson, A. and Morris, J. In vivo measurements of intradiscal pressure. *J. Bone Joint Surg.,* 46A:1077-1092, 1964.
28. Naylor, A. Intervertebral disc prolapse and degeneration. *Spine.* 1:108, 1976.
29. Naylor, A. The biophysical and biochemical aspects of intervertebral disc herniation and degeneration. *Ann. Roy. Coll. Surg.,* 31, 91, 1962.
30. Naylor, A. The structure and function of the intervertebral disc. *Orthop. Oxford,* 3:3-22, 1970.
31. Neuman, P.H. The etiology of spondylolisthesis. *J. Bone Joint Surg.* 45-B: 36-59, 1963.
32. Perry, O. Fracture of vertebral endplates in the lumbar spine. An experimental biomechanical investigation. *Acta. Orthop. Scan. Supp.* 25, 1957.
33. Rothman, R.H. and Simeone, F. *The Spine.* Philadelphia: Saunders, 1982.
34. Rowe, M.L. Low back pain in industry: A position paper. *J. Occup. Med.* 11:161-164, 1969.
35. Scham, S., and Taylor, T. Tension signs in lumbar disc prolapse. *Clin. Orthop.,* 44:163-170, 1966. compression. *Acta. Orthop. Scan.,* 43:234-238, 1972.

36. Smyth, M.J. and Wright, V.J. Sciatica and the intervertebral disc. An experimental study. *J. Bone Joint Surg.* 40A:1401, 1958.
37. Snook, S.H., Campanelli, R.A., Hart, J.W. A study of the three preventive approaches to low back injury. *J. Occup. Med.* 20:478, 481, 1978.
38. Urban, J.P.G., Holm, S., Maroudas, A. and Nachemson, A. Nutrition of the intervertebral disc. An in vivo study of solute transport. *Clin. Orthop. Rel. Res.* 129:101-114, 1977.
39. Valkenburg, H.A. and Haanen, H.C.M. Symposium on idiopathic low back pain. Section I-2. The epidemiology of low back pain, C.V. Mosby Co., p. 9-22, 1982.
40. Verbiest, H. Pathomorphologic aspects of developmental lumbar stenosis. *Orthop. Clin. N. Amer.*, 6, #1:177-196, January 1975.
41. Vernon-Roberts, B., and Pirie, C.J. Degenerative changes in the intervertebral discs of the lumbar spine and their sequelae. *Rheumatol. Rehabil.* 16:13-21, 1977.
42. Weber, H. Lumbar Disc Herniations: A prospective study of prognostic factors including a controlled trial. *J. Oslo City Hosp.*, 28:33-64, 89-120, 1978.
43. White, A.A. Symposium on idiopathic low back pain, C.V. Mosby Co., p. 1, 1982.
44. Wiesel, S.W., and Rothman, R.H. Acute low back pain: An objective analysis of conservative therapy. *Spine,* 5:4:324-330.
45. Wilberg, G. Back pain in relation to nerve supply of the intervertebral disc. *Acta. Orthop. Scan.* 10:211-221, 1949.
46. Wilson, C.B., Ehni, G., and Grollmus, J. Neurogenic intermittent claudication. *Clin. Neurosurg.,* 18:62-85, 1971.
47. Wiltse, L.L. Spondylolisthesis: Classification and etiology. *American Academy Orthopaedic Surgeons: Symposium on the Spine.* St. Louis: C.V. Mosby, 1969, pp. 143-167.

TABLE I

Etiologic Factors to Be Considered in Differential Diagnosis of Low Back Pain *

I. Congenital Disorders
  A. Facet Tropism (asymmetry)
  B. Transitional Vertebra
    1. Sacralization of a lumbar vertebra
    2. Lumbarization of a sacral vertebra

II. Tumors
  A. Benign
    1. Tumors involving nerve roots or meninges (*e.g.,* neurinoma or meningioma)
    2. Tumors involving vertebrae (*e.g.,* osteoid-osteoma, Paget's disease, benign osteoblastoma)
  B. Malignant
    1. Primary bone tumors (*e.g.,* multiple myeloma)
    2. Primary neural tumors
    3. Secondary tumors (*e.g.,* metastases from breast, prostate, lung, kidney, and thyroid)

III. Trauma
  A. Lumbar Strain
    1. Acute
    2. Chronic
  B. Compression Fracture
    1. Fracture of vertebral body
    2. Fracture of transverse process
  C. Subluxated Facet Joint (facet syndrome)
  D. Spondylolysis and Spondylolisthesis

IV. Toxicity
  A. Heavy Metal Poisoning (*e.g.,* radium)

V. Metabolic Disorders (*e.g.,* osteoporosis)

VI. Inflammatory Diseases (*e.g.,* rheumatoid arthritis, Marie-Strümpell disease)

VII. Degenerative Diseases (*e.g.,* spondylosis, osteoarthritis, herniated disc — HNP or herniated nucleus pulposus)

VIII. Infections
  A. Acute (*e.g.,* pyogenic disc space infections)
  B. Chronic (*e.g.,* tuberculosis, chronic osteomyelitis, fungal infections)

IX. Circulatory Disorders (*e.g.,* abdominal aortic aneurysm, vascular insufficiency such as varicose veins)

X. Mechanical Causes
  A. Intrinsic (*e.g.,* poor muscle tone, chronic postural strain, myofascial pain, unstable vertebrae)
  B. Extrinsic (*e.g.,* uterine fibroids, pelvic tumors or infections, hip diseases, prostate disease, sacroiliac joint infections and sprains, untreated lumbar scoliosis)

XI. Psychoneurotic Problems (*e.g.,* hysteria, malingering, compensatory low back pain — the green poultice syndrome)

* Keim, H.A., LOW BACK PAIN, from CIBA Clinical Symposia, 25:3, p. 8, 1973.

TABLE II

CLASSIFICATION OF SPONDYLOLISTHESIS

    I - Dysplastic

   II - Isthmic

       A. Lytic

       B. Acute Fracture

 III - Degenerative

 IV - Pathologic

Figure 1

Cephalad View of a Typical Lumbar Vertebra

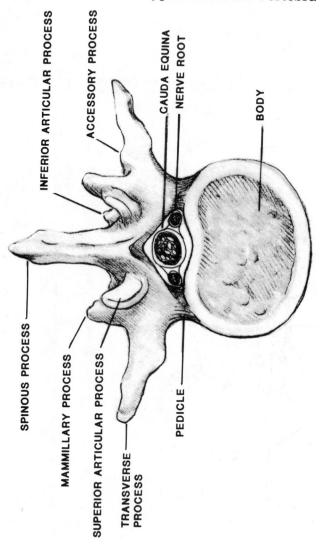

Figure 2A

An Anterior View of the Sacrum and Coccyx

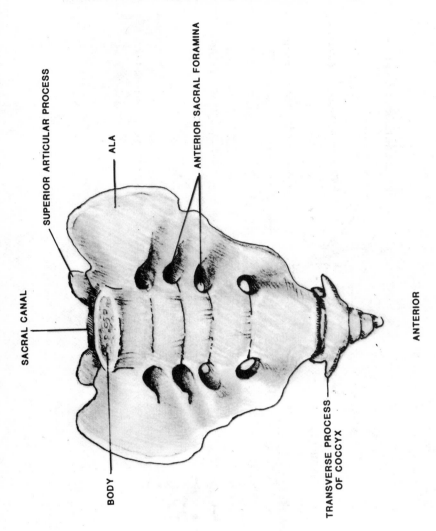

Figure 2B
A Posterior View of the Sacrum and Coccyx

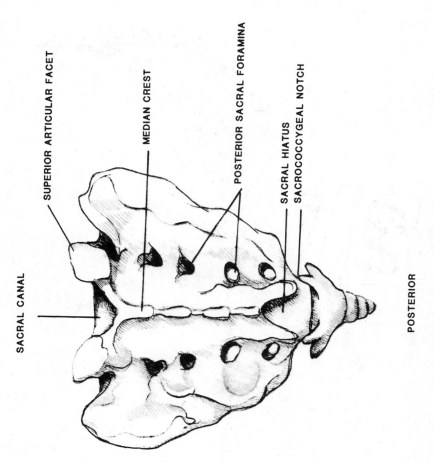

Figure 3
Lumbosacral Articulation

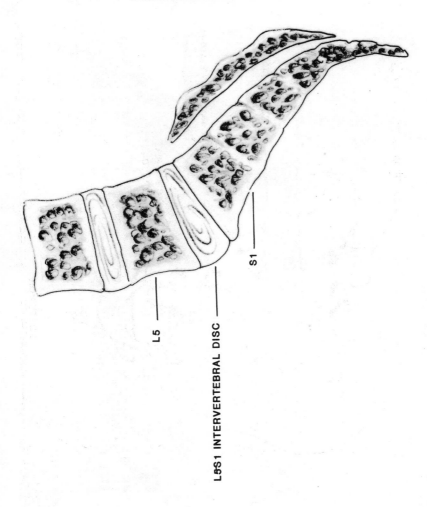

Figure 4

Sectioned Lumbar Intervertebral Disc

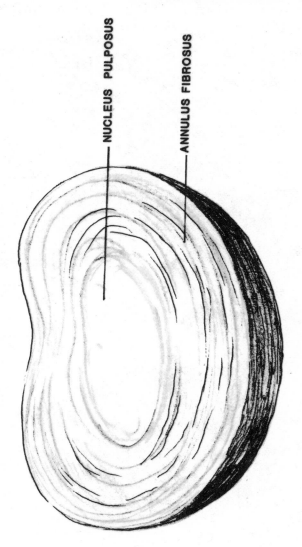

Figure 5

The Ligamenta Flava Attach to the Anterior Surface of the Lamina Above and Extend to the Posterior Surface and Upper Margin of the Lamina Below

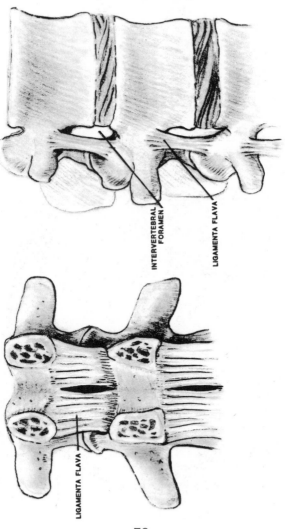

Figure 6

Spinal Branch of the Lumbar Artery

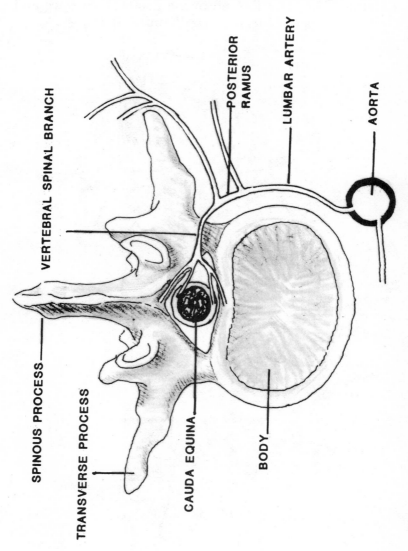

Figure 7

Branches of the Recurrent Sinuvertebral Nerve

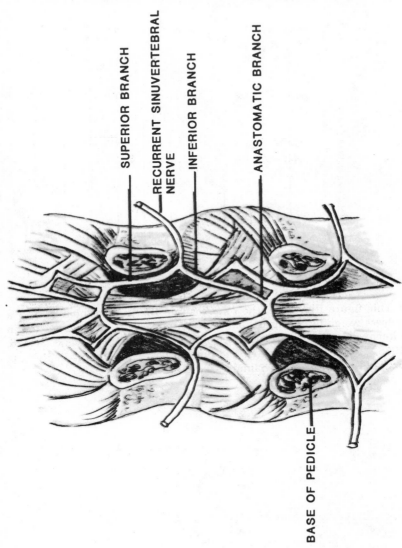

Figure 8

Total Load on the Third Lumbar Disc

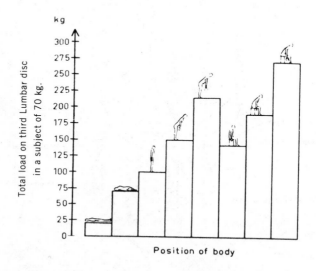

Position of body

This figure illustrates the total load on the third lumbar disc in a subject weighing 70 kg. (From Nachemson, A.: In vivo discometry in lumbar discs with irregular nucleograms. Acta Orthop. Scand. 36:426, 1965.) (From Rothman, R.H. and Simeone, F.A.: The Spine 2nd Ed., p. 523, Philadelphia, W.B. Saunders Co., 1982.)

Figure 9

Axial View of the Spine

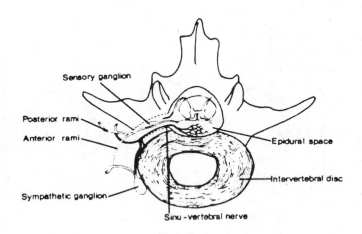

An axial view of the spine with all of the neural pathways that can be involved with neural encroachment or an inflammatory process occurring within either the spinal canal or the neural foramina.

(From Wiesel, S.W., Bernini, P., and Rothman, R.H.: The Aging Lumbar Spine, p. 15, Philadelphia, W.B. Saunders Co., 1982.)

Figures 10(A), 10(B), 10(C)

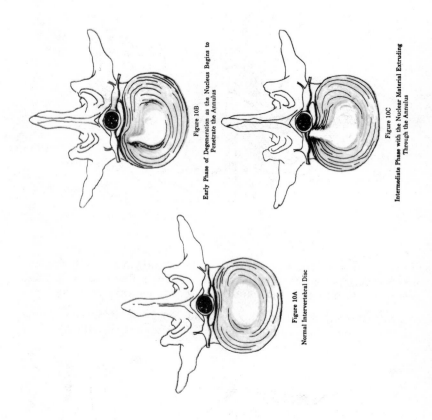

Figure 10B
Early Phase of Degeneration as the Nucleus Begins to Penetrate the Annulus

Figure 10C
Intermediate Phase with the Nuclear Material Extruding Through the Annulus

Figure 10A
Normal Intervertebral Disc

Figure 11
The Normal and Stenotic Lumbar Spinal Canal

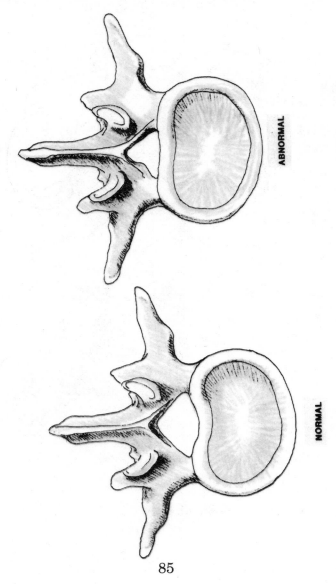

Figure 12A

Degrees of Spondylolisthesis (Slipping) of the Fifth Lumbar Vertebra on the Sacrum

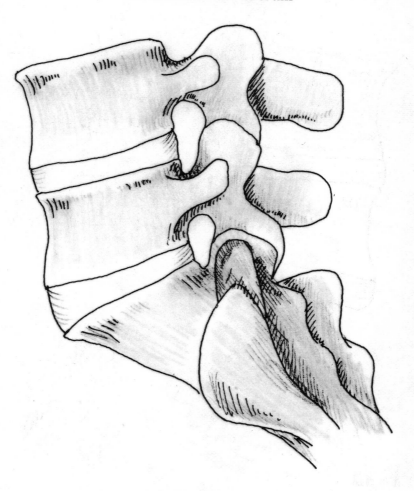

NORMAL

Figure 12B

Degrees of Spondylolisthesis (Slipping) of the Fifth Lumbar
Vertebra on the Sacrum

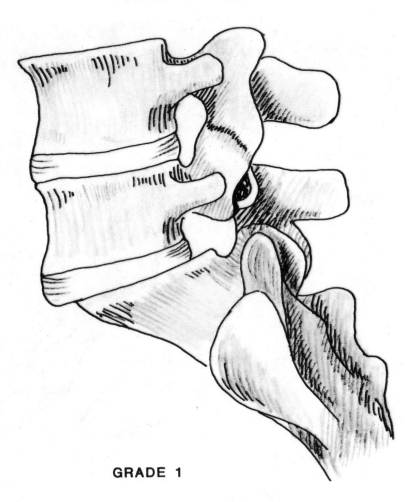

GRADE 1

Figure 12C

Degrees of Spondylolisthesis (Slipping) of the Fifth Lumbar
Vertebra on the Sacrum

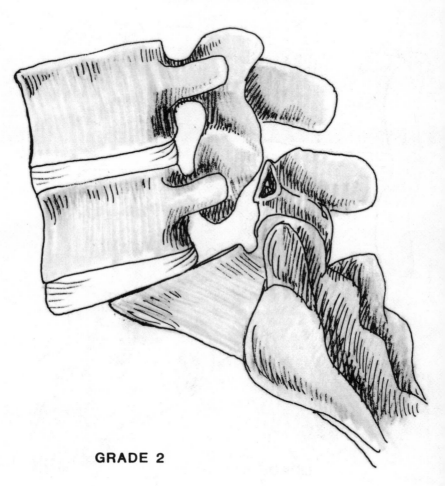

GRADE 2

Figure 12D

Degrees of Spondylolisthesis (Slipping) of the Fifth Lumbar Vertebra on the Sacrum

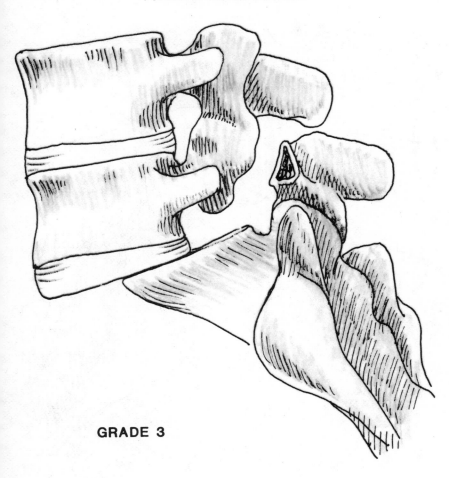

GRADE 3

Figure 12E

Degrees of Spondylolisthesis (Slipping) of the Fifth Lumbar Vertebra on the Sacrum

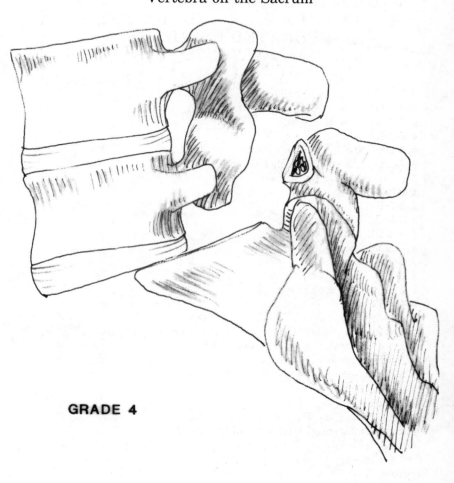

GRADE 4

CHAPTER 3

STANDARDIZED APPROACHES TO THE DIAGNOSIS AND TREATMENT OF LOW BACK PAIN AND MULTIPLY OPERATED LOW BACK PATIENTS

Sam W. Wiesel, M.D.

When patients with low back pain present themselves to the physician, they arrive complaining of a symptom. They do not come labeled with a specific diagnosis such as back strain. The problem confronting the examining physician is to integrate the patient's symptoms, physical signs, and x-ray findings into a logical diagnosis and treatment plan. This can be quite confusing because often the symptom complex is not straightforward.

The purpose of this chapter is, first, to present a standardized approach to the diagnosis and treatment of low back pain patients. This will give insight into the physician's thought process in assessing low back problems. Next, several general treatment regimens such as drug therapy and traction will be discussed, with attention given to their efficacy and current role in the treatment of low back pain. Finally, there will be a review of patients who have had one or more low back operations and continue to have significant complaints. A diagnostic and treatment protocol of this group also will be presented.

Each low back patient will have an associated set of unique circumstances surrounding his case. There are, however, a number of common objectives a physician should keep in mind when managing this population. These goals will be reviewed first, and then the diagnostic and treatment protocol will be presented in detail.

§ 3-1. Goals.

The first and most important objective facing the physician is the prompt return of the individual to his normal function. Total relief of pain is not always achieved. However, even for those patients whose pain cannot be eradicated, a return to fruitful endeavor must be sought. Patient education is germane to this goal. Many people refrain from work, recreation, or their household chores simply because increased activity produces mild pain. In their minds any pain signals do harm to their backs. A word of reassurance often is the key to successful management of these cases.

A second goal should be to devise a treatment format that will make therapy available at an acceptable cost to society. In a disease with the enormous economic impact of low back pain, it is vital that the individual, and society, can afford the diagnostic tests and treatment. In this day of the $600 body scan and the $6,000 surgical procedure, one must be careful to avoid unnecessary and cost-ineffective measures.

The third goal is to minimize unnecessary surgery. Surgical intervention which is premature and thus, by definition, ineffective in attaining the desired goal, has been a terrible burden to the spinal surgeon, the luckless patient, and society. The sorrowful saga of the thousands of patients who have undergone multiple spine operations with poor results bears witness to the disastrous overuse of spinal surgery in the past. Conversely, when conservative management has failed and the patient demonstrates pathology that lends itself to surgical intervention, the operation should not be delayed too long. Surgery to relieve a ruptured disc becomes less effective when delayed more than three months and almost worthless after a year or two. There is an optimal time for surgical intervention, and this must be clearly defined and understood (see entries 39, 40 in Bibliography at end of this chapter).

The final objective is the efficient and precise use of diagnostic studies. With the availability of computer-assisted tomography, epidural venography, psychological profiles, and specialty consultants — each with his own battery of tests — one must resist the impulse to utilize every test available and to meet the often insistent demands of the patient for "the latest study." There is a proper time and indication for each of these diagnostic measures. Decision

93

making actually is more difficult and less accurate when too much data are made available too early in the treatment process.

§ 3-2. Diagnostic and Treatment Protocol.

The task of the physician, when confronted with the low back pain patient, is to integrate his complaints into an accurate diagnosis and to prescribe appropriate therapy. Achieving this idealized goal depends on the accuracy of the physician's decision making. Although hard data for every aspect of low back care are not available, there is a large body of information to guide us in handling these patients. Using this knowledge, which has been presented in Chapters 2 through 5, an algorithm for low back pain has been designed. Webster defines an algorithm as "a set of rules for solving a particular problem in a finite number of steps." It is, in effect, an organized pattern of decision making and thought processes found useful in approaching the universe of low back pain patients. This algorithm can be followed in sequence (Figure 1) and is presented in table form also (Figure 2).

In the first interview and examination of the patient, the physician should try to rule out other medical causes of low back pain. We will assume that after this initial evaluation the patient's symptoms are felt to be due to a musculoskeletal problem in the low back.

The first major decision is to rule in or out the presence of C.E.C. — cauda equina compression syndrome. Compression of the cauda equina, or truly progressive motor weakness, is the only surgical emergency in lumbar spine disease (entry 12). This usually is due to pressure on the caudal sac, through which pass the nerves to the lower extremity, bowel, and bladder. The signs and symptoms are a complex of low back pain, bilateral motor weakness of the lower extremities, bilateral sciatica, saddle anesthesia, and even frank paraplegia with bowel and bladder incontinence. It can be caused by either bone or soft tissue, the latter generally a ruptured or herniated disc in the midline. These patients should undergo an immediate myelogram and surgical decompression. The principal reason for prompt surgical intervention is to stop the progression of neurologic loss; actual return of lost neurologic function following surgery is very low. Although incidence of cauda equina compression syndrome in the total low back population is very small, it is the only entity that requires immediate operative intervention.

The remaining patients comprise the overwhelming majority. They should be started on a course of conservative (non-operative) therapy regardless of the diagnosis. At this stage the specific diagnosis, whether a herniated disc or simple back strain, is not important because the entire population is treated the same way. Some of these patients eventually will need an invasive procedure (surgery), but at this point there is no way to predict which individuals will respond to conservative therapy and which will not.

The early stage of the treatment of low back pain (with or without leg pain) is a waiting game. The passage of time, the use of anti-inflammatory medication such as salicylates, and bedrest are the modalities proven safest and most effective (entry 44). The vast majority of these patients will respond to this approach within the first ten days, although a small percentage will not. In today's society with its emphasis on quick solutions, many patients are pushed too rapidly toward more complex — surgical — management. This "quick fix" approach has no place in the treatment of low back pain patients, especially in workers' compensation cases. Surgery does not return people to heavy work, and in the long run the best chance of getting a patient back to full duty is the non-operative approach.

95

The physician should treat the patient conservatively and wait up to six weeks for a response. As already stated, most of these patients will improve within ten days — a few will take longer. Of these latter few, the problem arises in distinguishing those who have a back problem from those using their backs as an excuse to stay out of work and collect compensation.

If there are no objective findings to substantiate a patient's subjective complaints, he should be strongly encouraged to return to some type of work as quickly as possible, no later than the two-week mark. In spite of this, some patients will complain of continued pain and inability to work. The primary treating physician is in a ticklish situation: on the one hand, as the patient's representative he accepts what the patient tells him (synonymous with a lawyer representing a client who may not be telling the complete truth); on the other hand, he has a responsibility to society to keep medical costs down, return people to work as quickly as possible, and not prescribe unnecessary treatments such as long periods of physical therapy.

The solution for the primary physician is to recommend an independent medical examiner (I.M.E.). This does not destroy the patient-physician relationship. If the I.M.E. concurs with the primary physician that no serious pathology exists, the primary physician — with reinforcement from the "expert" I.M.E. — can recommend that the patient return to work as soon as possible. It cannot be overemphasized that the sooner an I.M.E. is consulted the more effective this is; if there are no objective findings at ten days but the patient continues to complain of pain, an I.M.E. should be obtained. If two or three months pass before this is done, the patient will have his pain pattern firmly fixed in his mind and he will be difficult to convince that there is nothing seriously wrong with him.

Continuing with the overall treatment plan, once the patient has achieved approximately eighty percent relief he should be mobilized with the help of a lightweight, flexible corset. After he is more comfortable and has increased his activity level, he should begin a program of isometric lumbar exercises and return to his normal lifestyle (entries 7, 28). The pathway along this section of the algorithm is a two-way street: should regression occur with exacerbation of symptoms, the physician can resort to more stringent conservative measures. The patient may require further bedrest. Most acute low back pain patients will proceed along this pathway, returning to their normal life patterns within two months of onset of symptoms. The vast majority in this initial group have non-radiating low back pain, termed lumbago or back strain. The etiology of lumbago is not clear. There are several possibilities, including ligamentous or muscular strain, continuous mechanical stress from poor posture, or a small tear in the annulus fibrosis (the covering that surrounds the disc). Patients usually complain of pain in the low back, often localized to a single area. On physical examination they demonstrate a decreased range of lumbar spine motion, tenderness to palpation over the involved area, and muscle spasm. Their roentgenographic examinations usually are normal, but if therapy is not rapidly successful, films should be obtained to rule out other possible etiological factors such as an infection or tumor. The mainstay of successful treatment for lumbago is bedrest. Anti-inflammatory and muscle relaxant medications really do not alter its course (entry 44).

If the initial conservative treatment regimen fails and six weeks have passed, symptomatic patients are sorted into four groups. The first group is comprised of people with low back pain predominating. The second group complain mainly of leg pain, defined as pain radiating below the knee and commonly referred to as sciatica. The third group have posterior thigh pain. The fourth group have anterior thigh pain. Each group follows a separate diagnostic path.

§ 3-2(A). Low Back Pain.

Those patients who continue to complain predominantly of low back pain for six weeks should have their plain x-rays carefully examined for abnormalities. Spondylolysis with or without spondylolisthesis is the most common structural abnormality to cause significant low back pain. Spondylolysis can be defined as a break in the continuity of the pars interarticularis in the lamina. Approximately five percent of the population has this defect, thought to be caused by a combination of genetics and environmental stress. If the defect permits displacement of one vertebra on another, it is termed spondylolisthesis. In spite of this defect, most of these people are able to perform their activities of daily living with little discomfort. These patients usually will respond to non-operative measures, including a thorough explanation of the problem, a back support, and exercises. In a small percentage of such cases, conservative treatment fails and a fusion of the involved spinal segments becomes necessary (entry 33). This is one of the few times primary fusion of the lumbar spine is indicated, and it must be stressed that it is a relatively infrequent occurrence. Most patients with spondylolisthesis do not need surgery.

The vast majority of patients with pain predominantly in the low back will have normal plain x-rays. The diagnosis at this point is back strain. Before there is any additional work-up, a local injection of steroids and Xylocaine may be tried at the point of maximum tenderness. This can be quite successful, and if there is a good response the patient is begun on exercises with gradual resumption of normal activity. In some instances, if there are no objective

98

findings, an injection can be considered as early as the third week.

Should the patient not respond to local injection, other pathology must be seriously considered. A bone scan, along with a general medical evaluation, should be obtained. The bone scan is an excellent tool, often identifying early bone tumors or infections not visible on routine radiographic examination. A thorough medical search also frequently reveals problems missed earlier such as a posterior penetrating ulcer, pancreatitis, renal disease, or an abdominal aneurysm. If these diagnostic studies are positive, the patient should be transferred into a non-orthopaedic treatment mode and would no longer be in the therapeutic algorithm.

Those patients who have no abnormality on their bone scans and do not show other medical disease as cause for their back pain are then referred for another type of therapy — the low back school (entry 47). This concept has as its basis the belief that patients with low back pain, given proper education and understanding of their disease, often can return to a productive and functional life. Ergonomics, the proper and efficient use of the spine in work and recreation, is stressed. Back school need not be an expensive proposition. It can be a one-time classroom session with a review of back problems and a demonstration of exercises with patient participation. This type of educational process has proved very effective. It is most important, however, that before a patient is referred to this type of facility, he be thoroughly screened. One does not want to be in the position of treating a metastatic tumor in a classroom.

If the low back school is not successful, the patient should undergo a thorough psychosocial evaluation in an attempt to explain the failure of the previous treatment. This is predicated on the knowledge and belief that a patient's disability is related not only to his pathologic anatomy but also to his perception of pain and his stability in relation to his social environment. It is quite common to see a patient with a frank herniated disc continue working while regarding his disability as only a minor problem, while at the other end of the spectrum the hysterical patient takes to his bed at the slightest twinge of low back discomfort. Drug habituation, depression, alcoholism, and other psychiatric problems are seen frequently in association with back pain. If the evaluation suggests any of these problems, proper measures should be insitituted to overcome the disability. There are a surprising number of ambulatory patients addicted to commonly prescribed medications who use back pain as an excuse to obtain these drugs. Percodan and Valium, alone or in combination, are the two most popular offenders. Percodan is truly addictive; Valium is both habituating and depressing. Since the complaint of low back pain may be a common manifestation of depression, it is counterproductive to treat such patients with Valium.

§ 3-2(B). Sciatica.

The next group of patients consists of those with sciatica, which is pain radiating below the knee. These people usually experience their symptoms secondary to mechanical pressure and inflammation of the nerve roots that originate in the back and then travel down the leg. The etiology of the mechanical pressure can be soft tissue (herniated disc), bone, or a combination of the two.

At this juncture, these patients have had up to six weeks of rest and oral anti-inflammatory medication but still have persistent leg pain. The next therapeutic step is an epidural steroid injection. This is performed on an outpatient basis. The anti-inflammatory medication (steroids) is placed directly into the area where compression of the nerve is taking place. Epidural injections have proved forty percent effective in relieving leg pain. The maximum benefit from a single injection is achieved at two weeks. The injection may have to be repeated on one or two occasions, and another four to six weeks should pass before its success or failure is determined. It should be pointed out that all sorts of alternative conservative treatments are available (such as traction, passive physical therapy, and manipulation). Unfortunately, these methods have not stood the test of scientific scrutiny and should not become a routine part of the treatment program.

If epidural steroids are effective in alleviating the patient's leg pain or sciatica, he is started on a program of isometric lumbar flexion exercises and encouraged to return promptly to a normal lifestyle. Heavy work (lifting more than fifty pounds regularly) is *not* recommended because these patients will be prone to recurrences. Those who have sustained a disc herniation should be placed on some type of permanent light duty. If the employee-patient is told from the beginning that a permanent light duty job is available and that the only reason to undergo surgery is to relieve pain, he generally will respond to non-operative therapy. Surgery does not return people to heavy work, it only controls pain. This treatment pathway usually will be complete in less than three months, and most patients with sciatica will not have to undergo any major invasive treatment.

Should the epidural steroids prove ineffective, and if three months have passed since the initial injury without relief of pain, some type of invasive treatment should be considered. The patient group is then divided into those with probable herniated discs and those with symptoms secondary to spinal stenosis.

Patients with herniated discs have symptoms secondary to the nucleus pulposus herniating through the annulus fibrosis and causing pressure and inflammation of an individual nerve root. The fifth lumbar and first sacral nerve

101

roots are those most commonly involved. The pain will radiate along the anatomic pathway of the nerves which travel below the knee and into the foot. The highest incidence of this entity occurs in the fourth decade of life.

The physician must now carefully reevaluate the patient for a neurologic deficit and for a positive tension sign or straight leg-raising test. For those who have either a neurologic deficit or positive tension sign along with continued leg pain, a metrizamide myelogram should be considered. If the myelogram is clearly positive, an invasive procedure is a viable option. The computerized axial tomogram also can be helpful in this situation, and as more experience is gained in using it, it may eventually replace the myelogram. If all criteria are present, and an invasive procedure is indicated, one must choose between the new treatment of chemonucleolysis and the standard surgical removal of the herniated disc.

Chemonucleolysis — the injection of medicine into the nucleus pulposus to dissolve it and thus relieve the pressure on the nerve root — has been used for several years. Its results have been controversial, and studies to date have failed to convince many physicians that it is as effective and as safe as surgery. Two types of drugs are currently available: chymopapain and collagenase. Each dissolves a different substance within the nucleus pulposus. Chymopapain dissolves the proteoglycans, while collagenase works on the collagen. At present, the exact indications for chemonucleolysis are not clear, and there remain questions about its safety. The argument will not be resolved until more surgeons gain experience with chemonucleolysis and adequate scientific research has been completed.

For now, surgery still seems the safest and most efficient way to handle these patients. There is repeated documentation that for surgery to be effective in the treatment of a herniated disc, the surgeon must find unequivocal operative evidence of a nerve root compression (entries 6, 15). This mechanical root compression must be firmly substantiated, not only by neurologic examination but also by radiographic data, before the laminectomy. There is no place for "exploratory" back surgery. If the patient has neither a neurologic deficit nor a positive straight leg-raising test, then, regardless of radiographic findings, there is not enough evidence of root compression to proceed with surgery. Twenty-five percent of asymptomatic patients have positive myelograms, and thirty-five percent have positive CAT scans (entry 18). These patients without objective findings are the ones who have bad results and have given back surgery such a bad name.

If there are no objective findings, the physician should avoid surgery and proceed to the psychosocial evaluation. Exceptions should be few and far between. When sympathy for the patient's complaints outweighs the objective evaluation, surgery is fraught with difficulties. For those who meet these specific criteria for lumbar laminectomy, results will be satisfactory: ninety-five percent of them can expect a good to excellent result.

The second group of patients whose symptoms are based on mechanical pressure on the neural elements are those with spinal stenosis. This is a narrowing of the spinal canal secondary to increased bone formation, a natural occurrence with age. However, if the spinal canal is small to start with and then decreases further, pressure develops on the nerve and may cause radiation of pain into the leg. These patients may or may not have a positive neurologic examination or straight leg-raising test and occasionally have to ambulate until symptoms are reproduced; this stress test can cause the neurologic and tension signs which were negative initially to become positive. The older age group, i.e., those over sixty, are the people most affected by spinal stenosis.

The diagnosis of spinal stenosis usually can be made from the plain x-rays which will demonstrate facet degeneration, disc degeneration, and decreased interpedicular and sagittal canal diameter. A computerized axial tomogram or myelogram will better define the involved areas. If symptoms are severe enough and there is radiographic evidence of spinal stenosis, surgery is appropriate. Chemonucleolysis is not indicated since it will not dissolve the bone which is the major offending agent in stenosis. Age alone is no deterrent to surgery; many elderly people who are in good health except for a narrow spinal canal will benefit greatly from adequate decompression of the lumbar spine.

§ 3-2(C). Anterior Thigh Pain.

A small percentage of patients will have pain which radiates from the back into the anterior thigh. This usually is relieved by rest and anti-inflammatory medications. If, after six weeks of treatment, the discomfort persists, a work-up should be initiated to search for underlying pathology, Several entities must be considered.

A hip problem or a hernia can be ruled out with a good physical examination. If the hip examination is positive, x-rays should be done. An intravenous pyelogram (IVP) may be considered to evaluate the urinary tract, because a kidney stone often presents as anterior thigh pain. Peripheral neuropathy, most commonly secondary to diabetes, also can present initially with anterior thigh pain; a glucose tolerance test will reveal this problem. Finally, a retroperitoneal tumor can cause symptoms by mechanically pressing on the nerves which innervate the anterior thigh. A CAT scan or sonogram of the retroperitoneal area will eliminate or confirm this possibility.

If any of the entities listed above is uncovered, the patient is treated accordingly. If no physical cause can be found for the anterior thigh pain, the patient is treated for recalcitrant back strain by the method already outlined in the first part of this chapter.

§ 3-2(D). Posterior Thigh Pain.

This final group of patients will complain of back pain with radiation into the buttocks and posterior thigh. Most of them will be relieved of their symptoms with six weeks of conservative therapy. However, if their pain persists after the initial treatment period, they can be considered to have back strain and given a local injection of steroids and Xylocaine in the area of maximum tenderness in the lumbosacral area. If the injection is unsuccessful, it will be necessary to distinguish between referred and radicular pain.

Referred pain is pain in mesodermal tissues of the same embryologic origin. The muscles, tendons, and ligaments of the buttocks and posterior thighs have the same embryologic origin as those of the low back. When the low back is injured, pain may be referred to the posterior thigh where it is perceived by the patient. Referred pain cannot be cured with a surgical procedure.

Radicular pain is caused by compression of an inflamed nerve root along the anatomic course of the nerve. A herniated disc or spinal stenosis in the high lumbar areas (e.g., at the L-2/3 or L-3/4 interspace) could cause radiation of pain into the posterior thigh. An electromyogram (EMG) and a CAT scan are used in this situation to differentiate the two types of pain. If both studies are within normal limits, the patient is considered to have back strain and treated according to the algorithm. If either test is abnormal, the patient is diagnosed as having mechanical root compression from either a herniated disc or spinal stenosis. Epidural steroids should be tried first; if these do not give adequate relief, a myelogram — with subsequent surgery — is recommended.

This group of patients is very difficult to treat. The biggest mistake is the performance of surgery on people thought to have radicular pain who actually have referred pain. Again, referred pain is not responsive to surgery.

In most instances the treatment of low back pain is no longer a mystery. The algorithm presents a series of easy to follow and clearly defined decision making processes. Use of this algorithm provides patients with the most helpful therapeutic measures at optimal times and neither denies them helpful surgery nor subjects them to procedures that are useless technical exercises.

§ 3-3. Treatment Modalities.

As the algorithm indicates, all low back pain patients, regardless of diagnosis (except those with cauda equina compression syndrome), require an initial period of conservative therapy. At present there are many modalities available; unfortunately, most of them are based on empiricism and tradition. Few are scientifically valid because of the difficulty in performing a prospective double-blind study in this field. Each treatment plan in popular use today is surrounded by conflicting claims for its indication and efficacy.

The purpose of this section is to discuss the rationale behind the use of some of the more common therapeutic measures. Each treatment modality will be described first, with available scientific evidence for or against its use then being presented.

§ 3-3(A). Bedrest.

Bedrest has evolved over the years as the most important element in the treatment of low back pain. Patients should be asked to remain in bed, except for going to the bathroom, preferably at home. There is no real need to admit these people to a hospital unless they have no family or supportive friends to help them. The patient's position in bed should be one of comfort. For most people this means lying on the back with the hips and knees flexed to a moderate degree; several pillows can be placed under the knees to help maintain this position comfortably. Lying on either side, with the legs drawn up in the "fetal position" also can be quite comfortable. The only position to avoid is lying prone (face down), which will cause hyperextension of the lumbar spine; this can be uncomfortable for the patient and, theoretically, lead to further extrusion of an already herniated disc. Finally, the amount of bedrest prescribed varies for each patient; these people should not be mobilized until reasonably comfortable. The type of pathology will determine the duration of bedrest required. A patient with an acute herniated disc will require a minimum of two weeks of complete bedrest with a further ten days for gradual mobilization. However, most patients with acute back strain will need only three to seven days of bedrest before they can ambulate. Each patient should be followed carefully and not allowed complete mobility until his objective signs, such as a list and/or paravertebral muscle spasm, disappear.

The purpose of bedrest is to allow any inflammatory reaction that is present to subside. For example, inflammation is associated with a herniated disc, and bedrest will not result in the disc's return to its original position. But as the disc herniates, it causes a secondary inflammatory process responsible for the patient's pain, and if *this* reaction can be brought under control, the patient's symptoms will disappear. This relief may or may not be permanent.

Scientific evidence that bedrest is effective is available. Biomechanically, Nachemson has demonstrated that the supine position significantly reduces the pressure on the intervertebral discs, compared with sitting or standing positions (entries 27, 28). Assuming that increased pressure on the structures in the low back will increase symptoms, bedrest is a rational approach.

Clinically, Wiesel and Rothman compared bedrest with ambulation for treatment of low back strain (entry 44). Bedrest patients were shown to experience less pain and a faster return to duty than those required to remain ambulatory. Thus, from the evidence available, bedrest appears to be a most effective conservative treatment for low back pain.

§ 3-3(B). Drug Therapy.

The judicious use of drug therapy is an important adjunct in the treatment of low back pain. There are three major categories of drugs in common use: anti-inflammatories, analgesics, and muscle relaxants.

Anti-inflammatory agents are employed because it is felt that inflammation within the affected tissues is a major contributor to pain production in the low back. This is especially true for those patients with symptoms secondary to a herniated disc. The leg pain these people experience is due not only to the mechanical pressure from the ruptured disc but also to the inflammation around the involved nerve roots. If one can get rid of the inflammation, the patient's pain usually will subside.

There are a variety of anti-inflammatory agents available. On the basis of several scientific studies, none of these appear superior to the others (entries 2, 16, 19, 37, 44). It should be noted that all these studies looked at the various agents in combination with bedrest. Anti-inflammatory drugs should be considered only an adjunct and not a primary treatment.

The author's preference is to begin the patient on adequate doses of aspirin, which is effective and inexpensive. If the response is not satisfactory, other anti-inflammatories such as Naprosyn, Motrin, or Indocin can be tried. Most patients will get significant relief from one of the agents presently available. Again, all these anti-inflammatory medications are utilized in conjunction with bedrest for acute pain; they do not replace adequate rest. Occasionally, after an initial recovery, a patient will experience intermittent recurrent attacks or complain of a chronic low backache; in many cases these patients will be helped by a maintenance dose of an anti-inflammatory drug.

Analgesic medication is very important during the acute phase of low back pain (entry 44). The goal is to keep the patient comfortable while in bed. Most patients will respond to 30-60 mg. of codeine every four to six hours. If stronger medication is necessary, it is the author's feeling that the patient should be admitted to the hospital and given parenteral narcotics as required. As pain decreases, non-narcotic analgesics may be substituted for the more potent drugs.

The biggest mistake seen is treatment with very strong narcotics such as Demerol or Percodan on an outpatient basis. Many of these patients become addicted to the medication. In other cases, patients try to shortcut the bedrest and use analgesic medication instead. This, of course, will not work, and when the patient tries to stop the drug, the back pain returns.

Analgesic medication must be prescribed with great care for the low back pain patient. The treating physician must maintain control of the patient's drug use at all times. There are too many instances of patients taking undue advantage of the situation if left unattended. Finally, addicting narcotics have no place in the treatment regimen of patients with chronic low back pain problems.

Muscle relaxants are not generally recommended for the treatment of low back pain (entry 34). In most cases the muscle spasm is secondary to a primary problem such as a herniated disc. If the pain from the ruptured disc can be controlled, the muscle spasm will subside.

Occasionally, muscle spasm will be so severe that some type of treatment is required. Methocarbamol or Carisoprodol are the drugs recommended. Diazepam (Valium) should be discouraged since it is actually a physiological depressant, and depression is often an integral feature of the patient's low back problem (entry 25). Administering Diazepam to a depressed patient only increases the problem. If anxiety is prominent and a sedative is needed, phenobarbital will alleviate the symptoms.

In summary, drug therapy for low back pain should be viewed as an adjunct to adequate bedrest. Anti-inflammatory medication is the primary agent to be employed. Analgesic medication should be used judiciously in a controlled environment and not over a long period. Muscle relaxants are not routinely recommended and if employed should be carefully monitored. Diazepam, as a muscle relaxant, is not the drug of choice for low back pain because of its depressant qualities.

§ 3-3(C). Traction.

The application of traction to the lumbar spine is a very popular treatment for patients with herniated discs. The theory is that stretching the lumbar spine distracts the vertebrae so that the protruded disc is allowed to return to a more nearly normal anatomic position. In fact, the disc material probably does not change position at all. Scientific evidence indicates that a distraction force equal to sixty percent of body weight is needed just to reduce the intradiscal pressure at the third lumbar vertebra by twenty-five percent (entry 29). Such force could not practically be applied to a patient. There has never been any proof that disc material returns to its normal position following herniation.

Double-blind studies on the effect of traction on sciatica patients have not demonstrated any benefit (entries 41, 42). Weber studied two groups of patients with proven herniated discs (by myelogram) by applying traction apparatus to each group. However, for one group he used weights in the traction bags; for the other, he used no weights. He found no significant statistical difference between the two groups in terms of relief of symptoms. Traction had no effect on spinal mobility, tension signs, deep tendon reflexes, paresis, or sensory deficit, and while it usually was well tolerated, it made some patients worse.

In spite of the above, traction is used frequently as an excuse for admitting patients to the hospital. Thousands of health care dollars are spent each year for this treatment modality. It is the author's feeling that traction may benefit the patient by keeping him in bed and creating a positive psychologic effect on his expectations for recovery. It is recommended that patients can be given a home traction unit rather than being admitted to the hospital routinely.

§ 3-3(D). Manipulation.

Spinal manipulation is another very popular conservative modality in treating low back pain. It is somewhat controversial in the United States because it is performed mostly by chiropractors. The principle involved is that any malalignment of the spinal structures can be corrected by manipulation; the assumption here is that the malalignment is the etiology of the patient's pain. Unfortunately, there is "no scientific proof for or against either the efficacy of this spinal manipulation therapy or the pathophysiological foundation from which it is derived." (Entry 1).

There have been several clinical randomized trials studying the efficacy of manipulation. The two best studies demonstrated that patients felt some immediate relief of

111

pain but had no long-term benefit (entries 8, 14). The author's experience is that some patients do have short periods of symptomatic relief after manipulation but must keep returning for repeat sessions to maintain it. There is rarely a lasting improvement in symptoms.

At present it is felt that manipulation is not indicated for the routine treatment of low back pain. There is no scientific evidence to support its use. Some patients in fact may be harmed if pathologic bone disease such as a tumor or osteopenia is present when manipulation is performed (entry 34).

§ 3-3(E). Braces and Corsets.

External support of the lumbar pine with a corset or brace is indicated for only a short period in the average patient's recovery period. As his acute symptoms subside, a properly fitted corset or brace will aid him in regaining mobility sooner. As his recovery progresses, he usually should abandon the brace in favor of an exercise program. With continued long-term use of a brace, soft tissue contractures and muscle atrophy will occur. The young patient should rely on his brace only to hasten ambulation. In theory, strong, flexible lumbar and abdominal muscles function as an excellent "internal brace" because they are adjacent to the structures (vertebrae) they are supporting.

There are two situations in which long-term bracing is a reasonable approach. One is for the obese patient with weak abdominal muscles. A firm corset with flexible metal stays will reinforce the abdominal muscles (entry 26). It has been demonstrated that if a lumbosacral corset is properly applied, the intradiscal pressure in the lumbar area will decrease by approximately thirty percent (entry 30).

The aging patient with multi-level degenerative disease of the lumbar spine is the second type of patient for whom long-term bracing may be beneficial. These older people do not tolerate exercise very well, and in some cases exercise will aggravate their back condition. They can attain significant relief of pain with a well fitted brace.

Finally, a note specifically on compensation patients and bracing — Most of these patients are young, and long-term bracing is not indicated. However, a corset can be of great benefit to a patient who is returning to heavy work; it keeps the worker aware of his back and prevents him from applying maximum stress to this area as he resumes his normal activities. As soon as he can perform a regular set of exercises comfortably, he should be weaned quickly from the support.

§ 3-3(F) Exercises.

Some form of exercise is probably the most commonly prescribed therapy for patients recovering from low back pain. There are two regimens commonly advocated: isometric flexion exercises and hyperextension exercises. These programs are purported to reduce the frequency and intensity of low back pain episodes, although there is no scientific evidence to support this contention.

The isometric flexion exercises are the most popular (entries 5, 20, 21). They are based on Williams' theory that by reducing the lumbar lordosis, back pain is decreased. This goal is achieved by strengthening both the abdominal and lumbar muscles, thereby creating a corset of muscles to support the lumbar spine.

Hyperextension exercises strengthen the paravertebral muscles. These exercises generally are used after a patient has satisfactorily performed a course of isometric flexion exercises. The goal is to have the paravertebral muscles act as an internal support for the lumbar spine. McKenzie also feels that extending the spine moves the nucleus pulposus anteriorly (entry 23). Theoretically, these exercises can be used to keep the discs in anatomic position. Unfortunately, again there is no proof that this occurs.

The author feels that an exercise regimen is very important for the rehabilitation of low back patients. This regimen (Figure 3) should not be instituted while the patient is experiencing acute pain but may be started after his symptoms have subsided to the point where no list or paravertebral spasm is present. The number of repetitions is increased gradually; if the patient has any recurrence of acute symptoms, the exercises are stopped. The patient is then closely monitored; when his symptoms again decrease, the exercises can be resumed. The author's preference is for the isometric flexion exercises because of extensive experience with them. It should be stressed that there is no proof that exercises decrease recovery time or reduce the frequency of recurrences. Empirically, they appear to have a positive psychologic effect and give the patient an active part in his treatment program.

§ 3-3(G). Back School.

The concept of a back school was originated in 1970 by Zachrissen-Forcel in Sweden (entry 47). The concept is that if a patient understands the anatomic, epidemiologic, and biomechanical factors that give rise to low back pain, he can handle his own problem better than without this knowledge.

An audiovisual approach to teaching this material was developed. The patient is first introduced to the goals of the program. Then the basic anatomy and physiology involved in various back disorders are presented in a classroom setting, so that each patient realizes his problem is not unique. Next the patient is taught proper biomechanics of the spine and how to apply these to his everyday activities so as to reduce the forces applied to the spine. Finally, an exercise program is outlined for him. This multi-faceted approach has been quite successful and has yielded more encouraging results than routine physiotherapeutic modalities.

114

Many individuals have produced variations on the above program (entry 45). Some of these back school programs are complex and costly, involving as much as fifteen hours of classroom time. The author feels that the simpler the program, the better accepted it will be by the patient. There is no evidence that back school can prevent or decrease incidence of back pain. Its aim is to give the patient a basic understanding of his problem so he will not be afraid of his back when dealing with everyday situations.

§ 3-3(H). Physical Therapy.

There are many other treatment modalities used for low back pain. These include local heat, light massage, and ultrasound. They are well tolerated and pleasant. Most patients experience some immediate relief of symptoms but, unfortunately, not long-lasting relief. There is no evidence that any of these treatment regimens offers any long-term benefit or even adds to the efficacy of bedrest alone.

§ 3-4. The Multiply Operated Low Back Patient.

The patient who has undergone one or more back operations and continues to have significant discomfort is becoming an ever-increasing problem. It is estimated that 300,000 new laminectomies are performed each year in the United States alone and that fifteen percent of these patients will continue to be disabled. The inherent complexity of these cases necessitates a method of problem solving that is precise and unambiguous.

The best possible solution for preventing recurrent symptoms after spine surgery is to prevent inappropriate spine surgery whenever possible (entries 17, 35, 48). This stresses the earlier point that proper surgical indications for the initial procedure should be strictly followed. The idea of "exploring" the low back when the necessary objective criteria are not present is no longer acceptable. In fact, even when there are objective findings but the patient is psychologically unstable or there are compensation-litigation factors, the outcome of low back surgery is uncertain (entry 35). Thus, the initial decision to operate is the most important one. Once the situation of recurrent pain after surgery arises, the potential for a solution is limited at best.

The problem confronting the physician is to distinguish the patient with a mechanical lesion from the patient whose symptoms are secondary to a non-mechanical problem. The types of mechanical lesions include recurrent herniated disc, spinal instability, and spinal stenosis. These three entities produce symptoms by causing direct pressure on the neural elements and are amenable to surgical intervention. The non-mechanical lesions consist of scar tissue, whether arachnoiditis or perineural fibrosis, psychosocial instability, or a medical disease. These problems will not be helped by additional lumbar spine surgery.

The keystone for successful treatment is to obtain an accurate diagnosis. Although a seemingly obvious need, this essential step often is not taken. Consequently the rehabilitation of this patient group has been fraught with difficulty. The goals of this section are to analyze the significant points in the evaluation of the multiply operated back patient and to present an algorithm (flow chart) for obtaining a specific diagnosis.

§ 3-4(A). Evaluation.

Each multiply operated back patient has an involved history. Many patients want to relate their entire story to the evaluating physician, and it is best to let them do so. After the patient finishes talking, however, there are several important points in the history that must be elucidated so that the proper decision making process can be followed.

First, it should be determined if the patient's complaint is based on a non-orthopaedic cause such as pancreatitis or an abdominal aneurysm. Thus, a thorough general medical examination should be obtained routinely. If this early examination reveals anything significant, it should be treated appropriately. In addition, if there is any indication of psychosocial instability, evidenced by alcoholism, drug dependence, or depression, a thorough psychiatric evaluation is necessary. It has been clearly demonstrated that persons with profound emotional disturbances do not derive any observable benefit from additional surgery (entry 34). In many cases, once a patient's underlying psychosocial problem has been treated successfully, his somatic back complaints and disability will disappear.

If it is determined that the lumbar spine is the probable source of the patient's complaints, three specific historical points need clarification. The first factor is the number of previous lumbar spine operations the patient has undergone. It has been shown that with every subsequent operation, regardless of the diagnosis, the percentage of good results decreases. Statistically, the second operation has a fifty percent chance of success, and beyond two operations, patients are more likely to be made worse than better (entries 11, 38).

The next important historical point is determination of the pain-free interval following the patient's previous operation. If the patient awoke from surgery with pain still present, the nerve root may not have been properly decompressed, or the wrong level may have been explored. If the pain-free interval was longer than six months, the patient's recent pain may be on the basis of a recurrent herniated disc at the same or a different level. If the pain-free interval was between one and six months, the diagnosis most often is arachnoiditis (entry 11).

117

Finally, the patient's pain pattern must be evaluated. If leg pain predominates, a herniated disc or spinal stenosis is most likely. If back pain predominates, instability, tumor, infection, or arachnoiditis are the major considerations. If both back and leg pain are present, spinal stenosis or arachnoiditis are the possibilities.

Physical examination is the next major step in the evaluation of the multiply operated back patient. The neurologic findings and existence of a tension sign, such as the straight leg-raising test or sitting root test, are noted. It is helpful to have the results of a dependable previous examination so a comparison between the preoperative and postoperative states can be made. If the neurologic picture is unchanged from before the previous surgery, and the tension is negative, mechanical compression is unlikely. If, however, there is a new neurologic deficit, or the tension sign is positive, pressure on the neural elements is possible.

Roentgenographic studies are the last major part of the patient's workup. Again, it is most helpful to have a previous set of plain x-rays, CAT scan, and myelogram studies for comparison of the pre- and postoperative situations. The plain x-rays are evaluated for the extent and level of previous laminectomy(ies) and for evidence of spinal stenosis. Weightbearing lateral flexion-extension x-rays of the lumbar spine are examined to see if instability is present. An unstable spine may be the result of the patient's intrinsic disease or secondary to a previous surgical procedure.

Computerized axial tomography (CAT scan) is employed to evaluate for spinal stenosis. Bony encroachment in the lateral recesses and foramina can be specifically observed with this study. With the addition of intravenous contrast, the CAT scan can help to distinguish between scar tissue and a recurrent herniated disc. As the technology of the CAT scan advances, it will become an even more important part of this evaluation.

Contrast studies (myelogram) are reserved for those patients who have either a new neurologic deficit or a positive tension sign. Also, preoperative patients with a diagnosis of spinal stenosis or spinal instability should undergo a myelogram to make sure there are no additional pathologic features. A myelogram is an invasive procedure. Pain in and of itself is not an adequate justification for myelography. Metrizamide, a water-soluble contrast agent, is now employed when the problem is confined to the lumbar spine.

§ 3-4(B). An Algorithm for the Multiply Operated Back.

A specific diagnosis is necessary if an additional operation on the spine is to succeed. Basically, the physician is trying to separate the patients whose symptoms are on a mechanical basis from those who have pain secondary to scar tissue.

When a patient first arrives at the multiply operated back clinic, an information form is completed, as shown in Figure 4. This is most important, for these patients are difficult to evaluate. If the information is obtained in an organized manner there is less chance of missing important details. A physician's or hospital's summary form and the x-rays, CAT scans, and myelograms (not the reports) completed before the initial and most recent surgery must be obtained. This information is invaluable in the decision making process; in many cases one will discover that the initial operation was not indicated.

Figure 5 is a summary of the various pathologic entities with their associated signs, symptoms, and x-rays. Figure 6 is a graphic display using the information form Figure 5 in algorithmic form. Each of the pathologic problems that respond to surgery can be differentiated from arachnoiditis which is not amenable to surgery. It must be stressed that

119

the incidence of surgical problems is *much* lower than that of scar tissue.

As already discussed, the first two steps in the work-up of a failed back patient are the medical and psycosocial evaluations. If anything is uncovered, it should be appropriately treated; if not, the various mechanical etiologies (herniated disc, spinal instability, spinal stenosis) need to be considered.

§ 3-4(C). Herniated Intervertebral Disc.

Three possibilities exist if the patient's pain is caused by a herniated disc. First, the original disc may not have been satisfactorily removed. This can happen if the wrong level was operated on, the laminectomy performed was not large enough to free the neural elements, or a fragment of disc material was left behind. Such patients will have leg pain predominating, and their neurologic findings, tension signs, and myelographic patterns will remain unchanged from the preoperative state. The differentiating feature is that they will have had no pain-free interval; they will have awakened from the operation complaining of their preoperative pain. Patients in this category will be aided by a technically correct laminectomy.

A second possibility is that there is a recurrent herniated intervertebral disc at the previously decompressed level. These patients complain of sciatica and have unchanged neurologic findings, tension signs, and myelograms. The distinguishing characteristic here is that their pain-free interval is greater than six months. Another operative procedure is indicated in patients with this situation.

Finally, a recurrent disc may occur at a different level. Such patients will have a pain-free interval greater than six months, sciatica predominating, and positive tension signs. However, a neurologic deficit, if present, and the myelogram or CAT scan findings, will occur at a different level. A repeat operation for these patients will be beneficial.

§ 3-4(D). Lumbar Instability.

Lumbar instability is another major condition causing pain in the multiply operated back patient on a mechanical basis. The etiology may be the patient's intrinsic back disease or an excessively wide bilateral laminectomy. Instability is the abnormal movement of one vertebra on another, causing pain. Pseudarthrosis, a failed spine fusion, is included in this category, since its pain would be caused by the instability created by the failed fusion.

Patients with instability will complain predominatly of back pain, and their physical examinations may be negative. The key to diagnosis of these patients is the standing lateral flexion-extension x-ray. Reversal of the normal lordotic curve between two vertebral bodies and/or anterior-posterior translation of one vertebral body on another are the radiographic criteria for diagnosing segmental instability; either would constitute a strong indication for a spinal fusion or repair of the pseudarthrosis.

§ 3-4(E). Spinal Stenosis.

Spinal stenosis also is a major diagnostic entity that can mechanically produce pain. Patients with this problem will complain of both back and leg pain. Their physical examinations often are inconclusive, although a neurological deficit may occur following exercise; this phenomenon is termed a positive stress test. The plain x-rays can be suggestive, displaying facet degeneration, decreased interpedicular distance, decreased sagittal canal diameter, and disc degeneration. A CAT scan will demonstrate bony encroachment upon the neural elements; this is especially helpful in evaluating the lateral recesses. A metrizamide myelogram will show a block or significant narrowing of the dye column at the involved levels. It should be appreciated that spinal stenosis and arachnoiditis can co-exist (entry 9). Current feeling is that if there is definite evidence of bony compression, a laminectomy is indicated. However, if scar tissue is present, the degree of pain relief the patient may anticipate is uncertain.

§ 3-4(F). Arachnoiditis.

Arachnoiditis as a postoperative occurrence constitutes an area of considerable concern to the spinal surgeon. At present there is no effective cure for symptomatic arachnoiditis.

Arachnoiditis is strictly defined as an inflammation of the pia-arachnoid membrane surrounding the spinal cord or cauda equina (entry 3). The condition may be present in varying degrees of severity, from mild thickening of the membranes to solid adhesions. The scarring may be severe enough to obliterate the subarachnoid space, blocking the flow of contrast agents.

Statistically, the histories of these patients will reveal more than one previous operation, and their pain-free intervals will fall between one and six months. They may complain of both back and leg pain. Their physical examinations are not conclusive. The myelogram will be definitive for the diagnosis, displaying non-visualization of the nerve roots within the dural sac and narrowing or blockage of the contrast column at the involved levels. There is no bony pressure on the neural elements, since the symptoms originate from the scar tissue formation.

Arachnoiditis is not amenable to surgical intervention. Patients will have chronic pain and are very difficult to take care of; in most cases, with time their symptoms will decrease in severity, but how much decrease and over how long a time are unpredictable.

These people need encouragement, however, and there are various non-operative measures which can be tried (entries 3, 4, 13, 22, 24, 31, 32, 36, 4). Epidural steroids, transcutaneous nerve stimulation, operant conditioning, bracing, antidepressant medication, and patient education are some of the modalities employed to deal with arachnoiditis. However, for most patients these treatment regimens are at best palliative and frequently offer only short-term relief.

In conclusion, it should be stressed that the physician must take an organized approach to evaluation of the multiply operated low back patient. The origin of the problem in most cases is a faulty decision to perform the original surgical procedure. Further surgery on an "exploratory" basis is not warranted and will lead only to further disability. Another operative procedure is indicated only when there are objective findings.

The etiology of each patient's complaint must be accurately localized and identified. In addition to the orthopaedic evaluation, the patient's psychosocial and general medical status need thorough investigation. Once the spine is identified as the source of the patient's symptoms, specific features should be sought in the patient's clinical history, physical examination, and radiographic studies. The number of previous operations, characteristics of the pain-free interval, and predominance of leg pain versus back pain are the major historical points. The most important aspects of the physical examination are the neurologic findings and the presence of a tension sign. Plain x-rays, motion films, myelography, and computerized axial tomography all have specific meaning in the work-up. When all of the information is integrated, the physician usually can separate the patients with arachnoiditis from those with mechanical problems of the spine.

123

The chance of returning a multiply operated low back patient to any kind of heavy work is non-existent. Depending on the type of previous surgery and the patient's active complaints, there will be a permanent disability involved, with significant work restrictions. It is felt that these patients need job retraining as quickly as possible. As already stated, low back surgery does not return patients to heavy work.

BIBLIOGRAPHY

1. Anonymous. The scientific status of the fundamentals of chiropractic analysis and recommendations. *National Institute of Neurological and Communicative Disorders and Stroke.* April 8, 1975.
2. Barretter, R.R. A double blind comparative study of carisoprodol, propoxyphene and placebo in the management of low back syndromes. *Curr. Ther. Res.,* 20:233-240, 1976.
3. Burton, C. Lumbosacral arachnoiditis. *Spine,* 3:24, 1978.
4. Coventry, M.B., and Stauffer, R.N. The multiply operated back. In *American Academy of Orthopaedic Surgeons: Symposium on the Spine.* St. Louis: C.V. Mosby, 1969, pp. 132-142.
5. Davies, J.E., Gibson, T., and Tester, L. The value of exercises in the treatment of low back pain. *Rheum. Rehabil.,* 18:243-247, 1979.
6. DePalma, A., and Rothman, R.H. *The Intervertebral Disc.* Philadelphia: W.B. Saunders Co., 1970.
7. DePalma, A., and Rothman, R.H. Surgery of the lumbar spine. *Clin. Orthop.,* 63:162-170, 1969.
8. Doran, D.M., and Newell, D.J. Manipulation in treatment of low back pain: a multi-center study. *Br. Med. J,* 7:161-164, 1975.
9. Epstein, B.S. *The Spine.* Philadelphia: Lea and Febiger, 1962.
10. Fager, C.A., and Friedler, S.R. Analysis of failures and poor results of lumbar spine surgery. *Spine,* 5:87, 1980.
11. Finnegan, W.J., Tenlin, J.M., Marvel, J.P., Nardine, R.J., and Rothman, R.H. Results of surgical intervention in the symptomatic multiply-operated back patient. *J. Bone Joint Surg.,* 61-A:1077, 1979.
12. Floman, Y., Wiesel, S.W., and Rothman, R.H. Cauda equina syndrome presenting as a herniated lumbar disk. *Clin. Orthop.* 147:234-237, 1980.

13. Ghormley, R.K. The problem of multiple operations on the back. In *American Academy of Orthopaedic Surgeons Instructional Course Lectures, Volume XIV.* Ann Arbor: J.W. Edwards Co., 1957, p. 56.

14. Glover, J.R., Morris, J.G., and Khosla, T. Back pain: a randomized clinical trial of rotational manipulation of the trunk. *Br. J. Ind. Med.,* 31:59-64, 1974.

15. Hakelius, A. Long term follow-up in sciatica. *Acta. Orthop. Scand, Supp.* 129:33-41, 1972.

16. Hingorani, K., and Templeton, J.S. A comparative trial of azapropazone and ketoprofen in the treatment of acute backache. *Curr. Med. Res. Opin.,* 3:407-412, 1975.

17. Hirsch, C. Efficiency of surgery in low back disorders. *J. Bone Joint Surg.,* 47-A:991, 1965.

18. Hitselberger, W.E., and Witten, R.M. Abnormal myelograms in asymptomatic patients. *J. Neurosurg.,* 28:204-206, 1968.

19. Jaffe, G. A double blind multi-center comparison of naproxen and indomethacin in acute musculoskeletal disorders. *Curr. Med. Res. Opin.,* 4:373-380, 1976.

20. Kendall, P.H., and Jenkins, J.M. Exercises for backache: a double-blind controlled trial. *Physiotherapy,* 54:154-157, 1968.

21. Lidstrom, A., and Zachrissen, M. Physical therapy on low back pain and sciatica: an attempt at evaluation. *Scand. J. Rehab. Med.,* 2:37-42, 1970.

22. Loeser, J.D., Black, R.G., and Christman, A. Relief of pain by transcutaneous stimulation. *J. Neurosurg.,* 42:308, 1975.

23. McKenzie, R.A. *The Lumbar Spine.* New Zealand: Spinal Publications, Ltd., 1981.

24. Mooney, V. Innovative approaches to chronic back disability. Given as an Instructional Course Lecture at the American Academy of Orthopaedic Surgeons' annual meeting, Dallas, 1974.

25. Mooney, V., and Cairns, D. Management of patients with chronic low back pain. *Orthop. Clin. North Am.,* 9:543-557, 1978.

26. Morris, J.M. Low back bracing. *Clin. Orthop.,* 103:120-132, 1974.

27. Nachemson, A. The load on lumbar disks in different positions of the body. *Acta. Orthop. Scand,* 36:426, 1965.

28. Nachemson, A. The lumbar spine — An orthopaedic challenge. *Spine,* 1:59-71, 1976.

29. Nachemson, A., and Elfstrom, G. Intravital dynamic pressure measurements in lumbar disc. *Scand. J. Rehabil. Med. Supp.* 7:5-40, 1970.

30. Nachemson, A., and Morris, J.M. In vivo measurements of intradiscal pressure: discometry, a method for determination of pressure in the lower lumbar disc. *J. Bone Joint Surg.*, 46-A:1077, 1964.
31. Oudenhover, R.C. The role of laminectomy, facet rhizotomy and epidural steroids. *Spine*, 4:145, 1979.
32. Rose, D.L. The decompensated back. *Arch. Phys. Med. Rehabil.*, 56:51, 1975.
33. Rothman, R.H. Indications for lumbar fusion. *Clin. Neurosurg.*, 71: 215-219, 1973.
34. Rothman, R.H., and Simeone, F.A. *The Spine*, 2nd ed. Philadelphia: W.B. Saunders Co., 1982.
35. Spengler, D.M., and Freeman, D.W. Patient selection for lumbar discectomy: An objective approach. *Spine*, 4:129, 1979.
36. Tibodeau, A.A. Management of the problem postoperative back (in Proceedings of the American Orthopaedic Association). *J. Bone Joint Surg.*, 55-A:1766, 1973.
37. Vignon, G. Comparative study of intravenous ketoprofen versus aspirin. *Rheum. Rehabil.*, 15:83-84, 1976.
38. Waddell, G., Kummel, E.G., Lotto, W.N., Graham, J.D., Hall, H., and McCulloch, J.A. Failed lumbar disc surgery and repeat surgery following industrial injuries. *J. Bone Joint Surg.*, 61-A:201, 1979.
39. Weber, H. The effect of delayed disc surgery on muscular paresis. *Acta. Orthop. Scand.*, 46:631-642, 1975.
40. Weber, H. An evaluation of conservative and surgical treatment of lumbar disc protrusion. *J. Oslo City Hosp.*, 20:81-93, 1970.
41. Weber, H. Lumbar disc herniation: a prospective study of prognostic factors including a controlled trial. *J. Oslo City Hosp.*, 28:36-61, 89-103, 1978.
42. Weber, H. Traction therapy in sciatica due to disc prolapse. *J. Oslo City Hosp.*, 23:167-179, 1973.
43. White, A.H., Derby, R., and Wynne, G. Epidural injections for the diagnosis and treatment of low back pain. *Spine*, 5:78, 1980.
44. Wiesel, S.W., Cuckler, J.M., DeLuca, F., et al. Acute low back pain: An objective analysis of conservative therapy. *Spine*, 5:324-330, 1980.
45. Williams, S.J. Back school. *Physiotherapy*, 63:590, 1977.
46. Wilson, J. Low back pain and sciatica. *JAMA*, 200:129, 1967.
47. Zachrissen, M. *The Low Back Pain School.* Danderyd, Sweden: Danderyd's Hospital, 1972.

Figure 1
Low Back Pain Algorithm

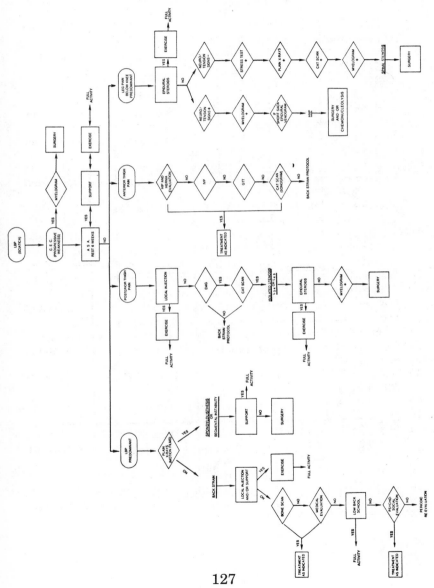

127

Figure 2

Differential Diagnosis of Low Back Pain

EVALUATION	BACK STRAIN	HNP	SPINAL STENOSIS	SPONDYLO-LISTHESIS /INSTABILITY	TUMOR	SPONDYLO-ARTHROPATHY	METABOLIC	INFECTION
PREDOMINANT PAIN (LEG vsBACK)	BACK	LEG	LEG	BACK	BACK	BACK	BACK	BACK
CONSTITUTIONAL SYMPTOMS					+	+		+
TENSION SIGN		+						
NEUROLOGIC EXAM		+	+ AFTER STRESS					
PLAIN X-RAYS			+	+	+/-	+	+	+/-
LATERAL MOTION X-RAYS				+				
C.A.T. SCAN		+	+		+			+
MYELOGRAM		+	+					
BONE SCAN					+	+	+	+
E.S.R.					+	+		+
Ca/P/ALK PHOS					+		+	

128

Figure 3

The George Washington University
Low Back Exercise Program

Do your exercise routine on the floor once a day. Do not start your exercises for the first hour after arising; your spinal discs need that time to develop their protective function.

Begin by doing the first six exercises (nos. 1 through 6), which are flexion exercises, _____ (repetitions) once a day. Increase your repetitions by one (for each exercise) every other day. You may stop the increases when you are doing each of these six exercises _____ times.

Do not begin the last two exercises (nos. 7 and 8), which are extension exercises, until you can do the first six (flexion) exercises comfortably.

Start the extension exercises (nos. 7 and 8) by doing each one the lowest number of repetitions with which you started the flexion exercises. Gradually increase the repetitions until you can do all eight exercises (flexion and extension) the same number of repetitions comfortably.

129

Low Back Exercises

1. **Pelvic Tilt:** Lie on back with knees bent and feet flat on the floor. Squeeze buttocks together and pull stomach in, flattening low back against floor. Hold to count of 5; relax to count of 5.

2. **Pelvic Tilt:** Lie on back with knees straight. Squeeze buttocks together and pull stomach in, flattening low back against floor. Hold to count of 5, relax to count of 5.

3. **To Increase Back Flexibility:** **a)** Bring right knee to chest and give a gentle stretch with arms. **b)** Bring left knee to chest and give a gentle stretch with arms. **c)** Bring right knee to chest; then, while holding that position, bring left knee to chest also. Give both a gentle stretch. Then lower one leg at a time.

4. **To Strengthen Abdominal Muscles:** **a)** Lie on floor with knees bent; tuck chin in and raise upper body slowly until shoulderblades clear the floor; hold to count of 5. **b)** Same position as before; raise upper body, reaching hands toward right side of knees; hold to count of 5. **c)** Same position; reach hands to left sides of knees; hold to count of 5. Progression: Grade I — arms held straight out. Grade II — arms crossed in front of chest. Grade III — hands behind head.

130

5. **To Stretch Hamstrings:** Lie on back with both knees bent. **a)** Straighten one leg up in the air for the count of 5. **b)** Repeat for the other side. **c)** Extend both legs at the same time.

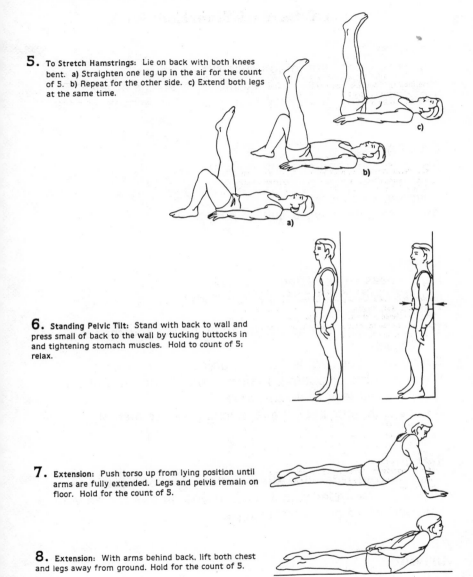

6. **Standing Pelvic Tilt:** Stand with back to wall and press small of back to the wall by tucking buttocks in and tightening stomach muscles. Hold to count of 5; relax.

7. **Extension:** Push torso up from lying position until arms are fully extended. Legs and pelvis remain on floor. Hold for the count of 5.

8. **Extension:** With arms behind back, lift both chest and legs away from ground. Hold for the count of 5.

131

General Care of Your Low Back

Posture
Active posture: standing pelvic tilt
Passive posture: rest one leg higher than the other
Never perform an activity that makes you arch your back

Sitting
Sit no longer than one-half hour at a time
Keep one or both legs elevated
Keep back supported
Avoid leaning forward and arching back

Lifting
Bend knees while lifting
Always hold object close to your body; do not lift above
    chest level
Try to maintain the pelvic tilt while lifting and carrying
    objects
Keep knees bent while carrying objects
While pushing or pulling, maintain the pelvic tilt and use
    hips and legs to do the work
Do not arch your back while reaching — use a stool

Sleeping
Sleep on your side with knees pulled up or on your back
with a pillow under your knees. If you must sleep on your
stomach, place pillow under hips.

Driving
Pull seat close to the pedals

132

Figure 4

The Multiply Operated Back

Name _____ Age _____ Date _____
Weight _____ Height _____ Social Security No. _____

II. History:
    A. Number of back operations ___
    B. First operation: Date _____
        1. Pre-operative Complaints: Back Pain ___ Leg Pain ___
           Combined ___
        2. Pain-free interval after first operation: None ___ Up to
           4 Weeks ___ One month to one year ___ Greater than
           one year ___
        3. Review of physician's report: Neurologic Exam ___
                              Straight Leg Raising Test ___
        4. Review of first pre-operative myelogram (Type) _____
    C. Last Operation: Date _____
        1. Pre-operative complaints: Back Pain ___ Leg Pain ___
           Combined ___
        2. Pain-free interval after last operation: None ___ Up to
           4 weeks ___ One month to one year ___ Greater than
           one year ___
        3. Review of physician's report: Neurologic Exam ___
                              Straight Leg Raising Test ___
        4. Review of last-operative myelogram (Type) _____
    D. Present complaints: Back Pain ___ Leg Pain ___ Combined ___

III. Physical Exam:
    A. Back: Range of Motion (%) ___ Point Tenderness _____
    B. Neurologic Exam: Knee Jerk ___ Ankle Jerk ___ Quadriceps
                        Weakness ___ E.H.L. Weakness _____
    C. Straight Leg Raising ___ Crossed SLR ___

133

IV. X-rays:
   Plain _____ Motion _____ Bone Scan _____ CAT Scan _____
   Myelogram (Type) _____

V. Medical and GYN Evaluation

VI. Psycho-social Assessment: Divorce _____ Alcohol _____ Drugs _____
   Multiple Jobs _____ Depression _____ Compensation _____
   Litigation _____

VII. Diagnosis: HNP (Level)   Spinal Stenosis   Spinal Instability
                 Arachnoiditis   Other

VIII. Treatment Plan

Figure 5

Differential Diagnosis of the Multiply Operated Back

History-Physical Radiographs	Original Disc Not Removed	Recurrent Disc At Same Level	Recurrent Disc At Different Level	Spinal Instability	Spinal Stenosis	Arachnoiditis
# Previous Operations						>1
Pain-Free Interval	None	>6 months	>6 months			>1 month but < 6 months
Predominant Pain (Leg Versus Back)	Leg Pain	Leg Pain	Leg Pain	Back Pain	Back and Leg Pain	Back and Leg Pain
Tension Sign	+	+	+			May Be Positive
Neurologic Exam	+ Same Pattern	+ Same Pattern	+ Different Level		+ after Stress	
Plain X-rays	+ If Wrong Level				+	
Lateral Motion X-rays				+		
Metrizamide Myelogram	+ But Unchanged	+ Same Level	+ Different Level		+	+
C.A.T. Scan					+	

135

Figure 6

Salvage Spine Surgery Algorithm

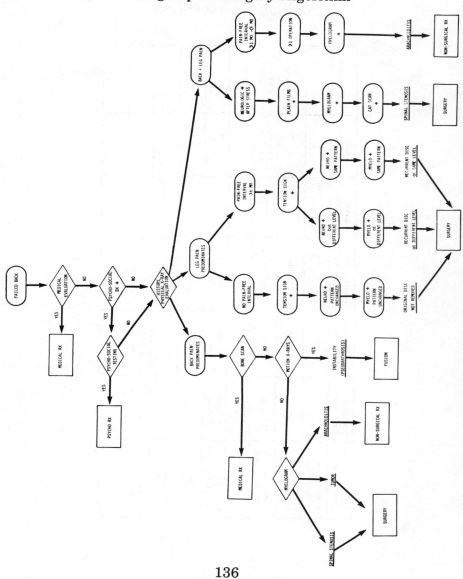

136

CHAPTER 4

NON-MECHANICAL CAUSES OF LOW BACK PAIN

David G. Borenstein, M.D.

PART I. INTRODUCTION

PART II. RHEUMATOLOGIC DISORDERS OF THE LUMBOSACRAL SPINE

139

§ 4-12. Vertebral Osteochondritis.
  § 4-12(A). Prevalence and Etiology.
  § 4-12(B). Clinical Findings.
  § 4-12(C). Physical Findings.
  § 4-12(D). Laboratory Findings.
  § 4-12(E). Radiographic Findings.
  § 4-12(F). Differential Diagnosis.
  § 4-12(G). Treatment.
  § 4-12(H). Course and Outcome.
Bibliography.

§ 4-13. Osteitis Condensans Ilii.
  § 4-13(A). Prevalence and Etiology.
  § 4-13(B). Clinical Findings.
  § 4-13(C). Physical Findings.
  § 4-13(D). Laboratory Findings.
  § 4-13(E). Radiographic Findings.
  § 4-13(F). Differential Diagnosis.
  § 4-13(G). Treatment.
  § 4-13(H). Course and Outcome.
Bibliography.

§ 4-14. Polymyalgia Rheumatica.
  § 4-14(A). Prevalence and Etiology.
  § 4-14(B). Clinical Findings.
  § 4-14(C). Physical Examination.
  § 4-14(D). Laboratory Findings.
  § 4-14(E). Radiologic Findings.
  § 4-14(F). Differential Diagnosis.
  § 4-14(G). Treatment.
  § 4-14(H). Course and Outcome.
Bibliography.

§ 4-15. Fibrositis.
  § 4-15(A). Prevalence and Etiology.
  § 4-15(B). Clinical Findings.
  § 4-15(C). Physical Findings.
  § 4-15(D). Laboratory Findings.
  § 4-15(E). Radiographic Findings.

141

145

146

147

PART I. INTRODUCTION

Many systemic and local disorders of the lumbosacral spine may be associated with pain in the absence of prior trauma or mechanical factors. Diseases which affect bone, cartilage, synovium, nerves, blood vessels, muscle, or connective tissue systemically, have the potential to cause damage in these same structures in the lumbosacral spine. In addition, abnormalities in visceral organs which border the lumbosacral spine (vascular, genitourinary, gastrointestinal) may cause pain which is referred to the back.

The nonmechanical causes of low back pain make up a wide range of diseases including rheumatologic, infectious, infiltrative, endocrinologic, hematologic, and referred pain disorders. Rheumatologic disorders of the spine cause inflammation of the joints of the spine resulting in decreased motion and pain. Infections of the spine may occur in discs, bones, or joints and are associated with both localized and systemic symptoms. Neoplasms, both benign

150

and malignant, may arise in the spine in association with severe, progressive pain. Abnormalities of the lumbosacral spine can relate to a systemic disease when endocrinologic and hematologic disorders are the cause of pain. Referred pain to the lumbosacral spine has its origin in the vascular, genitourinary or gastrointestinal systems and is usually associated with symptoms characteristic of dysfunction in the diseased organ. A careful history, physical examination and laboratory evaluation can differentiate the specific findings which are associated with these heterogeneous disease processes which can cause low back pain.

The diagnostic process, at times, is complicated by events which are related to a patient's work. In most circumstances, the onset of symptoms of diseases associated with low back pain occurs in the absence of any job related event. Occasionally, however, the onset of symptoms may coincide with trauma from a work-related injury and the worker may ascribe all subsequent medical symptoms to that accident. In fact, the trauma is an incidental event, not associated with the true pathologic process causing the pain. Unfortunately, by ascribing the symptoms to trauma, the patient often delays his medical evaluation, allowing the disease process to progress to his own disadvantage.

A patient's work may at times aggravate back symptoms primarily due to non-mechanical disorders and vice versa. Heavy labor may in particular have an adverse effect on them. For example, patients with osteoporosis may be at greater risk for vertebral body compression fractures when they engage in heavy work. Patients with ankylosis of the axial skeleton may react adversely to strenuous activity; poor body positioning may make them vulnerable to extremity injuries.

On the other hand, the non-mechanical disorders of the spine, either because of local or systemic factors, may limit work potential. Patients with ankylosing spondylitis limited to the spine may continue to work, while those with associated hip arthritis may be severely limited in walking and may become disabled. Those with enteropathic arthritis may be minimally limited by back symptoms, but may have occupational difficulties secondary to their inflammatory bowel disease.

This chapter will describe those diseases associated with non-mechanical low back pain. The historical, physical, laboratory, radiographic findings, and therapy of each illness is reviewed. The course of the illnesses and their effect on work potential is discussed.

PART II. RHEUMATOLOGIC DISORDERS OF THE LUMBOSACRAL SPINE

§ 4-1. Introduction.

Rheumatologic disorders of the lumbosacral spine are common causes of lumbar pain. These disorders affect the bone, joints, ligaments, tendons and muscles which are components of the lumbosacral spine. While muscle strain, disease of the intervertebral discs, and osteoarthritis of the lumbosacral spine are frequent causes of low back pain, there are a number of other inflammatory and non-inflammatory disorders associated with pain in the lumbosacral spine. The most important rheumatic disorders which cause inflammation of the joints of the axial skeleton are the seronegative spondyloarthropathies. This group of diseases is characterized by involvement of the sacroiliac joints, peripheral large joint disease and the absence of rheumatoid factor. The seronegative spondyloarthropathies include ankylosing spondylitis, Reiter's syndrome, psoriatic arthritis, enteropathic disease, familial Mediterranean fever, Behcet's syndrome, Whipple's disease, and arthritis associated with hidradenitis suppurativa. They are closely associated with genetic factors which predispose patients to these illnesses. Environmental factors play a role as the triggers of the inflammatory response in genetically predisposed individuals, but they have been only partially identified. Bacterial infection is associated with the onset of Reiter's syndrome and reactive arthritis. The role of trauma as an environmental trigger in ankylosing spondylitis, Reiter's syndrome and psoriatic arthritis remains controversial.

In spondyloarthropathy, the history of pain, which is most severe in the morning and improves with activity, is characteristic. Physical examination demonstrates localized tenderness over the sacroiliac joints and vertebral column, with limitation of motion in all directions. Laboratory abnormalities are consistent with systemic inflammatory disease, but are non-specific. Radiographic evaluation is very useful in identifying characteristic joint space narrowing, sclerosis and fusion in the sacroiliac joints, vertebral body squaring, and ligamentous calcification in the axial skeleton.

Rheumatoid arthritis, an inflammatory arthropathy, may involve the facet joints of the lumbar spine, but it more frequently affects the cervical spine and occurs in the setting of diffuse long standing joint disease. Non-inflammatory lesions affecting bone in the lumbosacral spine include diffuse idiopathic skeletal hyperostosis, osteochondritis and osteitis condensans ilii. Muscle syndromes associated with low back pain include polymyalgia rheumatica and fibrositis.

In addition to drug therapy, treatment for these rheumatologic disorders involves a number of therapeutic modalities which include patient education, and physical and occupational therapy. Although there are no cures for these illnesses, medical therapy can be very effective in controlling symptoms.

The prognosis of patients in regard to their work potential is related to their specific ailment, sites of involvement, and intensity of symptoms. Some rheumatologic disorders such as osteitis condensans ilii and diffuse idiopathic skeletal hypertrophy, are unassociated with any significant disability. The seronegative spondyloarthropathies may affect work capacity to a varying degree. Patients with limitation of motion of the axial skeleton alone may be able to participate in all occupational tasks except heavy lifting. Patients with axial skelton and hip joint involvement are frequently severely limited in motion and have great difficulty working.

§ 4-2. Ankylosing Spondylitis.

Ankylosing spondylitis is a chronic inflammatory disease characterized by a variable symptomatic course and progressive involvement of the sacroiliac and axial skeletal joints. It is the prototype of the seronegative spondyloarthropathies. This disease complex is characterized by axial skeletal arthritis, the absence of rheumatoid factor in serum (seronegative), the lack of rheumatoid nodules, and the presence of a tissue factor on host cells, HLA-B27. Ankylosing spondylitis is a disease of antiquity, having been found in the remains of mummies from Egypt and having been known to Hippocrates (entry 20 in Bibliography at end of this section). It has also had many names, including rheumatoid spondylitis, Marie-Strumpell disease, von Bechterew's disease, and rheumatoid variant.

§ 4-2(A). Prevalence and Etiology.

Ankylosing spondylitis affects about 1% of the Caucasian population, a number equal to the prevalence for rheumatoid arthritis. Initially, the male to female ratio was thought to be 10:1. More recent studies have demonstrated the ratio to be in the range of 3:1 (entry 6). Women tend to be less symptomatic and develop less severe disease, and this may explain their small representation in earlier studies.

The etiology of ankylosing spondylitis is unknown. In the past, infections, trauma, and heredity were thought to be involved in the pathogenesis of this disease. Bacteria in the urinary tract or gut were considered to be the initiators of an inflammatory response which results in ankylosing spondylitis. However, the findings in studies which demonstrated the increased colonization of the gut and heightened lymphocyte sensitivity to Klebsiella organisms in ankylosing spondylitis patients have not been borne out in subsequent investigations (entries 9, 26).

The role of incidental trauma to the lumbosacral spine in the initiation of the inflammatory process of ankylosing spondylitis is unproven. Patients with ankylosing spondylitis may present with radicular back pain which is thought to be secondary to a herniated lumbar disc. Some of these patients have laminectomies and have gone on to develop classic changes of ankylosing spondylitis in the axial skeleton. One of the potential complications of total hip joint replacement in a patient with ankylosing spondylitis is myositis ossificans, or the calcification of soft tissues surrounding a joint. It may be possible that in the patients predisposed to this disease, significant tissue injury to the lumbosacral spine or peripheral joints can result in an inflammatory process which promotes tissue calcification and joint ankylosis (entry 19).

The difficulty substantiating the role of lumbosacral spine trauma and the development of ankylosing spondylitis is illustrated by two histories. In the case of fraternal twin brothers, one was asymptomatic with no evidence of ankylosing spondylitis. The other brother, who was asymptomatic before a fall at work, developed low back pain. He had a decompression procedure (laminectomy) for a suspected herniated lumbar disc. Subsequently, he developed progressive fusion of the lumbar, thoracic, and cervical spine and peripheral arthritis of the hip (entry 3). In another case, an Egyptian soldier who had symptoms of low back pain compatible with ankylosing spondylitis, was wounded with schrapnel in the lumbar spine and thighs. He was immobilized for an extended period and experienced marked limitation of axial skeleton motion. Radiographic evaluation demonstrated increased spondolytic changes in his axial skeleton (entry 3). Did significant tissue damage to the axial skeleton or peripheral joints in those patients with active ankylosing spondylitis result in an exacerbation of symptoms and progression of disease in the injured area? Would ankylosing spondylitis have developed to the same degree in the absence of spinal surgery or tissue injury? The scientific data are not available to answer these important questions.

A genetic predisposition to ankylosing spondylitis and to the seronegative spondyloarthropathies in general does exist. HLA (Human Leukocyte Antigen) antigens are cell surface markers which are present on all nucleated mammalian cells. The portion of the 6th chromosome of man which determines the expression of the HLA antigen is the major histocompatibility complex (MHC) which is associated with control of the immune response of the host. In the MHC region are loci which code for the A, B, C, and D HLA antigens. Using immunologic methods, antibody for A, B, and C loci, and lymphocytes for the D locus, investigators have identified specific antigens associated with each locus. In ankylosing spondylitis, a strikingly high association between HLA-B27 and the disease has been demonstrated. HLA-B27 is present in over 90% of white patients with ankylosing spondylitis compared to a frequency of 8% in a normal white population (entry 23). HLA-B27 is present in 50% of American blacks with ankylosing spondylitis with a prevalence of 4% in the normal black population (entry 12). African blacks do not have the HLA-B27 antigen and rarely develop ankylosing spondylitis. Other ethnic groups such as the Haida and Pima Indian tribes have a large proportion of individuals affected by ankylosing spondylitis associated with an increased prevalence of HLA-B27. Approximately 20% of individuals who are HLA-B27 have evidence of spondylitis (entry 6). However, genetic factors alone will not result in the expression of the disease. Many individuals who are HLA-B27 positive have no evidence of a spondyloarthropathy. Identical twins who are HLA-B27 positive may be discordant for ankylosing spondylitis. One twin may have ankylosing spondylitis, while the other may be normal or develop symptoms and signs of a non-ankylosing spondylitis spondyloarthropathy (entry 13). Environmental factors must play a role in the development of the disease.

Ankylosing spondylitis is a disease of the synovial and cartilaginous joints of the axial skeleton: sacroiliac joints, spinal apophyseal joints, and symphysis pubis. The large appendicular joints, hips, shoulders, knees, elbows, and ankles are also affected in 30% of patients. The inflammatory process is characterized by chondritis, inflammation of cartilage; or osteitis, inflammation of bone, at the junction of the cartilage and bone in the spine. An inflammatory granulation tissue forms and erodes the vertebral body margins. As opposed to rheumatoid arthritis which is associated with osteoporosis as an early manifestation of disease, the inflammation of ankylosing spondylitis is characterized by ankylosis of joints and ossification of ligaments surrounding the vertebrae (syndesmophytes) and other musculotendinous structures such as the heels and pelvis.

§ 4-2(B). Clinical Features.

The classical picture is of a man between the ages of 15 and 40 with intermittent low back pain and stiffness slowly progressing over a period of months (entry 18). Back pain is greatest in the morning and is increased by periods of inactivity. Patients may have difficulty sleeping because of pain and stiffness. The back pain improves with exercise. The mode of onset is variable with a majority of the patients developing pain in the lumbosacral region. In a small number, peripheral joints (hips, knees, and shoulders) are initially involved, but occasionally acute iridocyclitis (eye inflammation) or heel pain may be the first manifestation of disease. Rarely, patients may develop ankylosis of the spine without any back pain (entry 14). At the other end of the spectrum, back pain may be severe with radiation into the lower extremities, mimicking an acute lumbar disc herniation. The usual patient has a moderate degree of intermittent aching pain which is localized to the lumbosacral area. Paraspinal musculoskeletal spasm may also contribute to the discomfort. With progression of the disease, pain develops in the dorsal and cervical spine and rib joints.

Spinal involvement is manifested by flattening of the lumbar spine and loss of normal lordosis. Thoracic spine disease causes decreased motion at the costovertebral joints, reduced chest expansion, and impaired pulmonary function. Involvement of the cervical spine causes the head to protrude forward, making it difficult to look straight ahead.

Peripheral joint arthritis (hips, knees, ankles, shoulders, elbows) occurs in 30% of patients within the first ten years of disease (entry 7). Joint disease appears as pain and stiffness. The inflammatory process may proceed to joint space narrowing and contractures. Fixed flexion contractures of the hips give rise to difficulty with walking and cause a rigid gait. Hip disease is the most frequent limiting factor in mobility rather than spinal stiffness. In fact, peripheral joint disease, particularly of the hips, which appears in the first ten years of disease, is associated with greater disease activity and more extensive restriction of spinal motion.

Ankylosis may also occur in cartilaginous joints such as the symphysis pubic, sternomanubrial and costosternal joints. Erosions of the plantar surface of the calcaneus at the attachment of the plantar fascia results in an enthesopathy (inflammation of an enthesis-attachment of tendon to bone) (entry 1). This inflammation causes a fasciitis and periosteal reaction with heel pain and the formation of heel spurs. Achilles tendonitis is another enthesopathy associated with heel pain and ankylosing spondylitis.

Ankylosing spondylitis is also associated with many non-articular abnormalities. Constitutional manifestations of disease, such as fever, fatigue, and weight loss are seen in a small number of patients with active disease, particularly in those with peripheral joint manifestations. Iritis, inflammation of the anterior uveal tract of the eye, may be the presenting complaint in 25% of the patients with ankylosing spondylitis and is present in up to 40% of cases over the course of the illness. It is usually unilateral and recurrent and is most often independent of the severity of the joint disease. Iritis is normally treated with topical corticosteroids along with the occasional use of systemic corticosteroids. Mild visual loss may be associated with iritis, but blindness is rare.

Neurologic complications of ankylosing spondylitis are secondary to nerve impingement or trauma to the spinal cord. Patients with long standing ankylosing spondylitis who develop new leg pain, urinary or bowel incontinence may be developing impingement of the nerves at the caudal end of the spinal cord, the cauda equina (entry 22). As impingement progresses, sensory loss in the perineum develops along with thigh and leg weakness. There is no effective therapy.

Ankylosing spondylitis causes two significant changes to occur in the spine over time. Although the disease is associated with calcification of ligaments and joints, the loss of motion in the spine causes the vertebrae to become osteoporotic. Osteoporosis is a process in which there is a loss of bone calcium. Osteoporotic bones are weaker and are at greater risk of fracture (entry 15). The other change is the loss of normal flexibility because of ankylosis of the spinal joints and ligaments. The spine in this ankylosed state is much more brittle and is prone to fracture even with minimal trauma. The most common location for fracture is in the cervical spine though dorsal and lumbar spine fractures have also been described. Patients who develop fractures may complain of nothing more than localized pain and decreased spinal motion, but severe sensory and motor functional loss corresponding to the location of the lesion may develop. Therapy usually consists of external fixation with a brace when neurologic symptoms are minimal, but surgical decompression and fusion for severe neurologic abnormalities such as paraplegia may be required. Fractures which mend with external bracing or with surgical fusion heal with normal bone formation.

Another complication of long standing ankylosing spondylitis is spondylodiscitis, a destructive lesion of the disc and its surrounding vertebral bodies (entry 27). This lesion is associated with a fresh onset of localized pain in the spine, which uncharacteristically for the ankylosing spondylitis patient, is improved with bed rest. The etiology of these lesions may be localized inflammation or minor trauma. In one study, patients with ankylosing spondylitis who did heavy manual labor were at greater risk of developing this abnormality than those who had sedentary occupations (entry 8). In most cases, external immobilization was effective in controlling symptoms while surgical fusion was reserved for the more severely affected patients.

Cardiac involvement occurs in 10% of patients with disease durations of 30 years or longer. Mild features include tachycardia, conduction defects, and pericarditis. The most serious cardiac abnormality is proximal aortitis which results in aortic valve insufficiency, heart failure, and death (entry 5). Prosthetic valve replacement may forestall cardiac deterioration.

Pulmonary involvement is manifested by decreased chest expansion which limits lung capacity. In addition, a fibrotic process affects the apical segments of the lung as a late and rare complication.

Amyloid, a deposition of a protein-like material in a number of visceral organs, is a very rare complication of ankylosing spondylitis.

§ 4-2(C). Physical Findings.

Physical examination shows tenderness on palpation over the sacroiliac joints and diminished spinal motion in all planes. Measurements of spinal motion including Schober's test (lumbar spine motion), lateral bending of lumbosacral spine, occiput to wall (cervical spine motion) and chest expansion are important in ascertaining limitations of motion and following the progression of the disease. Peripheral joint examination is also indicated. Examination of the eyes, heart, lungs and nervous system may uncover unsuspected extra-articular disease.

§ 4-2(D). Laboratory Findings.

Laboratory results are non-specific and add little to the diagnosis of ankylosing spondylitis. A minority of patients have a mild anemia. Erythrocyte sedimentation rate is raised in 80% of patients with active disease. The rheumatoid factor and antinuclear antibody are characteristically absent.

Histocompatibility testing (HLA) is positive in 90% of patients with ankylosing spondylitis but is also present in an increased percentage of patients with other spondyloarthropathies (Reiter's syndrome, psoriatic spondylitis, spondylitis with inflammatory bowel disease). It is not a diagnostic test for ankylosing spondylitis. HLA testing may be useful in the young patient with early disease where the differential diagnosis may be narrowed by the presence of HLA-B27 positivity.

§ 4-2(E). Radiologic Findings.

Characteristic changes of ankylosing spondylitis in the sacroiliac joints and lumbosacral spine are very helpful in making a diagnosis but may be difficult to determine in the early stages of the disease (entry 17). Both sacroiliac joints are involved with a loss of subchondral bone, joint widening, reactive sclerosis and eventual ankylosis. Typical radiographic abnormalities in the spine include "squaring" of the vertebral bodies, symmetrical marginal syndesmophyte formation and apophyseal joint fusion which proceeds from the sacrum to the neck. Scintigraphic evaluation by bone scan may demonstrate increased activity over the sacroiliac joints in early disease prior to any detectable radiographic changes in the spinal joints (entry 21).

§ 4-2(F). Differential Diagnosis.

The Rome clinical criteria, used in studies of ankylosing spondylitis, include bilateral sacroiliitis on radiologic examination plus low back pain for more than three months which is not relieved by rest, pain in the thoracic spine,

162

limited motion in the lumbar spine, and limited chest expansion or iritis (entry 16). In the office setting, a physician makes the diagnosis of ankylosing spondylitis when the patient is a young male, with bilateral sacroiliac pain, lumbar spine stiffness improved with activity, recurring radicular pain which alternates side to side, a history of iritis, radiologic changes of spondylitis, and HLA-B27 positivity. However, in the early stages of disease, clinical symptoms may be mild or atypical and radiologic changes nonexistent, preventing early diagnosis.

The differential diagnosis of low back pain in the young patient includes other spondyloarthropathies, herniated lumbar disc, DISH, osteitis condensans ilii, osteoarthritis, rheumatoid arthritis, fibrositis, and tumors. Characteristics of these specific diseases are listed in Table I at end of § 4-14(H), and in other portions of this chapter.

§ 4-2(G). Treatment.

The goals of therapy, as with other forms of inflammatory arthritis, are to control pain and stiffness, reduce inflammation, maintain function and prevent deformity. Patients require a comprehensive program of education, physiotherapy, medications and other measures. Patients are educated about their disease and are encouraged to continue with as normal a life style as possible. Patients are taught proper posture and mobilizing and breathing exercises to prevent the tendency to stoop forward and lose chest motion. The importance of a firm upright chair for sitting and a hard mattress with no pillows for sleeping is stressed. The genetic implications of the disease are placed in perspective. Offspring are at 10% risk of developing ankylosing spondylitis if a parent is HLA-B27 positive.

Medications to control pain and inflammation are useful in the patient with ankylosing spondylitis (Table 2, at end of § 4-13(H)) (entry 11). In the patient with mild spinal or

163

peripheral joint disease, aspirin may be somewhat effective. However, in patients with more severe disease, aspirin is usually ineffective. Other nonsteroidal anti-inflammatory drugs, which decrease pain and inflammation have been demonstrated to be effective in ankylosing spondylitis. The two most commonly used drugs for ankylosing spondylitis have been indomethacin and phenylbutazone. These drugs are effective in controlling spinal and peripheral joint disease but are associated with potentially serious side effects (entry 10). In one retrospective radiographic study phenylbutazone decreased the rate of progression of axial skeletal disease in ankylosing spondylitis (entry 2). Newer anti-inflammatory drugs, such as ibuprofen, tolmetin, fenoprofen, piroxicam, sulindac, and mefenamic acid may also be effective in ankylosing spondylitis (entry 25). However, these drugs have only been available for clinical use for a relatively short period of time. Data on their chronic use for this disease are relatively scanty.

Systemic corticosteroids are rarely needed and are ineffective for the articular disease of ankylosing spondylitis. Occasionally systemic corticosteroids are needed to control persistent iritis. Intra-articular injection of a long-acting steroid preparation is indicated if the patient has peripheral joint disease with persistent effusions.

Intramuscular gold salt injections are not indicated for the axial skeletal disease of ankylosing spondylitis, but they may be helpful in the rare patient with peripheral joint disease who demonstrates persistent synovitis and joint destruction. Antimalarials and immunosuppressive medications are not indicated.

Orthopaedic appliances, such as a heel cup for plantar fasciitis, or a temporary spinal brace to prevent forward flexion, are useful in appropriate patients. In the patient with fixed flexion deformity of the hip, total joint replacement relieves pain and increases mobility. Surgical procedures on the spine, such as a lumbar spine osteotomy, are limited to patients who have such a degree of forward flexion as to prevent them from looking up from the ground (entry 24). This operation has multiple potential complications and is viewed as a last resort procedure.

Radiotherapy was used frequently in patients with ankylosing spondylitis up to the 1960's and it was effective at controlling pain and stiffness in the lumbosacral spine (entry 4). The effects of this therapy, however, were transient and did not prevent progression of the disease. In addition, long term follow-up studies on these patients demonstrated an increased mortality secondary to leukemia. The therapy has been abandoned except for the rarest of patients who is intolerant of all other treatment.

§ 4-2(H). Course and Outcome.

The general course of ankylosing spondylitis is benign and is characterized by exacerbations and remissions. Many patients with ankylosing spondylitis may have sacroiliitis with mild involvement of the lumbosacral spine. Limitation of lumbosacral motion may be mild. The disease can become quiescent at any time. Patients who go on to develop total fusion of the spine may feel better since ankylosis of the spinal joints is associated with decreased pain.

Most studies report that the majority of patients remain functional and employed over the course of the illness (entry 7). The prime predictor of more severe dysfunction is the presence of peripheral joint involvement, particularly in the hips, and this usually appears within the first ten years of disease. A majority of these patients developed severe spinal restriction. Patients with fixed flexion contractures of the hips and ankylosis of the spine are severely limited in their functional capacity; however, total hip joint replacement may improve their mobility. Patients with spinal rigidity, but normal hip function, have minimal disability. However, they should avoid heavy labor, such as lifting objects heavier than 40 pounds.

BIBLIOGRAPHY

1. Ball, J. Enthesopathy of rheumatoid and ankylosing spondylitis. *Ann. Rheum. Dis.,* 30:213-223, 1971.
2. Boersma, J.W. Retardation of ossification of the lumbar vertebral column in ankylosing spondylitis by means of phenylbutazone. *Scand. J. Rheumatol.,* 5:60-64, 1976.
3. Borenstein, D. Unpublished observation.
4. Brown, W.M.C., and Doll, R. Mortality from cancer and other causes after radiotherapy for ankylosing spondylitis. *Br. Med. J.,* 2:1327-1332, 1965.
5. Bulkley, B.H., and Roberts, W.C. Ankylosing spondylitis and aortic regurgitation: description of the characteristic cardiovascular lesion from study of eight necropsy patients. *Circulation,* 48:1014-1027, 1973.
6. Calin, A., and Fries, J.F. Striking prevalence of ankylosing spondylitis in "healthy" W27 positive males and females: a controlled study. *N. Eng. J. Med.,* 293:835-839, 1975.
7. Carette, S., Graham, D., Little, H., Rubenstein, J., and Rosen, P. The natural disease course of ankylosing spondylitis. *Arthritis Rheum.,* 26:186-190, 1983.
8. Cawley, M.I.D., Chalmers, T.M., Kellgren, J.H., and Ball, J. Destructive lesions of vertebral bodies in ankylosing spondylitis. *Ann. Rheum. Dis.,* 31:345-358, 1972.
9. Edmonds, J., Macauley, D., Tyndall, A., Liew, M., Alexander, K., Geczy, A., and Baskin, H. Lymphocytotoxicity of anti-Klebsiella antesera in ankylosing spondylitis and related arthropathies: patient and family studies. *Arthritis Rheum.,* 24:1-7, 1981.
10. Fowler, P. Phenylbutazone and indomethacin. *Clin. Rheum. Dis.,* 1:267-283, 1975.
11. Godfrey, R.G., Calabro, J.J., Mills, D., and Matty, B.A. A double blind crossover trial of aspirin, indomethacin and phenylbutazone in ankylosing spondylitis. *Arthritis Rheum.,* 15:110, 1972.
12. Good, A.E., Kawaniski, H., and Schultz, J.S. HLA-B27 in blacks with ankylosing spondylitis or Reiter's disease. *N. Eng. J. Med.,* 294:166-167, 1976.
13. Hochberg, M.C., Bias, W.B., and Arnett, F.C., Jr. Family studies in HLA-B27 associated arthritis. *Medicine,* 57:463-475, 1978.

14. Hochberg, M.C., Borenstein, D.G., and Arnett, F.C. The absence of back pain in classical ankylosing spondylitis. *Johns Hopkins Med. J.,* 143:181-183, 1978.
15. Hunter, T., and Dubo, H. Spinal fractures complicating ankylosing spondylitis. *Ann. Intern. Med.,* 88:546-549, 1978.
16. Kellgren, J.H. Diagnostic criteria for population studies. *Bull. Rheum. Dis.,* 13:291-292, 1962.
17. McEwen, C., DiTata, D., Ling, G.C., Porini, A., Good, A., and Rankin, T. Ankylosing spondylitis and spondylitis accompanying ulcerative colitis, regional enteritis, psoriasis and Reiter's disease: A comparative study. *Arthritis Rheum.,* 14:291-318, 1971.
18. Neustadt, D.H. Ankylosing spondylitis. *Postgrad Med.,* 61:124-135, 1977.
19. Resnick, D., Dwosh, I.L., Goergen, T.G., Shapiro, R.F., and D'Ambrosia, R. Clinical and radiographic "reankylosis" following hip surgery in ankylosing spondylitis. *Am. J. Roentgenol.,* 126:1181-1188, 1976.
20. Ruffer, A. Arthritis deformans and spondylitis in ancient Egypt. *J. Pathol.,* 22:159-196, 1918.
21. Russel, A.S., Lentle, B.C., and Percy, J.S. Investigation of sacroiliac disease comparative evaluation of radiological and radionuclide techniques. *J. Rheumatol.,* 2:45-51, 1975.
22. Russell, M.L., Gordon, D.A., Ogryzlo, M.A., and McPhedran, R.S. The cauda equina syndrome of ankylosing spondylitis. *Ann. Intern. Med.,* 78:551-554, 1973.
23. Schlosstein, L., Terasaki, P.I., Bluestone, R., and Pearson, C.M. High association of an HL antigen, W27, with ankylosing spondylitis. *N. Eng. J. Med.,* 288:704-706, 1973.
24. Scudese, V.A., and Calabro, J.J. Vertebral wedge osteotomy: correction of rheumatoid (ankylosing) spondylitis. *JAMA,* 186:627-631, 1963.
25. Simon, L.S., and Mills, J.A. Nonsteroidal anti-inflammatory drugs. *N. Eng. J. Med.,* 302:1179-1185, 1237-1243, 1980.
26. Warren, R.E., and Brewerton, D.A. Faecal carriage of Klebsiella by patients with ankylosing spondylitis and rheumatoid arthritis. *Ann. Rheum. Dis.,* 39:37-44, 1980.
27. Wholey, M.H., Pugh, D.G., and Bickel, W.H. Localized destructive lesions in rheumatoid spondylitis. *Radiology,* 74:54-56, 1960.

§ 4-3. Reiter's Syndrome.

Reiter's syndrome is a disease associated with the triad of urethritis (inflammation of the lower urinary tract), arthritis, and conjunctivitis. Reiter's syndrome is the most common cause of arthritis in young men and primarily affects the lower extremity joints and the low back. The disease results from the interaction of an environmental factor, usually a specific infection, and a genetically predisposed host. The course of the illness, while usually benign, may be chronic and remitting, resulting in significant disability.

Many physicians, including Hippocrates, have written about the apparent relation between venereal and gastrointestinal infections and the development of arthritis. In 1916, Reiter, Fiessinger, and Leroy, described young soldiers with an acute febrile illness that included conjunctivitis, urethritis and polyarthritis appearing after a dysenteric illness (entries 25, 11). However, the triad and the term Reiter's syndrome were not associated until Bauer and Engleman in 1942, referred to the findings of urethritis, conjunctivitis, and arthritis in World War II soldiers as Reiter's disease (entry 5).

§ 4-3(A). Prevalence and Etiology.

Reiter's syndrome occurs in patients throughout the world with no racial or ethnic predisposition. Approximately 1% of patients with the common infection, non-gonococcal urethritis, develop the syndrome. The syndrome develops in between 0.2% and 3% of all patients with enteric infections secondary to Shigella, Salmonella, Campylobacter, and Yersinia. The male to female ratio in venereal infection is in the range of 10:1, while the ratio is 1:1 in large outbreaks secondary to enteric infection. The ratio may also be 1:1 in patients with no antecedent infection.

The etiology of Reiter's syndrome is unknown. The initiation of the disease is thought to be related to dysenteric (epidemic) or venereal (endemic) infections. The post-dysenteric form of Reiter's syndrome follows infections by Shigella dysenteriae and flexneri, Salmonella enteriditis, Yersinia enterocolitica and Campylobacter jejuni. The relationship of sexually acquired genital infections with Chlamydia trachomatis and Ureaplasma urealyticum and the precipitation of joint disease is not clearly established. The distinction of urethritis as an initiating event and as an integral manifestation of the syndrome is a difficult one to make. Patients with enteric infections develop urethritis without urethral infection (entry 23). Chlamydia trachomatis is cultured in only 50% of patients with Reiter's syndrome at the onset of joint disease (entry 17).

Reiter's syndrome may develop in patients who deny enteric or venereal infection. Some of these patients associate the onset of their disease with an episode of joint trauma (entry 1). The trauma is associated with swelling, stiffness and pain in the traumatized joint, and this is followed by the emergence of additional joint symptoms, urethritis, conjunctivitis, or cutaneous lesions (entry 6). Another possible manifestation of a response to trauma in these patients is the presence of bony bridging or nonmarginal syndesmophytes in the spine in the absence of sacroiliitis or back symptoms (entries 7, 8). However, there is no scientific evidence to substantiate the role of trauma as an initiator of Reiter's syndrome.

Characteristic of the spondyloarthropathies in general, Reiter's syndrome is associated with the HLA-B27 antigen. Between 60% and 80% of the patients are positive for HLA-B27. In addition, a majority of those who are negative for this antigen have HLA-B antigens which cross-react with B27, including B7, BW22, B40, and BW42 (entry 2).

§ 4-3(B). Clinical Features.

The classical picture of Reiter's syndrome is in a young man about 25 years old who develops urethritis and a mild conjunctivitis, followed by the onset of a predominantly lower extremity oligoarthritis. The symptoms of urethritis are usually mild with a mucopurulent discharge and dysuria. Men may also develop acute or chronic prostatitis. Women may have vaginitis or cervicitis, although many with these manifestations of genitourinary tract involvement in Reiter's syndrome may be asymptomatic. This paucity of symptoms from the genitourinary tract may in part explain the infrequency of the diagnosis of Reiter's syndrome in women (entry 21). Urethritis occurs in both epidemic and endemic forms of the disease. Up to 93% of patients with Reiter's syndrome will have genitourinary symptoms during the course of their illness (entry 12).

The conjunctivitis is usually mild and is manifested by an erythema (redness) and crusting of the lids. Conjunctival inflammation is usually bilateral and gradually resolves over a few days, but it may recur spontaneously. Acute iritis occurs in 20% of Reiter's syndrome patients and is marked by severe pain, photophobia and scleral injection.

Arthritis may occur one to three weeks after the initial infection. In many patients, arthritis is their only manifestation of disease (entry 3). The term reactive arthritis is used for patients who develop only the arthritis of Reiter's syndrome after an enteric or genitourinary infection. The weight-bearing joints—knees, ankles, and feet—are most frequently affected in an asymmetrical manner. A minority of patients have a persistent monoarthritis as their only articular abnormality. The involved joints are acutely inflamed with some joints developing very large effusions.

Back pain is a frequent symptom of patients with Reiter's syndrome. During the acute course, between 31% and 92%

of the patients may develop pain in the lumbosacral region (entries 12, 22, 24). The pain is of an aching quality and is improved with activity. Occasionally the pain will radiate into the posterior thighs, but rarely below the knees; it may be unilateral. This finding corresponds to the asymmetrical involvement of the sacroiliac joints and it contrasts with the symmetrical involvement of ankylosing spondylitis (entry 27). Sacroiliitis is the cause of back pain in a majority of patients in the acute phase of the disease as measured by increased activity over the sacroiliac joints on scintiscan. Radiographic evidence of sacroiliitis is usually restricted to those patients with severe disease. In retrospective studies, sacroiliitis can be detected radiographically in 9% of patients in the acute phase of illness and in up to 71% of patients who have had disease activity longer than five years (entry 9). Spondylitis affecting the lumbar, thoracic and cervical spine occurs less commonly than sacroiliitis with up to 23% of patients with severe disease showing such involvement (entry 15).

Another musculoskeletal manifestation of Reiter's syndrome is inflammation of the insertion of tendons and fascia. This anatomic structure is called an enthesis and a process which results in its inflammation is an enthesopathy. Heel pain, "lover's heel," secondary to plantar fascia inflammation, is a common finding (entry 4). Other manifestations of enthesopathy in Reiter's syndrome patients are Achilles tendinitis, chest wall pain, dactylitis or "sausage digit," and low back pain with no evidence of active sacroiliitis.

Although not part of the classic triad of the disease, mucocutaneous lesions are very characteristic of Reiter's syndrome. Keratodermia blennorrhagica is a skin rash characterized by waxy, macular lesions, which become vesicular and scale. They are found predominantly on the palms and soles. Histologically, they are indistinguishable

171

from pustular psoriasis. Keratodermia which appears on the glans penis is referred to as circinate balanitis. It occurs in up to 31% of patients. Oral ulcers occur on the palate, tongue, buccal mucosa and lips. These lesions are shallow and painless and occur in 33% of patients. Nail involvement is characterized by opacification and hyperkeratosis.

Constitutional symptoms occur in about a third of patients and are characterized by fever, anorexia, weight loss, and fatigue. Cardiac complications, including heart block and aortic regurgitation, occur as a late manifestation of disease and in 2% of patients (entry 26). Neurologic disease occurs in 1% of patients and is associated with peripheral neuropathy, hemiplegia, and cranial nerve abnormalities (entry 14). Amyloidosis is also a rare and late complication of Reiter's syndrome.

§ 4-3(C). Physical Findings.

Physical examination should include all organ systems which may be involved in the disease process. Many of the manifestations of Reiter's syndrome may be overlooked by patients. Important findings, such as oral ulcers, circinate balanitis, and limitation of lumbosacral spine motion may be missed if not looked for by the physician. Conjunctivitis is manifested by erythema of the conjunctivae and crusting of the lids. Urethritis may be detected only by "milking" the urethra before urination for the presence of a mucopurulent discharge.

A complete musculoskeletal examination should include both upper and lower extremities as well as the axial skeleton. Men tend to have involvement in the knees, ankles, and feet, while women have more upper extremity disease (entry 21). The usual patient will have six or fewer joints affected. Percussion tenderness over the sacroiliac joints may be unilateral as in the asymmetrical involvement in Reiter's syndrome. The mobility of the lumbosacral spine should be measured in all planes of motion. A search for evidence of enthesopathy, heel or Achilles tendon tenderness, is also required.

An examination of the oropharynx, genitals, palms, soles, and nails will cover the areas which are associated with the mucocutaneous lesions of Reiter's syndrome.

§ 4-3(D). Laboratory Findings.

Laboratory results are non-specific and not helpful in making the diagnosis of Reiter's syndrome. A mild anemia of chronic disease, an elevated white blood cell count (leukocytosis) and elevated platelet count (thrombocytosis) is demonstrated in about a third of the patients (entry 1). The erythrocyte sedimentation rate is elevated in 70% to 80% of patients but it does not follow the course of the disease. Synovial fluid analysis demonstrates an inflammatory fluid with no specific abnormalities. Rheumatoid factor and antinuclear antibodies are not present in this illness. HLA-B27 or one of the cross-reactive antigens is present in 80% of patients. HLA-B27 positivity is helpful in differentiating Reiter's syndrome from rheumatoid arthritis (this is more easily done by clinical examination). The test, however, is not helpful in differentiating Reiter's syndrome from the other spondyloarthropathies.

§ 4-3(E). Radiologic Findings.

In patients who do not manifest the complete triad of Reiter's syndrome, radiographic changes are helpful in confirming the diagnosis (entry 19). Joint destruction is most severe in the feet. The hips and shoulders are usually spared. The radiologic correlate of the enthesopathy of Reiter's syndrome is periosteal new bone formation at the attachments of the plantar fascia and Achilles tendon into the calcaneus. Sacroiliac involvement may mimic ankylosing spondylitis (symmetric disease) or may be asymmetric. Spondylitis is discontinuous in its involvement of the axial skeleton (skip lesions) and is characterized by nonmarginal bony bridging of vertebral bodies. These vertebral hyperostoses are markedly thickened compared to the thin syndesmophytes of ankylosing spondylitis.

§ 4-3(F). Differential Diagnosis.

Preliminary criteria for the diagnosis of acute Reiter's syndrome have recently been reported by the American Rheumatism Association (entry 28). Patients with Reiter's syndrome were distinguished from patients with other spondyloarthropathies and gonococcal arthritis by an episode of peripheral arthritis of more than one month's duration, occurring in association with urethritis and/or cervicitis. The differential diagnosis of the patient who presents with low back pain and no other manifestation of Reiter's syndrome is the same as that presented for ankylosing spondylitis. This differential diagnosis includes the other spondyloarthropathies, herniated lumbar disc, DISH, and tumors. In the patient who presents with acute monarticular disease after sexual intercourse, the diagnosis of infectious arthritis secondary to Neisseria gonorrhoeae must not be overlooked (entry 20). Bacterial cultures of the urethra, rectum, pharynx and synovial fluid are necessary to detect the presence of the bacterium.

§ 4-3(G). Treatment.

The therapeutic regimen includes patient education, medications and physical therapy. Patients are confused by the association of Reiter's syndrome and sexual relations. Many develop guilt or anxiety over sexual intercourse. The fact that urethritis may recur without any obvious cause must be made clear to the patient.

Acute joint symptoms are treated symptomatically with nonsteroidal anti-inflammatory drugs and the modalities of physical therapy. The drugs which are most effective for Reiter's syndrome are listed in Table II at end of § 4-14(H). The joint and enthesopathic manifestations of Reiter's syndrome appear to respond better to indomethacin, phenylbutazone, or the other newer nonsteroidal anti-inflammatory drugs than to aspirin. The drugs are continued as long as the patient remains symptomatic. The effect of these agents on the long term course of the disease is unknown.

The role of antibiotic therapy in the acute phase of Reiter's syndrome remains controversial. A recent study demonstrated that antibiotic therapy did not influence the appearance of Reiter's syndrome in patients with non-specific urethritis (entry 16).

Gold salt therapy may be helpful in the patient with progressive, destructive, peripheral joint disease. The immunosuppressive, methotrexate, is reserved for the patient with uncontrolled progression of joint disease and unresponsive, extensive skin involvement (entry 10). Corticosteroids are used as drops for iritis and as long acting preparations for intra-articular injection. Systemic corticosteroids have less of an effect in Reiter's syndrome than they do in rheumatoid arthritis and are rarely used.

§ 4-3(H). Course and Outcome.

Reiter's syndrome has no cure. The course of the illness is unpredictable. About 30% to 40% of patients have a self-limited illness, lasting three months to one year. Another 30% to 50% develop a relapsing pattern of illness with periods of complete remission. The final 10% to 25% develop chronic, unremitting disease associated with significant disability (entries 12, 13). Patients who develop significant disability from Reiter's syndrome typically have painful or deformed feet or visual loss from iritis. A five-year follow-up study of 131 consecutive patients with Reiter's syndrome revealed that 83% of the patients had some disease activity (entry 12). Fifty-one percent of patients had continued low back pain while 45% had persistent foot or heel pain. Thirty-four percent of patients had disease activity which interfered with their job while 26% had to change jobs or were unemployed. Heel involvement at the time of diagnosis was the finding most closely associated with a poor functional outcome. The presence of HLA-B27 did not correlate with functional outcome; however, a report from Finland suggested that the HLA-B27 status of the patient was associated with disease severity (entry 18). Patients who were HLA-B27 positive had more frequent back pain, mucocutaneous and genitourinary symptoms. They also had a longer duration of disease and more frequent chronic low back pain and sacroiliitis. Chronic joint symptoms continued in 68% of 140 Reiter's syndrome patients studied and 41% had chronic back pain. Most patients were able to lead normal lives although 16% of the group had chronic destructive peripheral arthritis.

Reiter's syndrome is no longer considered a benign disease. Early therapy with nonsteroidal anti-inflammatory drugs, physical therapy, and appropriate shoes may have beneficial effects on patient function. Unfortunately, even aggressive therapy is unable to prevent disease activity and progressive disability in many of Reiter's syndrome patients.

BIBLIOGRAPHY

1. Arnett, F.C., Jr. Reiter's syndrome. *Johns Hopkins Med. J.,* 150:39-44, 1982.
2. Arnett, F.C., Hochberg, M.D., and Bias, W.B. Cross-reactive HLA antigens in B27-negative Reiter's syndrome and sacroiliitis. *Johns Hopkins Med. J.,* 141:193-197, 1977.
3. Arnett, F.C., McClusky, E., Schacter, B.Z., and Lordon, R.E. Incomplete Reiter's syndrome: discriminating features and HL-A W27 in diagnosis. *Ann. Intern. Med.,* 84:8-12, 1976.

4. Ball, J. Enthesopathy of rheumatoid and ankylosing spondylitis. *Ann. Rheum. Dis.,* 30:213-223, 1971.
5. Bauer, W., and Engleman, E.P. A syndrome of unknown etiology characterized by urethritis, conjunctivitis, and arthritis (so-called Reiter's disease). *Trans. Assoc. Am. Phys.,* 57:307-313, 1942.
6. Borenstein, D. Unpublished observation.
7. Calin, A., and Fries, J.F. Striking prevalence of ankylosing spondylitis in "healthy" W27 positive males and females: a controlled study. *N. Eng. J. Med.,* 293:835-839, 1975.
8. Cohen, L.M., Mittal, K.K., Schmid, F.R., Rogers, L.F., and Cohen, K.L. Increased risk for spondylitis stigmata in apparently healthy HL-A W27 men. *Ann. Intern. Med.,* 84:1-7, 1976.
9. Csonka, G.W. The course of Reiter's syndrome. *Br. Med. J.,* 1:1088-1090, 1958.
10. Farber, G.A., Forshner, J.G., and O'Quinn, S.E. Reiter's syndrome: treatment with methotrexate. *JAMA,* 200:171-173, 1967.
11. Fiessinger, N., and Leroy, E. Contribution a l'elude d'une epidemic de dysenterie dans le somme. *Bull. Soc. Med. Hop. Paris,* 40:2030-2057, 1916.
12. Fox, R., Calin, A., Gerber, R.C., and Gibson, D. The chronicity of symptoms and disability in Reiter's syndrome: an analysis of 131 consecutive patients. *Ann. Intern. Med.,* 91:190-193, 1979.
13. Good, A.E. Involvement of the back in Reiter's syndrome: follow-up study of thirty-four cases. *Ann. Intern. Med.,* 57:44-59, 1962.
14. Good, A.E. Reiter's disease: a review with special attention to cardiovascular and neurologic sequelae. *Semin. Arthritis Rheum.,* 3:253-286, 1974.
15. Good, A.E. Reiter's syndrome: long-term follow-up in relation to development of ankylosing spondylitis. *Ann. Rheum. Dis.,* 38:39-45, 1979.
16. Keat, A.C., Maini, R.N., Nkwazi, G.C., Pegrum, G.D., Ridgway, G.L., and Scott, J.T. Role of chlamydia trachomatis and HLA-B27 in sexually acquired reactive arthritis. *Br. Med. J.,* 1:605-607, 1978.
17. Keat, A.C., Thomas, B.J., Taylor-Robinson, D., Pegrum, G.D., Maini, R.N., and Scott, J.T. Evidence of chlamydia trachomatis infection in sexually acquired reactive arthritis. *Ann. Rheum. Dis.,* 39:431-437, 1980.
18. Leirisalo, M., Skylv, G., Kousa, M., Voipio-Pulkki, L., Suoranta, H., Nissila, M., Hvidman, L., Nielsen, E.D., Svejgaard, A., Tilikainen, A., and Laitinen, O. Follow-up study on patients with Reiter's disease and reactive arthritis, with special reference to HLA-B27. *Arthritis Rheum.,* 25:249-259, 1982.

19. Martel, W., Braunstein, E.M., Borlaza, G., Good, A.E., and Griffin, P.E., Jr. Radiologic features of Reiter's disease. *Radiology,* 132:1-10, 1979.

20. McCord, W.C., Nies, K.M., and Louie, J.S. Acute venereal arthritis: comparative study of acute Reiter syndrome and acute gonococcal arthritis. *Arch. Intern. Med.,* 137:858-862, 1977.

21. Neuwelt, C.M., Borenstein, D.G., and Jacobs, R.P. Reiter's syndrome: a male and female disease. *J. Rheumatol.,* 9:268-272, 1982.

22. Oates, J.K., and Young, A.C. Sacroiliitis in Reiter's disease. *Br. Med. J.,* 1:1013-1015, 1959.

23. Paronen, I. Reiter's disease: a study of 344 cases observed in Finland. *Acta. Med. Scand.* (Suppl.), 212:1-112, 1948.

24. Popert, A.J., Gill, A.J., and Laird, S.M. A prospective study of Reiter's syndrome: an interim report on the first 82 cases. *Br. J. Vener. Dis.,* 40:160-165, 1964.

25. Reiter, H. Uber eine bisher unerkannte Spirochaeteninfektion (Spirochaetosis arthritica) *Dtsch. Med. Wochenschr,* 42:1535-1536, 1916.

26. Ruppert, G.B., Lindsay, J., and Barth, W.F. Cardiac conduction abnormalities in Reiter's syndrome. *Am. J. Med.,* 73:335-340, 1982.

27. Russel, A., Davis, P., Percy, J.S., and Lentle, G.C. The sacroiliitis of acute Reiter's syndrome. *J. Rheumatol.,* 4:293-296, 1977.

28. Willkens, R.F., Arnett, F.C., Bitter, T., Calin, A., Fisher, L., Ford, D.K., Good, A.E., and Masi, A.T. Reiter's syndrome: evaluation of preliminary criteria for definite disease. *Bull. Rheum. Dis.,* 32:31-34, 1982.

§ 4-4. Psoriatic Arthritis.

Patients with psoriasis who develop a characteristic pattern of joint disease have psoriatic arthritis. The characteristic patterns of psoriatic arthritis include asymptomatic oligoarthritis, symmetric polyarthritis, and spondylitis. The arthritis is slowly progressive and is rarely associated with significant disability.

French physicians, Bagin and Bouidillon, in the last century were the first to name the disease and describe it in detail (entries 5, 7). In the United States, there was hesitancy in ascribing joint disease to psoriasis. Many physicians thought that two common diseases were occurring in patients simultaneously, psoriasis and rheumatoid arthritis. More recent studies, however, have clearly demonstrated the association of psoriasis and arthritis (entry 32).

§ 4-4(A). Prevalence and Etiology.

Precise data concerning the prevalence of psoriasis are not available. Many patients with mild disease may never be seen by physicians. Therefore, only estimates of prevalence have been made and have suggested that 1% to 3% of the population is affected by psoriasis. Psoriasis does, however, occur more commonly in people from temperate climate zones. Prevalence of psoriasis in the United States and Japan is similar. People from eastern and northern Africa are also similarly affected. Psoriasis is rare in southern and western Africa, and this is reflected in the low percentage of American blacks affected, most of whom originated from western Africa. Psoriatic arthritis occurs in 5% to 7% of individuals with psoriasis, and in 0.1% of the general population (entry 16). Psoriasis and psoriatic arthritis occur in equal frequency in both sexes.

The basic abnormality, which results in the increased metabolic activity of the skin, is unknown. Some investigators believe this abnormality resides in the most superficial layers of the skin (epidermis), while others believe in the inner layers of the skin (dermis) as the source of the increased metabolic activity. A psoriatic diathesis exists in patients who are at risk for the disease. Abnormalities in protein synthesis, blood flow, and metabolism are present in normal appearing skin, hair, and nails in these individuals.

A genetic predisposition for the development of psoriasis and psoriatic arthritis does exist. Although a positive family history is obtained in about a third of patients with psoriasis, a definite pattern of inheritance has not been established. The skin disease of psoriasis has been associated with HLA-B13, and HLA-BW17 antigens. Genetic factors also play a role in psoriatic arthritis. Patients with psoriatic arthritis have an increased frequency of HLA-BW38, HLA-DR4 and HLA-DR7 antigens (entry 14). Psoriatic arthritis occurs more frequently in family members (entry 26).

Like ankylosing spondylitis and Reiter's syndrome, psoriatic arthritis may develop after exposure to a number of environmental factors. Trauma has been reported by a number of investigators as the initiator of arthritis or osteolysis (bone loss) in psoriasis (entries 9, 30). Chronic arthritis has developed after trauma to normal joints in patients with psoriasis uncomplicated by arthritis. Synovial joints in psoriasis patients may be more liable to damage due to an enzyme deficiency of synovial fluid (entry 11). Other environmental factors including infections with Staphylococcus aureus and Clostridium perfrigens, and delayed hypersensitivity have also been suggested as important in the development of psoriatic arthritis (entries 23, 21).

§ 4-4(B). Clinical Features.

Psoriatic arthritis has more than one clinical form and this initially caused confusion in the description of the illness (entry 27). Classical psoriatic arthritis is described as involving distal interphalangeal (DIP) joints and associated nail disease alone. This pattern occurs in 5% of patients. The most common form of disease, affecting 70% of patients with psoriatic arthritis, is an asymptomatic oligoarthritis; a few large or small joints are involved. Dactylitis, diffuse swelling of a digit, is most closely associated with this form of the disease. Skin activity and joint symptoms do not correlate since patients with little skin activity may experience continued joint pain and stiffness. As opposed to rheumatoid arthritis, the clinical appearance of an involved joint does not necessarily correlate with patient symptoms. Patients with severely affected joints may be asymptomatic. Symmetric polyarthritis, which affects the small joints of the hands and feet and resembles rheumatoid arthritis, occurs in 15% of patients. Arthritis mutilans, characterized by extensive destruction of bone in the hands, is found in 5% of patients. Spondylitis with or without peripheral joint disease occurs in 5% of patients.

In a recent report, clinical forms of psoriatic arthritis were simplified to three major types — asymmetric, oligoarticular arthritis (54%); symmetric arthritis (25%); and spondyloarthritis (21%) (entry 17). DIP involvement occurred most commonly in the asymmetric oligoarthritis group and rarely in the spondyloarthropathy group. Arthritis mutilans occurred rarely in all groups.

Patients who develop axial skeletal disease, sacroiliitis or spondylitis are men who have the onset of psoriasis later in life (entry 20). Low back pain, which is indistinguishable from the pain associated with the other spondyloarthropathies, is present in the vast majority of patients with axial skeletal disease. These patients may have back pain or peripheral joint symptoms as their initial complaint. Asymmetric or symmetric peripheral joint involvement may antedate the development of axial skeletal disease.

The typical patient with the symmetric or asymmetric form of psoriatic arthritis is a man or woman between the age of 35 and 45. Patients with more severe disease have an onset of symptoms at an earlier age, manifested by inflammation in a few joints or by diffuse selling of an entire digit. Psoriasis antedates the arthritis in a majority of patients. Between 10% and 20% of patients will have characteristic arthritis before the appearance of psoriatic lesions. Patients with severe skin involvement are more likely to develop arthritis (entry 22). However, even patients with minimal skin involvement still have some risk of developing joint disease. Nail involvement, characterized by pitting, horizontal ridging, onycholysis (opacification of the nail bed), and discoloration, occurs in 80% of patients with psoriatic arthritis in contrast to a 30% incidence in patients with uncomplicated psoriasis. The activity of skin and nail disease does not necessarily correlate with joint symptoms since any of the forms of psoriatic skin disease, whether guttate, pustular, seborrheic or others, may be associated with joint involvement.

Although constitutional symptoms of fever, anorexia, and weight loss are rare in patients with psoriatic arthritis, fatigue and morning stiffness are common. Ocular involvement includes conjunctivitis in 20%, iritis in 10%, and scleritis in 2% of patients with psoriatic arthritis (entry 19). Iritis is more commonly seen in patients with axial skeletal disease. Cardiac complications, such as aortic insufficiency as seen in ankylosing spondylitis, are very rare and are usually associated with spondylitis.

§ 4-4(C). Physical Findings.

An extensive examination of the skin is an essential part of the investigation of a patient with suspected psoriatic arthritis. The diagnosis of psoriasis is not difficult to make when the patient has the characteristic erythematous, raised, circumscribed, dry scaling lesions over the elbows, knees and scalp, and pitting of the nails. They may be hidden in the scalp, gluteal folds, perineum, rectum, or umbilicus, and may remain undetected unless a complete skin exam is done. Nails are examined for the presence of pitting, ridges, opacification and hyperkeratosis.

A complete musculoskeletal examination is essential in determining the extent of joint involvement. Patients may be asymptomatic in a specific joint although physical examination demonstrates decreased function. They may have dactylitis of a toe and may be unaware of the change until pointed out by the physician. Examination of the axial skeleton should be completed even in the asymptomatic patient. A loss of spinal motion may be a manifestation of axial skeleton disease. Sacroiliac involvement may be unilateral or bilateral. Percussion over the sacroiliac joints can elicit symptoms over the affected side. Patients may develop spondylitis in the absence of sacroiliitis and these have maximal tenderness with percussion over the spine above the sacrum.

§ 4-4(D). Laboratory Findings.

The findings of anemia, mild leukocytosis, and elevated erythrocyte sedimentation rate occur in a minority of patients (entry 2). An elevated uric acid, hyperuricemia, is detected in 20% of patients. This may be secondary to an increased metabolic rate and protein breakdown in patients with extensive skin involvement. These patients may develop secondary gout. Psoriatic synovial fluid is inflammatory but has no diagnostic features. Rheumatoid factor and antinuclear antibody are usually absent. If they are present, the rheumatoid factor and ANA occur in the same frequency as found in age-matched controls. HLA-B27 is detected in approximately 35% to 60% of patients with axial skeleton disease (entry 8). In one study, patients with sacroiliitis and spondylitis had 90% positivity for HLA-B27, while in another study patients with spondylitis and normal sacroiliac joints had 43% positivity (entries 20, 25). Peripheral joint disease is associated with HLA-BW38, while psoriasis alone is associated with HLA-B13 and B17 (entry 15).

183

§ 4-4(E). Radiologic Findings.

While radiologic features of psoriatic arthritis and rheumatoid arthritis may be similar, certain features of DIP and PIP joints in psoriatic arthritis are distinctive (entries 31, 1). The joint involvement is oligoarticular with erosive changes in the DIP joint and terminal phalanx, especially the big toe. The "pencil-in-cup" deformity, osteolysis of the proximal phalanx and widening of the distal phalanx, is characteristic of psoriatic arthritis. Periosteal reaction occurs along the shafts of the long bones, as opposed to the periosteal changes in Reiter's syndrome which are localized to the metatarsals and phalanges of the feet and hands.

Axial skeleton involvement is manifested by sacroiliitis which can be unilateral or bilateral (entry 24). Spondylitis is characterized by asymmetric involvement of the vertebral bodies and non-marginal syndesmophytes. Spondylitis with normal sacroiliac joints may show up radiographically. Some patients have been described with axial skeletal disease which mirrors the involvement characteristic of ankylosing spondylitis. Paravertebral ossification separated from the vertebral body may occur in the thoraco-lumbar region (entry 10).

In psoriatic spondylitis, the cervical spine may also be affected with joint space sclerosis and narrowing and anterior ligamentous calcification (entry 18). As with the other spondyloarthropathies, a scintiscan may demonstrate increased activity over the sacroiliac joints or axial skeleton before radiographic changes are detectable (entry 3).

§ 4-4(F). Differential Diagnosis.

The diagnosis of psoriatic arthritis is easily made when the patient has characteristic skin lesions and joint changes. The diagnosis is more difficult in the patient who presents with joint symptoms before the appearance of skin lesions. The differential diagnosis for such a patient should include Reiter's syndrome, gout, erosive osteoarthritis, and rheumatoid arthritis. Reiter's syndrome may be differentiated by urethritis, predominantly lower extremity involvement, and periosteal changes. Acute gout is confirmed by the detection of monosodium urate crystals in synovial fluid. Erosive osteoarthritis occurs in post-menopausal women and is characterized by inflammation in the DIP and PIP joints and radiographic findings of osteophytes, sclerosis and cysts in these joints. The erythrocyte sedimentation rate remains normal. The clinical appearance of symmetrical polyarthritis in psoriatic arthritis and rheumatoid arthritis is similar. The absence of rheumatoid nodules, rheumatoid factor and the presence of DIP involvement and periostitis helps differentiate the patient with psoriatic arthritis from the one with rheumatoid arthritis.

§ 4-4(G). Treatment.

The goals of therapy are the maintenance and improvement in function by the reduction of inflammation. Therapy includes patient education, non-steroidal anti-inflammatory drugs, immunosuppressives and physical therapy. The importance of appropriate skin care must be stressed. While in the past no correlation between improvement in skin and joint symptoms could be demonstrated, a recent study reported the improvement of non-spondylitic psoriatic arthritis in patients who responded to photochemotherapy for their skin disease (entry 28). Non-steroidal anti-inflammatory drugs, particularly indomethacin and phenylbutazone, are useful in controlling pain and stiffness in the peripheral and axial joints. Mefenamic acid has also been suggested as a useful drug for the control of joint symptoms.

Drugs with skin rash as a potential toxicity have been contraindicated in the past; however, recent studies have demonstrated the efficacy and lack of skin toxicity for hydroxychloroquine and gold salts (entries 17, 12). These drugs are most helpful for the patient with refractory peripheral arthritis.

In patients with severe and extensive skin disease and destructive arthritis, immunosuppressive therapy with methotrexate, 6-mercaptopurine and azathioprine (entries 6, 4, 13) is indicated. These drugs have the potential for severe liver toxicity and leukopenia and should be reserved for the most severely affected patients. Systemic corticosteroids are rarely used for skin or joint disease since a rebound phenomenon appears when the drug is discontinued. Intra-articular corticosteroids are useful in the patient with psoriatic monarthritis and persistent effusion.

§ 4-4(H). Course and Outcome.

The course of psoriatic arthritis is unpredictable (entry 29). A small percentage of patients develop destructive, disabling disease while a majority have less pain and disability than seen in rheumatoid arthritis. In one large study, 97% of psoriatic patients were able to work at their jobs, missing less than 12 months of work during a minimum 10-year follow-up period. Another survey found that women who developed symmetric large and small joint disease developed more destructive arthritis (entry 17).

Patients who develop psoriatic spondylitis develop varying degrees of restriction of spinal motion. There is no consistent correlation between the severity of peripheral joint disease and axial skeletal disease.

Patients become disabled when destructive lesions in their hands and feet result in persistent pain and loss of function, but patients with spondylitis are limited by loss of

186

motion of the spine. Patients with documented psoriatic spondylitis should be prohibited from engaging in heavy labor. As in ankylosing spondylitis, fusion of the axial skeleton makes it more vulnerable to fracture from minimal trauma.

BIBLIOGRAPHY

1. Avila, R., Pugh, D.G., Slobumb, C.H., and Winkelman, R.K. Psoriatic arthritis: a roentgenologic study. *Radiology,* 75:691-701, 1960.
2. Baker, H., Golding, D.H., and Thompson, M. Psoriasis and arthritis. *Ann. Intern. Med.,* 58:909-925, 1963.
3. Barraclough, D., Russell, A.S., and Percy, J.S. Psoriatic spondylitis: a clinical radiological and scintiscan survey. *J. Rheumatol.,* 4:282-287, 1977.
4. Baum, J., Hurd. E., Lewis, D., Ferguson, J.L., and Ziff, M. Treatment of psoriatic arthritis with 6-mercaptopurine. *Arthritis Rheum.,* 16:139-147, 1973.
5. Bazin, P. Lecons Theoriques et Cliniques sur les Affections Cutane es de Nature Arthritique et Dartreux. *Paris, Delahaye,* 1860, pp. 154-161.
6. Black, R.L., O'Brien, W.M., Van Scott, E.J., Auerbach, R., Eisen, A.Z., and Bunim, J.J. Methotrexate therapy in psoriatic arthritis: double-blind study in 21 patients. *JAMA,* 189:743-747, 1964.
7. Bourillon, C. Psoriasis et Arthropathies. *These, Paris,* 1888.
8. Brewerton, D.A., Coffrey, M., Nicholls, A., Walters, D., and James D.C.O. HLA-B27 and arthropathies associated with ulcerative colitis and psoriasis. *Lancet* I:956-958, 1974.
9. Buckley, W.R., and Raleigh, R.L. Psoriasis with acro-osteolysis. *New Eng. J. Med.,* 261:539-541, 1959.
10. Bywaters, E.G.L., and Dixon, A.S.J. Paravertebral ossification in psoriatic arthritis. *Ann. Rheum. Dis.,* 24:313-331, 1965.
11. Cotton, D.W.K., and Mier, P.D. An hypothesis on the aetiology of psoriasis. *Br. J. Dermatol.,* 76:519-528, 1969.
12. Dorwart, B.B., Gall, E.P., Schumacher, H.R., and Krauser, R.E. Chrysotherapy in psoriatic arthritis: efficacy and toxicity compared to rheumatoid arthritis. *Arthritis Rheum.,* 21:513-515, 1978.
13. DuVivier, A., Munro, D.D., and Verbov, J. Treatment of psoriasis with azathioprine. *Br. Med. J.,* 1:49-51, 1974.

14. Espinoza, L.R., Vasey, F.B., Gaylord, S.W., Dietz, C., Bergan, L., Bridgeford, P., and Germain, B.F. Histocompatibility typing in the seronegative spondyloarthropathies: a survey. *Semin. Arthritis Rheum.*, 11:375-381, 1982.

15. Espinoza, L.R., Vasey, F.B., Oh, J.H., Wilkinson, R., and Osterland, C.K. Association between HLA-BW38 and peripheral psoriatic arthritis. *Arthritis Rheum.*, 21:72-75, 1978.

16. Hellgren, L. Association between rheumatoid arthritis and psoriasis in total populations. *Acta. Rheum. Scand.*, 15:316-326, 1969.

17. Kammer, G.M., Soter, W.A., Gibson, D.J., and Schur, P.H. Psoriatic arthritis: a clinical, immunologic and HLA study of 100 patients. *Semin. Arthritis Rheum.*, 9:75-97, 1979.

18. Kaplan, D., Plotz, C.M., Nathanson, L., and Frank, L. Cervical spine in psoriasis and in psoriatic arthritis. *Ann. Rheum. Dis.*, 23:50-55, 1964.

19. Lambert, J.R., and Wright, V. Eye inflammation in psoriatic arthritis. *Ann. Rheum. Dis.*, 35:354-356, 1976.

20. Lambert, J.R., and Wright, V. Psoriatic spondylitis: a clinical radiological description of the spine in psoriatic arthritis. *Quar. J. Med.*, 46:411-425, 1977.

21. Landau, J.W., Gross, B.G., Newcomer, V.D., and Wright, E.T. Immunologic response of patients with psoriasis. *Arch. Dermatol.*, 91:607-610, 1965.

22. Leczinsky, C.G. The incidence of arthropathy in a ten-year series of psoriasis cases. *Acta. Dermatol. Vener.*, 28:483-487, 1948.

23. Mansson, I., and Olhagen, B. Intestinal clostridium pertrigens in rheumatoid arthritis and other connective tissue disorders: studies of fecal flora, serum antitoxin levels, and skin hypersensitivity. *Acta. Rheum. Scand.*, 12:167-174, 1966.

24. McEwen, C., DiTata, D., Lingg, C., Porini, A., Good, A., and Rankin, T. Ankylosing spondylitis and spondylitis accompanying ulcerative colitis, regional enteritis, psoriasis and Reiter's disease: a comparative study. *Arthritis Rheum.*, 14:291-318, 1971.

25. Metzger, A.L., Morris, R.I., Bluestone, R., and Terasaki, P.I. HLA-A W27 in psoriatic arthropathy. *Arthritis Rheum.*, 18:111-115, 1965.

26. Moll, J.M.H., and Wright, V. Familial occurrence of psoriatic arthritis. *Ann. Rheum. Dis.*, 32:181-199, 1973.

27. Moll, J.M.H., and Wright, V. Psoriatic arthritis. *Semin. Arthritis Rheum.*, 3:55-78, 1973.

28. Perlman, S.G., Gerber, L.H., Roberts, R.M., Nigra, T.P., and Barth, W.F. Photochemotherapy and psoriatic arthritis: a prospective study. *Ann. Intern. Med.*, 91:717-722, 1979.

29. Roberts, M.E.T., Wright, V., Hill, A.G.S., and Mehra, A.C. Psoriatic arthritis: follow-up study. *Ann. Rheum. Dis.*, 35:206-212, 1976.
30. Williams, K.A., and Scott, J.T. Influence of trauma on the development of chronic inflammatory polyarthritis. *Ann. Rheum. Dis.*, 26:532-537, 1967.
31. Wright, V. Psoriatic arthritis: a comparative study of rheumatoid arthritis and arthritis associated with psoriasis. *Ann. Rheum. Dis.*, 20:123-131, 1961.
32. Wright, V. Rheumatism and psoriasis: a re-evaluation. *Am. J. Med.*, 27:454-462, 1959.

§ 4-5. Enteropathic Arthritis.

Ulcerative colitis and Crohn's disease are inflammatory bowel diseases. Ulcerative colitis is limited to the colon while Crohn's disease, or regional enteritis, may involve any part of the gastrointestinal tract. Inflammation of the gut results in numerous gastrointestinal symptoms including abdominal pain, fever, and weight loss. These inflammatory diseases are also associated with extra-intestinal manifestations including arthritis. Articular involvement in these inflammatory bowel diseases includes both peripheral and axial skeleton joints. Peripheral arthritis is generally non-deforming and follows the activity of the underlying bowel disease. Axial skeleton disease is similar to ankylosing spondylitis and follows a course independent of activity of bowel inflammation.

The association of arthritis and ulcerative colitis was first elucidated by Bargen in the 1920's (entry 3). Crohn's disease was described by Crohn, Ginzberg and Oppenheimer in 1932. The association of arthritis and Crohn's disease was commented on by Van Patter in the 1950's (entry 21).

§ 4-5(A). Prevalence and Etiology.

Ulcerative colitis occurs four times more commonly in whites than non-whites and more commonly among Jews than non-Jews. In a white population, the annual incidence of disease is up to 10 of 100,000 people. Symptomatic ulcerative colitis usually occurs between 25 and 45 years of age, and the disease is more common among women than among men.

Crohn's disease occurs in all races and is distributed worldwide. In the United States, the annual incidence of the disease is 4 per 100,000 individuals (entry 16). The disease appears most often between the ages of 15 and 35. Men and women are equally affected. Patients from urban backgrounds and with high levels of education are at greater risk.

The frequency of peripheral arthritis is 11 percent in ulcerative colitis and in Crohn's disease, 20 percent, and it is equal in both sexes (entries 9, 11). Spondylitis occurs in 4 percent of both diseases (entry 7) with women being equally affected in Crohn's disease and half as often as men in ulcerative colitis.

The etiology of both of these inflammatory bowel diseases is unknown. Specific infections with bacteria, overproduction of enzymes, vascular disorders, and hypersensitivity to foods are but a few of the unproven theories suggested as possible causes. No specific genetic predisposition for these illnesses has been discovered although there may be a familial predilection (entries 1, 20).

§ 4-5(B). Clinical Features.

The early symptoms of ulcerative colitis are frequent bowel movements with blood or mucous. Mild disease is associated with some abdominal pain and a few bowel movements per day. Severe disease is characterized by fatigue, weight loss, fever, and extracolonic involvement. Crohn's disease is frequently an indolent illness characterized by generalized fatigue, mild non-bloody diarrhea, anorexia, weight loss and cramping lower abdominal pain.

Patients may have symptoms for years before the diagnosis is made.

Articular involvement in these inflammatory bowel diseases is divided into two forms: peripheral and spondylitic. In ulcerative colitis, peripheral arthritis starts as an acute monoarticular or oligoarticular arthritis affecting the knee, ankle, elbow, PIP, wrist or shoulder joints in patients with active bowel disease (entry 22). The attacks are painful, sometimes associated with effusions. They subside after six to eight weeks and are non-deforming (entry 19). Joint symptoms follow the activity of the bowel disease. Patients with arthritis frequently demonstrate other extracolonic manifestations of ulcerative colitis such as erythema nodosum, pyoderma gangrenosum and uveitis, as well as extensive and chronic bowel disease with pseudopolyps and perianal disease (entry 15).

The peripheral joint involvement in Crohn's disease is similar in onset and distribution to that described in ulcerative colitis (entry 2). However, the correlation of bowel inflammation and joint symptoms is more variable. In children, arthritis may antedate the symptoms of Crohn's disease, while this is unusual for adults. In some patients with Crohn's disease, joint symptoms may continue while gastrointestinal complaints subside. The lack of association is probably secondary to the more scattered nature of Crohn's disease throughout the gut compared to the colonic involvement of ulcerative colitis and the difficulty in ascertaining complete remission of the illness (entry 10). Patients with colonic involvement with Crohn's disease may be at greater risk of developing peripheral arthritis (entry 6).

Axial skeleton involvement in ulcerative colitis and Crohn's disease is similar. Three groups of patients with inflammatory bowel disease and spondylitis have been described by Dekher-Saeys (entry 8). In about a third of

191

patients, spondylitis antedated bowel disease. Seventy percent were HLA-B27 positive, 68% had radiographic changes of spondylitis, and 25% had iritis. One-fourth of the patients developed bowel symptoms before the onset of spondylitis. Thirteen percent were HLA-B27 positive, 36% had radiographic changes of spondylitis and none had iritis. These patients had severe disease of the gut. The remaining fraction of patients had simultaneous onset of gut and spine disease. The spondylitis of inflammatory bowel disease has a course totally independent of that of the bowel disease. The clinical and radiographic findings are similar to ankylosing spondylitis, including involvement of shoulders and hips, although some have suggested that it is a milder disease and has an increased proportion of women with spondylitis (entry 14). Patients with spondylitis complain of aching low back pain with stiffness which is maximal in the early morning (entry 4). Occasional patients with Crohn's disease have been reported with severe back pain, limitation of lumbar motion and decreased chest expansion even though they have no radiographic changes of spondylitis. In these patients, improvement in bowel disease was associated with improved musculoskeletal function (entry 8). Other musculoskeletal complications of inflammatory bowel disease include clubbing and periosteitis, avascular necrosis of bone, septic arthritis of the hips, and granulomatous inflammation of synovium, bone, and muscle (entries 18, 13).

§ 4-5(C). Physical Findings.

The general physical examination concentrates on the gastrointestinal tract including inspection of the oropharynx, perineum and rectum. Examination for extraintestinal disease such as aphthous ulcers, iritis, erythema nodosum, and pyoderma gangrenosum is indicated. The musculoskeletal examination must include both peripheral joints and axial skeleton.

§ 4-5(D). Laboratory Findings.

Usual laboratory tests, such as hematocrit, white blood cell count, platelet count, ESR, may be abnormal, but are nonspecific. The inflammatory findings of increased WBC with poor mucin clot in the synovial fluid is also non-diagnostic. Rheumatoid factor and antinuclear antibodies are absent. No specific HLA antigen has been associated with ulcerative colitis, Crohn's disease, or peripheral joint disease. Approximately 50% of patients with spondylitis and inflammatory bowel disease are HLA-B27 positive (entry 17).

§ 4-5(E). Radiographic Findings.

Radiographs of peripheral joints demonstrate soft tissue swelling and occasionally joint effusions (entry 5). Findings of joint destruction, joint space narrowing, erosions and periosteal proliferation are rare. The radiographic changes of the spondylitis in inflammatory bowel disease are indistinguishable from those with classic ankylosing spondylitis. They include "squaring" of vertebral bodies, erosions, widening and fusion of the sacroiliac joints, symmetric involvement of sacroiliac joints, and marginal syndesmophytes (entry 14). Asymptomatic radiographic sacroiliitis is demonstrated in up to 15% of patients with inflammatory bowel disease. In contrast to classic ankylosing spondylitis, these patients have no association with HLA-B27 and women and men have equal involvement (entry 12).

§ 4-5(F). Differential Diagnosis.

A specific diagnosis of ulcerative colitis or Crohn's disease is made from the histologic examination of biopsy material from the gut. The inflammatory abnormalities of ulcerative colitis are limited to the superficial layers of the gut,

193

mucosa, and submucosa. The most characteristic findings in ulcerative colitis are atrophy of mucosal glands and inflammatory cells in the crypts of the colon causing crypt abscesses. Crohn's disease is characterized by a transluminal granulomatous inflammation. While biopsy material yields supportive evidence for one or the other illness, the final differentiation rests heavily on the history, clinical course, and barium contrast studies of the gut. The differential diagnosis for the bowel disease includes bacterial colitis, diverticulitis, pseudomembranous colitis and ischemic colitis. The diagnosis of enteropathic arthritis is not difficult to make in the patient with ulcerative colitis or Crohn's disease with a non-deforming peripheral arthritis. Since most patients develop peripheral arthritis after the onset of gastrointestinal symptoms, the possibilities for the cause of their arthritis are limited. The real difficulty arises in the patient with back symptoms before the onset of bowel disease and their differential diagnosis includes the other spondyloarthropathies including ankylosing spondylitis and Reiter's syndrome. The lower frequency of HLA-B27 may help differentiate enteropathic arthritis from ankylosing spondylitis. The absence of conjunctivitis, urethritis, and of periostitis, particularly in the heels, may help to distinguish it from Reiter's syndrome.

§ 4-5(G). Treatment.

Treatment for peripheral arthritis must be directed toward control of the underlying bowel disease. Therapy might include asulfasalazine (Azulfidine, 4-6 gm. per day), corticosteroid enemas, oral corticosteroids and colectomy for the severe patient with ulcerative colitis. Colectomized patients do not have recurrences of peripheral joint disease (entry 22). Surgery is not nearly as effective in Crohn's disease since it may have more extensive distribution through the gastrointestinal tract.

The therapeutic program for peripheral joint disease may also include non-steroidal anti-inflammatory drugs, intra-articular injections of corticosteroids and physical therapy. There is no evidence of increased gastrointestinal adverse effects with non-steroidal anti-inflammatory drugs in these patients with inflammatory bowel disease.

Therapy for enteropathic spondylitis is similar to that of classic ankylosing spondylitis. This program includes patient education, non-steroidal anti-inflammatory drugs, and physical therapy. Control of bowel disease does not necessarily correlate with improvement in back symptoms. In ulcerative colitis, colectomy should not be done with the expectation of resolution of spondylitic symptoms.

§ 4-5(H). Course and Outcome.

The ultimate course and outcome of these patients is dependent on the severity of their bowel disease. Patients with severe ulcerative colitis have a mortality rate of 10% to 20% over five years. Patients with a severe initial attack, continuous clinical activity, involvement of the entire colon, and disease for ten years or longer have a higher risk of developing cancer of the colon. These patients may require colectomy. Although Crohn's disease is associated with frequent recurrences, the overall mortality rate of 5% for the first five years of disease is much lower than in ulcerative colitis.

Patients with peripheral enteropathic arthritis have non-deforming disease, of short duration. These patients experience little disability from the arthritis and are able to work. The disability associated with enteropathic spondylitis is similar to that of ankylosing spondylitis. The association of hip disease and spondylitis results in a marked decrease in mobility. Patients with spinal rigidity are at risk for fracture. These patients should not perform heavy labor associated with lifting.

BIBLIOGRAPHY

1. Almy, T.P., and Sherlock, P. Genetic aspects of ulcerative colitis and regional enteritis. *Gastroenterology,* 51:757-763, 1966.
2. Ansell, B.M., and Wigley, R.A.D. Arthritic manifestations in regional enteritis. *Ann. Rheum. Dis.,* 23:64-72, 1964.
3. Bargen, J.A. Complications and sequelae of chronic ulcerative colitis. *Ann. Intern. Med.,* 3:335-352, 1929.
4. Bowen, G.E., and Kirsner, J.B. The arthritis of ulcerative colitis and regional enteritis ("intestinal arthritis"). *Med. Clin. North Am.,* 49:17-32, 1965.
5. Clark, R.L., Muhletaler, C.A., and Margulies, S.I. Colitic arthritis: clinical and radiographic manifestations. *Radiology,* 101:585-594, 1971.
6. Cornes, J.S., and Stecher, M. Primary Crohn's disease of the colon and rectum. *Gut,* 2:189-201, 1961.
7. Dekher-Saeys, B.J., Meuwissen, S.G.M., van den Berg-Loonen, E.M., DeHaas, W.H.D., Agenant, D., and Tytgat, G.N.J. Prevalence of peripheral arthritis, sacroiliitis and ankylosing spondylitis in patients suffering from inflammatory bowel disease. *Ann. Rheum. Dis.,* 37:33-35, 1978.
8. Dekher-Saeys, B.J., Meuwissen, S.G.M., van den Berg-Loonen, E.M., DeHaas, W.H.D., Meijers, K.A.F., and Tytgat, G.N.J. Clinical characteristics and results of histocompatability typing (HLA-B27) in 50 patients with both ankylosing spondylitis and inflammatory bowel disease. *Ann. Rheum. Dis.,* 37:36-41, 1978.
9. Greenstein, A.J., Janowitz, H.D., and Sachar, D.B. The extra-intestinal complications of Crohn's disease and ulcerative colitis: a study of 700 patients. *Medicine,* 55:401-412, 1976.
10. Haslock, I. Arthritis and Crohn's disease: a family study. *Ann. Rheum. Dis.,* 32:479-486, 1973.
11. Haslock, F., and Wright, V. The musculoskeletal complications of Crohn's disease. *Medicine,* 52:217-225, 1973.
12. Hyla, J.F., Franck, W.A., and Davis, J.S. Lack of association of HLA-B27 with radiographic sacroiliitis in inflammatory bowel disease. *J. Rheumatol,* 3:196-200, 1976.
13. London, D., and Fitton, J.M. Acute septic arthritis complicating Crohn's disease. *Brit. J. Surg.,* 57:536-537, 1970.
14. McEwen, C., DiTata, D., Lingg, C., Porini, A., Good, A., and Rankin, T. Ankylosing spondylitis and spondylitis accompanying ulcerative colitis, regional enteritis, psoriasis and Reiter's disease: a comparative study. *Arthritis Rheum.,* 14:291-318, 1971.

15. McEwen, C., Lingg, C., Kirsner, J.B., and Spencer, J.A. Arthritis accompanying ulcerative colitis. *Am. J. Med.,* 33:923-941, 1962.

16. Monk, M., Mendeloff, A.I., Siegel, C.I., and Lilienfeld, A. An epidemiological study of ulcerative colitis and regional enteritis among adults in Baltimore: hospital incidence and prevalence, 1960 to 1963. *Gastroenterology,* 53:198-210, 1967.

17. Morris, R.I., Metzger, A.L., Bluestone, R., and Terasaki, P.I. HL-A-W27 — a useful discriminator in the arthropathies of inflammatory bowel disease. *N. Engl. J. Med.,* 290:1117-1119, 1974.

18. Neale, G., Kelsall, A.R., and Doyle, F.H. Crohn's disease and diffuse symmetrical periostitis. *Gut,* 9:383-387, 1968.

19. Palumbo, P.J., Ward, L.E., Sauer, W.G., and Scudamore, H.H. Musculoskeletal manifestations of inflammatory bowel disease: ulcerative and granulomatous colitis and ulcerative proctitis. *Mayo Clin. Proc.,* 48:411-416, 1973.

20. Paulley, J.W. Ulcerative colitis: a study of 173 cases. *Gastroenterology,* 16:566-576, 1950.

21. Van Patter, W.N., Bargen, J.A., Dockerty, M.B., Feldman, W.H., Mayo, C.W., and Waugh, J.M. Regional enteritis. *Gastroenterology,* 26:347-450, 1954.

22. Wright, V., and Watkinson, G. The arthritis of ulcerative colitis. *Br. Med. J.,* 2:670-680, 1965.

§ 4-6. Familial Mediterranean Fever.

Familial Mediterranean fever is a hereditary disorder characterized by recurrent, brief episodes of fever, serosal inflammation (peritonitis or pleuritis), and arthritis. Back symptoms and sacroiliitis have been described in a minority of patients with the disorder. Major disability occurs in the patient with familial Mediterranean fever who develops persistent, destructive hip disease and it can be fatal for those who develop amyloidosis and associated renal failure. Familial Mediterranean fever was first described in 1945 by Siegal (entry 9) and the name of familial Mediterranean fever was associated with the illness in the 1950's (entry 4). The disease has also been called benign paroxysmal peritonitis and recurrent polyserositis.

197

§ 4-6(A). Prevalence and Etiology.

The most commonly affected people are those from eastern Mediterranean countries including Sephardic Jews, Armenians, Turks, and some Arabs but the disease may occur sporadically in other nationalities. A recent review stated that 1,327 patients with familial Mediterranean fever had been reported in the literature (entry 6). Large family studies suggest that the disease is transmitted by a single recessive gene, but no specific HLA antigen has been associated with the illness (entry 13). Men are more frequently affected by a ratio of 3:2. The cause of the illness is unknown.

§ 4-6(B). Clinical Features.

The disease is characterized by acute attacks of fever associated with peritonitis, pleuritis or arthritis. The episodes of abdominal or chest pain are limited to from hours to days, while arthritis may persist for a longer period of time. Areas of painful erythema may also occur on the lower extremities below the knee. The onset of this illness is usually during childhood or adolescence.

The most frequent manifestation of familial Mediterranean fever, occurring in 95% of patients, is peritonitis. Patients develop severe adbominal pain with absent bowel sounds suggestive of an acute abdominal crisis. Not uncommonly, they undergo a laparotomy which reveals no specific pathology. Within 24 to 48 hours, symptoms abate, leaving the patient in his usual state of health. Pleural pain, with difficulty breathing, and a minor effusion occurs in 40% of familial Mediterranean fever patients. Recurrences of abdominal or pleural pain occur irregularly, with remissions lasting months to years. Factors which have been suggested as possible initiators of attacks include menstruation, stress, heavy activity and exposure to cold (entry 8).

198

Articular manifestations of familial Mediterranean fever occur in 75% of patients and are a presenting feature in a third (entry 3). The joints most commonly affected are knee, ankle, hip, shoulder and rarely the sacroiliac. The usual joint attack has an abrupt onset, with rapidly intensifying pain affecting a single joint. An effusion may accompany the development of joint pain and the attacks may last from a few days to a month. Most episodes of joint pain and swelling are unassociated with residual joint damage. Occasionally a patient with familial Mediterranean fever may develop a protracted episode of arthritis affecting a single joint, such as a knee or hip, and this may last for a year or more (entry 3). These joints develop marked swelling and surrounding muscle atrophy. Complete recovery of function may be expected in the patient with resolution of a protracted attack of arthritis in a knee. The outcome of hip arthritis, however, is more ominous with residual limitation of motion and pain being the issue. Hip joint destruction may also occur.

Sacroiliitis, frequently asymptomatic, has been described in patients with familial Mediterranean fever with between 10% and 17% of patients having either unilateral or bilateral disease (entries 3, 1). Sacroiliitis has also been described in children (entry 5). Lumbar spine changes consistent with spondylitis occur less commonly. Amyloidosis is the fatal complication of familial Mediterranean fever because these patients have a deposition of this protein in the kidneys. Nephrotic syndrome, characterized by proteinuria and peripheral edema, ensues which results in renal failure (entry 11). Most patients who develop amyloidosis die before they reach 40 years of age (entry 12).

§ 4-6(C). Physical Findings.

Patients with familial Mediterranean fever may have an entirely normal examination between attacks. During attacks, examination of the abdomen, chest, skin and joints including the back are essential. Examination of the lower extremities for peripheral edema is helpful in detecting the presence of a nephrotic syndrome or the erysipelas-like erythema associated with acute attacks.

§ 4-6(D). Laboratory Findings.

Laboratory findings are non-specific in familial Mediterranean fever (entry 10). During attacks, WBC may be markedly elevated along with increases in ESR. Abnormal values quickly return to normal with resolution of the attack. Urinalysis may demonstrate protein and red blood cells indicative of renal amyloidosis. Rheumatoid factor and antinuclear antibody are absent. Synovial fluid may include increased WBC to one million, good mucin clot, increased proteins and normal or low glucose (entry 12). HLA testing demonstrates no increased frequency of any specific antigen.

§ 4-6(E). Radiographic Findings.

Soft tissue swelling and osteoporosis are seen in patients during short attacks, but these changes are rapidly reversible with remission. Radiographic changes are more severe in patients with protracted attacks. When the knee is affected, osteoporosis may be widespread throughout the limb. On at least two occasions, resumption of weight bearing has resulted in fractures of the tibia or femur secondary to the osteoporosis (entry 12). Other findings include sclerosis, joint space narrowing and erosions. Marked joint space narrowing is a common finding after protracted attacks in a hip. Sacroiliac joint changes include loss of cortical definition and sclerosis on both sides of the joint with or without erosions and fusions (entry 1). The changes in the lumbar spine include bony bridging between lumbar vertebrae (entry 3).

§ 4-6(F). Differential Diagnosis.

The diagnostic criteria for familial Mediterranean fever include recurrent short attacks of fever with peritonitis, pleuritis, arthritis, erythema and absence of data suggesting an alternative diagnosis. A Mediterranean ancestry is a helpful piece of data when a patient has his initial attack.

Sacroiliac joint involvement is similar to that associated with the sero-negative spondyloarthropathies. Abdominal symptoms differentiate familial Mediterranean fever from ankylosing spondylitis, Reiter's syndrome and psoriatic arthritis. Patients with inflammatory bowel disease develop diarrhea while familial Mediterranean fever patients have normal bowel habits even during attacks. The absence of HLA-B27 antigen in familial Mediterranean fever patients suggests that the pathogenesis of lumbosacral spine changes is different than that of the HLA-B27 positive spondyloarthropathies. The sacroiliitis is not related to a second disease, ankylosing spondylitis, but rather to some unspecified mechanism associated with familial Mediterranean fever.

§ 4-6(G). Treatment.

An effective therapy for familial Mediterranean fever is colchicine. A regimen of colchicine 0.6 mg. orally twice a day helps in both ameliorating the attacks and in reducing their frequency (entry 2). Non-narcotic analgesics are useful in controlling pain. Bedrest, anti-inflammatory drugs and corticosteroids may have no benefit for joint disease in familial Mediterranean fever (entry 12). Prolonged immobilization may aggravate the osteoporosis and muscle atrophy. Joint replacement may be necessary for advanced disease of the hip. Therapy for renal failure is necessary for the patient who develops renal amyloidosis.

§ 4-6(H). Course and Outcome.

The major morbidity from familial Mediterranean fever, other than the frequency of acute attacks, is protracted disease of the hip. In one study, 16 of 18 hips affected developed severe limitation of motion and pain (entry 12). Eight of these patients required joint arthroplasty and had a good outcome.

The major cause of mortality from familial Mediterranean fever is amyloidosis and renal failure. This complication may be more common in patients in Israel than in patients in the United States. Studies suggest that daily colchicine therapy may prevent the development or progression of amyloidosis in these patients (entry 7).

BIBLIOGRAPHY

1. Brodey, P.A., and Wolff, S.M. Radiographic changes in the sacroiliac joints in familial Mediterranean fever. *Radiology,* 114:331-333, 1975.
2. Dinarello, C.A., Wolff, S.M., Goldfinger, S.E., Dale, D.C., and Alling, D.W. Colchicine therapy for familial Mediterranean fever: a double-blind trial. *N. Eng. J. Med.,* 291:934-937, 1974.
3. Heller, H., Gafni, J., Michaeli, D., Shanin, N., Sohar, E., Ehrlich, G., Karten, I., and Sokoloff, L. The arthritis of familial Mediterranean fever. *Arthritis Rheum.,* 9:1-17, 1966.
4. Heller, H., Sohar, E., Kariv, I., and Sherf, L. Familial Mediterranean fever. *Harefuah,* 48:91-94, 1955.
5. Lehman, T.J.A., Hanson, V., Kornreich, H., Peters, R.S., and Schwabe, A.D. HLA-B27—negative sacroiliitis: a manifestation of familial Mediterranean fever in childhood. *Pediatrics,* 61:423-426, 1978.
6. Meyerhoff, J. Familial Mediterranean fever: report of a large family, review of the literature and discussion of the frequency of amyloidosis. *Medicine,* 59:66-77, 1980.
7. Ravid, M., Robson, M., and Kedar, I. Prolonged colchicine treatment in four patients with amyloidosis. *Ann. Intern. Med.,* 87:568-570, 1977.
8. Schwartz, J. Periodic peritonitis, onset simultaneously with menstruation. *Ann. Intern. Med.,* 53:407-411, 1960.

9. Siegal, S. Benign paroxysmal peritonitis. *Ann. Intern. Med.,* 23:1-21, 1945.
10. Siegal, S. Familial paroxysmal peritonitis: analysis of fifty cases. *Am. J. Med.,* 36:893-918, 1964.
11. Sohar, E., Gafni, J., Pras, M., and Heller, H. Familial Mediterranean fever: a survey of 470 cases and review of the literature. *Am. J. Med.,* 43:227-253, 1967.
12. Sohar, E., Pras, M., and Gafni, J. Familial Mediterranean fever and its articular manifestations. *Clin. Rheum. Dis.,* 1:195-209, 1975.
13. Sohar, E., Pras, M., Heller, J., and Heller, H. Genetics of familial Mediterranean fever. A disorder with recessive inheritance in non-ashkenazi jews and armenians. *Arch. Intern. Med.,* 107:529-538, 1961.

§ 4-7. Behcet's Syndrome.

Behcet's syndrome is a chronic relapsing systemic disease characterized by the triad of oral and genital ulcers and iritis. Additional features of the disease include vasculitis with aneurysms, erythema nodosum, meningoencephalitis and arthritis with sacroiliitis. Disability from Behcet's syndrome occurs most commonly with loss of vision secondary to iritis and neurologic impairment secondary to meningoencephalitis and stroke. The disease was first described by Hulusi Behcet, a Turkish dermatologist, in 1937 (entry 1). The triad of this syndrome has also been referred to as the mucocutaneous ocular syndrome (entry 16).

§ 4-7(A). Prevalence and Etiology.

Behcet's syndrome occurs most commonly in people from eastern Mediterranean countries and Japan (entries 2, 13). The disease usually affects men more commonly than women except in North America where one study demonstrated an increased frequency in women (entry 11).

The etiology of Behcet's syndrome is unknown, although a number of immunologic abnormalities of a cellular and humoral variety have been described. These abnormalities include reduced T-lymphocyte reactivity and increased concentration of immunoglobulins (entries 11, 9). However, no coherent pattern to these immunologic abnormalities has been ascertained. Genetic factors may play a role in Japanese and Turkish patients where an increased frequency of HLA-B5 has been associated with the disease (entry 12). HLA-B27 positivity is seen in patients with sacroiliitis.

§ 4-7(B). Clinical Features.

The usual patient is 30 years old who presents with multiple, painful oral ulcers which resolve over weeks. Genital ulcers, which are also painful, occur over the vulva, penis or scrotum, and are present in 80% of patients. Eye lesions, predominantly unilateral or bilateral iritis, occur in 66% of patients and may present with blurred vision and little ocular pain. Iritis may lead to blindness in these patients (entry 4). Skin manifestations, including ulceration, vasculitis, erythema nodosum, erythema multiforme, and the formation of pustules after the trauma of venopuncture, occur in a majority of patients (entry 3). Central nervous system involvement occurs in 24% of patients. A multitude of neurologic manifestations have been described including meningoencephalitis, characterized by fever, headache, stiff neck, cerebrovascular accident or stroke, hemiparesis, seizures, loss of speech, and profound confusion (entry 14). Neurologic involvement is a bad prognostic sign with a mortality rate of 31% to 41% in two studies (entries 21, 22). Vascular disease is associated with thrombophlebitis and arterial aneurysm. Gastrointestinal involvement includes colonic disease similar to ulcerative colitis (entry 3).

Articular involvement occurs in a majority of patients with Behcet's syndrome (entries 23, 10). Peripheral arthritis is usually polyarticular involving the knees, ankles, wrists, or elbows. Monoarticular disease occurs in the knees or ankles, but is less common. Joint symptoms frequently occur after the onset of oral ulcerations (entry 18). The arthritis is asymmetric, recurrent and non-deforming. Axial skeleton disease, back pain and sacroiliitis, has been reported in Behcet's syndrome (entry 15). In one study 10 of 79 patients demonstrated radiographic changes of marked sacroiliitis. Spondylitis has also been described with the disease (entry 22).

§ 4-7(C). Physical Findings.

Examination of the oropharynx and genitourinary system is essential in Behcet's syndrome. Ophthalmological examination is frequently necessary to demonstrate iritis which may be mildly symptomatic. A complete neurologic examination is also indicated to document any abnormalities in mentation or neurologic function.

§ 4-7(D). Laboratory Findings.

Laboratory results are non-specific in Behcet's syndrome. With active disease, ESR is elevated along with an increase in peripheral WBC. Synovial fluid is inflammatory in type with an increase in WBC from 80,000 to 250,000, normal glucose and poor mucin clot (entry 23).

§ 4-7(E). Radiographic Findings.

Bone and joint findings are usually mild and are characterized by osteoporosis and soft tissue swelling. Osseous erosions and joint space narrowing are rarely encountered (entry 19). Sacroiliitis, both unilateral and bilateral, have been noted in patients with Behcet's syndrome along with the rarer occurrence of spondylitis (entry 5). Patients who are HLA-B27 positive may be at greater risk for developing sacroiliitis and spondylitis. Patients with inflammatory disease of the bowel may develop sacroiliitis similar to that seen with ulcerative colitis or Crohn's disease (entries 3, 6).

§ 4-7(F). Differential Diagnosis.

The diagnosis of Behcet's syndrome is based on clinical features since there are no pathognomonic laboratory findings. Diagnostic criteria that have been suggested include oral ulcers, genital ulcers, iritis, skin lesions as a major group; and gastrointestinal, vascular, musculoskeletal, central nervous system lesions and family history a minor group (entry 17). The diagnosis is difficult to make since the different manifestations of the illness may take years to appear. The diagnosis should be considered in a patient with recurrent oral ulcerations and one other manifestation of the disease. The differential diagnosis is very small when a patient has multisystem involvement with Behcet's syndrome. In patients with single organ system disease, the differential diagnosis is directed toward diseases which affect that organ system. For example, the differential diagnosis of oral ulcers would include herpes simplex infection, gastrointestinal disease, ulcerative colitis or Crohn's disease, aseptic meningitis, and Mollaret's meningitis. In patients with oral ulcers, back pain and lower extremity arthritis, the diagnosis of Reiter's syndrome must be excluded. Patients with Reiter's syndrome usually have painless oral ulcerations.

§ 4-7(G). Treatment.

No specific therapy has been demonstrated to be effective in controlling the manifestations of Behcet's syndrome on a continued basis. Systemic corticosteroids suppress skin and joint inflammation. They are less effective on oral and genital ulcers and iritis. Transfer factor, a component of blood which stimulates the immune system, has been effective in a small group of patients (entry 20). Cytotoxic drugs, particularly chlorambucil, is reserved for patients with severe iritis and central nervous system dysfunction (entry 8).

§ 4-7(H). Course and Outcome.

Behcet's syndrome is characterized by frequent attacks early in the course of the illness. After three to seven years the frequency of attacks decrease. The disease is rarely disabling when the illness is limited to ulcerations, skin disease and arthritis. Blindness is a major disability in patients with iritis. Severe ocular disease is more characteristic of Japanese patients with Behcet's syndrome than it is of patients from other geographic areas (entry 17). Central nervous system disease is associated with increased mortality. Death may be secondary to cranial nerve involvement, paraplegia or encephalitis. Vascular disease is a late complication of Behcet's syndrome and is associated with severe disease. Patients with arterial disease are at risk for the development of aneurysms and rupture, leading to death (entry 7).

BIBLIOGRAPHY

1. Behcet, H. Uber rezidivierende aphthose, durch ein Virus verursachte Geschwure am Mund, am Auge und an den Genitalien. *Dermat Wchnschr.*, 105:1152-1157, 1937.
2. Chajek, T., and Fainam, M. Behcet's syndrome: report of 41 cases and a review of the literature. *Medicine*, 54:179-196, 1975.
3. Chamberlain, M.A. Behcet's syndrome in 32 patients in Yorkshire. *Ann. Rheum. Dis.*, 36:491-499, 1977.
4. Colvard, D.M., Robertson, D.M., and O'Duffy, J.D. The ocular manifestations of Behcet's disease. *Arch. Ophthal.*, 95:1813-1817, 1977.
5. Dilsen, A.N. Sacroiliitis and ankylosing spondylitis in Behcet's syndrome. *Scand. J. Rheum. Supp.*, 8:20, 1975.
6. Empey, D.W., and Hale, J.E. Rectal and colonic ulceration in Behcet's syndrome. *Proc. R. Soc. Med.*, 65:163-164, 1972.
7. Enoch, B.A., Castillo-Olivares, J.L., Khoo, T.C.L., Grainger, F.G., and Henry, L. Major vascular complications in Behcet's syndrome. *Postgrad. Med. J.*, 44:453-459, 1968.
8. Mamo, J.G., and Azzam, S.A. Treatment of Behcet's syndrome with Chlorambucil. *Arch. Ophthalmol.*, 84:446-450, 1970.

9. Marquardt, J.L., Synderman, R., and Oppenheim, J.J. Depression of lymphocyte transformation and exacerbation of Behcet's syndrome by ingestion of English walnuts. *Cell Immunol.*, 9:263-272, 1973.

10. Mason, R.M., and Barnes, C.G. Behcet's syndrome with arthritis. *Ann. Rheum. Dis.*, 28:95-103, 1969.

11. O'Duffy, J.D., Carney, J.A., and Deadhar, S. Behcet's syndrome: report of 10 cases, 3 with new manifestations. *Ann. Inter. Med.*, 75:561-570, 1971.

12. Ohno, S., Nakarayama, E., Sugiura, S., Itakura, K., Aoki, K., and Aizawa, M. Specific histocompatibility antigens associated with Behcet's syndrome. *Am. J. Ophth.*, 80:636-641, 1975.

13. Oshima, Y., Shimizu, T., Yokakari, R., Matsumoto, T., Kano, K., Kagami, T., and Nagaya, H. Clinical studies on Behcet's syndrome. *Ann. Rheum. Dis.*, 22:36-45, 1963.

14. Pallis, C.A., and Fudge, B.J. The neurological complications of Behcet's syndrome. *AMA Arch. of Neur. and Psych.*, 75:1-14, 1956.

15. Perkins, E.S. Behcet's disease. Ophthalmological aspects. *Proc. R. Soc. Med.*, 54:106-107, 1961.

16. Robinson, H.M., Jr., and McCrumb, F.R., Jr. Comparative analysis of the mucocutaneous-ocular syndromes: report of eleven cases and review of the literature. *Arch. Derm. Syph.*, 61:539-560, 1950.

17. Shimizu, T., Ehrlich, G.E., Inaba, G., and Hayashi, K. Behcet disease (Behcet syndrome). *Semin. Arthritis Rheum.*, 8:223-260, 1979.

18. Strachen, R.W., and Wigzell, F.W. Polyarthritis in Behcet's multiple symptom complex. *Ann. Rheum. Dis.*, 22:26-35, 1963.

19. Vernon-Roberts, B., Barnes, C.G., and Revell, P.A. Synovial pathology in Behcet's syndrome. *Ann. Rheum. Dis.*, 37:139-145, 1978.

20. Wolf, R.E., Fudenberg, H.H., Welch, T.M., Spitlier, L.E., and Ziff, M. Treatment of Behcet's syndrome with transfer factor. *JAMA*, 238:869-871, 1977.

21. Wolff, S.M., Schotland, D.L., and Phillips, L.L. Involvement of nervous system in Behcet's syndrome. *Arch. Neurol.*, 12:315-325, 1965.

22. Wright, V.A., Chamberlain, M.A., and O'Duffy, J.D. Behcet's syndrome. *Bull. Rheum. Dis.*, 29:972-979, 1978-79.

23. Zizic, T.M., and Stevens, M.B. The arthropathy of Behcet's disease. *Johns Hopkins Med. J.*, 136:243-250, 1975.

§ 4-8. Whipple's Disease.

Whipple's disease, or intestinal lipodystrophy, is a rare disease, first described by G.H. Whipple in 1907, characterized by multiorgan system dysfunction (entry 13). Abnormalities of the musculoskeletal, gastrointestinal, cardiovascular, pulmonary, and nervous systems characterize the illness. Early diagnosis is essential since antibiotic therapy is effective for this illness of bacterial origin.

§ 4-8(A). Prevalence and Etiology.

Whipple's disease is a rare illness with approximately 200 cases reported in the world literature as of 1970 (entry 2). The disease occurs most commonly in Caucasian men between the ages of 40 to 60. The male to female ratio is 10:1. Familial clustering of the disease has been reported (entry 12).

Whipple's disease may have an infectious etiology. A poorly characterized bacterium, which stains pink with periodic acid-Schiff (PAS) is found in the organ systems affected by this illness. The inflammation caused by these bacteria is granulomatous, manifested by macrophages filled with PAS positive organisms in biopsy specimens. The source of these organisms is unknown.

§ 4-8(B). Clinical Features.

Musculoskeletal symptoms, arthralgias, joint pain without inflammation, or arthritis occur as the earliest manifestations of disease in a majority of patients (entry 11). Arthritis may antedate other manifestations of disease by as long as 20 to 35 years (entry 5). Peripheral joint involvement is marked by a migratory oligoarticular or polyarticular arthritis affecting the knees, ankles, elbows, or fingers (entry 4). The joint disease is episodic and recurrent, and rarely causes damage to articular structures

209

although more cases of joint destruction have been reported (entry 1). Back pain and axial skeleton involvement, spondylitis or sacroiliitis may occur in up to 19% of patients (entry 8). Many patients with axial skeleton arthritis have concomitant peripheral joint disease.

The classic triad of malabsorption, diarrhea, and weight loss occurs at unspecified intervals after the onset of joint symptoms. In some patients, arthritic complaints remit after the onset of intestinal symptoms (entry 6). Some of the other protean manifestations of this illness include hypotension, lymphadenopathy, hyperpigmentation, fever and peripheral edema (entry 11).

§ 4-8(C). Physical Examination.

Patients may be febrile with hyperpigmented skin and lymphadenopathy. The abdominal examination may be normal. Patients with axial skeleton disease have tenderness over the spine with limitation in all planes of motion. Peripheral joints may be normal.

§ 4-8(D). Laboratory Findings.

The hematologic findings are non-specific with 90% developing anemia. Intestinal absorption studies, serum carotene, 5-hour D-xylose absorption, 72-hour fecal fat, demonstrate values consistent with malabsorption. Synovial fluid analysis shows an inflammatory exudate with a white blood cell count up to 36,000. The synovial tissue is hyperplastic with PAS positive bodies in synoviocytes (entry 7). Rheumatoid factor and antinuclear antibody are absent. HLA-B27 is demonstrated to be positive in 33% of patients with Whipple's disease.

§ 4-8(E). Radiographic Findings.

Radiographic findings are infrequent in patients with peripheral joint disease. Rare instances of joint destruction and ankylosis have been reported (entry 1). Patients with spondylitis have changes in the sacroiliac joints and lumbar spine similar to those of ankylosing spondylitis (entries 3, 9).

§ 4-8(F). Differential Diagnosis.

The diagnosis of Whipple's disease is made by biopsy of the mucosa of the jejunum of the small intestine. The biopsy is accomplished by an oral route. The histologic material is stained with PAS to demonstrate the bacteria-like structures in granulomas. The differential diagnosis includes the other spondyloarthropathies, particularly Reiter's syndrome and inflammatory bowel disease, Addison's disease and lymphoma.

§ 4-8(G). Treatment.

Prolonged antibiotic therapy with tetracycline has a favorable effect on the disease process. Joint pain, diarrhea, and lymphadenopathy resolve within a few months. The effect of antibiotic therapy on the progression of spondylitis is unknown.

§ 4-8(H). Course and Outcome.

Early diagnosis is essential to the favorable outcome of this treatable illness. Patients may be ill for a number of years before it is correctly diagnosed. Antibiotic therapy has a beneficial effect on the manifestations of this illness. The joint disease is non-deforming and does not cause disability, but axial skeletal involvement is associated with some limitation of motion. Occasionally, severe manifestations, such as central nervous system disease, do occur in patients who are on continuous antibiotic therapy (entry 10).

BIBLIOGRAPHY

1. Ayoub, W.T., Davis, D.E., Torretti, I.D.E., and Viozzi, F.J. Bone destruction and ankylosis in Whipple's disease. *J. Rheumatol,* 9:930-931, 1982.

2. Bayless, T.M. Whipple's disease: newer concepts of therapy. *Adv. Intern. Med.,* 16:171-189, 1970.

3. Canoso, J.J., Saini, M., and Hermos, J.A. Whipple's disease and ankylosing spondylitis simultaneous occurrence in HLA-B27 positive male. *J. Rheumatol,* 5:79-84, 1978.

4. Caughey, D.E., and Bywaters, E.G.L. The arthritis of Whipple's syndrome. *Ann. Rheum. Dis.,* 22:327-355, 1963.

5. DeLuca, R.F., Silver, T.S., and Rogers, A.I. Whipple disease: occurrence in a 76-year-old man with a 20-year prodrome of arthritis. *JAMA,* 233:59-60, 1975.

6. Hargrove, M.D., Jr., Verner, J.V., Smith, A.G., Horswell, R.R., and Ruffin, J.M. Whipple's disease: report of two cases with intestinal biopsy before and after treatment. *Gastroenterology,* 39:619-624, 1960.

7. Hawkins, C.F., Farr, M., Morris, C.J., Hoare, A.M., and Williamson, N. Detection by electron microscope of rod-shaped organisms in synovial membrane from a patient with the arthritis of Whipple's disease. *Ann. Rheum. Dis.,* 35:502-509, 1976.

8. Kelley, J.J., and Weisiger, B.B. The arthritis of Whipple's disease. *Arthritis Rheum.,* 6:615-632, 1963.

9. Khan, M.A. Axial arthropathy in Whipple's disease. *J. Rheumatol,* 9:928-929, 1982.

10. Knox, D.L., Bayless, T.M., and Pittman, F.E. Neurological disease in patients with treated Whipple's disease. *Medicine,* 55:467-476, 1976.

11. LeVine, M.E., and Dobbins, W.O. Joint changes in Whipple's disease. *Semin. Arthritis Rheum.,* 3:79-93, 1973.

12. Puite, R.H., and Tesluk, H. Whipple's disease. *Am. J. Med.,* 19:383-400, 1955.

13. Whipple, G.H. A hitherto undescribed disease characterized anatomically by deposits of fat and fatty acids in the intestinal and mesenteric lymphatic tissues. *Bull. Johns Hopkins Hosp.,* 18:382-391, 1907.

212

§ 4-9. Hidradenitis Suppurativa.

Hidradenitis suppurativa and acne conglobata are chronic suppurative disorders of the skin. A recent report has described a group of patients with these skin diseases who have developed peripheral and axial skeleton arthritis similar to that of the seronegative spondyloarthropathies (entry 1).

§ 4-9(A). Prevalence and Etiology.

These skin diseases are relatively uncommon conditions and their exact prevalence is unknown. Hidradenitis suppurativa is an infection of the apocrine sweat glands located in the axillae and inguinal regions. The disease is characterized by the development of recurrent inflammatory, suppurative nodules and sinus tracts. Acne conglobata is a form of acne characterized by the development of large abscesses and interconnecting sinuses in the skin and they form cysts which are similar to those of hidradenitis suppurativa.

The etiology of the arthritis associated with these skin conditions is unknown. Chronic cutaneous infections are components of the inflammatory process in hidradenitis suppurativa and acne conglobata. The joint disease which develops in these patients may be a reactive arthritis secondary to chronic infection similar to the arthritis which develops in patients after a genitourinary or enteric infection; however, as opposed to other patients with reactive arthritis patients with hidradenitis suppurativa or acne conglobata with arthritis do not have an increased frequency of HLA-B27 positivity.

§ 4-9(B). Clinical Features.

The usual patient is 32 years of age and is black (entry 1). Most have both hidradenitis suppurativa and acne conglobata. In the majority of patients, the skin disease

213

precedes the onset of arthritis by one to twenty years. Peripheral arthritis affects the knees, elbows, wrists, and ankles most commonly, but small joints of the hands and feet may also be involved (entry 2). Attacks of arthritis last from weeks to months and axial skeleton disease, particularly of the lumbosacral spine, is very common; disease of the cervical and thoracic spine occurs less often. Joint symptoms frequently mirror the activity of skin disease.

§ 4-9(C). Physical Findings.

Skin examination is essential to document the extent and activity of the cutaneous disease. Musculoskeletal examination may show swelling, effusion and warmth in peripheral joints, along with limitation of spinal movement and tenderness to percussion over the sacroiliac joints.

§ 4-9(D). Laboratory Findings.

A mild anemia occurs in a majority of patients along with occasional leukocytosis. The ESR is usually elevated. Rheumatoid factor and antinuclear antibody are not present. Cultures of skin lesions are frequently sterile. There is no increased frequency of HLA-B27 or its crossreactive antigens.

§ 4-9(E). Radiographic Findings.

Radiographic findings in patients with peripheral arthritis include swelling, periarticular osteoporosis, periosteal new bone formation and joint space erosions of finger joints. Axial skeletal abnormalities include sacroiliitis with narrowing, sclerosis, erosion and fusion and they are unilateral in most circumstances. Axial skeleton disease is associated with squaring of vertebral bodies, ligamentous calcification and asymmetric syndesmophytes.

§ 4-9(F). Differential Diagnosis.

The diagnosis of hidradenitis suppurativa or acne conglobata is based on the appearance and distribution of the skin lesions. These lesions must be differentiated from other suppurative infections of the skin including Barthlin abscesses and actinomycosis. The associated arthritis must be differentiated from the other seronegative spondyloarthropathies.

§ 4-9(G). Treatment.

The therapy for hidradenitis suppurativa includes antibiotics and incision and drainage of skin tracts. Occasionally, corticosteroids or low dose X-ray therapy is given to those patients with persistent skin disease. Acne conglobata is effectively treated with corticosteroids but the drug is limited in its utility because of side effects. Antibiotic therapy, tetracycline, is worthwhile on a long term basis. Joint disease is responsive to non-steroidal anti-inflammatory drugs. Rarely are corticosteroids needed for control of joint symptoms. Surgery for control of active skin disease may also improve joint disease.

§ 4-9(H). Course and Outcome.

Both skin diseases tend to be chronic and recurrent. They cause disfigurement in their severe forms but no physical disability. The joint disease associated with hidradenitis suppurativa and acne conglobata is associated with loss of range of motion in affected joints. No significant disability has been reported; however, patients with severe axial skeleton and hip disease may be at the same risk for disability as reported in patients with other spondyloarthropathies.

BIBLIOGRAPHY

1. Rosner, I.A., Richter, D.E., Huettner, T.L., Kuffner, G.H., Wisnieski, J.J., and Burg, C.G. Spondyloarthropathy associated with hidradenitis suppurativa and acne conglobata. *Ann. Intern. Med.*, 97:520-525, 1982.
2. Windom, R.E., Sanford, J.P., and Ziff, M. Acne conglobata and arthritis. *Arthritis Rheum.*, 4:632-635, 1961.

§ 4-10. Rheumatoid Arthritis.

Rheumatoid arthritis is a chronic systemic inflammatory disease which causes pain, heat, swelling and destruction in synovial joints. The joints characteristically affected by rheumatoid arthritis are small joints of the hands and feet, wrists, elbows, hips, knees, ankles, and cervical spine. The lumbar spine is rarely involved in the rheumatoid process. Patients who develop low back pain secondary to rheumatoid arthritis have longstanding, extensive disease in the usual joint locations associated with the illness. Rheumatoid arthritis of the lumbosacral spine may be associated with apophyseal joint erosions and secondary intervertebral disc narrowing. Sacroiliac joint involvement is characterized by narrowing of the joint space without erosions or bony sclerosis. Rheumatoid arthritis of the lumbosacral spine responds to the same therapy which is effective for joint disease in other locations. The disability of rheumatoid arthritis is related to the extent of joint destruction in all locations rather than the presence or absence of lumbosacral spine involvement.

§ 4-10(A). Prevalence and Etiology.

The prevalence of rheumatoid arthritis is approximately 1% to 3% of the United States population (entry 5). Rheumatoid arthritis is found in all racial and ethnic groups. The condition occurs in all age groups. The male to female ratio is approximately 1:3. One study suggested that

5% of males and 3% of females with rheumatoid arthritis may have lumbar spine involvement (entry 4).

The etiology of rheumatoid arthritis is probably multifactorial, including both genetic and environmental factors. Clinical symptoms and signs, laboratory abnormalities, and characteristic histologic patterns in the synovium of patients with rheumatoid arthritis suggests that immunologic factors play an important role in its pathogenesis (entry 7).

§ 4-10(B). Clinical Symptoms.

Patients with rheumatoid arthritis develop joint pain, heat, swelling, and tenderness. The joint involvement is additive and symmetrical. The joints at greatest risk of being affected by the disease process include the proximal interphalangeal, metacarpal-carpal, wrist, elbow, hip, knee, ankle, and metatarsal-phalangeal joints. In the axial skeleton, the cervical spine is most frequently affected. Patients have joint pain and stiffness which is most severe in the morning. Activity improves symptoms. This phenomenon, stiffness of a joint with rest, occurs frequently with active disease. As a component of systemic inflammation, afternoon fatigue is a common complaint.

Patients with rheumatoid arthritis of the lumbosacral spine usually have a history of extensive involvement of long duration. Low back pain appears along with increased synovitis in peripheral joints. The pain may be localized to the low back or may radiate into the thighs (entry 6). The sacroiliac joints are asymptomatic.

§ 4-10(C). Physical Findings.

Physical examination of a rheumatoid arthritis patient with lumbar spine involvement would discover diffuse joint involvement characterized by heat, swelling, tenderness, and loss of motion. Examination of the lumbar spine may show tenderness with palpation over the bony skeleton and limitation of all spinal movements. Neurologic examination, including straight leg raising test, is normal.

§ 4-10(D). Laboratory Findings.

Abnormal laboratory findings include anemia, elevated erythrocyte sedimentation rate, and increases in serum globulins. Rheumatoid factors (antibodies directed against host antibodies) are present in 80% of patients with rheumatoid arthritis. Antinuclear antibodies are present in 30% of rheumatoid arthritis patients. Synovial fluid analysis demonstrates an inflammatory fluid characterized by poor viscosity, increased numbers of white blood cells, decreased glucose, and increased protein.

Histologic examination of the synovium from affected joints demonstrates an inflammatory, hyperplastic tissue characterized by mononuclear cell infiltration, synovial cell proliferation, fibrin deposition and necrosis. Examination of lumbar spine apophyseal joints from rheumatoid arthritis patients at autopsy has shown similar hyperplastic changes (entry 4).

§ 4-10(E). Radiographic Findings.

Characteristic radiographic changes of rheumatoid arthritis include soft tissue swelling, bony erosion without reactive sclerotic bone, joint space narrowing, and periarticular osteopenia. Radiographic examination of patients with lumbar spine rheumatoid arthritis demonstrates apophyseal joint erosion without sclerosis, secondary disc space narrowing, and malalignment (entries 4, 6, 2). Sacroiliac joint changes include asymmetric involvement with mild narrowing without any erosions, sclerosis or fusion (entry 1).

§ 4-10(F). Differential Diagnosis.

Rheumatoid arthritis is a clinical diagnosis based upon history of joint pain, distribution of joint involvement and characteristic laboratory abnormalities (rheumatoid factor). In the patient who develops back pain in the setting of active disease of long duration, the diagnosis of rheumatoid arthritis of the lumbar spine is probable. However, since involvement of the lumbar spine in rheumatoid arthritis is unusual, other possibilities must be considered. These diseases include herniated intervertebral disc, spondyloarthropathies, and local infection.

§ 4-10(G). Treatment.

Therapy for rheumatoid arthritis of the lumbar spine is similar to that given for generalized disease. Patient education, physical therapy, non-steroidal anti-inflammatory drugs, remittive agents (gold, penicillamine), corticosteroids, and immunosuppressive agents may be indicated (entry 3).

§ 4-10(H). Prognosis and Course.

The course of rheumatoid arthritis cannot be predicted at time of onset. Some patients develop sustained disease which is associated with joint destruction and resistance to therapy. Patients with long term disease are at risk of developing lumbar spine involvement. Although lumbar spine disease occurs in patients with a more active process, these particular symptoms are responsive to therapeutic measures. Patients with rheumatoid arthritis are not disabled specifically because of lumbar spine involvement. However, patients with lumbar spine disease have extensive generalized involvement which usually affects their functional status and their ability to function in a job.

BIBLIOGRAPHY

1. Baggenstoss, A.H., Bickel, W.H., and Ward, L.E. Rheumatoid granulomatous nodules as destructive lesions of vertebrae. *J. Bone Joint Surg.*, 344:601-609, 1952.
2. Dixon, A.S.J., and Lience, E. Sacro-iliac joint in adult rheumatoid arthritis and psoriatic arthropathy. *Ann. Rheum. Dis.*, 20:247-257, 1961.
3. Jacobs, R.P. Update on the treatment of rheumatoid arthritis. *Primary Care*, 6:483-503, 1979.
4. Lawrence, J.S., Sharp, J., Ball, J., and Bier, F. Rheumatoid arthritis of the lumbar spine. *Ann. Rheum. Dis.*, 23:205-217, 1964.
5. O'Sullivan, J.B., and Cathcart, E.S. The prevalence of rheumatoid arthritis: follow-up evaluation of the effect of criteria on rates in Sudbury, Massachusetts. *Ann. Intern. Med.*, 76:573-577, 1972.
6. Sims-Williams, H., Jayson, M.I.V., and Baddeley, H. Rheumatoid involvement of the lumbar spine. *Ann. Rheum. Dis.*, 36:524-531, 1977.
7. Stastny, P. Immunogenetic factors in rheumatoid arthritis. *Clin. Rheum. Dis.*, 3:315-332, 1977.

§ 4-11. Diffuse   Idiopathic   Skeletal   Hyperostosis (DISH).

Diffuse idiopathic skeletal hyperostosis is a disease characterized clinically by spinal stiffness and pain, and radiographically by exuberant calcification of spinal and extraspinal structures. Despite impressive radiographic abnormalities, patients rarely have significant loss of function or disability from the illness except for the rare individual who develops difficulty swallowing (dysphagia) secondary to cervical spine involvement. This disease has been known by many different names including spondylitis ossificans ligamentosa, vertebral osteophytosis, ankylosing hyperostosis of Forestier and Rotes-Querol and Forestier's disease. Diffuse idiopathic skeletal hyperostosis was suggested in 1975 by Resnick as a more appropriate name in light of the diffuse bone growth which develops in both spinal and extraspinal locations (entry 4).

§ 4-11(A). Prevalence and Etiology.

Diffuse idiopathic skeletal hyperostosis is a common entity found in 6% to 12% of an autopsy population (entries 5, 9). The usual patient is a man between the ages of 48 to 85 years (entry 8). The ratio of men to women is 2:1 (entry 1). The disease occurs most commonly in Caucasians and rarely in blacks.

The etiology of diffuse idiopathic skeletal hyperostosis is unknown. In one series, occupational stress or spinal trauma was reported in 57% of patients with this condition (entry 4). The patients usually had occupations which included construction, ranching, and roofing, and required a moderate degree of physical activity. Other individuals in the same study had no history of occupational or accidental trauma. Endocrinologic abnormalities associated with bony hyperostosis, acromegaly and hypoparathyroidism have been suggested as causes of diffuse idiopathic skeletal hyperostosis. No abnormalities in growth hormone (acromegaly) or parathormone (hypoparathyroidism) have been found (entry 8). Diabetes mellitus occurs in 30% of patients with the disease; however, this frequency of diabetes mellitus may be related to the age of the population rather than a true association of the two disorders (entry 4).

A specific genetic predisposition to the development of the problem has not been identified. HLA-B27 positivity was found in 34% of diffuse idiopathic skeletal hyperostosis patients in one study (entry 6), but a subsequent study found no significant association with HLA antigens (entry 7). Until further data are obtained to the contrary, diffuse idiopathic skeletal hyperostosis should not be classified with the HLA-B27 positive, seronegative spondyloarthropathies (i.e., ankylosing spondylitis, Reiter's syndrome).

§ 4-11(B). Clinical Findings.

The principle musculoskeletal complaint in 80% of patients is spinal stiffness (entry 8). The duration of back stiffness may be 10 to 20 years, with onset when the patient is in the 40's. Morning stiffness dissipates within an hour, only to recur in the late evening (entry 1). Back pain in the thoracolumbar spine occurs in 57% of patients as their initial complaint (entry 4). Back pain is usually mild, intermittent and rarely radiating. Occasionally patients will have cervical spine pain as their initial complaint. Dysphagia is seen in 17% to 28% of patients (entry 8). Dysphagia occurs secondary to constriction of the esophagus by anteriorly located cervical osteophytes.

Extraspinal manifestations of diffuse idiopathic skeletal hyperostosis occur in 37% of patients. In 20%, extraspinal pain was the initial or predominant complaint. The most common extraspinal skeletal areas involved include shoulders, knees, elbows, and heels.

§ 4-11(C). Physical Findings.

Physical examination usually reveals little limitation of motion in the lumbar spine. Occasionally, a slight decrease in lumbar lordosis and a small increase in dorsal kyphosis may be present. Limitation of motion in the thoracic and cervical spine may also be found (entry 4). A minority of the patients will be tender to percussion over the sacroiliac joints (entry 8). Patients with extraspinal disease may have diminished range of motion and pain on palpation over affected areas. Areas which may show abnormalities include the hips, subtalar joints, shoulders, knees, elbows and ankle joints.

§ 4-11(D). Laboratory Findings.

Laboratory parameters are essentially normal in patients

with diffuse idiopathic skeletal hyperostosis (entry 2). Occasionally a mildly elevated ESR is noted. Since patients who develop the disease are elderly, laboratory abnormalities may be secondary to another illness affecting the patient. Therefore, elevations of fasting and two hour postprandial glucoses are probably secondary to impending diabetes rather than to diffuse idiopathic skeletal hyperostosis.

§ 4-11(E). Radiographic Findings.

The diagnosis of diffuse idiopathic skeletal hyperostosis is a radiographic one. It is made, not uncommonly, in asymptomatic people who happen to have characteristic bony changes in the thoracic spine on a chest radiograph. The three criteria for spinal involvement include flowing calcification along the anterolateral aspect of four contiguous vertebral bodies, preservation of intervertebral disc height, absence of apophyseal joint bony ankylosis and sacroiliac joint sclerosis, erosion or fusion (entry 5). These criteria help differentiate it from spondylosis deformans, intervertebral disc degeneration and ankylosing spondylitis. The fact that the posterior spinal elements are not affected permits almost normal range of motion on physical examination. Radiographic abnormalities are seen most frequently in the thoracic and lumbar spine (entry 4). A majority of patients also develop cervical spine involvement.

Extraspinal radiographic changes include bony proliferation or "whiskering" ligamentous, calcification, and para-articular osteophytes (entry 1). Common locations for these changes are in the pelvis, heel, foot, patella, elbow, shoulder, and wrist.

§ 4-11(F). Differential Diagnosis.

The diagnosis of diffuse idiopathic skeletal hyperostosis is based upon the presence of characteristic radiographic

changes and an absence of clinical abnormalities suggestive
of another illness. A number of diseases cause bony
outgrowths of the spine, including spondyloarthropathies,
acromegaly, hypoparathyroidism, fluorosis, ochronosis,
neuropathic arthropathy and trauma. Specific abnormalities associated with each of these entities, which are
reviewed in other portions of the chapter, help differentiate
these illnesses from diffuse idiopathic skeletal hyperostosis.

§ 4-11(G). Treatment.

Treatment is directed to relieving pain and maximizing
function. In patients with back pain and stiffness,
non-steroidal anti-inflammatory agents may be helpful.
Exercise programs are designed to encourage maximum
ranges of motion throughout the axial skeleton. Local injections of lidocaine and corticosteroid are used in areas of
bony overgrowth, such as the heel, for pain relief.

Patients with severe dysphagia may require removal of
the offending hyperostosis (entry 3). A possible complication of surgical excision of exostoses is recurrence of bony
overgrowth. Some patients with diffuse idiopathic skeletal
hyperostosis who have had hip joint replacements have
developed post-operative heterotopic ossification.

§ 4-11(H). Course and Outcome.

The course of diffuse idiopathic skeletal hyperostosis is
usually benign. Confusion may occur in patients who
present in their 40's with back pain with no physical
findings or early radiographic changes. The differential
diagnosis for such a patient requires investigation for a
number of illnesses which affect the axial skeleton. The
patients with diffuse idiopathic skeletal hyperostosis will
have a very slow progressive course. They may have aching
low back pain and stiffness for 20 years or longer, but will

rarely develop any limitations in their activities or morbidity from their illness.

BIBLIOGRAPHY

1. Forestier, J., and Lagier, R. Ankylosing hyperostosis of the spine. *Clin. Orthop.,* 74:65-83, 1971.
2. Harris, J., Carter, A.R., Glick, E.N., and Storey, G.O. Ankylosing hyperostosis: 1. clinical and radiological features. *Ann. Rheum. Dis.,* 33:210-215, 1974.
3. Meeks, L.W., and Renshaw, T.S. Vertebral osteophytosis and dysphagia. *J. Bone Joint Surg.,* 55A:197-201, 1973.
4. Resnick, D., and Niwayama, G. Radiographic and pathologic features of spinal involvement in diffuse idiopathic skeletal hyperostosis (DISH). *Radiology,* 119:559-568, 1976.
5. Resnick, D., Shaul, S.R., and Robins, J.M. Diffuse idiopathic skeletal hyperostosis (DISH): Forestier's disease with extraspinal manifestations. *Radiology,* 115:513-524, 1975.
6. Shapiro, R.F., Utsinger, P.D., Wiesner, K.B., Resnick, D., Bryan, B.L, and Castles, J.L. The association of HLA-B27 with Forestier's disease (vertebral ankylosing hyperostosis). *J. Rheumatol,* 3:4-8, 1976.
7. Spagnola, A.M., Bennett, P.H., and Terasaki, P.I. Vertebral ankylosing hyperostosis (Forestier's disease) and HLA antigens in Pima Indians. *Arthritis Rheum.,* 21:467-472, 1978.
8. Utsinger, P.D., Resnick, D., and Shapiro, R. Diffuse skeletal abnormalities in Forestier disease. *Arch. Intern. Med.,* 136:763-768, 1976.
9. Vernon-Roberts, B., Pirie, C.J., and Trenwith, V. Pathology of the dorsal spine in ankylosing hyperostosis. *Ann. Rheum. Dis.,* 33:281-288, 1974.

§ 4-12. Vertebral Osteochondritis.

Vertebral osteochondritis is a condition associated with an irregularity of ossification and endochondral growth, with pathologic changes developing at the junction of the vertebral body and intervertebral disc. This abnormality results in increasing wedging of vertebral bodies and progressive forward flexion of the spine (kyphosis).

225

Vertebral osteochondritis primarily affects the thoracic spine in teenagers; however, a similar process has also been described as affecting the lumbar spine (entry 4). If left untreated, the illness may result in progressive kyphosis, persistent back pain and signs of spinal cord compression. Scheuermann's disease, juvenile kyphosis, osteochondritis deformans juvenile dorsi are other names used interchangeably for this illness. Scheuermann first demonstrated the radiographic changes of these wedged vertebrae in 1920 (entry 5).

§ 4-12(A). Prevalence and Etiology.

The prevalence of vertebral osteochondritis has been reported to vary from 0.4% to 8.3% of the general population, depending upon whether the diagnosis was based upon radiographic or clinical criteria. The male to female ratio is 1:2 to thoracic involvement, while the exact opposite ratio may be seen in disease of the lumbar spine (entries 1, 3).

The etiology of vertebral osteochondritis is unknown. Factors suggested as playing a possible role in the development of this illness include avascular necrosis of bone, herniation of disc material into vertebral bodies, excessive growth hormone and heavy labor. Scheuermann suggested that heavy labor at a young age may be associated with the development of vertebral osteochondritis (entry 6). However, vertebral osteochondritis usually occurs in young individuals with no history of heavy labor or trauma. Therefore, the affect of trauma or heavy labor on the development of vertebral osteochondritis is unknown.

§ 4-12(B). Clinical Findings.

Patients with thoracic vertebral osteochondritis may present in three ways. They may complain of back pain, increasing kyphosis, or may be discovered to have vertebral

226

osteochondritis by chance on a radiograph of the spine. Between 20% to 60% of the patients have back pain and the pain is concentrated over the area of kyphosis. The pain may radiate down the back and is usually relieved with bed rest.

Vertebral osteochondritis of the lumbar spine may occur independently or in conjunction with disease in the thoracic spine. The main symptoms of lumbar involvement include local pain and tenderness gradually increasing over weeks, radiation of pain to the hip, paravertebral muscle spasm and limitation of motion.

§ 4-12(C). Physical Findings.

Physical findings include a kyphotic deformity occasionally associated with scoliosis. The kyphosis occurs in the thoracic (75%), thoracolumbar (20%), and lumbar (5%) regions. Thoracic kyphosis is frequently associated with increased lumbar and cervical lordosis. Pain on palpation over the affected portion of axial skeleton is not unusual and paravertebral muscle spasm and tightness of the hamstrings does occur. Rarely, neurologic signs of cord compression may be elicited (entry 2).

§ 4-12(D). Laboratory Findings.

Laboratory parameters, such as hematocrit, WBC, ESR, chemistries, rheumatoid factor and antinuclear antibody are normal.

§ 4-12(E). Radiographic Findings.

Radiographic abnormalities of vertebral osteochondritis include wedging of vertebral bodies, Schmorl's nodules (invasion of vertebral body by an intervertebral disc), irregular end plates and increased kyphosis. Lumbar spine changes are similar. In adult patients, the healed lesion results in characteristic radiographic changes which

227

include increased anteroposterior diameter of a vertebral body, disc space narrowing, loss of normal spine curvature, large Schmorl's nodules, and persistence of a separate fragment of bone anterosuperior to the front edge of a vertebral body (limbus vertebra).

§ 4-12(F). Differential Diagnosis.

The diagnosis of thoracic vertebral osteochondritis is easily made when three contiguous vertebral bodies are wedged by 5 degrees or more (entry 7); however, this criterion eliminates patients with fewer affected vertebrae or those with vertebral irregularity without wedging. No specific criteria have been suggested for lumbar involvement. Therefore, the diagnosis of vertebral osteochondritis should be suspected in a patient with characteristic radiographic changes of one or more thoracic or lumbar vertebrae. The lack of any constitutional symptoms, fever, weight loss, loss of appetite, and abnormal laboratory tests help differentiate vertebral osteochondritis in its early stages from infections such as tuberculosis, neoplasm, hyperparathyroidism, Paget's disease and rheumatoid arthritis.

§ 4-12(G). Treatment.

Treatment is directed at preventing deformity during the patient's rapid growth years and avoiding the development of poor body posture, associated back pain and neurologic deficits as an adult. If the patient has a mild, reversible kyphosis, bed rest and daily exercises to strengthen the back extensor muscles are usually adequate to prevent further deformity. However, progression of kyphosis or radiographic changes of vertebral wedging require body casting or a Milwaukee brace. Milwaukee brace treatment, continued until growth is complete, has been effective at preventing the deformity in 40% of patients with vertebral osteochondritis (entry 1). Surgical intervention is reserved for patients with severe kyphosis, intolerable back pain, visceral compromise (respiratory insufficiency) or progressive neurologic deficit.

§ 4-12(H). Course and Outcome.

Patients with greater than 70 degrees of kyphosis may develop increasing curvature after skeletal maturation is complete. Curvature of this degree can be associated with a marked increase in lumbar lordosis and persistent low back pain. Severe deformities are also at a greater risk of neurologic compromise.

Lumbar spine osteochondritis may be associated with premature degeneration of intervertebral discs (entry 3). Butler has suggested that vertebral osteochondritis which results in loss of normal lumbar lordosis and mobility of the upper lumbar discs may make the lower lumbar discs more vulnerable to degeneration and protrusion. He believes that low back pain in these middle aged adults is secondary to osteochondritis and disc degeneration which occurred in their youth.

BIBLIOGRAPHY

1. Bradford, D.S., Moe, J.H., Montalvo, J.F., and Winter, R.B. Scheuermann's kyphosis and roundback deformity: results of Milwaukee brace treatment. *J. Bone Joint Surg.*, 56A:740-758, 1974.

2. Bradford, D.S. Neurological complications in Scheuermann's disease: a case report and review of the literature. *J. Bone Joint Surg.*, 51A:567-572, 1969.

3. Butler, R.W. The nature and significance of vertebral osteochondritis. *Proc. Roy. Soc. Med.*, 48:895-902, 1955.

4. Lamb, D.W. Localized osteochondritis of the lumbar spine. *J. Bone Joint Surg.*, 36B:591-596, 1954.

5. Scheuermann, H.W. Kyfosis dorsalis juvenilis. *Ugesk Laeger*, 82:385-393, 1920.

6. Scheuermann, H.W. Kyphosis juvenilis (Scheuermann's Krankheit). *Fortschr Geb Rontgens*, 53:1-16, 1936.

7. Sorenson, K.H. *Scheuermann's Juvenile Kyphosis*. Munksgaard, Copenhagen, 1964.

§ 4-13. Osteitis Condensans Ilii.

Osteitis condensans ilii is a disease characterized by mild back pain and unilateral or bilateral bony sclerosis of the lower ilium with sparing of the sacral portion of the sacroiliac joints. The illness is not progressive and is not associated with functional disability. The major difficulty with osteitis condensans ilii is that it is frequently confused with ankylosing spondylitis. The course of osteitis condensans ilii is benign while progressive disease is the usual outcome in ankylosing spondylitis. The illness has also been referred to as osteitis condensans, condensans ilii, and sacroiliac osteosclerosis.

§ 4-13(A). Prevalence and Etiology.

The prevalence of osteitis condensans ilii has been estimated to be 1.6% in the Japanese and 3% in Scandanavians (entries 2, 4). The usual patient is a woman in the age range of 30 to 40 years. The ratio of men to women is 1:9 or greater. The etiology of osteitis condensans ilii is unknown. Trauma, epiphysitis, chronic urinary tract infections, pelvic obliterative endarteritis, and pregnancy have all been entertained as etiologic factors in this illness (entry 5). Series of patients with osteitis condensans ilii have been reported with no history of trauma, epiphysitis, or chronic urinary infection (entry 3). The objection to either endarteritis or pregnancy being etiologic factors is based on the absence of endarteritis in other pelvic articulations and the presence of osteitis condensans ilii in men and nulliparous women. No specific genetic predisposition to developing osteitis condensans ilii has been identified. The prevalence of HLA-B27 in patients with this illness is no greater than that in control populations (entry 1).

§ 4-13(B). Clinical Findings.

The major symptoms of osteitis condensans ilii is low back pain which occurs in 30% of patients. The pain is dull, localized to one or the other side of the midline, with radiation into the buttock on occasion. The pain is not exacerbated by coughing, sneezing, or straining at stool, but may be increased with menstruation in women. Not uncommonly, women notice the onset of pain during pregnancy or the post-partum period. Morning stiffness is usually mild, lasting less than an hour. The episodes of pain may have a duration of weeks to months. The disease may then go into a complete or partial remission which may last for years. A small proportion of patients may complain of fibrositic symptoms characterized by widespread musculoskeletal aching and local point tenderness (entry 1).

§ 4-13(C). Physical Findings.

Physical examination may demonstrate tenderness on sacroiliac joint percussion, pain with sacroiliac joint motion and mild limitation of motion. The rest of physical examination is normal.

§ 4-13(D). Laboratory Findings.

Laboratory values are generally normal in patients with osteitis condensans ilii. Hematocrit, WBC, platelets, urinalysis and chemistry studies are normal. Rheumatoid factor and antinuclear antibody are negative.

§ 4-13(E). Radiographic Findings.

The radiographic findings include an area of triangular sclerosis on the iliac aspect of the sacroiliac joint. The bony sclerosis is unassociated with joint erosions or extensive involvement of the sacrum. The radiographic changes may resolve over time. There are no characteristic abnormalities in other portions of the lumbar, thoracic or cervical spine (entry 2).

§ 4-13(F). Differential Diagnosis.

The diagnosis of osteitis condensans ilii is based upon the presence of radiographic changes on the iliac side of the sacroiliac joint and absence of findings consistent with spondylitis. Patients with spondyloarthropathy have more persistent low back pain associated with more stiffness and limitation of motion. Radiographic changes of spondyloarthropathy are characterized by erosion on both sides of the sacroiliac joint. Other processes which might cause confusion in diagnosis include septic arthritis with bacteria or tuberculosis, Paget's disease, or tumor. The clinical features of these illnesses and the associated radiographic changes help distinguish the specific diseases.

§ 4-13(G). Treatment.

The majority of patients benefit from a conservative regimen of a firm mattress for sleeping, local wet or dry heat, and exercises. Nonsteroidal anti-inflammatory drugs are rarely required.

§ 4-13(H). Course and Outcome.

The course of osteitis condensans ilii is benign. In many circumstances the radiographic changes may reverse to normal (entry 2). Low back pain may persist for months, but is responsive to therapy. Low back pain of osteitis condensans ilii does not cause decreased motion of the lumbosacral spine. This illness is not associated with disability and patients are able to continue to work even though they experience symptoms of the illness.

BIBLIOGRAPHY

1. DeBosset, P., Gordon, D.A., Smythe, H.A., Urowitz, M.B., Koehler, B.E., and Singal, D.P. Comparison of osteitis condensans ilii and ankylosing spondylitis in female patients: clinical, radiological and HLA typing characteristics. *J. Chron. Dis.*, 31:171-181, 1978.

2. Numaguchi, Y. Osteitis condensans ilii, including its resolution. *Radiology,* 98:1-8, 1971.

3. Thompson, M. Osteitis condensans ilii and its differentiation from ankylosing spondylitis. *Ann. Rheum. Dis.,* 13:147-156, 1954.

4. Wassman, K. Osteitis condensans ilii. *Acta. Med. Scand.,* 151:151-154, 1955.

5. Wells, J. Osteitis condensans ilii. *Am. J. Roentgenol.,* 76:1141-1143, 1956.

§ 4-14. Polymyalgia Rheumatica.

Polymyalgia rheumatica is a clinical syndrome characterized by severe stiffness, tenderness and aching of the proximal musculature of the upper and lower extremities. Patients who are 50 years of age or older are most commonly affected by this syndrome and they have an elevated sedimentation rate as the primary abnormal laboratory finding. There is no pathognomonic pathological abnormality which helps physicians diagnose this illness; therefore, other illnesses which may be associated with proximal muscle pain must be eliminated as possible causes of pain before a diagnosis of polymyalgia rheumatica is made. Corticosteroids in small doses are very effective at controlling the symptoms and patients with polymyalgia rheumatica on corticosteroids return to their baseline state of health with no associated disability. Barber in 1957, was the first to propose the name of polymyalgia rheumatica for the clinical syndrome which had numerous names in the past, including senile rheumatic gout, polymyalgia arteritica and anarthritic rheumatoid disease (entry 1).

§ 4-14(A). Prevalence and Etiology.

The incidence of polymyalgia rheumatica in one Caucasian population over 50 years of age was 5.4/10,000 (entry 4). The prevalence of the disease increases in older age groups, with a majority of patients being over the age of 60 (entry 15). The male/female ratio is 1:4. Blacks rarely develop the illness (entry 2) and the etiology is unknown.

§ 4-14(B). Clinical Findings.

The classical picture for polymyalgia rheumatica is of a woman over 50 years of age who develops pain and stiffness symmetrically in the muscles of the shoulder girdle (entry 7). Discomfort in the low back, pelvic girdle, thighs and neck is also commonly experienced. The pain is worse in the morning, such that getting out of bed is very difficult. Activity improves the pain. Symptoms reappear when the patient becomes inactive. The onset of symptoms may be abrupt or gradual. There also may be constitutional symptoms including fever, malaise, fatigue, anorexia and weight loss.

§ 4-14(C). Physical Findings.

Physical examination demonstrates muscle tenderness on palpation and muscle pain with motion but atrophy and weakness are not present. Although active range of motion of joints may be limited by pain, passive motion is normal. Back motion is not limited. Occasionally joint swelling, particularly in the sternoclavicular area, may be seen (entry 14).

§ 4-14(D). Laboratory Findings.

The characteristic laboratory finding is an elevated erythrocyte sedimentation rate and it is present in almost every case of polymyalgia rheumatica (entry 11). Patients with active disease may also develop a hypochromic anemia (entry 8). Chemical studies are usually normal except for about a third of patients with abnormal liver function tests, particularly alkaline phosphatase (entry 13). Rheumatoid factor and antinuclear antibody are usually negative or present in the same proportion as found in normal age matched controls. Muscle enzyme levels are normal.

Pathological examination of muscle biopsy specimens from patients with polymyalgia rheumatica are unremarkable (entry 3). Synovial biopsy demonstrates non-specific inflammation of the synovium (entry 9).

§ 4-14(E). Radiologic Findings.

Plain radiographs demonstrate typical changes in the skeleton that might be expected in patients of this age group. Polymyalgia rheumatica is unassociated with any specific radiographic abnormality but joint scans with technetium pertechnetate may demonstrate increased uptake in the shoulder joints (entry 16).

§ 4-14(F). Differential Diagnosis.

Polymyalgia rheumatica is a diagnosis made after a number of other diseases with similar symptoms are excluded. Other diseases associated with proximal muscle pain include viral infections, subacute bacterial endocarditis, malignancy, osteoarthritis, rheumatoid arthritis, polymyositis, giant cell arteritis, fibrositis, thyroid and parathyroid dysfunction. A complete history, physical examination and laboratory evaluation can usually differentiate these diseases. In some instances where clinical symptoms or laboratory abnormalities are not absolutely characteristic, a tentative diagnosis of polymyalgia rheumatica and therapy is initiated. These patients are then continuously observed to be sure that another illness is not causing their muscle symptoms. When other diseases have been excluded and a patient demonstrates shoulder pain of a month's duration, is 50 years of age or older, has an elevated sedimentation rate, and responds to corticosteroid therapy, a diagnosis of polymyalgia rheumatica can be made.

Patients with giant cell arteritis, an inflammation of blood vessels, frequently have symptoms of polymyalgia rheumatica (entry 10) but in addition to the symptoms of polymyalgia these patients also have symptoms of headache, pain in the jaw with chewing (jaw claudication) and visual changes. The major complication of giant cell arteritis is blindness caused by the occlusion of the artery which supplies the retina. The diagnosis of giant cell arteritis is suspected in a patient with polymyalgia rheumatica who has headache, visual symptoms or jaw claudication and has the diagnosis confirmed by biopsy of the temporal artery.

Polymyositis is an inflammatory disease of muscle associated with muscle weakness. Muscle pain occurs in polymyositis in 50% of patients. Muscle weakness and pain occur in the shoulder and pelvic girdle but low back pain can rarely be an associated symptom. A muscle biopsy showing inflammatory changes helps differentiate polymyositis from polymyalgia rheumatica.

§ 4-14(G). Treatment.

The generally accepted treatment for polymyalgia rheumatica is daily low-dose (15 mg.) corticosteroids (entry 5). They are usually continued at as low a dose as needed to control symptoms for at least two years. Steroids discontinued before that time may result in a risk of relapse of 30% (entry 6). Non-steroidal anti-inflammatory drugs have been used in patients with polymyalgia rheumatica but are usually not as effective in controlling symptoms. Patients who have giant cell arteritis require high dose (60 mg.) corticosteroids to control the disease and prevent blindness (entry 12).

§ 4-14(H). Course and Outcome.

Patients with polymyalgia rheumatica have a resolution of their symptoms within five days of starting corticosteroid therapy. The sedimentation rate gradually returns to normal. These patients suffer little disability from their disease once they are on appropriate therapy.

BIBLIOGRAPHY

1. Barber, H.S. Myalgic syndrome with constitutional effects: Polymyalgia rheumatica. *Ann. Rheum. Dis.*, 16:230-237, 1957.
2. Bell, W.R., and Klinefelter, H.F. Polymyalgia rheumatica. *Johns Hopkins Med. J.*, 121:175-187, 1967.
3. Brooke, M.H., and Kaplan, H. Muscle pathology in rheumatoid arthritis, polymyalgia rheumatica and polymyositis: A histochemical study. *Arch. Pathol.*, 94:101-118, 1972.
4. Chuang, T.Y., Hunder, G.G., Ilbtrup, D.M., and Kurland, L.T. Polymyalgia rheumatica: A 10-year epidemiologic and clinical study. *Ann. Intern. Med.*, 97:672-680, 1982.
5. Davison, S., and Spiera, H. Concepts and treatment in polymyalgia rheumatica. *J. Mt. Sinai Hosp. N.Y.*, 35:473-478, 1968.
6. Fauchald, P., Rygvold, O., and Ystese, B. Temporal arteritis and polymyalgia rheumatica: clinical and biopsy findings. *Ann. Intern. Med.*, 77:845-852, 1972.
7. Fernandez-Herlihy, L. Polymyalgia rheumatica. *Semin. Arth. Rheum.*, 1:236-245, 1971.
8. Gordon, I. Polymyalgia rheumatica: A clinical study of 21 cases. *Quart. J. Med.*, 29:473-488, 1960.
9. Gordon, I., Rennie, A.M., and Branwood, A.W. Polymyalgia rheumatica: Biopsy studies. *Ann. Rheum. Dis.*, 23:447-455, 1964.
10. Hamilton, C.R., Shelley, W.M., and Tumulty, P.A. Giant cell arteritis including temporal arteritis and polymyalgia rheumatica. *Medicine*, 50:1-27, 1971.
11. Healy, L.A., Parker, F., and Wilske, K.R. Polymyalgia rheumatica and giant cell arteritis. *Arthritis Rheum.*, 14:138-141, 1971.

12. Hunder, G.G., and Allen, G.L. Giant cell arteritis: A review. *Bull. Rheum. Dis.,* 29:980-987, 1978.

13. Hunder, G.G., Sheps, S.G., Allen, G.L., and Joyce, J.W. Daily and alternate day corticosteroid regimens in treatment of giant cell arteritis: Comparison in a prospective study. *Ann. Intern. Med.,* 82:613-618, 1975.

14. Miller, L.D., and Stevens, M.B. Skeletal manifestations of polymyalgia rheumatica. *JAMA,* 240:27-29, 1978.

15. Mowat, A.G., and Hazelman, B.L. Polymyalgia rheumatica: A clinical study with particular reference to arterial disease. *J. Rheum.,* 1:190-202, 1974.

16. O'Duffy, J.D., Wahner, H.W., and Hunder, G.G. Joint imaging in polymyalgia rheumatica. *Mayo Clin. Proc.,* 51:519-524, 1976.

§ 4-15. Fibrositis.

Fibrositis is a soft tissue, pain amplification syndrome. It is characterized by chronic pain in discrete trigger point areas, and specific sleep disturbance which occurs in a perfectionist, compulsive individual. Fibrositis is unassociated with any structural abnormalities of muscle, bone or cartilage. However, the persistent pain and chronic fatigue associated with the illness prevents patients from achieving their full work potential (entry 10). Fibrositis occurs in two forms. Primary fibrositis or fibromyalgia is unassociated with any specific cause or associated disorder. Patients with an underlying disease process such as trauma, osteoarthritis, rheumatoid arthritis, hypothyroidism, and malignancy who develop characteristic musculoskeletal symptoms have secondary fibrositis. This disease has been known by many different names including fibromyalgia, myofibrositis, and scapulohumeral syndrome. The term fibrositis was first used by Sir William Gowers in 1904, in an article on low back pain (entry 3).

§ 4-15(A). Prevalence and Etiology.

The prevalence and incidence of fibrositis is unknown;

however, many primary care physicians believe fibrositis to be a very common ailment with over 10 million Americans affected (entry 9). The disease occurs most commonly in Caucasian women with a mean age of 29 years (entry 11).

The exact etiology of the disease is unknown. Moldofsky has suggested that specific disturbances in sleep may result in patients developing fibrositis. The abnormality in sleep is the superimposition of light stages of sleep, characterized by alpha waves on electro-encephalogram, on deep stages of sleep, characterized by delta rhythm (non-REM sleep) (entry 5). In a second study, Moldofsky produced fibrositic symptoms in healthy volunteers when their deep sleep was interrupted by loud noises over a three-day period (entry 6).

Fowler has suggested that patients with fibrositis have increased muscle tone (entry 2). When performing a standardized task, they have 50% more electrical activity on electromyogram than normal controls. Individuals with increased muscle tension and abnormal sleep patterns may be at risk of developing fibrositis. Smythe has suggested that trauma in a single incident or repeated episodes may be a cause of fibrositis (entry 8). In the anxious, perfectionist type individual, trauma in areas of increased sensitivity may result in perpetuation of pain long after the injury and associated pain should have subsided. For example, cab drivers may develop cervical fibrositis, while bus drivers may develop fibrositis of the back. The association of trauma and fibrositis remains a conjecture at this time, since no prospective study has been completed which demonstrates this association.

§ 4-15(B). Clinical Findings.

Patients with fibrositis complain of generalized aching pain associated with profound stiffness and fatigue. Areas of pain are confined to articular and periarticular structures, including ligaments, tendons, muscles, and bony prominences. The pain may be unremitting with durations of 20 years or longer. Generalized stiffness is most notable in the morning, usually lasting up to an hour. A smaller percent of patients have evening stiffness, while some patients have day-long stiffness. Fatigue is also a prominent feature of the patient with fibrositis. Characteristically, these patients arise in the morning after a restless sleep feeling exhausted. Some patients complain of unrelenting fatigue. Other clinical features include polyarthralgias, occasionally associated with hand swelling, numbness, headaches, and anxiety.

Cold or humid weather, overactivity, total inactivity, and poor sleep exacerbate fibrositis. Factors which improve symptoms include moderate activity, warm dry weather, and massage.

§ 4-15(C). Physical Findings.

Physical examination demonstrates specific areas which are tender to palpation. These areas are referred to as trigger or tender points and are localized to certain anatomical sites. The most commonly affected areas include the upper border of the trapezius, medial part of the knees, lateral border of the elbows, posterior iliac crest and the lumbar spine. Pressure over the area results in a pattern of local and referred pain; however, there is no tenderness outside the local area. In one study, patients had between 4 to 33 tender points (entry 11). Usually patients with primary fibrositis have 12 or more discrete areas.

Patients with fibrositis may have "fibrositic" nodules located about the sacrum and posterior iliac crest. The nodules are mobile, firm and tender to palpation. Biopsy of these nodules demonstrates fibrofatty tissue without inflammation. Besides the findings of tender points and nodules, the physical examination is normal, with full range of motion of the lumbosacral spine.

§ 4-15(D). Laboratory Findings.

Laboratory parameters are normal. Blood chemistries, hematologic parameters, the erythrocyte sedimentation rate, rheumatoid factor, and antinuclear antibody are all normal.

§ 4-15(E). Radiographic Findings.

Radiographic findings are normal in patients with fibrositis.

§ 4-15(F). Differential Diagnosis.

The diagnosis of primary fibrositis is based upon the presence of characteristic musculoskeletal abnormalities in the absence of stigmata of other diseases. Patients with diffuse aching and fatigue of at least three months' duration, at least 12 tender points, and disturbed sleep patterns probably have fibrositis. There are no specific physical or laboratory findings which are pathognomonic for this illness. Therefore, the diagnosis of fibrositis is one of exclusion and these patients require constant re-evaluation. This is necessary to detect early physical or laboratory findings which are indicative of an underlying disease process, such as rheumatoid arthritis or a malignancy since these patients have secondary fibrositis. Other illnesses associated with secondary fibrositis include rheumatoid arthritis, osteoarthritis, spondyloarthropathies, connective tissue diseases, malignancies, hypothyroidism, hyperparathyroidism, chronic infections, and sarcoidosis (entry 1). Characteristic physical findings, inflammatory joint signs, laboratory abnormalities, elevated ESR, and positive rheumatoid factor, help differentiate other illnesses from primary fibrositis.

Fibrositis must also be differentiated from psychogenic rheumatism which is characterized by significant anxiety, depression, or neurosis (entry 7). Patients with psychogenic rheumatism have severe pain of a burning or cutting quality. The pain is excruciating in intensity and without any recognizable anatomic boundaries. The patients usually deny stiffness. On examination, they have a marked response to minimal pressure on palpation. Their areas of tenderness do not correspond to the tender points of fibrositis. The laboratory evaluation of these patients is normal and their complaints are resistant to all forms of therapy.

§ 4-15(G). Treatment.

Treatment of fibrositis requires a multifactorial approach. In many circumstances, educating the patient about his illness is reassuring and relieves some symptoms. Many patients are encouraged by having a specific diagnosis for their ailments. Their symptoms are no longer "in their head."

Rest and relaxation are important for patients who are overworked. Patients are encouraged to remain at work, but they should not become excessively fatigued (entry 11). Some patients require a change in their job status to lighter duty work. Range of motion and stretching exercises encourage improved muscle function. Heat treatments are also useful. Drug therapy in the form of aspirin or other non-steroidal anti-inflammatory agents is helpful in reducing the pain associated with fibrositis (entry 1). Injection of trigger points with a combination of anesthetic agent and a long-acting corticosteroid is helpful in controlling localized pain (entry 4). Systemic corticosteroids are not used in patients with primary fibrositis. Antidepressant medications, such as amitriptyline in small doses taken six hours before bedtime may regularize sleeping patterns and allow the patient to feel rested in the morning.

The combination of all these modes of therapy does have a beneficial effect on the symptoms of fibrositis; however, fibrositis is a chronic illness which may be exacerbated by a number of factors including increased tension, cold exposure, or sleep disturbances. Compliance with a program is essential for the long-term management of these individuals. Appropriate rest, exercise, and stress management may control disease symptoms without the need to use medications or injections.

§ 4-15(H). Course and Outcome.

The course of fibrositis is one of exacerbations and remissions. The patients have low back, neck, knee and chest pain which hinders their ability to perform up to their potential. Wood, in a study of English workers, reported that non-articular rheumatism, which included patients with fibrositis, accounted for 10.9% of absences from work, and corresponded to 10.5% of lost days from work (entry 10). Fibrositis may go unrecognized for an extended period. In such circumstances, the patients may be thought of as malingerers who are unwilling to do a full day's work.

Although fibrositis is not disabling in the same sense as the spondyloarthropathies, it is associated with reduced productivity and absenteeism. Recognition of the disease and institution of appropriate therapy can have a beneficial effect on the patient's outlook and work performance.

BIBLIOGRAPHY

1. Beetham, W.P., Jr. Diagnosis and management of fibrositis syndrome and psychogenic rheumatism. *Med. Clin. North Am.,* 63:433-439, 1979.

2. Fowler, R.S., Jr., and Kraft, G.H. Tension perception in patients having pain associated with chronic muscle tension. *Arch. Phys. Med. Rehabil.,* 55:28-30, 1974.

3. Gowers, W.R. Lumbago: Its lessons and analogues. *Br. Med. J.,* 1:117-121, 1904.

4. Kraus, H. Triggerpoints. *N.Y. State J. Med.,* 73:1310-1314, 1973.

5. Moldofsky, H., Scarisbrick, P., England, R., and Smythe, H. Musculoskeletal symptoms and non-REM sleep disturbance in patients with "fibrositis syndrome" and healthy subjects. *Psychosom. Med.,* 37:341-351, 1975.

6. Moldofsky, H., and Scarisbrick, P. Induction of neurasthenic musculoskeletal pain syndrome by selective sleep stage deprivation. *Psychosom. Med.,* 38:35-44, 1976.

7. Rotes-Querol, J. The syndromes of psychogenic rheumatism. *Clin. Rheum. Dis.,* 5:797-805, 1979.

8. Smythe, H.A. Non-articular rheumatism and the fibrositis syndrome. Hollander, J.L., McCarty, D.J., Jr., eds., *Arthritis and Allied Conditions,* 8th ed., Phila: Lea and Febiger, 1972, pp. 874-884.

9. The American Rheumatism Association Committee on Rheumatologic Practice: A description of rheumatology practice. *Arthritis Rheum.,* 20:1278-1281, 1977.

10. Wood, P.H.N. Rheumatic complaints. *Br. Med. Bull.,* 27:82-88, 1971.

11. Yunus, M., Masi, A.T., Calabro, J.J., Miller, K.A., and Feigenbaum, S.L. Primary fibromyalgia (fibrositis): Clinical study of 50 patients with matched normal controls. *Semin. Arthritis Rheum.,* 11:151-171, 1981.

TABLE I
RHEUMATIC DISEASES ASSOCIATED WITH LOW BACK PAIN

	ANKYLOSING SPONDYLITIS	REITER'S SYNDROME	PSORIATIC SPONDYLITIS	ENTEROPATHIC ARTHRITIS	OSTEOARTHRITIS OF THE SPINE	HERNIATED NUCLEUS PULPOSUS
Sex	Male	Male	=		=	=
Age at Onset	15-40	20-30	30-40	15-45	40-50	20-40
Presentation	Back Pain	Arthritis Urethritis Conjunctivitis	Back Pain	Abdominal Pain	Back Pain	Radicular Pain
Sacroiliitis	Symmetrical	Asymmetrical	Asymmetrical	Symmetrical	-	-
Axial Skeleton	+	+/-	+/-	+	+	-
Peripheral Joints	Lower	Lower	Upper	Lower	Lower	-
Enthesopathy	+	+	+			
Erythrocyte Sedimentation rate	Elevated	Elevated	Elevated	Elevated	Normal	Normal
Rheumatoid Factor	-	-	-	-	-	
HLA-B27	90%	80%	60%	50%	8%	8%
Course	Continuous	Relapsing	Continuous	Continuous	Relapsing	Episodic
Therapy	Nonsteroidals Exercises	Nonsteroidals Gold Methotrexate	Nonsteroidals Gold Methotrexate	Nonsteroidals Corticosteroids Antibiotics	Nonsteroidals	Bed Rest Nonsteroidals Muscle Relaxants Surgery
Disability	Hip	Lower Extremity	Lower Extremity	Hip	Neurologic Dysfunction	Neurologic Dysfunction

245

TABLE II
NON-STEROIDAL ANTI-INFLAMMATORY DRUGS

	Brand Name	Size (mg)	Dose mg/day	Frequency/day	Loading	Major Toxicities
Salicylates						
Aspirin	Bayer	325	5300	4		GI, renal
Enteric-Coated	Ecotrin	325	5300	4		GI, renal
	Easprin	975	3900	4		GI, renal
Diflunisol	Dolobid	250, 500	500-1000	2	+	GI, renal
Phenylalkanoic Acids						
Ibuprofen	Motrin	300, 400, 600	1200-2400	4		GI, renal
Fenoprofen Calcium	Nalfon	200, 300, 600	600-2400	3-4		GI, renal
Naproxen	Naprosyn	250, 375, 500	500-1000	2		GI, renal
Naproxen Sodium	Anaprox	275	550-825	2		GI, renal
Indoles						
Sulindac	Clinoril	150, 200	300-400	2		GI
Tolmetin Sodium	Tolectin	200, 400	600-1600	4		GI, renal
Indomethacin	Indocin	25, 50, 75 (SR)	75-200	2-4		GI, Renal, hematologic
Anthranilic Acid						
Meclofenamate Sodium	Meclomen	50, 100	200-400	4		GI
Pyrazolone						
Phenylbutazone	Butazolidin	100	300-400	3	+	GI, renal, hematologic
Oxicam						
Piroxicam	Feldene	10, 20	10-20	1		GI, renal

246

PART III. INFECTIONS OF THE LUMBOSACRAL SPINE

§ 4-16. Introduction.

Infections of the lumbar spine are uncommon causes of low back pain; however, physicians who evaluate patients with low back pain must include them in the differential diagnoses. This is particularly important because the outcome is excellent if the disease process is recognized early and treated appropriately. When spinal infections are not promptly recognized, however, they can lead to catastrophic complications, including spinal deformity and spinal cord compression with its associated paralysis and incontinence.

The clinical symptoms and course of spinal infections is dependent on the organism involved. Bacterial infections cause acute, toxic symptoms, while tuberculous and fungal diseases are more indolent in onset and course. The primary symptom of patients with spinal infection is back pain which tends to be localized over the anatomic structure involved. Physical examination demonstrates decreased motion, muscle spasm and percussion tenderness over the involved area. Results of common laboratory tests are non-specific. Radiographic abnormalities including vertebral body subchondral bone loss, disc space narrowing and erosions of contiguous bony structures are helpful when present, but often lag behind clinical symptoms by weeks to months.

The definitive diagnosis of spinal infection requires identification of the offending organism by culturing aspirated and/or biopsied material from the lesion. Treatment consists of antimicrobial drugs directed against the specific organism causing the infection, immobilization with bed rest to relieve pain, a cast if spinal instability is present, and surgical drainage of the abscess to relieve spinal cord

247

compression. Patients who have a prompt diagnosis and appropriate antimicrobial therapy are able to combat the infection without residual disability. Significant disability from persistent pain, spinal instability, and spinal cord compression may occur when there has been a delay in diagnosing a persistent osteomyelitis or epidural abscess.

Paget's disease of bone, a disease associated with abnormal bone resorption and formation is usually inappropriately included in the group of illnesses associated with metabolic bone diseases. The cause of Paget's disease, however, is not known but there is increasing evidence to suggest that it is a viral infection of bone and it therefore has been included in § 4-20.

§ 4-17. Vertebral Osteomyelitis.

Vertebral osteomyelitis is a disease process caused by the growth of a potentially wide variety of organisms in the bones which comprise the axial skeleton. These organisms include bacteria — Staphylococcus aureus, Escherichia coli, and Brucella abortus; mycobacteria — Mycobacteria tuberculosis; fungi — Coccidioides immitis; spirochetes — Treponema pallidum; and parasites — Echinococcus granulosus. Vertebral osteomyelitis develops most commonly from hematogenous spread through the blood stream. The clinical symptoms and course are dependent on the infecting organism and the associated host inflammatory response. Bacterial infections are generally associated with an acute, toxic reaction, while granulomatous infections caused by tuberculous or fungal organisms are more indolent in onset and course.

The diagnosis of vertebral osteomyelitis is frequently missed because patient symptoms are ascribed to more common causes of low back pain such as muscle strain, and radiographic changes lag behind the evolution of the infection. The definitive diagnosis can be confirmed by culturing the offending organism from fluid obtained from aspiration of the bone lesion or from blood cultures. Vertebral osteomyelitis can be cured with antibiotics alone if these agents are used in adequate doses and are begun early in the course of the illness. Surgical therapy may be required when there is extensive destruction of vertebral bodies, spinal cord compression or recurrence of infection. Long term disability does not occur if there are no neurological complications and the infection is expeditiously brought under control.

§ 4-17(A). Prevalence and Etiology.

Over the past decade, 348 cases of vertebral osteomyelitis have been reported in the medical literature (entry 35) and it has been found to account for 2% to 4% of all cases of osteomyelitis (entry 23). The mean age of adults reported with vertebral osteomyelitis ranges from 45 to 62. The male to female ratio is up to 3:1.

Vertebral body bone is most frequently infected by hematogenous spread through the blood stream. It is supplied by the paravertebral venous system and nutrient arteries (entries 3, 38). The venous plexus of Batson is a network of valveless veins which lines the vertebral column and the flow in this venous system is modified by changes in intra-abdominal pressure. Increased pressure tends to force blood from infected areas in the pelvis (bladder, gastrointestinal tract) into these veins of the vertebral column. This fact might explain the predisposition of patients with urinary tract infections, rectosigmoid disease, or postpartum infection to vertebral osteomyelitis (entries 26, 24, 32). On the other hand, the localization of early foci of osteomyelitis in vertebral bodies in the subchondral region corresponds to an area richly supplied by nutrient arteries (entry 38). The spread of infection from extraspinal foci is associated with the constitutional symptoms of septicemia. It is probable that both routes may be involved in individual patients.

Infection from contiguous sources, direct implantations by lumbar puncture or disc operations, are relatively rare compared to those caused by hematogenous spread. Occasional patients with vertebral osteomyelitis may have a history of prior trauma to the spine but, trauma usually does not play a role in the pathogenesis of hematogenous vertebral osteomyelitis (entry 12). Studies of patients with vertebral osteomyelitis and epidural abscesses, however, report a significant percentage with a history of substantial trauma prior to the development of the infection (entry 2). A hematoma associated with trauma can become a culture medium for hematogenously spread organisms. The role of minor trauma such as muscle strains as a predisposing factor for vertebral osteomyelitis is uncertain. The importance of unrelated and coincident minor trauma may be overestimated in patients with vertebral osteomyelitis, thus confusing the diagnosis and delaying the initiation of appropriate antibiotic therapy to the disadvantage of the patient.

Approximately 40% or more of patients with vertebral osteomyelitis have an unequivocal extraspinal primary source for infection. The usual locations for these infections include the genitourinary tract, skin and respiratory tract (entry 29). Parenteral drug abusers also develop vertebral osteomyelitis, particularly with Pseudomonas aeruginosa (entry 22). Any patient with a chronic disease which decreases host immunity, such as diabetes mellitus, chronic alcoholism, malignancy, and sickle cell anemia are also at risk of developing vertebral osteomyelitis.

§ 4-17(B). Clinical Findings.

The primary symptom of patients with vertebral

osteomyelitis is low back pain. The pain may develop over 8 to 12 weeks before the diagnosis is established in patients (entry 29). A history of a recent primary infection, or an invasive diagnostic procedure, is common. The pain may be intermittent or constant, may be present at rest and it is exacerbated by motion. This group of symptoms is one of the four clinical syndromes of pyogenic vertebral osteomyelitis described by Guri (entry 16). Patients may develop a hip joint syndrome characterized by acute pain in the hip with limited motion and flexion contracture. The patient with the abdominal syndrome presents with symptoms easily confused with appendicitis. Signs of acute meningitis, including positive Kernig and Brudzinski tests and positive straight leg raising tests, are components of the meningeal syndrome. Paraplegia without back pain is a very rare occurrence (entry 27).

Patients with certain underlying illnesses may develop vertebral osteomyelitis secondary to specific organisms. Those with diabetes mellitus or chronic urinary tract infections develop vertebral osteomyelitis secondary to gram negative bacteria (entry 12). Parenteral drug abusers develop pseudomonas and candida osteomyelitis (entry 22). Patients with sickle cell anemia may get salmonella osteomyelitis (entry 33).

Brucellosis, a disease caused by the Brucella organism, affects patients who ingest unpasteurized milk products, or more commonly since the advent of pasteurization, workers involved with meat processing (entry 9). The infection, passed from lower animals, cattle or hogs, occurs in government meat inspectors, veterinarians, farmers, stockmen, rendering-plant workers and laboratory personnel. The Brucella organism, B. melitensis, the most virulent, B. suis, or B. abortus, penetrates the mucous membranes of the oropharynx or enters through breaks in the skin, traverses

251

the lymph nodes and enters the blood stream where it spreads to the reticuloendothelial system. The disease has an insidious onset with intermittent fever, chills, weakness, weight loss and headache. Patients commonly complain of tenderness or pain over the vertebral bodies which is worse with activity and relieved by rest (entries 21, 20).

Elderly patients are at greatest risk of developing vertebral osteomyelitis secondary to Mycobacterium tuberculosis. Prior to antibiotic therapy, children were most frequently affected but more recent data from patients with skeletal tuberculosis in the United States show average ages of 40 to 51 (entries 4, 11). Skeletal tuberculosis occurs as a result of hematogenous spread from another source, usually pulmonary, during an acute infection; or as a reactivation of a quiescent focus present in bone for many years after initial seeding (entries 13, 5). Fifty to sixty per cent of patients with skeletal tuberculosis have axial skeletal disease (entry 13). This may be explained in part by the affinity of this organism to relatively high oxygen concentrations which exist in the cancellous bone of the vertebral bodies. Tuberculous spondylitis begins in the subchondral area of the vertebral body adjacent to the intervertebral disc and the organism creates an inflammatory process characterized by the formation of granulomas and caseation necrosis of bone. Initially, only the vertebral body is. affected; however, the infection can spread to involve contiguous structures which include the intervertebral disc, other vertebral bodies, soft tissues such as muscle and ligaments to form a paravertebral abscess, or the spinal cord and meninges. The more extensive the destruction, the greater the potential for deformity of the axial skeleton with kyphoscoliosis and associated spinal cord compression.

The clinical presentation of a patient with tuberculous spondylitis consists of pain over the involved vertebrae with radiation into a buttock or lower extremity, low grade fever and weight loss of varying duration. Patients with more advanced disease may present with neurologic symptoms and angular deformities of the spine with loss of height. The onset of symptoms is gradual and the time before presentation to a physician may be as long as three years (entry 14). Paraplegia may be the first manifestation of tuberculous spondylitis even before any deformity of the spine is apparent (entry 31).

Infections secondary to Actinomyces israeli and fungal organisms, Coccidioides immitis and Blastomyces dermatidis, are very rare causes of vertebral osteomyelitis. The clinical course of the infections is very similar to that of the tuberculosis. These patients present with a history of constitutional symptoms over an extended period of time and complain of localized pain over the affected vertebral body (entries 25, 37, 30).

Before the advent of penicillin, syphilis was a common cause of axial skeleton infection but it currently is an extremely rare complication. Syphilis may affect the axial skeleton by direct infection of bone or by the loss of normal sensation (entry 17). Both forms are a result of tertiary syphilis which is the form of the disease which occurs after the initial infection (primary) and hematogenous spread (secondary) of the organism. The growth of the organism in bone results in the formation of a gumma and bony destruction. These patients develop local symptoms similar to those of tuberculous spondylitis patients and they may develop neurologic deficits (entry 18). Patients with neurosyphilis, on the other hand, lose normal sensation and without this protective awareness, fractures and destruction of the bony skeleton result. This is called a neuropathic arthropathy or Charcot disease. These patients feel little pain and are frequently symptom-free.

Echinococcus granulosa is a cestode worm of the dog (entry 1). Intermediate hosts for the ova of this worm are sheep, cattle, hogs, and man. The ova attach to the intestinal mucosa and gain entrance to the blood stream where they disseminate, particularly to the liver and occasionally to bone where they form cysts (hydatid disease). The disease produces a slow destructive lesion of bone which in the spine erodes through the vertebral bodies and can rupture into the neural canal. The cysts then migrate up and down the canal. Patients with hydatid disease of the spine have symptoms which can last from a few weeks to five years. Pain is a common symptom associated with swelling. Occasionally painless paraplegia of sudden onset may be the presenting sign of disease. Patients who develop paraplegia either die from the disease or are chronically disabled (entry 34).

§ 4-17(C). Physical Findings.

Physical findings in patients with vertebral osteomyelitis include a decreased range of motion, muscle spasm, and percussion tenderness over the involved bone. Those with psoas muscle irritation may demonstrate decreased hip motion along with a flexion contracture. Patients with thoracolumbar vertebral osteomyelitis may have abdominal tenderness on palpation. Some of the patients with bacterial vertebral osteomyelitis have a fever (entries 8, 6). Neurologic abnormalities, including paraplegia, are reported in a number of series of patients with vertebral osteomyelitis (entries 12, 15, 10). Patients with the more indolent infections of tuberculosis and coccidioidomycosis usually have less fever, but greater spinal deformity than those with pyogenic vertebral osteomyelitis.

§ 4-17(D). Laboratory Findings.

The commonly ordered blood tests (CBC, ESR and serum

254

chemistries) yield results which are normal or non-specific. The erythrocyte sedimentation rate is abnormal in the vast majority of patients with vertebral osteomyelitis, particularly during the acute phase (entries 12, 6). Hematocrits may be normal and there is a normal or slightly elevated white blood cell count.

The most useful laboratory test for the diagnosis of vertebral osteomyelitis is the direct culture of blood and bone lesions. This may be positive in 50% of patients with acute osteomyelitis and obviate the need for bone biopsy (entry 28). In patients with negative blood cultures, bone aspiration or surgical biopsy produces material which on culture is often positive for the offending organism (entry 29). Staphylococcus aureus is the bacterium associated with vertebral osteomyelitis in up to 60% of patients (entries 8, 6). Gram negative organisms, E. coli, Proteus, and Pseudomonas, are often grown from elderly patients and parenteral drug abusers with vertebral osteomyelitis. Cultures from peripheral sources of infection, urinary tract, skin and respiratory tract, may be positive and should be obtained from the patient with suspected vertebral osteomyelitis. The diagnosis of brucellosis is associated with a positive bone culture or elevations in brucellar agglutinin titers.

The PPD test is usually positive for patients with tuberculous spondylitis unless they are anergic secondary to miliary disease. The number of organisms in an infected spine is less than one million bacteria. Therefore, it is appropriate to culture both purulent material and biopsy specimens in order to improve the potential for a positive culture (entry 7). Histologic evidence of granulomas suggests tuberculous or fungal infection. The erythrocyte sedimentation rate is rarely elevated in tuberculous spondylitis. As far as fungal infections are concerned, antibody titers may be raised or skin tests reactive, but none of these tests are as specific as the growth of organisms from either aspirated material or biopsy specimens. The laboratory abnormalities of tertiary syphilis include the presence of antibodies to nontreponemal (VDRL test) and treponemal (FTA-ABS test) antigens. Spirochetal organisms are not usually found in tertiary lesions on histologic examination. The histologic changes include a granulomatous necrotizing process with a prominent obliterative endarteritis.

§ 4-17(E). Radiographic Findings.

Radiographic changes follow the symptomatic onset of disease by one to two months. The early abnormalities in pyogenic vertebral osteomyelitis include subchondral bone loss, narrowing of the disc space, loss of definition of vertebral end plates and erosions of the neighboring vertebral body. Continued dissemination of infection may produce soft tissue swelling associated with paravertebral abscesses (loss of psoas shadow). Once the lesion starts to heal, bony regeneration appears and is characterized by osteosclerosis which may finally result in bony fusion across the disc space. The lumbar vertebrae, particularly the first and second, are the vertebral bodies in the axial skeleton most commonly affected (entries 12, 29). The radiographic changes of Brucella spondylitis are similar to those of other pyogenic vertebral osteomyelitis (entry 21).

The vertebral body is more commonly affected than posterior elements in tuberculous spondylitis. The infection causes erosion of the subchondral bone and invades the disc space. These change much less rapidly with tuberculous spondylitis as compared with pyogenic spondylitis. The infection may also spread to soft tissue, forming paraspinal abscesses. Severe angular deformities occur from marked destruction of vertebral bodies. The reactive sclerosis characteristic of healing pyogenic vertebral osteomyelitis does not occur with tuberculous spondylitis.

Vertebral coccidioidomycosis differs from tuberculous spondylitis by sparing the intervertebral discs, involving of anterior and posterior elements of the vertebral body and rarely causing vertebral collapse (entry 14).

Radiographic changes in the spine may include marked destruction and dissolution of bone in Charcot arthropathy of the spine and may show lysis and sclerosis in bone from gummatous osseous lesions. Periosteal changes may also be present (entry 17).

Hydatid disease is associated with single or multiple expansile osteolytic lesions containing trabeculae. Soft tissue calcification may also be seen.

Bone scintography demonstrates abnormalities in the area of infection at an earlier stage of disease than does plain radiographs. A bone scan may also demonstrate areas of involvement other than the one which is symptomatic. It must be kept in mind that false positives and negatives do occur and that increased uptake on bone scan may be caused by tumor, trauma, or arthritis as well.

Computerized tomography may show bony changes prior to their appearance on routine radiographs. The extent of soft tissue abscesses are also more easily visualized by this method (entry 19).

§ 4-17(F). Differential Diagnosis.

The definitive diagnosis of vertebral osteomyelitis is based upon the recovery and identification of the causative organism from aspirated material or biopsy. Clinical history, physical examination, and laboratory investigation are too non-specific to assure an accurate diagnosis. Conditions confused with vertebral osteomyelitis include discitis, metastatic tumors, multiple myeloma, eosinophilic granuloma, aneurysmal bone cyst, giant cell tumor of bone and sarcoidosis. Osteomyelitis may occur in bones in the pelvis and may be associated with severe low back pain or abdominal pain (entry 36).

§ 4-17(G). Treatment.

Therapy of vertebral osteomyelitis includes antibiotics, bed rest, and immobilization. The choice of antibiotic therapy is decided by the organism causing the infection and its sensitivity to specific agents. Gram positive bacteria such as Staphylococcus require penicillin or semi-synthetic penicillin (Nafcillin), while gram negative bacteria are sensitive to aminoglycosides for eradication. Patients are treated with four to six weeks of parenteral antibiotics followed by a course of oral antibiotics which may have a duration of as long as six months. Bed rest is helpful in decreasing pain by limiting motion. Some patients require plaster casts or braces when there is major bone destruction or instability.

Brucellosis may be treated with oral tetracycline at a dose of 2 gm. per day. Therapy should continue for three to four weeks. In severe cases, 0.5 gm. of streptomycin is injected twice a day for a similar period.

Tuberculous spondylitis requires a three-drug regimen initially (isoniazid, ethambutol, rifampin or streptomycin) for two to three months. Two drugs are usually continued for a total of 18 months. Immobilization is necessary for healing. Patients are either placed on bed rest or in plaster body casts to insure limitation of motion at the site of infection.

Fungal osteomyelitis requires parenteral amphotericin B therapy at a dose that the patient can tolerate without developing renal dysfunction. A total dose of 2.5 gm. is usually employed as the initial course of therapy. Bone lesions, however, may be resistant to chemotherapy and may require surgical debridement. Immobilization is required for these patients as well.

Patients with tertiary syphilis without neurologic involvement are effectively treated with penicillin. Long acting penicillin, Benzathine penicillin G at 2.4 million units intramuscularly weekly for three weeks, is usually used. Patients who are allergic to penicillin may take tetracycline or erythromycin.

Hydatid disease has been resistant to most antihelminthic therapy. Mebendazole has shown some efficacy in preliminary studies. Definitive therapy requires surgical excision of cysts with care being taken to remove them in toto.

§ 4-17(H). Course and Outcome.

The improvement in vertebral osteomyelitis may be monitored by following the decrease in patients' symptoms and pain, fever and the return of the erythrocyte sedimentation rate to normal. With early diagnosis and the institution of appropriate therapy, vertebral osteomyelitis will resolve with minimal disability to the patient. However, when the diagnosis is delayed and the infection spreads to involve the spinal cord by compression (epidural abscess, granulation tissue) or direct extension (meningitis) potentially life threatening complications may occur (entries 2, 14). Surgical intervention, which may be a biopsy and/or debridement, is indicated when the diagnosis of vertebral osteomyelitis is suspected but the causative organism is not known or when a paravertebral abscess is present with signs of spinal cord compression, persistent nerve root compression or a lack of response (continued fever, elevated ESR) to antibiotic therapy. An anterior approach is used to limit the possibility of increasing spinal instability. Paraplegia is also an indication for surgical decompression since patients who have a surgical decompression soon after the onset of paraplegia have a better chance of full recovery than those who have had their neurologic deficits for a month or more (entry 6).

BIBLIOGRAPHY

1. Alldred, A.J., and Nisbet, N.W. Hydatid disease of bone in Australia. *J. Bone Joint Surg.,* 46B:260-267, 1964.
2. Baker, A.S., Ojemann, R.G., Swartz, M.N., and Richardson, E.P., Jr. Spinal epidermal abscess. *N. Eng. J. Med.,* 293:463-468, 1975.
3. Batson, O.V. The function of the vertebral veins and their role in the spread of metastases. *Ann. Surg.,* 112:138-149, 1940.
4. Brashear, H.R., Jr., and Rendleman, D.A. Pott's paraplegia. *South. Med. J.,* 71:1379-1382, 1978.
5. Chapman, M., Murray, R.O., and Stoker, D.J. Tuberculosis of the bones and joints. *Semin. Roentgenol.,* 14:266-282, 1979.
6. Collert, S. Osteomyelitis of the spine. *Acta. Orthop. Scand.,* 48:283-290, 1977.
7. Davidson, P.T., and Horowitz, T. Skeletal tuberculosis: A review with patient presentations and discussion. *Am. J. Med.,* 48:77-84, 1970.
8. Digby, J.M., and Kersley, J.B. Pyogenic non-tuberculous spinal infection: An analysis of thirty cases. *J. Bone Joint Surg.,* 61B:47-55, 1979.
9. Fox, M.D., and Kaufman, A.F. Brucellosis in the United States. *J. Infect. Dis.,* 136:312-316, 1977.
10. Freehafer, A.A., Furey, J.G., and Pierce, D.S. Pyogenic osteomyelitis of the spine: Resulting in spinal paralysis. *J. Bone Joint Surg.,* 44A:710-716, 1962.
11. Friedman, B. Chemotherapy of tuberculosis of the spine. *J. Bone Joint Surg.,* 48A:451-474, 1966.
12. Garcia, A., Jr., and Grantham, S.A. Hematogenous pyogenic vertebral osteomyelitis. *J. Bone Joint Surg.,* 42A:429-436, 1960.
13. Goldblatt, M., and Cremin, B.J. Osteoarticular tuberculosis: Its presentation in the coloured races. *Clin. Radiol.,* 29:669-677, 1978.
14. Gorse, G.J., Pais, M.J., Kusske, J.A., and Cesario, T.C. Tuberculous spondylitis: A report of six cases and a review of the literature. *Medicine,* 62:178-193, 1983.
15. Griffiths, H.E.D., and Jones, D.M. Pyogenic infection of the spine: A review of twenty-eight cases. *J. Bone Joint Surg.,* 53B:383-391, 1971.
16. Guri, J.P. Pyogenic osteomyelitis of the spine: Differential diagnosis through clinical and roentgenographic observations. *J. Bone Joint Surg.,* 28A:29-39, 1946.
17. Johns, D. Syphilitic disorders of the spine: Report of two cases. *J. Bone Joint Surg.,* 52B:724-731, 1970.

18. Karaharju, E.O., and Hannuksela, M. Possible syphilitic spondylitis. *Acta. Orthop. Scand.,* 44:289-295, 1973.
19. Kattapuram, S.V., Phillips, W.C., and Boyd, R. CT in pyogenic osteomyelitis of the spine. *Am. J. Roentgenol.,* 140:1199-1201, 1983.
20. Keenan, J.D., and Metz, C.W., Jr. Brucella spondylitis: A brief review and case report. *Clin. Orthop.,* 82:87-91, 1972.
21. Kelley, P.J., Martin, W.J., Schirger, A., and Weed, L.A. Brucellosis of the bones and joints: experience with 36 patients. *JAMA,* 174:347-353, 1960.
22. Kido, D., Bryan, D., and Halpern, M. Hematogenous osteomyelitis in drug addicts. *Am. J. Roentgenol.,* 118:356-63, 1973.
23. Kulowski, J. Pyogenic osteomyelitis of the spine: An analysis and discussion of 102 cases. *J. Bone Joint Surg.,* 18A:343-364, 1936.
24. Lame, E.L. Vertebral osteomyelitis following operation on the urinary tract or sigmoid: The third lesion of an uncommon syndrome. *Am. J. Roentgenol.,* 75:938-952, 1956.
25. Lane, T., Goings, S., Fraser, D., Ries, K., Pettrozzi, J., and Abrutyn, E. Disseminated actinomycosis with spinal cord compression: Report of two cases. *Neurology,* 29:890-893, 1979.
26. Leigh, T.F., Kelley, R.P., and Weens, H.S. Spinal osteomyelitis associated with urinary tract infections. *Radiology,* 65:334-342, 1955.
27. Ling, C.M. Pyogenic osteomyelitis of the spine. *Orthop. Rev.,* 4:23-32, 1975.
28. Musher, D.M., Thorsteinsson, S.B., Minuth, J.N., and Luchi, R.J. Vertebral osteomyelitis: Still a diagnostic pitfall. *Arch. Intern. Med.,* 136:105-110, 1976.
29. Ross, P.M., and Fleming, J.L. Vertebral body osteomyelitis: spectrum and natural history: A retrospective analysis of 37 cases. *Clinc. Orthop.,* 118:190-198, 1976.
30. Sarosi, G.A., and Davies, S.F. Blastomycosis. *Am. Rev. Resp. Dis.,* 120:911-938, 1979.
31. Seddon, H.J. Pott's paraplegia: Prognosis and treatment. *Br. J. Surg.,* 22:769-799, 1935.
32. Sherman, M., and Schneider, G.T. Vertebral osteomyelitis complicating postabortal and postpartum infection. *South. Med. J.,* 48:333-338, 1955.
33. Specht, E.E. Hemoglobinopathic Salmonella osteomyelitis. *Clinc. Orthop.,* 79:110-118, 1971.
34. Unger, H.S., Schneider, L.H., and Sher, J. Paraplegia secondary to hydatid disease: Report of a case. *J. Bone Joint Surg.,* 45A:1479-1484, 1963.

35. Waldvogel, F.A., and Vasey, H. Osteomyelitis: The past decade. *N. Eng. J. Med.*, 303:360-370, 1980.

36. Weld, P.W. Osteomyelitis of the ilium masquerading as acute appendicitis. *JAMA*, 173:634-636, 1960.

37. Wesselius, L.J., Brooks, R.J., and Gall, E.P. Vertebral coccidiodomycosis presenting as Pott's disease. *JAMA*, 238:1397-1398, 1977.

38. Wiley, A.M., and Trueta, J. The vascular anatomy of the spine and its relationship to pyogenic vertebral osteomyelitis. *J. Bone Joint Surg.*, 41B:796-809, 1959.

§ 4-18. Intervertebral Disc Space Infection.

Infection of the intervertebral disc is an uncommon but potentially disabling cause of low back pain. While once thought to be exclusively a complication of vertebral osteomyelitis, intervertebral disc infection also can develop secondary to hematogenous invasion through the bloodstream and by direct penetration during disc surgery. A significant clinical feature of this illness is the long delay between the onset of symptoms of low back pain, muscle spasm and limitation of motion and the establishment of the diagnosis. The diagnosis of an intervertebral disc infection is confirmed by the growth of organisms on culture media from fluid obtained by aspiration of the disc space. Antibiotics, chosen by susceptibility testing, and immobilization with a body cast are effective in eradicating the infection and relieving symptoms. The major potential complication of this infection is compression of the spinal cord with subsequent paraplegia. Most patients have full recovery and are able to return to their usual work once fusion of the disc space occurs.

§ 4-18(A). Prevalence and Etiology.

Intervertebral disc infection is an uncommon illness, occurring with an incidence of two patients per year at an orthopaedic hospital in one study (entry 9). Approximately

2.8% of patients who have lumbar disc surgery develop disc space infections (entry 11). Men are more frequently affected and range in age from 17 to 63 (entry 9).

The intervertebral discs may be infected by hematogenous spread through the bloodstream. Although it is generally believed that the intervertebral discs are avascular in adult life, a number of investigators have demonstrated some blood flow in adults (entries 16, 14). While there is a decrease in the number of vessels which enter the nucleus pulposus with aging, an adequate circumferential supply is maintained from the periphery (entry 8). The blood supply to intervertebral disc is greater in children and this may explain the increased frequency of disc space infection in the pediatric age group when compared to adults (entry 2).

There is evidence supporting the importance of both the venous and arterial systems in the development of disc infections. Patients with pelvic and urinary tract infections may develop involvement of disc spaces with the identical organism and it is postulated that these organisms may spread by means of the vertebral venous plexus of Batson (entries 1, 3, 7). Batson's plexus is a network of veins which surrounds the vertebral column and it is connected to the major veins which return blood to the heart, the inferior and superior vena cava. However, other investigators believe that the reversal of flow in the venous system, which would allow infected blood to enter Batson's plexus, does not occur under usual circumstances. Instead, they suggest hematogenous spread through the arterial system as the source of infection (entries 16, 6). It is probable that either route may be involved in appropriate circumstances.

A few patients have been reported to have developed intervertebral disc infection following trauma, such as lifting objects at work or heavy physical labor (entries 4, 10). Trauma in the area of the intervertebral disc may result in the formation of a hematoma which is then infected by blood-borne organisms but the likelihood of this occurrence is not great. In fact, most patients with intervertebral disc infections deny previous trauma (entry 9). Therefore, the association of this infection and trauma is unlikely. Patients who undergo operative procedures, needle biopsy, discography or puncture of a disc during a lumbar puncture can develop infection by direct inoculation of the organism at the time of the procedure (entry 11).

§ 4-18(B). Clinical Findings.

Patients with intervertebral disc infections of the lumbar spine have symptoms of localized low back pain. In some, the onset of the pain is acute and very severe. In other patients, the pain may be more insidious and milder. It becomes chronic and may be associated with radiation into the flanks, abdomen, testes, or lower extremities. The pain is exacerbated by movement and relieved by absolute rest. Motion may initiate paroxysms of paravertebral muscle spasm and the patients usually have great difficulty walking. Loss of motor strength or sensory symptoms may suggest spinal cord compression.

§ 4-18(C). Physical Examination.

Physical findings in patients with disc infection include localized tenderness on palpation, limitation of motion of the lumbar spine, and paravertebral muscle spasm. Fever is present in very few of them (entry 10). Examination of the skin, respiratory system, gastrointestinal system, and genitourinary system may demonstrate the primary source of infection.

§ 4-18(D). Laboratory Findings.

The laboratory findings are non-specific in disc space infections. The most commonly abnormal test is the erythrocyte sedimentation rate. It is elevated in up to 75% of patients (entry 9). A mild leukocytosis with a normal differential is also seen (entry 10). Occasionally blood cultures are positive during the acute phases of the illness (entry 10) but culture of fluid obtained by needle aspiration or at surgery is usually positive (entries 4, 10). The most frequent cultured organism causing disc infection is Staphylococcus aureus. Gram negative organisms have also been implicated, particularly Pseudomonas aeruginosa in intravenous drug abusers (entries 12, 13).

§ 4-18(E). Radiographic Findings.

The radiographic features of disc infection are distinctive and help differentiate it from that of vertebral osteomyelitis; however, they may lag behind clinical symptoms by six weeks or more (entry 9). The earliest change is a decrease in the height of the affected intervertebral disc space. Two months after the appearance of disc space narrowing, reactive sclerosis of subchondral bone appears in the adjoining vertebral bodies. Subsequently, progressive irregularity of the vertebral end plates develops and indicates a local extension of the inflammatory process and an osteomyelitis. At this juncture, the loss of vertebral bone may be associated with a "ballooning" of the intervertebral disc space (entry 9). This increase in apparent disc space is usually associated with involvement of the vertebral body posteriorly. Repair may occur at any stage of infection and it is manifested by bony proliferation about the outside margin of the disc. The process may heal with bony ankylosis across the affected disc space of the adjacent vertebral bodies.

Bone scintiscans are very useful in rapidly identifying increased bone activity in areas contiguous to infected discs. The bone scan may be positive in an area of infection which appears normal on plain radiographs (entry 5).

§ 4-18(F). Differential Diagnosis.

The diagnosis of intervertebral disc infection is often overlooked because of non-specific symptoms and its relative infrequency as a cause of back pain compared to mechanical disorders and the spondyloarthropathies. Nonetheless, physicians should be aware of this treatable entity so that the appropriate diagnosis will be made as early as possible. The definitive diagnosis of a disc infection depends upon the aspiration or biopsy of the infected site with subsequent confirmation by culture of the causative organism. A bone scan may help identify the patient with this entity who has normal radiographs.

Other diseases which may be associated with intervertebral disc changes in the low back include osteomyelitis of a vertebral body, spinal cord tumor and metastasis.

§ 4-18(G). Treatment.

Therapy of intervertebral disc infections includes antibiotics and immobilization by bed rest, casts, or bracing. There is little agreement in the literature as to just what component of therapy is most effective in this disease. Some authors suggest bed rest and immobilization are adequate and antibiotics are not necessary (entry 15); however, most patients receive a four to six-week course of antibiotic therapy (entry 10). The response to treatment is monitored by a relief of pain, return of the sedimentation rate to normal and radiographic evidence of disc space restoration or bony ankylosis of adjacent vertebral bodies. Surgical exploration is not indicated unless there are signs of spinal cord compression. Surgical fusion is unnecessary. Patients who develop a disc space infection after lumbar disc surgery may be treated with immobilization and antibiotics; surgical exploration is usually unnecessary (entry 11).

§ 4-18(H). Course and Outcome.

Most patients with disc infections have a benign course and fully recover without disability. Back bracing may be required for a year or longer until ankylosis of adjacent vertebral bodies occurs. Patients usually have full range of motion with healing and are asymptomatic (entry 10).

The most serious complication of disc space infection was reported by Kemp (entry 9). In his group of thirteen patients, three of them were hemiplegic and three were paraplegic, and all required surgical decompression. At operation direct extension of inflammatory granulation tissue was identified growing posteriorly and involving the meninges and spinal cord. Cord damage was secondary to compression of the cord by edema, inflammatory tissue or from thrombosis of spinal cord vessels. Of the six patients with neurologic symptoms, two had complete recovery, two had partial recovery and two had no recovery after surgical decompression. The appearance of neurologic symptoms was rapid in these patients although most of them had back pain for greater than four months. Accurate diagnosis and the prompt institution of appropriate therapy should prevent the emergence of this complication of disc infection.

BIBLIOGRAPHY

1. Batson, O.V. The function of the vertebral veins and their role in the spread of metastasis. *Ann. Surg.,* 112:138-149, 1940.
2. Boston, H.C., Bianco, A.J., Jr., and Rhodes, K.H. Disc space infections in children. *Orthop. Clin. North Am.,* 6:953-964, 1975.
3. Doyle, J.R. Narrowing of the intervertebral-disc space in children. *J. Bone Joint Surg.,* 42A:1191-1200, 1960.
4. Ettinger, W.H., Jr., Arnett, F.C., Jr., and Stevens, M.B. Intervertebral disc space infections: Another low back syndrome of the young. *Johns Hopkins Hosp. Med. J.,* 141:23-27, 1977.
5. Garroway, R., and Healy, W.A., Jr. Importance of bone scan in disc space infection. *Orthop. Review,* 11:87-92, 1982.

6. Ghormley, R.K., Bickel, W.H., and Dickson, D.D. A study of acute infectious lesions of the intervertebral discs. *South. Med. J.,* 33:347-353, 1940.

7. Griffiths, H.E.D., and Jones, D.M. Pyogenic infection of the spine. *J. Bone Joint Surg.,* 53B:383-391, 1971.

8. Hassler, O. The human intervertebral disc. *Acta. Orthopaedica Scand.,* 40:765-772, 1970.

9. Kemp, H.B.S., Jackson, J.W., Jeremiah, J.D., and Hall, A.J. Pyogenic infections occurring primarily in intervertebral discs. *J. Bone Joint Surg.,* 55B:698-714, 1973.

10. Onofrio, B.M. Intervertebral discitis: Incidence, diagnosis and management. *Clin. Neurosurgery,* 27:481-516, 1980.

11. Pilgaard, S. Discitis (closed space infection) following removal of lumbar intervertebral disc. *J. Bone Joint Surg.,* 51A:713-716, 1969.

12. Scherbel, A.L., and Gardner, J.W. Infections involving the intervertebral discs: Diagnosis and management. *JAMA,* 174:370-374, 1960.

13. Selby, R.C., and Pillay, K.V. Osteomyelitis and disc infection secondary to Pseudomonas aeruginosa in heroin addiction: Case report. *J. Neurosurg.,* 37:463-466, 1972.

14. Smith, N.R. The intervertebral discs. *Brit. J. Surg.,* 18:358-375, 1931.

15. Sullivan, C.R. Diagnosis and treatment of pyogenic infections of the intervertebral disc. *Surg. Clin. North Am.,* 41:1077-1086, 1961.

16. Wiley, A.M., and Trueta, J. The vascular anatomy of the spine and its relationship to pyogenic vertebral osteomyelitis. *J. Bone Joint Surg.,* 41B:796-809, 1959.

§ 4-19. Pyogenic Sacroiliitis.

Septic arthritis is a disease process caused by the direct invasion of a joint space by infectious agents, usually bacteria. Joints become infected by direct penetration, spread from contiguous structures, or more often by hematogenous invasion through the bloodstream. Septic arthritis occurs more commonly in large peripheral joints (knees, shoulders) than in the lumbosacral spine but when a joint of the axial skeleton is involved, the sacroiliac is the one most commonly affected. The diagnosis of pyogenic sacroiliitis is frequently delayed, with the back pain that patients experience ascribed to hip disease, herniated intervertebral discs, or malignancy; but it can be confirmed by culturing the offending organism from fluid obtained by aspiration of the joint space. Antibiotic therapy is effective in eradicating the infection. Long term disability is usually not a problem.

§ 4-19(A). Prevalence and Etiology.

Pyogenic sacroiliitis is an uncommon illness accounting for 0.07% of hospital admissions in one study (entry 3). Approximately 80 patients with pyogenic sacroiliitis have been reported in the medical literature (entries 6, 10, 9, 7). The disease occurs most commonly in young adult men. The range of ages is 20 to 66 (entry 3) and the male to female ratio is 3:2 (entry 6).

Infectious agents reach the sacroiliac joint by travelling through the bloodsteam. They lodge in the vascular synovial membrane which lines the lower portion of the sacroiliac joint. The infectious agents grow in the synovium and invade the joint space. Once an infection is established in a joint, rapid destruction may occur because of both the direct toxic effects of products of organisms on joint structures and the host's inflammatory response to them.

Any factor, such as intravenous drug abuse, skin infections, bone and urinary tract infections and endocarditis (entries 3, 9, 2) which promotes blood-borne infection or inhibits the normal defense mechanisms of the synovial joint predisposes the host to infection. Although the role of trauma in the pathogenesis of pyogenic sacroiliitis is unclear, buttock or hip injuries have been reported in patients prior to developing pyogenic sacroiliitis (entries 9, 4). Most patients, however, deny a history of trauma and its importance as a direct cause of infection is in question. The histocompatibility typing associated with seronegative

269

spondyloarthropathies, HLA-B27, is not associated with pyogenic sacroiliitis (entry 13).

§ 4-19(B). Clinical Findings.

The typical patient with pyogenic sacroiliitis is a young man with fever who develops acute pain over the buttock (entry 6). Those who have subacute or chronic symptoms are frequently afebrile. The pain is severe and may radiate to the low back, hip, thigh, abdomen, or calf. Abdominal pain associated with nausea and vomiting can confuse the diagnosis (entry 8). Patients also complain of severe leg pain that prevents them from bearing weight on the affected limb. Pyogenic sacroiliitis is usually unilateral although bilateral disease can occur.

§ 4-19(C). Physical Findings.

Physical examination demonstrates tenderness over the sacroiliac joint and pain on compression of the joint. Hyperextension of the hip, Gaenslen maneuver, stresses the sacroiliac joint capsule and causes pain in the buttock. Patrick's test or the Fabere maneuver — flexion, abduction, external rotation and extension — of the hip, may be associated with sacroiliac joint pain. The test is done by placing the ankle of the affected side on the opposite knee and exerting downward pressure on the bent knee. If the maneuver is performed slowly, the hip may be taken through a full range of motion without producing sacroiliac pain. Therefore, the test should be done rapidly so that the sacroiliac joint will be stressed and painful. Straight leg raising may be limited due to stretching of inflamed sacral nerve roots over the anterior of the sacroiliac joint. A subgluteal abscess with swelling of the buttock and obliteration of the gluteal fold may also be seen. Infection may track along different fascial planes, presenting as masses in the inner thigh, posterior thigh, hip, lumbar spine or abdomen (entries 8, 1). Muscle spasm of the gluteal and lumbar paravertebral muscles may be severe. The severity of the spasm may cause severe pain and scoliosis of the lumbar spine (entry 9). Fever and chills may be seen in the patient with an acute onset of disease (entry 10).

§ 4-19(D). Laboratory Findings.

Laboratory results demonstrate mild anemia, leukocytosis and elevation in the erythrocyte sedimentation rate. Blood cultures may be positive in up to 50% of patients, eliminating the need for joint aspiration. Arthrocentesis of the sacroiliac joint under general anesthesia with fluoroscopic guidance is the most effective method of diagnosing pyogenic sacroiliitis. Aspirations without fluoroscopy do not obtain adequate specimens (entry 6).

Staphylococcus aureus and Staphylococcus epidermidis account for a majority of infections of the sacroiliac joint (entry 3). Pseudomonas aeruginosa is frequently cultured from sacroiliac joints in parenteral drug abusers (entry 5). Aerobic and anaerobic streptococci may also be cultured (entries 3, 10).

Tuberculosis is also associated with pyogenic sacroiliitis. Sacroiliac tuberculosis was a more common disease before the development of effective antituberculous antibiotic therapy. Strange reported 329 cases of sacroiliac tuberculosis (entry 12) and the patients developed this complication as a manifestation of a generalized tubercular infection. The onset of disease is insidious, the pain aching and the course indolent. The diagnosis of tuberculous sacroiliitis may be confirmed by aspiration of fluid. The yield is improved by the simultaneous culture of synovial tissue. Therefore, open surgical biopsy may be required to confirm this diagnosis (entry 14). Open biopsy may also be required from patients with fungal sacroiliitis as well.

§ 4-19(E). Radiographic Findings.

The duration of symptoms before the appearance of radiographic changes in the sacroiliac joints in bacterial sacroiliitis is two weeks (entry 3). The radiographic findings include blurring of joint margins, pseudowidening of the joint space, erosions and reactive sclerosis. Healing is associated with bony ankylosis.

Bone scintiscans are very useful in rapidly identifying increased joint activity associated with infection and localization is possible within two days of symptoms (entry 6). Shielding of radioactivity in the urinary bladder may allow detection of minimal asymmetry in tracer uptake early in the course of the illness. Computerized tomography may be more useful than plain radiographs in detection of early joint changes (entry 10).

§ 4-19(F). Differential Diagnosis.

The diagnosis of pyogenic sacroiliitis is frequently missed due to a lack of physician awareness of this infection and the non-specific low back symptoms of these patients, but it must be suspected in the appropriate host who presents with severe radiating low back pain. A bone scan will help localize the lesion to the sacroiliac joints. Positive synovial fluid cultures from the sacroiliac joint confirm the diagnosis if blood cultures are negative. Biopsy is necessary if the diagnosis remains in doubt, cultures are negative and granulomatous infection is being considered.

Unilateral sacroiliac joint infection must be differentiated from appendicitis, septic arthritis of the hip, intravertebral disc herniation, intraspinal tumors, and metastatic disease (entry 11). Osteomyelitis of the ilium may cause infection of a contiguous sacroiliac joint. The converse rarely occurs due to the tendency of primary infections of the sacroiliac joint to perforate the joint capsule anteriorly as opposed to invading surrounding bone. The rare occurrence of bilateral disease must be differentiated from the seronegative spondyloarthropathies including ankylosing spondylitis, enteropathic arthritis, psoriatic arthritis, and Reiter's syndrome.

§ 4-19(G). Treatment.

The therapy for septic arthritis of the sacroiliac joint consists of a high dose of parenteral antibiotics which are chosen on the basis of susceptibility testing. The duration of the therapy is variable but some studies have reported control of the infection in as little as seventeen days of antibiotic therapy (entry 3). Most studies have suggested a minimum of four and preferably six weeks of parenteral therapy (entries 6, 9) but oral antibiotics may be used if adequate bacteriocidal levels of antibiotic at 1:8 dilution against the patient's organism are achieved. Surgical drainage is required for periarticular abscesses and the debridement of necrotic bone and cartilage. Immobilization in a spica cast is unnecessary.

§ 4-19(H). Course and Outcome.

With early diagnosis and adequate antibiotic therapy, septic arthritis of the sacroiliac joint has a good outcome. With therapy, patients can expect resolution of their symptoms of pain, limitation of motion and fever and the erythrocyte sedimentation rate returns to normal. Many patients have no radiographic changes and return to normal function. Even patients who have fusion of the sacroiliac joint as a result of an infection should expect no functional disability.

BIBLIOGRAPHY

1. Avila, L. Primary pyogenic infections of the sacroiliac articulation. *J. Bone Joint Surg.,* 23:922-928, 1941.
2. Chandler, F.A. Pneumonoccic infection of the sacroiliac joint complicating pregnancy. *JAMA,* 101:114-116, 1933.
3. Delbarre, F., Rondier, J., Delrieu, F., Evrard, J., Cuyla, J., Menkes, C.J., and Amor, B. Pyogenic infection of the sacroiliac joint. *J. Bone Joint Surg.,* 57A:819-825, 1975.
4. Dunn, E.J., Bryan, D.M., Nugent, J.T., and Robinson, R.A. Pyogenic infections of the sacroiliac joint. *Clin. Orthop.,* 118:113-117, 1976.
5. Gifford, D.B., Patzakis, M., Ivler, D., and Swezey, R.L. Septic arthritis due to pseudomonas in heroin addicts. *J. Bone Joint Surg.,* 57A:631-635, 1975.
6. Gordon, G., and Kabins, S.A. Pyogenic sacroiliitis. *Am. J. Med.,* 69:50-56, 1980.
7. Iczkovitz, J.M., Leek, J.C., and Robbins, D.L. Pyogenic sacroiliitis. *J. Rheumatol.,* 8:157-160, 1981.
8. L'Episcopo, J.B. Suppurative arthritis of the sacroiliac joint. *Ann. Surg.,* 104:289-303, 1936.
9. Lewkonia, R.M., and Kinsella, T.D. Pyogenic sacroiliitis: Diagnosis and significance. *J. Rheumatol.,* 8:153-156, 1981.
10. Longoria, R.K., and Carpenter, J.L. Anaerobic pyogenic sacroiliitis. *South. Med. J.,* 76:649-651, 1983.
11. Murphy, M.E. Primary pyogenic infection of sacroiliac joint. *N.Y. State J. Med.,* 77:1309-1311, 1977.
12. Strange, F.G.S. The prognosis in sacro-iliac tuberculosis. *Brit. J. Surg.,* 50:561-571, 1963.
13. Veys, E.M., Govaerts, A., Coigne, E., Mielants, H., and Verbruggen, A. HLA and infective sacroiliitis. *Lancet* II:349, 1974.
14. Wallace, R., and Cohen, A.S. Tuberculous arthritis: A report of two cases with review of biopsy and synovial fluid findings. *Am. J. Med.,* 61:277-282, 1976.

§ 4-20. Herpes Zoster.

Herpes zoster (shingles) is a later complication of a varicella infection (chicken pox) during childhood. The disease is characterized by an erythematous, papular rash accompanied by pain in the distribution of a peripheral sensory nerve. The pain may antedate the skin lesions by four to seven days and confuse the diagnosis. The process may resolve without any residual symptoms but, in older patients, the infection may result in scarring and persistent pain which is resistant to treatment. In some patients, the pain is severe enough to cause considerable disability.

§ 4-20(A). Prevalence and Etiology.

Herpes zoster is a reactivation of a prior varicella infection. The occurrence of the disease depends on the number of individuals who have had varicella, their age, and status of their immune system. In people 16 to 60 years of age, the annual incidence ranges from 3.1 to 5.4 per 1,000. The incidence is two to three times greater in people over 60 years of age and up to ten times greater in immunocompromised individuals who have had cancer chemotherapy or organ transplants.

Patients with varicella (chicken pox—DNA virus) develop a viremia during their illness and the virus reaches the posterior spinal sensory ganglia in the spinal cord where it remains dormant for an unspecified time. During a period of low host resistance, the viruses grow in a ganglion or nerve, resulting in skin lesions and pain. These patients are infectious and can transmit a varicella infection to previously unexposed patients. Evidence suggests that zoster is a reactivation of a previous infection and not due to a new exposure since a varicella infection in a patient does not give zoster to someone who has previously had chicken pox. The disease is usually limited to a single spinal nerve and may be due to a resurgence of immune response.

The factors which initiate the resurgence of this infection are not known. Any lesion which irritates sensory ganglia may initiate the illness. Patients who have sustained trauma, infection or a hidden malignancy may be at risk of developing this illness. In the vast majority of circumstances, however, the initiating factor is not known (entries 3, 4).

§ 4-20(B). Clinical Findings.

Zoster presents with a 4 to 28-day prodrome of constitutional symptoms including fever, malaise, chills, and gastrointestinal symptoms, particularly in the elderly. The first local symptom is usually pain in the nerve segment, burning or shooting in character, and often associated with dysesthesias in the area of skin supplied by the affected nerve root. Within a week after the onset of pain, an eruption appears as a series of localized erythematous papules which develop into vesicles grouped together upon an erythematous base which follow a segmental distribution. Within days, the eruption fades and the vesicles dry with crusts, leaving small scars in the skin. The skin may become partially or completely analgesic. Pain usually resolves with the skin lesions; however, it may persist for years. This post-herpetic neuralgia is more likely to occur in elderly patients.

§ 4-20(C). Physical Findings.

The skin lesions may be found in any segmental sensory nerve distribution. They have a characteristic appearance which helps distinguish this lesion from other skin diseases. In the immunocompromised host, they may be present in more than one segment, suggesting a generalized zoster infection. The illness may also involve other portions of the spinal cord resulting in a segmental myelitis characterized by paralysis. Patients who develop zoster must be examined for the presence of a hidden malignancy.

§ 4-20(D). Laboratory Findings.

Cerebrospinal fluid examination may demonstrate an increase in white blood cells in up to one-third of patients with zoster. Viral cultures may be positive for virus if fluid from acute vesicular lesions is cultured. Zoster antibodies may show a four-fold rise during an infection.

§ 4-20(E). Radiographic Findings.

Radiographs of the lumbar spine may demonstrate abnormalities such as fracture-dislocation, metastatic disease or spinal tumor, which corresponds with the location of affected sensory root ganglion.

§ 4-20(F). Differential Diagnosis.

The diagnosis of herpes zoster offers little difficulty when the characteristic pain and vesicular eruption is present. The diagnosis is more difficult in the pre-eruptive stage. Zoster should be considered in the patient with the sudden onset of burning pain in a segmental distribution. Following these patients for a short period until the appearance of vesicles helps establish the correct diagnosis.

§ 4-20(G). Treatment.

The treatment of zoster is mainly supportive and directed at controlling pain; antibiotic therapy is indicated only for patients who develop bacterial superinfection. Patients with vesicles are infectious and should limit contact with individuals who have not had varicella but they are no longer infectious once the lesions crust over.

The use of corticosteroids in noncompromised hosts may reduce the likelihood of the development of post-herpetic neuralgia (entry 1); however, not all physicians believe that corticosteroids help prevent this disease (entry 2) and corticosteroids may occasionally encourage dissemination of the disease.

§ 4-20(H). Course and Outcome.

Zoster is usually a disease of short duration with little permanent disability; however, the patient who is older and has an underlying illness may be left with permanent residuals. These may include post-herpetic neuralgia, paralysis, and widespread dissemination. Those patients left with persistent and intractable post-herpetic pain may suffer significant disability. Available therapy to prevent the illness or to manage its after effects is inadequate.

BIBLIOGRAPHY

1. Eaglestein, W.H., Katz, R., and Brown, J.A. The effects of early corticosteroid therapy on skin eruption and pain of herpes zoster. *JAMA*, 211:1681-1683, 1970.
2. Editorial. Chemotherapy for varicella-zoster infections. *Brit. Med. J.*, 2:1466-1467, 1976.
3. Lipton, S. *The Control of Chronic Pain.* Yearbook Medical Publishers, Inc., Chicago, 1979, pp. 32-33.
4. Rash, M.R. Herpes zoster complicating a herniated-thoracic disc. *Orthop. Review*, 11:91-95, 1982.

§ 4-21. Paget's Disease of Bone.

Paget's disease of bone is a localized disorder of bone characterized by a remarkable degree of bone resorption and subsequent formation of disorganized and irregular new bone. In the past, Paget's disease has been classified with metabolic bone diseases; however, recent data suggest that Paget's disease is an infectious disease which may be caused by a virus. Most patients with Paget's disease are asymptomatic and have no disability; however, patients with extensive disease may develop a number of complications which include bone pain, skeletal deformity, pathologic fractures, hypercalcemia, increased cardiac output, deafness, and sarcomatous malignant transformation. The diagnosis of Paget's disease is based upon characteristic laboratory and radiographic abnormalities. Therapy with calcitonin, diphosphonates, or mithramycin suppresses the activity of the disease but does not offer a cure. The disease was first described by Sir James Paget in 1877 (entry 10).

§ 4-21(A). Prevalence and Etiology.

Paget's disease is a common disorder in certain areas of the world, affecting up to 3% of people over 40 years of age (entry 13). It is more common in Western Europe, Australia and New Zealand than in the United States. Most Americans with Paget's disease are of Western European or Mediterranean descent. Paget's disease is rare in blacks in Africa, but is reported in blacks in the United States (entry 16). Most of the patients are over 40 years of age. The incidence increases with increasing age and men are slightly more commonly affected than women with a rate of 1.3:1 (entry 1).

The etiology of Paget's disease is unknown. There is little evidence to support abnormalities in hormone secretion, vascular supply or connective tissue metabolism as a cause of this illness. The primary abnormality resides in osteoclasts, cells which resorb bone. Histologic examination of pagetic bone reveals increased numbers of osteoclasts with marked increase in size and number of nuclei in these cells. This change in osteoclasts results in increased bone resorption and inadequate new bone formation. New pagetic bone is less compact, more vascular and weaker than normal bone (entry 6).

Ultrastructural studies of pagetic bone has demonstrated the presence of nuclear and cystoplasmic inclusions that resemble portions of paramyxoviruses (entry 15). The characteristics of a slow virus infection, long latent period, single organ disease and lack of inflammatory response, match closely with those of Paget's disease. However, the infectious agent has not been isolated from cultured pagetic cells and the specific virus has not been identified. Therefore, a conclusion concerning the role of viruses as the etiologic agent of Paget's disease cannot be made at this time.

§ 4-21(B). Clinical Findings.

Most patients with Paget's disease are asymptomatic (entry 4). Frequently the possibility that a patient may have the disease is raised by the presence of an elevated serum alkaline phosphatase on screening chemistry tests or by the discovery of an area of bony change on radiographs. When patients become symptomatic, however, they frequently develop rheumatologic complaints. Back pain is common affecting 37% of 290 patients in one study (entry 1). It is of a deep boring quality and is not increased at night. The pain may radiate with a radicular pattern in the gluteal region, thighs, legs or feet. Cauda equina symptoms with saddle anesthesia, progressive weakness and bladder or bowel incontinence is present in a small percentage of patients (entry 1). The lumbosacral spine is the area of the axial skeleton most commonly symptomatic.

The pain of Paget's disease involving the axial skeleton may be of bone, joint or nerve origin. Vertebral bodies may fracture, facet joints may develop secondary osteoarthritis due to deformity and neural elements may become compressed by new growth (entry 7). Neurologic complications are relatively uncommon, but may cause significant disability including myelopathy, radiculopathy, and cranial nerve signs including optic atrophy, deafness, and cerebellar dysfunction (entry 3). Spinal cord compression occurs most commonly in the thoracic spine where vertebral width is narrowed (entry 12). Occasionally, cord compression develops suddenly as a result of a collapse of a vertebra (entry 14).

Other frequently affected sites are the skull, pelvis, femora and tibias. Patients may experience pain in weight bearing bones with walking. Paget's disease may also cause deformities in bones and result in increased size of the skull, hip joint disease (osteoarthritis) and bowing of the legs.

280

Bone softening may give rise to invagination of the base of the skull which may result in obstructive hydrocephalus (entry 16).

Other complications of Paget's disease include hyperuricemia with gout, hypercalcemia in the immobilized patient, and high output cardiac failure due to the increased blood flow to bones. Malignant degeneration develops in patients with polyostotic Paget's disease but is a rare occurrence. Less than 1% of Paget's disease patients develop osteosarcoma or fibrosarcoma (entry 19).

§ 4-21(C). Physical Findings.

Physical examination may be entirely normal in the patient with asymptomatic Paget's disease. Increasing disease activity, manifested by rapid bone growth, may correspond with physical findings consistent with rapid bone metabolism and the temperature over bones may be elevated from increased blood flow. Bone growth of the skull is manifested by increased skull circumference and dilated scalp veins. Angiod streaks are an occasional finding on funduscopic examination. Resting tachycardia is seen in patients with high cardiac output.

Musculoskeletal abnormalities may include a pagetic stature (dorsal kyphosis) and abnormal gait. In the weight bearing bones, new bone formation leads to lateral bowing in the femur and anterior bowing in the tibia. Decreased motion may be demonstrated in the lumbar spine and point tenderness may be elicited over vertebral bodies which have sustained fractures. Neurologic examination may be remarkable for cranial nerve, sensory, motor or cerebellar dysfunction if bone growth has resulted in nerve compression.

§ 4-21(D). Laboratory Findings.

The most characteristic laboratory abnormality is an elevation of the serum alkaline phosphatase since this mirrors the extent of new bone, osteoblastic, activity. A measure of bone resorption is 24-hour total urinary hydroxyproline. This amino acid is formed from collagen matrix during bone breakdown. Together the tests can provide an indication of bone disease, its progression and its response to therapy. From a practical standpoint, serum alkaline phosphatase is used routinely, since 24 urine collections are difficult to obtain on a regular basis.

Serum levels of calcium and phosphorus are normal since bone resorption and formation are closely linked. Elevations of calcium occur in pagetic patients who are immobilized, who have coexistent primary hyperparathyroidism, or who have metastatic disease to bone. Serum uric acid concentrations may be increased in men with extensive disease.

Pathologic abnormalities of early disease are characterized by an increased number of osteoclasts. Subsequently new bone is produced in a chaotic, mosaic pattern and woven bone as opposed to normal lamellar bone is produced. The amount of minerals in pagetic new bone appears normal.

§ 4-21(E). Radiographic Findings.

Radiographic abnormalities vary with the stage of disease. The osteolytic phase corresponds with localized lytic lesions of bone. There is a well demarcated area of lucency without associated bony reaction. These lesions are most commonly discovered in the skull and are referred to as osteoporosis circumscripta. Lytic lesions may be seen in long bones but are rarely seen in the spine.

The mixed phase consists of bone sclerosis combined with osseous demineralization. The bone increases in size with greatly thickened and widely spaced trabeculae, cortical thickening, and irregularly distributed zones of increased and reduced density. A "cotton wool" appearance is a term used to describe the patchy involvement.

The sclerotic phase of disease appears as areas of homogenous increase in bone density. This form of disease may be too difficult to clearly distinguish from the mixed form of disease.

In the spine, mixed or sclerotic radiographic changes are usually demonstrated (entry 17). Vertebral bodies may develop coarse, parallel, vertical striations which may simulate the appearance of a hemangioma. Increased cortical thickening at the inferior and superior vertebral borders may result in a "picture frame" appearance. Marked osteoblastic changes may result in a homogenously sclerotic "ivory" vertebra. The "ivory" vertebra is usually enlarged in Paget's disease and this helps differentiate it from metastatic disease. Compression fractures and large osteophytes may also be seen.

Bone scintiscans are sensitive in detecting increased bone activity even in locations where plain radiographs are normal. In a patient with localized disease and normal biochemical parameters, a bone scan may give the only objective indication of the presence and activity of disease (entry 18). In a patient with Paget's disease, a positive scan in a "normal" area of skeleton suggests involvement at that location (entry 9). Response to treatment is characterized by radiographic changes in an area of normal activity on scan.

§ 4-21(F). Differential Diagnosis.

In the great majority of patients, the history, physical examination, chemical, radiographic and bone scan abnormalities are adequate to confirm the diagnosis of Paget's disease. In an occasional patient with unusual clinical or radiological presentation, bone biopsy may necessary to confirm the diagnosis.

The differential diagnosis is broad since all lesions which may cause sclerosis of bone must be included. Diseases associated with sclerotic vertebral bodies, including metastatic tumor, lymphoma, myelofibrosis, fluorosis, mastocytosis, renal osteodystrophy, fibrous dysplasia, tuberous sclerosis, axial osteomalacia, and fibrogenesis imperfecta ossium may be confused with Paget's disease. Abnormalities on physical and radiological examination, such as hepatosplenomegaly (myelofibrosis) anemia (metastatic tumor), bowing deformities and "ground glass" appearance of bone (fibrous dysplasia) help differentiate these disorders from Paget's disease.

§ 4-21(G). Treatment.

Most patients with Paget's disease are asymptomatic and do not require therapy. Indications for treatment include disabling bone pain which is not relieved with nonsteroidal anti-inflammatory agents, progressive skeletal deformity with frequent fractures, vertebral compression, acetabular protrusion, neurologic complications, deafness, high output congestive heart failure, or immobilization.

None of the three classes of agents used for Paget's disease to produce symptomatic improvement and better control of bone metabolism cure the illness. Calcitonin, a polypeptide hormone from the parafollicular cells of the thyroid gland, slows osteoclastic bone resorption. Injection of 50 to 100 units of calcitonin three or more times per week results in a gradual decrease in serum alkaline phosphatase to about half of the initial elevated concentrations and resolution of bone pain (entry 5); however, patients may develop antibodies to calcitonin which abrogates its beneficial effects.

Diphosphonates are structural analogues of pyrophosphate which, when ingested by osteoclasts, decrease their ability to resorb bone. In addition, to a variable extent, they interfere with the mineralization of normal bone and this limits the dose and duration of a course of diphosphonate therapy. Diphosphonates are given at a dose of 5 mg/kg/day for six months of the year. Some physicians give a course for a six-month period and discontinue therapy for six months, while others give therapy every other month for six months and then re-evaluate the response. Diphosphonates may increase the tendency of Paget's patients to develop pathological fractures (entry 2).

Mithramycin is an antibiotic that binds to DNA and inhibits RNA synthesis. It has a cytotoxic effect on osteoclasts and is used as a treatment for hypercalcemia related to cancer. The drug is potent and rapidly effective but, since it is associated with toxicity, it cannot be used as a first line agent (entry 11). It is administered intravenously to patients with progressive neurologic compression syndromes secondary to Paget's disease.

Physicians usually use one agent at a time, following the patient's symptoms, serum alkaline phosphatase and activity on bone scan. Many have a remission of their disease after a course of therapy, and may not need to resume the drug for an extended period. Patients with persistent neurologic syndromes may require a combination of agents for adequate control (entry 8).

§ 4-21(H). Course and Outcome.

Most patients with Paget's disease have an asymptomatic illness. Others with more active disease are controlled with nonsteroidal anti-inflammatory drugs and agents which modify bone metabolism. The complications of Paget's disease which cause disability and mortality are rare. Occasionally they develop neurologic symptoms from spinal cord compression. Increased drug therapy or laminectomy can be effective in controlling these complications. Malignant transformation of Paget's disease is associated with a very poor prognosis but is fortunately a very rare occurrence.

BIBLIOGRAPHY

1. Altman, R.D. Musculoskeletal manifestations of Paget's disease of bone. *Arthritis Rheum.*, 23:1121-1127, 1980.

2. Canfield, R., Rosner, W., Skinner, J., McWhorter, J., Resnick, K., Feldman, F., Kammerman, S., Ryan, K., Kunigonis, M., and Bohne, W. Diphosphonate therapy of Paget's disease of bone. *J. Clin. Endocrinol Metab.*, 44:96-106, 1977.

3. Chen, J., Rhee, R.S.C., Wallach, S., Avramides, A., and Flores, A. Neurologic disturbances in Paget disease of bone: Response to calcitonin. *Neurology*, 29:448-457, 1979.

4. Collins, D.H. Paget's disease of bone. Incidence and subclinical forms. *Lancet*, 2:51-57, 1956.

5. DeRose, J., Singer, F.R., Avramides, A., Flores, A., Dziadiw, R., Baker, R.K., and Wallach, S. Response of Paget's disease to porcine and salmon calcitonins. Effects of long-term treatment. *Am. J. Med.*, 56:858-866, 1974.

6. Deuxchaines, C.N. de , and Krane, S.M. Paget's disease of bone: Clinical and metabolic observations. *Medicine*, 43:233-266, 1964.

7. Franck, W.A., Bress, N.M., Singer, F.R., and Krane, S.M. Rheumatic manifestations of Paget's disease of bone. *Am. J. Med.*, 56:592-603, 1974.

8. Hosking, D.J., Bijvoet, O.L.M., van Aken, J., and Will, E.J. Paget's bone disease treated with diphosphonate and calcitonin. *Lancet*, 1:615-617, 1976.

9. Khairi, M.R.A., Wellman, H.N., Robb, J.A., and Johnston, C.C., Jr. Paget's disease of bone (osteitis deformans). Symptomatic lesions and bone scan. *Ann. Intern. Med.*, 79:348-351, 1973.

10. Paget, J. On a form of chronic inflammation of bones (osteitis deformans). *Trans. Roy. Med. Chir. Soc.*, London, 60:37-63, 1877.

11. Ryan, W., Schwartz, T.B., and Perlia, C.P. Effects of mithramycin on Paget's disease of bone. *Ann. Intern. Med.*, 70:549-557, 1969.

12. Schmidek, H.H. Neurologic and neurosurgical sequelae of Paget's disease of bone. *Clin. Orthop.*, 127:70-77, 1977.

13. Schmorl, G. Ueber osteitis deformans Paget. *Virchows Arch. Pathol. Anat. Physiol.*, 283:694-699, 1932.

14. Schreiber, M.H., and Richardson, G.A. Paget's disease confined to one lumbar vertebra. *Am. J. Roentgenol.*, 90:1271-1276, 1963.

15. Singer, F.R., and Mills, B.G. Evidence for a viral etiology of Paget's disease of bone. *Clin. Orthop.*, 178:245-251, 1983.

16. Siris, E.S., Jacobs, T.P., and Canfield, R.E. Paget's disease of bone. *Bull. N.Y. Acad. Med.,* 56:285-304, 1980.
17. Steinbech, H.L. Some roentgen features of Paget's disease. *Am. J. Roentgenol.,* 86:950-964, 1961.
18. Waxman, A.D., Ducker, S., McKee, D., Siemsen, J.K., and Singer, F.R. Evaluation of 99mTc diphosphonate kinetics and bone scans in patients with Paget's disease before and after calcitonin treatment. *Radiology,* 125:761-764, 1977.
19. Wick, M.R., Siegal, G.P., Unni, K.K., McLeod, R.A., and Greditzer, H.G., III. Sarcomas of bone complicating osteitis deformans (Paget's disease). Fifty years' experience. *Am. J. Surg. Path.,* 5:47-59, 1981.

PART IV. TUMORS AND INFILTRATIVE LESIONS OF THE LUMBOSACRAL SPINE

§ 4-22. Introduction.

Tumors and infiltrative lesions of the lumbar spine are unusual causes of back pain; however, these diseases are associated with the highest morbidity, mortality, and dysfunction of all. Physicians who evaluate patients with low back pain must be aware of the existence of neoplastic disease of the spine and must include it in their differential diagnosis. Patients with tumors of the lumbosacral spine usually have back pain as their initial complaint and not infrequently, a traumatic event is thought to be the inciting cause. Only as the pain persists and increases in intensity does it become clear that the trauma was an incidental event unassociated with the underlying disease process.

A history of pain which increases with recumbency is a hallmark for tumors of the spine. Physical examination demonstrates localized tenderness as well as neurologic dysfunction if the spinal cord is compressed. In many circumstances, laboratory abnormalities are non-specific. In contrast, radiographic evaluation is very useful in identifying characteristic changes in the bony and soft tissue areas of the spine which help identify the location and type

287

of neoplastic lesion. The definitive diagnosis of a tumor must be derived from histologic examination of biopsy material obtained from the lesion (tissue diagnosis). The most effective therapy for both benign and malignant tumors is the removal of the lesions accessible to surgical excision. When excision is not possible, partial resection, radiation therapy, corticosteroids, or chemotherapy may be indicated to control symptoms and compression of the spinal cord and nerve roots. In general, patients with malignant tumors have a poorer prognosis than those with benign tumors.

§ 4-23. Osteoid Osteoma.

Osteoid osteoma is a benign neoplasm of bone characterized by pain which is worse at night, which is relieved by aspirin, and has radiographic features of a radiolucent central lesion, a nidus, surrounded by a variable area of bony sclerosis. Bone pain may be severe even before radiographic evidence of an osteoid osteoma is present. The diagnosis of osteoid osteoma must be considered in a young individual with persistent back pain dramatically relieved by aspirin. A thorough examination, utilizing a bone scan and computerized tomography may be necessary to discover the location, particularly if in the spine. In the past, this lesion has been referred to as sclerosing osteomyelitis, an osteoblastic disease, highlighting the confusion surrounding its pathogenesis. Jaffe in 1935, first described the osteoid osteoma and placed it in the category of benign osteoblastic tumors (entry 5).

§ 4-23(A). Prevalence and Etiology.

Osteoid osteomas comprise about 2.6% of excised primary bone tumors (entry 6). They are most frequently discovered in young adults between 20 and 30 years of age. The ratio

288

of men to women is 2:1. The pathogenesis of osteoid osteoma remains uncertain. There is no conclusive evidence that the origin of this tumor is related to trauma.

§ 4-23(B). Clinical Findings.

Pain is a characteristic feature of osteoid osteoma. The pain is intermittent and vague initially, but with time, it becomes constant with a boring quality. The pain frequently exacerbates at night, disturbing sleep. Spinal osteoid osteomas are associated with muscle spasm and scoliosis (spinal curvature). Approximately 7% of osteoid osteomas occur in the spine (entry 1). The appearance of marked paravertebral muscle spasm and the sudden onset of scoliosis in a young adult requires an evaluation for the presence of this lesion. The pain of osteoid osteoma is frequently relieved with aspirin and in some patients, symptoms may be present two years before the correct diagnosis is made (entry 6). It may be present for a considerable period of time before the radiographic findings become evident (entry 8). These patients may be inaccurately labeled as malingerers or neurotics. It should also be noted that some osteoid osteomas are painless (entry 7).

§ 4-23(C). Physical Findings.

In a patient with an osteoid osteoma of the spine, physical examination may demonstrate marked muscular spasm with resultant scoliosis. The spine may be curved without rotation (entry 6). The osteoid osteomas are usually located on the concave side of the scoliosis. Muscle atrophy may also be present if spasm has been of prolonged duration. When superficial, osteoid osteoma may be associated with swelling and erythema of overlying structures such as the skin.

289

§ 4-23(D). Laboratory Findings.

The laboratory findings in this benign neoplasm of bone are normal. Pathologically, osteoid osteomas are characterized by a nidus, usually less than 1 cm. in diameter and composed of osteoid and loose fibrovascular tissue, surrounded by thickened cortical bone (entry 4).

§ 4-23(E). Radiographic Findings.

Osteoid osteomas have characteristic radiographic findings; they produce an oval area of lucency, the nidus, surrounded by an area of sclerosis. Osteoid osteomas arising in the spine are commonly located in the posterior elements, transverse and spinal processes, pedicles, and laminae. These lesions may not be seen on plain radiographs and may require computed tomography or a bone scan for detection (entries 10, 11). When located in the vertebral bodies, they may affect surrounding bone and adjacent intervertebral discs (entry 3).

§ 4-23(F). Differential Diagnosis.

The diagnosis of an osteoid osteoma is suggested by characteristic clinical and radiographic features, and is confirmed by the histologic examination of biopsied material. Other abnormalities which may mimic the findings of an osteoid osteoma include osteoblastoma, osteosarcoma, osteomyelitis, Brodie's abscess, Ewing's sarcoma, eosinophilic granuloma, and metastasis (entry 6). Abnormalities in laboratory tests (elevated white blood cell count, erythrocyte sedimentation rate, bone chemistries) and characteristic histologic findings help differentiate these inflammatory, infectious, or malignant lesions from osteoid osteoma.

§ 4-23(G). Treatment.

Treatment of this benign lesion is simple excision of the

nidus and surrounding sclerotic bone. If the nidus is not entirely removed, recurrence of the lesion is possible (entry 9). This may be a particular problem in poorly accessible areas of the spine. It should also be noted that occasionally osteoid osteomas undergo spontaneous healing (entry 2).

§ 4-23(H). Course and Outcome.

The course of osteoid osteoma is benign once the diagnosis is made and the lesion is excised. It may be a difficult diagnostic problem, however, because clinical symptoms may appear before it is radiographically evident. Low back pain and associated limitation of activities at work may be inappropriately ascribed to malingering or psychoneurosis. In the young adult with low back pain which is exacerbated at night and is relieved with aspirin, a thorough evaluation for the presence of this lesion is mandatory.

BIBLIOGRAPHY

1. Freiberger, R.H. Osteoid osteoma of the spine: A cause of backache and scoliosis in children and young adults. *Radiology*, 75:232-235, 1960.
2. Golding, J.S.R. The natural history of osteoid osteoma with a report of 20 cases. *J. Bone Joint Surg.*, 36B:218-229, 1954.
3. Heiman, M.L., Cooley, C.J., and Bradford, D.S. Osteoid osteoma of a vertebral body: Report of a case with extension across the intervertebral disc. *Clin. Orthop.*, 118:159-163, 1976.
4. Huvos, A.G. *Bone Tumors: Diagnosis, Treatment and Prognosis.* W.B. Saunders, Phila: 1979, pp. 18-32.
5. Jaffe, H.L. Osteoid osteoma: A benign osteoblastic tumor composed of osteoid and atypical bone. *Arch. Surg.*, 31:709-728, 1935.
6. Mirra, J.M. *Bone Tumors: Diagnosis and Treatment.* J.B. Lippincott Co., Phila: 1980, pp. 97-108.
7. Sevitt, S., and Horn, J.S. A painless and calcified osteoid osteoma of the little finger. *J. Pathol.*, 67:571-574, 1954.
8. Silberman, W.W. Osteoid osteoma. *J. Inter. Coll. Surgeons*, 38:53-66, 1962.

9. Sim, F.H., Dahlin, D.C., and Beabout, J.W. Osteoid osteoma: Diagnostic problems. *J. Bone Joint Surg.*, 57A:154-159, 1975.
10. Wedge, J.H., Tchang, S., and MacFadyen, D.J. Computed tomography in localization of spinal osteoid osteoma. *Spine*, 6:423-427, 1981.
11. Winter, P.F., Johnson, P.M., Hilal, S.K., and Feldman, F. Scintigraphic detection of osteoid osteoma. *Radiology*, 122:177-178, 1977.

§ 4-24. Osteoblastoma.

Osteoblastoma is a rare benign neoplasm of bone which is frequently confused with an osteoid osteoma. The pain of an osteoblastoma is not increased at night, is not relieved by aspirin, and the lesion is larger, with no surrounding bony sclerosis. Back pain occurs commonly with osteoblastoma and may be associated with muscle spasm, scoliosis, and spinal cord compression. In the past, osteoblastoma has been referred to as an osteogenic fibroma, giant osteoid osteoma, spindle-cell variant of giant cell tumor, and osteoblastic osteoid tissue-forming tumor (entry 4).

§ 4-24(A). Prevalence and Etiology.

Osteoblastoma comprises 0.5% of biopsied bone tumors (entry 7). A majority of the lesions appear between the second and third decades of life. Nearly 90% of patients diagnosed with an osteoblastoma are 30 years of age or younger (entry 6). The male to female ratio is 2:1. The etiology of the osteoblastoma is unknown.

§ 4-24(B). Clinical Findings.

Approximately 30% to 40% of osteoblastomas occur in the spine (entry 5). The major clinical symptom is dull, aching, localized pain over the involved bone. The pain is insidious in onset and may have a duration of months to years before diagnosis. As opposed to osteoid osteoma, the pain of an osteoblastoma is less severe, not nocturnal, and is not relieved by salicylates. Osteoblastomas located in the lumbar spine may be associated with pain radiating into the legs associated with muscle spasm and limitations of motion. Scoliosis may also occur.

§ 4-24(C). Physical Findings.

Physical examination may demonstrate local tenderness on palpation with mild swelling over the spine. Osteoblastomas associated with spinal cord compression will result in abnormalities on sensory and motor examination of the lower extremities. Reflexes may also be abnormal.

§ 4-24(D). Laboratory Findings.

Laboratory findings are normal in osteoblastoma. On pathologic examination, osteoblastic lesions are larger than osteoid osteoma, ranging from 2 to 10 cm. in length and are well circumscribed. Histologic features include cellular osteoblastic tissue, with large amounts of osteoid material and the absence of cartilage cells. Multinucleated giant cells may also be present (entry 2).

§ 4-24(E). Radiographic Findings.

Radiographic findings of osteoblastoma are variable and non-specific. In the spine, lesions are most commonly located in the posterior elements of the vertebrae including pedicles, laminae, transverse and spinous processes (entry 1). Vertebral bodies are rarely the primary location. Osteoblastomas are expansile and may grow rapidly as measured by serial radiographic studies. Characteristically, the lesion is well delineated and is covered by a thin layer of periosteal new bone. The extent of reactive new bone formation is much less than that associated with osteoid osteoma (entry 8). The center of the lesion may be radiolucent or radio-opaque. A bone scan is helpful in localizing the exact site of an osteoblastoma that is difficult to visualize on plain radiographs.

§ 4-24(F). Differential Diagnosis.

The diagnosis of an osteoblastoma is made by the thorough examination of biopsied samples. Osteogenic sarcoma, at presentation, may be easily confused clinically, radiographically and histologically with osteoblastoma. The presence of an outer rim of bone on radiographic examination and the absence of cartilage and anaplastic cells on biopsy help differentiate it from a malignant process. Other entities which must be differentiated from osteoblastoma include osteoid osteoma, giant cell tumor of bone, and brown tumor of hyperparathyroidism.

§ 4-24(G). Treatment.

Local excision of the entire lesion is the treatment of choice if bone can be sacrificed without loss of function or excessive risk for neurologic dysfunction. Osteoblastoma in the posterior elements of the spine is usually inaccessible for complete excision. Partial curettage of these lesions may be associated with cessation of growth and relief of symptoms for an extended period of time (entry 5). Osteoblastomas which are rapidly expanding or recurrent lesions may be controlled with radiation therapy.

§ 4-24(H). Course and Outcome.

The course of osteoblastoma is usually benign. The lesion is responsive to partial curettage and low dose radiation therapy. Marsh reported a series where one of thirteen spinal osteoblastomas had a recurrence resulting in paraplegia seven months after surgery (entry 5). Laminectomy and post-operative radiation was the treatment for this complication. Recurrences occur in less than 5% of osteoblastomas; however, repeated recurrences have been described and rarely, malignant transformation has been reported (entries 3, 9).

BIBLIOGRAPHY

1. DeSouza-Dias, L., and Frost, H.M. Osteoblastoma of the spine: A review and report of eight new cases. *Clin. Orthop.*, 91:141-151, 1973.
2. Huvos, A.G. *Bone Tumors: Diagnosis, Treatment, and Prognosis.* W.B. Saunders, Phila: 1979, pp. 33-46.
3. Jackson, R.P. Recurrent osteoblastoma: A review. *Clinc. Orthop.*, 131:229-233, 1978.
4. Lichtenstein, L. Benign osteoblastoma — A category of osteoid and bone forming tumors other than classical osteoma, which may be mistaken for giant cell tumor or osteogenic sarcoma. *Cancer*, 9:1044-1052, 1956.
5. Marsh, B.W., Bonfiglio, M., Brady, L.P., and Enneking, W.F. Benign osteoblastoma: Range of manifestations. *J. Bone Joint Surg.*, 57A:1-9, 1975.
6. McLeod, R.A., Dahlin, D.C., and Beabout, J.W. The spectrum of osteoblastoma. *Am. J. Roentgenol.*, 126:321-335, 1976.
7. Mirra, J.M. *Bone Tumors: Diagnosis and Treatment.* J.B. Lippincott Co., Phila: 1980, pp. 108-122.
8. Pochaczevsky, R., Yen, Y.M., and Sherman, R.S. The roentgen appearance of benign osteoblastoma. *Radiology*, 75:429-437, 1960.
9. Schajowicz, F., and Lemos, C. Malignant osteoblastoma. *J. Bone Joint Surg.*, 58B:202-211, 1976.

§ 4-25. Osteochondroma.

Osteochondroma is a common benign tumor of bone which occurs in single or multiple locations in the skeleton. When multiple osteochondromas exist, the spine is more frequently affected. Although usually associated with mild clinical symptoms, the lesions may grow to the extent of causing spinal cord compression and associated neurologic dysfunction. Cooper was the first to describe this lesion in 1818 (entry 4).

§ 4-25(A). Prevalence and Etiology.

Solitary and multiple osteochondromas represent 11% of biopsied primary bone tumors (entry 8). Approximately 60% of patients develop the lesion between the second and third decades of life. The male to female ratio is 2:1 (entry 6).

The etiology of osteochondroma is postulated to be related to an abnormality of cartilage growth. Nests of cartilage in a periosteal location grow out from the epiphyseal growth plate and results in a bony prominence capped by a layer of cartilage which is contiguous with the cortex of the underlying bone. They may be thought of as slow-growing developmental anomalies which cease to enlarge once growth has stopped. This first awareness is usually during childhood or young adulthood (entry 8).

§ 4-25(B). Clinical Findings.

Osteochondromas occur most commonly at the ends of long tubular bones and develop in the spine in 1% of patients with this lesion (entry 3). The lesion is frequently symptomless and is only discovered as a painless prominence of bone or as a chance finding on a radiograph. If pain is present, it is mild and usually secondary to mechanical irritation of overlying soft tissue structures. An osteochondroma which continues to grow may cause loss of function and decreased motion. When attached to the spinal column, they have been associated with kyphosis and spondylolisthesis (entry 1). They may even grown large enough to cause nerve root or spinal cord compression, and may be associated with radicular pain, sensory abnormalities, motor weakness and urinary and fecal incontinence (entries 2, 9, 10).

§ 4-25(C). Physical Findings.

Physical examination may be normal without any

neurological deficit; however, osteochondromas near facet joints of the spine may cause some restriction in motion. Neurologic findings, when present, will correspond to the location of the lesion and related nerve root compression.

§ 4-25(D). Laboratory Findings.

This benign lesion is not associated with any abnormal laboratory findings. Pathologically, the lesion may vary in size up to 100 cm. and it is well-circumscribed. It may be sessile or stalked and is capped by a layer of cartilage. Histologic features include microscopic foci of cartilage intermixed with woven bone (entry 6).

§ 4-25(E). Radiographic Findings.

The radiographic features of an osteochondroma are diagnostic. The lesion protrudes from the underlying bone on a sessile or pedunculated bony stalk which is continuous with the cortex and spongiosa of the underlying bone. The outer surface of the lesion may be smooth or irregular but it is almost always well demarcated. If the cartilage cap is calcified, it may obscure the underlying stalk.

In the spine, osteochondromas are located close to centers of secondary ossification, including the spinous process, pedicle and neural arch (entry 7). Computerized tomography and myelography are helpful in localizing the site of the lesion in the spinal column, its size, and its relationship to the nerve roots and spinal cord.

§ 4-25(F). Differential Diagnosis.

The diagnosis of an osteochondroma is based upon its appearance on radiographs and the lack of clinical and laboratory findings. Chondrosarcomatous degeneration of osteochondroma occurs in less than 1% of patients, usually in adult life and patients with multiple lesions are at

297

greater risk in this respect. Malignant transformation is usually heralded by increasing pain, enlarging soft tissue mass, and loss of definition of the outer border of the lesion on radiographs. The onset of pain in a previously asymptomatic osteochondroma, however, is not always associated with malignant degeneration since infarction of the cartilage cap or fracture through the base of an osteochondroma may also cause pain and new bone growth (entry 8).

§ 4-25(G). Treatment.

Osteochondromas are usually asymptomatic and require no treatment but when they become symptomatic, surgical excision should be considered. Patients with neurologic symptoms can be helped by removal of the lesion and decompression of the affected nerve root or spinal cord (entry 5).

§ 4-25(H). Course and Outcome.

The course of the solitary osteochondroma is usually benign and asymptomatic, and is not associated with any dysfunction or inability to work. In the rare patient who has a vertebral osteochondroma and neurologic dysfunction, surgical decompression of the site should result in a return of function. Continuous observation of patients with osteochondromas is important, particularly for those with multiple lesions, because they can on occasion become malignant.

BIBLIOGRAPHY

1. Blaauw, G. Osteocartilagenous exostosis of the spine, in Vinken, P.J., Bruyn, G.W. (eds): *Handbook of Clinical Neurology Tumors of the Spine and Spinal Cord, Part I.* New York: American Elsevier Publishing Co., Inc., 1975, pp. 313-319.

2. Borne, G., and Payrot, C. Right lumbo-crural sciatica due to a vertebral osteochondroma. *Neurochirurgie*, 22:301-306, 1976.

3. Chrisman, O.D., and Goldenberg, R.R. Untreated solitary osteochondroma. Report of two cases. *J. Bone Joint Surg.*, 50A:508-512, 1968.

4. Cooper, A. Exostosis, in Cooper, A., Travers, B. (eds): *Surgical Essays*, 3rd ed. London: Cox and Son, 1818, pp. 169-226.

5. Gokay, H., and Bucy, P.C. Osteochondroma of the lumbar spine: Report of a case. *J. Neurosurg.*, 12:72-78, 1955.

6. Huvos, A.G. *Bone Tumors: Diagnosis, Treatment and Prognosis.* W.B. Saunders, Phila: 1979, pp. 139-149.

7. Inglis, A.E., Rubin, R.M., Lewis, R.J., and Villacin, A. Osteochondroma of the cervical spine: Case report. *Clin. Orthop.*, 126:127-129, 1977.

8. Mirra, J.M. *Bone Tumors: Diagnosis and Treatment.* J.B. Lippincott Co., Phila: 1980, pp. 520-532.

9. Palmer, F.J., and Blum, P.W. Osteochondroma with spinal cord compression. *J. Neurosurg.*, 52:842-845, 1980.

10. Twersky, J., Kassner, E.G., Tenner, M.S., and Camera, A. Vertebral and costal osteochondromas causing spinal cord compression. *Am. J. Roentgenol.*, 124:124-128, 1975.

§ 4-26. Giant Cell Tumor.

Giant cell tumor of bone is a common, locally aggressive lesion which may turn malignant. Patients who develop low back pain from a giant cell tumor have lesions in the sacrum, pelvis or lumbar spine and those located in the sacrum and lumbar spine often present with neurologic abnormalities from nerve root or spinal cord compression (entry 12). In the past, the giant cell tumor has been referred to as myeloid sarcoma, medullary sarcoma, and osteoclastoma. Cooper in 1818, was the first to describe this lesion and its usual benign characteristics (entry 3) but, Bloodgood was the first to refer to it as benign giant cell tumor (entry 1).

§ 4-26(A). Prevalence and Etiology.

Giant cell tumors comprise 4% to 5% of biopsied primary

299

bone tumors (entry 13). Approximately 70% of patients are diagnosed between the ages of 20 to 40. The average age of patients with malignant giant cell tumors is greater than those with benign tumors. Benign tumors predominate in women in a ratio of 3:2, while malignant tumors predominate in men in a ratio of 3:1 (entry 7). The etiology of giant cell tumor is not known. The tumor arises from non-bone-forming fibrous connective tissue in bones and, factors which make it inherently invasive and potentially malignant are unknown.

§ 4-26(B). Clinical Findings.

The vertebral column, including the sacrum, is involved in 8% of patients with giant cell tumor (entry 8). A smaller number of patients may also have involvement of the ilium and ischium (entry 11). Patients with this lesion usually present with intermittent, aching pain over the affected bone. The duration of symptoms may vary from a few weeks to six months but some patients have had pain for over two years prior to diagnosis (entry 4). Patients with sacral or vertebral involvement may describe neurologic dysfunction, including paresthesias with radiation of pain into the lower extremities, muscle weakness, and urinary or rectal incontinence (entries 4, 15).

§ 4-26(C). Physical Findings.

Physical examination may demonstrate tenderness on palpation over the spine and sacrum. Localized swelling may be noted if the location of the giant cell tumor is superficial, that is, in the spinous process. Kyphosis, muscle spasm and associated limitation of motion may also be noted. In a sacral lesion, an extracolonic mass may be found on rectal examination (entry 15). Neurologic findings may show sensory, motor or reflex abnormalities dependent on the level of nerve root compression.

§ 4-26(D). Laboratory Findings.

Laboratory results are normal in patients with benign giant cell tumors but serum calcium, phosphorus, and alkaline phosphatase tests should be obtained to differentiate it from hyperparathyroidism, Paget's disease, and malignant giant cell tumor. Patients with malignant giant cell tumors may show abnormalities such as anemia and elevated sedimentation rates.

Although the pathologic findings of giant cell tumor are characteristic, they may mimic other benign, and malignant lesions. It is composed of large numbers of osteoclast-like giant cells separated by inconspicuous mononuclear stromal cells. In a minority of lesions, small foci of osteoid and woven bone are seen. Thin walled vessels with abundant hemorrhages are also characteristic.

§ 4-26(E). Radiographic Findings.

The radiographic findings of giant cell tumor are characteristic but not pathognomonic. The lesion is expansile, with irregular thinning of the cortical margin. It is lytic but may contain a delicate trabecular meshwork. Little bony reaction occurs in response to this lesion. Extensive sclerotic borders or periosteal reaction are not seen. In the spine, the vertebral body is frequently affected but the spinous and transverse processes may also be involved. Radiographic changes of giant cell tumor in the sacrum may be subtle and large tumors may be missed on plain radiographs. The lesion may be eccentrically located in the sacrum and may spread across the sacroiliac joint to involve the ilium. When located in the superior portion of the sacrum, they may erode through the L5-S1 disc space (entry 15). If plain radiographs are unable to demonstrate abnormalities of the sacrum or spine, a bone scan may be helpful in demonstrating abnormalities. Computerized tomography

301

is useful in localizing the extent of a lesion in the sacrum. Myelography, however, is needed in patients with neurologic dysfunction if the level of spinal cord or nerve root compression is to be identified.

§ 4-26(F). Differential Diagnosis.

The diagnosis of giant cell tumor of bone can be made only after a thorough review of all clinical, laboratory, radiologic and pathologic data. This is necessary because of the similarity of this lesion to a number of benign and malignant conditions. The list of diseases which may be confused with giant cell tumor includes giant cell tumor of hyperparathyroidism, chondroblastoma, aneurysmal bone cyst, osteoblastoma, non-ossifying fibroma, fibrosarcoma, and osteosarcoma.

§ 4-26(G). Treatment.

The treatment of choice for a giant cell tumor is en-bloc excision if it is in an accessible location. Recurrence rates of 10% to 15% are reported in lesions that have been excised (entry 14). Curettage may control growth of the tumor, but the local recurrence rate is 50% within five years (entry 9). Radiation therapy is rarely curative, is associated with frequent recurrences and may promote malignant transformation (entry 7). It is reserved for lesions which are inaccessible for surgical removal or curettage (entry 5).

The ideal treatment for giant cell tumor of the spine is complete removal of accessible lesions (entries 2, 10, 16). Decompression of the spine is necessary once neurologic symptoms appear. A delay of greater than three months after the onset of nerve root symptoms may result in irreversible nerve deficits (entry 12). However, complete resection is frequently impossible because of the location and size of the lesions and the potential for critical blood loss with resection. In large sacral tumors, partial excision with irradiation or irradiation alone may be the only reasonable option (entry 6).

§ 4-26(H). Course and Outcome.

Giant cell tumors of bone are invasive, benign tumors which have a high local recurrence rate. Patients with benign giant cell tumors of the sacrum have died because of local invasion, malignant transformation, or secondary complications such as renal failure secondary to neurogenic bladder and obstruction (entry 15). Regardless of treatment, the patients must continue to be examined for signs of recurrence. In some circumstances up to five courses of therapy were needed to successfully eradicate the disease (entry 7).

BIBLIOGRAPHY

1. Bloodgood, J.G. Bone tumors. Central (medullary) giant cell tumor (sarcoma) of lower end of ulna, with evidence that complete destruction of the bony shell or perforation of the bony shell is not a sign of increased malignancy. *Ann. Surg.,* 69:345-359, 1919.
2. Chow, S.P., Leong, J.C.Y., and Yau, A.C.M.C. Osteoclastoma of the axis: Report of a case. *J. Bone Joint Surg.,* 59A:550-551, 1977.
3. Cooper, A., and Travers, B. *Surgical Essays,* 3rd ed., London: Cox and Son, 1818.
4. Dahlin, D.C. Giant cell tumor of vertebrae above the sacrum: A review of 31 cases. *Cancer,* 39:1350-1356, 1977.
5. Dahlin, D.C., Cupps, R.E., and Johnson, E.W. Giant-cell tumor: A study of 195 cases. *Cancer,* 25:1061-1070, 1970.
6. Harwood, A.R., Fornasier, V.L., and Rider, W.D. Supervoltage irradiation in the management of giant cell tumor of bone. *Radiology,* 125:223-226, 1977.
7. Hutter, R.V.P., Worcester, J.W., Jr., Francis, K.C., Foote, F.W., Jr., and Stewart, F.W. Benign and malignant giant cell tumors of bone: A clinicopathological analysis of the natural history of the disease. *Cancer,* 15:653-690, 1962.
8. Huvos, A.G. *Bone Tumors: Diagnosis, Treatment and Prognosis.* W.B. Saunders, Phila: 1979, pp. 265-291.

9. Johnson, E.W., Jr., and Dahlin, D.C. Treatment of giant cell tumor of bone. *J. Bone Joint Surg.*, 41A:895-904, 1959.

10. Johnson, E.W., Jr., Gee, V.R., and Dahlin, D.C. Giant cell tumors of the sacrum. *Am. J. Orthop.*, 4:302-305, 1962.

11. Kuritzky, A.S., and Joyce, S.T. Giant cell tumor in the ischium: A therapeutic dilemma. *JAMA*, 238:2392-2394, 1977.

12. Larrson, S.E., Lorentzon, R., and Boquist, L. Giant cell tumors of the spine and sacrum causing neurological symptoms. *Clin. Orthop.*, 111:201-211, 1975.

13. Mirra, J.M. *Bone Tumors: Diagnosis and Treatment.* J.B. Lippincott, Co., Phila: 1980, pp. 332-362.

14. Parrish, F. Treatment of bone tumors by total excision and replacement with massive autologous and homologous grafts. *J. Bone Joint Surg.*, 48A: 968-990, 1966.

15. Smith, J., Wixon, D., and Watson, R.C. Giant cell tumor of the sacrum: clinical and radiologic features in 13 patients. *J. Can. Assoc. Radiol.*, 30:34-39, 1979.

16. Stevens, W.W., and Weaver, E.W. Giant cell tumors and aneurysmal bone cysts of the spine: Report of 4 cases. *South. Med. J.*, 63:218-221, 1970.

§ 4-27. Aneurysmal Bone Cyst.

Aneurysmal bone cysts are benign, non-neoplastic, cystic vascular lesions of bone which are expansile and usually painful. In the spine, they may expand to the extent of producing nerve root or spinal cord compression. Aneurysmal bone cysts have been previously described as an ossifying hematoma, bone cyst and atypical giant cell tumor. Jaffe and Lichtenstein were the first to describe the distinctive characteristics of the lesion (entries 9, 10).

§ 4-27(A). Prevalence and Etiology.

Aneurysmal bone cyst is a rare lesion, comprising about 1% of primary bone tumors (entry 12). The vast majority of young adults who develop them are under the age of 30 (entry 6) and in contrast to other primary bone tumors, most series report a slight female predominance (entry 2).

The etiology of aneurysmal bone cyst remains uncertain although trauma may play a role in its initiation since injuries may induce the formation of arteriovenous malformations (AVM). These malformations consist of abnormal vascular channels and several reports have suggested that trauma which initiated an AVM, may have led to the development of an aneurysmal bone cyst (entries 1, 7).

In a third of the cases, aneurysmal bone cyst is superimposed upon another pathologic process which may be either a benign or malignant bone tumor. The basic abnormality in both circumstances is a local change in intraosseous blood flow. The blood pools in bone and results in increasing intraosseous pressure followed by resorption, expansion and cyst formation (entry 3).

§ 4-27(B). Clinical Findings.

Patients with aneurysmal bone cysts usually present symptoms with pain or swelling in the affected area. The pain is usually of acute onset which increases in severity over a short period of time. The duration of symptoms can range from months to several years. While a majority of aneurysmal bone cysts occur in the long bones of the extremities, 15% to 25% of cases occur in the spine (entries 11, 4). Neurologic symptoms and signs include a spectrum of abnormalities from sensory changes to paraplegia and these may occur if the expansion of the lesion results in nerve root or spinal cord compression (entries 7, 8, 5).

§ 4-27(C). Physical Findings.

Physical examination may demonstrate tenderness to palpation over the site of involvement. The overlying skin may be erythematous and warm if the aneurysmal bone cyst is close to the surface. Neurologic findings correlate with the location of nerve root or spinal cord compression.

§ 4-27(D). Laboratory Findings.

The laboratory findings are normal in this vascular lesion of bone. The pathologic findings include cystic cavities, composed of vascular channels filled with fibrous connective tissue, osteoid, granulation tissue, and multinucleated giant cells (entry 2).

§ 4-27(E). Radiographic Findings.

The radiographic features of an aneurysmal bone cyst consist of a solitary, eccentrically located, osteolytic, expansile lesion which is sharply demarcated by a thin subperiosteal shell of bone (entry 9). A soft tissue mass may also be associated with the bony lesion. When in the spine, they occur most commonly in the lumbar and thoracic area and affect the posterior elements of the vertebrae, including pedicles, laminae, spinous and transverse processes (entry 7). Vertebral bodies are only secondarily invaded by the markedly expansile lesion originating in the posterior elements.

§ 4-27(F). Differential Diagnosis.

The characteristic radiographic appearance of the aneurysmal bone cyst helps differentiate it from other benign and malignant lesions. In the spine, radiolucent defects of the posterior elements also occur with osteoblastoma; however, it tends to be smaller and shows intralesional calcification.

§ 4-27(G). Treatment.

Aneurysmal bone cysts can be treated by surgery, radiotherapy, or cryotherapy. Although they are benign lesions, they are highly prone to local recurrence after curettage. If the location of an aneurysmal bone cyst allows for removal of a section of bone without loss of function, en bloc resection is the treatment of choice and lesions in the posterior elements of the spine may be treated with resection and bone grafting. Lesions which are too large or involve a vertebral body are treated with radiotherapy (entry 13). Cryosurgery, freezing the lesion, has also been reported to halt expansion of the cyst and prevent recurrence (entry 2).

§ 4-27(H). Course and Outcome.

Aneurysmal bone cyst is a benign lesion but may cause severe dysfunction because of its expansile characteristics. If diagnosed early and treated appropriately, dysfunction may be kept to a minimum. However, if it is located in the spine and is allowed to expand unchecked, serious neurologic deficits may result. In addition, an aneurysmal bone cyst weakens the bone and increases the risk of pathologic fracture.

BIBLIOGRAPHY

1. Barnes, R. Aneurysmal bone cyst. *J. Bone Joint Surg.*, 38B:301-311, 1956.
2. Biesecker, J.L., Marcove, R.C., Huvos, A.G., and Mike, V. Aneurysmal bone cyst: a clinicopathologic study of 66 cases. *Cancer*, 26:615-625, 1970.
3. Clough, J.R., and Price, C.H.G. Aneurysmal bone cyst: pathogenesis and long term results of treatment. *Clin. Orthop.*, 97:52-63, 1973.
4. Dabska, M., and Buraczewski, J. Aneurysmal bone cyst: pathology, clinical course and radiologic appearance. *Cancer*, 23:371-389, 1969.
5. Dahlin, D.C., and McLeod, R.A. Aneurysmal bone cyst and other non-neoplastic conditions. *Skeletal Radiol.*, 8:243-250, 1982.
6. Dahlin, D.C., Besse, B.E., Pugh, D.G., and Ghormley, R.K. Aneurysmal bone cysts. *Radiology*, 64:56-65, 1955.
7. Donaldson, W.F. Aneurysmal bone cyst. *J. Bone Joint Surg.*, 44A:25-40, 1962.
8. Hay, M.C., Patterson, D., and Taylor, T.K.F. Aneurysmal bone cyst of the spine. *J. Bone Joint Surg.*, 60B:406-411, 1978.
9. Jaffe, H.L. Aneurysmal bone cyst. *Bull. Hosp. Joint Dis.*, 11:3-13, 1950.

10. Lichtenstein, L. Aneurysmal bone cyst: a pathological entity commonly mistaken for a giant cell tumor and occasionally for hemangioma and osteogenic sarcoma. *Cancer*, 3:279-284, 1950.

11. Lichtenstein, L. Aneurysmal bone cyst: observations on fifty cases. *J. Bone Joint Surg.*, 39A:873-882, 1957.

12. Mirra, J. *Bone Tumors: Diagnosis and Treatment*, J.B. Lippincott Co., Phila: 1980, pp. 478-492.

13. Slowick, F.A., Campbell, C.J., and Kettelkamp, D.B. Aneurysmal bone cyst. *J. Bone Joint Surg.*, 50A:1142-1151, 1968.

§ 4-28. Hemangioma.

Hemangiomas are benign vascular lesions composed of cavernous, capillary or venous blood vessels which may affect soft tissues or bone. They are frequently located in the spine, but are usually asymptomatic. Patients with symptomatic hemangiomas develop localized pain and muscle spasm. Neurologic complications, including compression of nerve roots, spinal cord, or cauda equina, may occur through vertebral body collapse or expansion secondary to a vertebral hemangioma. The first reference to a hemangioma was reported by Toynbee in 1845 (entry 9).

§ 4-28(A). Prevalence and Etiology.

Hemangiomas account for fewer than 1% of clinically symptomatic primary bone tumors (entry 6). However, necropsy studies by a number of investigators have demonstrated that asymptomatic vertebral lesions are found in 12% of autopsies (entries 4, 7). The prevalence of hemangioma increases with age. They are usually identified in patients between the fourth and fifth decades of life. When taking into account hemangiomas from all sites, women and men are equally affected.

The etiology of hemangioma remains unknown. They are considered congenital vascular malformations by some and as benign neoplasms by others.

308

§ 4-28(B). Clinical Findings.

The most common location for hemangiomas is in the spine, particularly in the thoracic area, and the skull. Pain and tenderness over the involved vertebrae are the usual presenting symptoms. The pain usually starts as vague and non-descript. It gradually increases with a constant and throbbing quality. There may be associated muscle spasm. Neurologic manifestations of cord compression by vertebral hemangioma may include sensory changes, motor weakness, radiculitis or transverse myelitis (entries 1, 5).

§ 4-28(C). Physical Findings.

Physical examination may demonstrate tenderness with palpation over the affected vertebral body. Limitation of motion may be present if related muscle spasm is severe.

§ 4-28(D). Laboratory Findings.

Solitary vertebral hemangiomas are not associated with any abnormal laboratory results. Pathologic examination demonstrates a well delineated cystic space which contains sparsely cellular osseous trabecular tissue and increased capillary, cavernous or venous vessels (entry 6).

§ 4-28(E). Radiographic Findings.

Vertebral hemangiomas primarily involve the vertebral bodies. In an affected vertebral body, the vertical striations are prominent while horizontal striations are absent because of absorption (entry 8). The alteration of vertebral striations is diffuse and vertebral body configuration is usually unchanged. Occasionally, vertebral hemangiomas may extend from the body to the laminae, pedicles, transverse or spinous processes. Rarely, expansion or enlargement of a vertebra may occur (entry 5).

§ 4-28(F). Differential Diagnosis.

The diagnosis of a vertebral hemangioma is uncomplicated when one vertebral body is affected with characteristic radiographic changes. It is more difficult when portions of a vertebra other than the body are affected. Bony resorption of a pedicle may mimic the destructive changes of metastatic cancer. A vertebral body fracture may occur with a hemangioma, but is more frequently seen in metastatic tumor. Course trabeculation of a vertebral body may also be seen in Paget's disease. Abnormal laboratory tests (elevated ESR, serum alkaline phosphatase) should differentiate tumor and Paget's disease from a hemangioma.

§ 4-28(G). Treatment.

The treatment of choice for vertebral hemangiomas is irradiation since they are radiosensitive. Radiation therapy effectively relieves symptoms even though the appearance of the lesion remains unchanged (entry 3). Surgical intervention in the form of laminectomy has an excessive morbidity and mortality due to profuse hemorrhage (entry 3); therefore, laminectomy should be reserved for those patients with neurologic deficits who require decompression of the spinal cord (entry 2).

§ 4-28(H). Course and Outcome.

Vertebral hemangiomas are usually asymptomatic and have a benign course; however, when they become symptomatic, they require therapy to prevent expansion of the lesion. The major complication of a vertebral hemangioma is neural compression. Compression fractures may occur more easily in vertebrae affected by a hemangioma. Hemangiomas may cause neural compression by compression fracture, expansion of an involved vertebra, direct extension of the hemangioma into the extradural space or extradural hemorrhage. Appropriate diagnosis and treatment may help prevent this potentially disabling complication of this vascular neoplasm.

BIBLIOGRAPHY

1. Barnard, L., and Von Nuys, R.G. Primary hemangioma of the spine. *Ann. Surg.,* 97:19-25, 1933.
2. Lozman, J., and Holmblad, J. Cavernous hemangiomas associated with scoliosis and a localized consumptive coagulopathy: A case report. *J. Bone Joint Surg.,* 58A:1021-1024, 1976.
3. Manning, J.H. Symptomatic hemangioma of the spine. *Radiology,* 56:58-65, 1951.
4. Marcial-Rojas, R.A. Primary hemangiopericytoma of bone: Review of the literature and report of the first case with metastasis. *Cancer,* 13:308-311, 1960.
5. McAllister, V.L., Kendall, B.E., and Bull, J.W.D. Symptomatic vertebral hemangiomas. *Brain,* 98:71-80, 1975.
6. Mirra, J. *Bone Tumors: Diagnosis and Treatment.* J.B. Lippincott Co., Phila: 1980, pp. 492-498.
7. Schmorl, G., and Junghanns, H. *The Human Spine in Health and Disease,* 2nd ed., New York: Grune and Stratton, 1971, pp. 325.
8. Sherman, R.S., and Wilner, D. The roentgen diagnosis of hemangioma of bone. *Am. J. Roentgenol.,* 86:1146-1159, 1961.
9. Toynbee, J. An account of two vascular tumors developed in the substance of bone. *Lancet,* 2:676, 1845.

§ 4-29. Eosinophilic Granuloma.

Eosinophilic granuloma occurs in solitary and multifocal forms and is characterized by the infiltration of bone with histiocytes, mononuclear phagocytic cells, and eosinopohils. Eosinophilic granuloma, Hand-Schuller-Christian disease, and Letterer-Siwe disease are thought to have the same pathogenesis and are referred to collectively as histiocytosis X. Eosinophilic granuloma is the mildest form, and Letterer-Siwe the most aggressive form, of histiocytosis X. Eosinophilic granuloma, when it is discovered in an adult, may involve the spine causing vertebral body collapse and neurologic symptoms. The course of this illness is benign and it is very responsive to therapy. It has been called pseudotuberculous granuloma, Taratynov's disease, traumatic myeloma, and histiocytic granuloma in the past. Lichtenstein and Jaffe were the first to use the term eosinophilic granuloma in 1944 (entry 5).

§ 4-29(A). Prevalence and Etiology.

Eosinophilic granuloma is a rare lesion occurring in less than 1% of biopsied primary infiltrative lesions of bone (entry 7). It occurs most commonly in children and adolescents with approximately 10% of patients being 20 years or older (entry 2). It has a higher incidence in males at a 2:1 ratio and is more frequent in caucasians than blacks. The etiology of this disease is unknown.

§ 4-29(B). Clinical Findings.

Pain is the predominant symptom and is constant. It is not relieved by rest or aspirin therapy. If the lesion is close to the skin, localized swelling may be noticed.

§ 4-29(C). Physical Findings.

A palpable mass may be present over the affected bone but it is neither tender nor associated with redness or heat. A low grade fever is present in some patients. Spinal cord and nerve root compression secondary to vertebral body collapse results in corresponding abnormalities on neurologic examination.

§ 4-29(D). Laboratory Findings.

Eosinophilic granuloma is associated with peripheral eosinophilia in 6% to 10% of patients. There is also an elevated erythrocyte sedimentation rate.

The histologic appearance of eosinophilic granuloma is characterized by collections of eosinophils and histiocytes without the formation of local, distinct granulomas. The pleomorphic appearance of the histiocytes may superficially resemble malignant cells of Hodgkin's disease. Histiocytes of eosinophilic granuloma are benign. With time, they may form giant cells and take on the appearance of foam cells after they have ingested necrotic tissue and convert it into cytoplasmic lipids. The presence of lipid in eosinophilic granuloma is a secondary phenomenon and is not of pathogenetic importance as it is in Gaucher's disease. Fibrous tissue appears as the lesion heals (entry 4).

§ 4-29(E). Radiographic Findings.

Eosinophilic granuloma in the spine is associated with a spectrum of radiographic abnormalities. The features of early lesions are those of a destructive, radiolucent oval area of bone lysis without peripheral sclerosis. Progressive destruction in a vertebral body results in a flattened vertebral body, termed vertebra plana, first described by Calve (entry 1). Eosinophilic granuloma is the most common cause of vertebra plana in children. It may erode the posterior elements of a vertebral body and spare the vertebral body. These lesions are not associated with vertebral collapse (entry 6). Rarely, eosinophilic granuloma can produce expansile lesions with extensive destruction of multiple vertebrae and paraspinal extension (entry 3).

§ 4-29(F). Differential Diagnosis.

The diagnosis of eosinophilic granuloma is made from the close inspection of biopsy material. Laboratory and radiographic features are too non-specific to assure an accurate diagnosis. Other diseases which must be considered in the differential diagnosis include osteomyelitis, Hodgkin's disease, giant cell tumor, granulomatous diseases, such as tuberculosis, fungal infections, sarcoidosis, and non-ossifying fibroma.

§ 4-29(G). Treatment.

The treatment of choice for eosinophilic granuloma is curettage, with or without packing of the lesion with bone chips. Inaccessible lesions in the spine which may lead to pathologic fractures are best treated with low dosage radiation therapy.

§ 4-29(H). Course and Outcome.

Eosinophilic granuloma is a benign lesion and the prognosis is good. Patients with vertebra plana may heal in time with partial reconstitution of the affected vertebral body. The prognosis of patients with multifocal eosinophilic granuloma may not be as good if their illness progresses to the diffuse involvement associated with other components of histiocytosis X. This is more of a concern for younger patients than for adults.

BIBLIOGRAPHY

1. Calve, J.A. Localized affection of spine suggesting osteochondritis of vertebral body, with clinical aspects of Pott's disease. *J. Bone Joint Surg.*, 7:41-46, 1925.

2. Cheyne, C. Histiocytosis X. *J. Bone Joint Surg.*, 53B:366-382, 1971.

3. Ferris, R.A., Pettrone, F.A., McKelvie, A.M., Twigg, H.L., and Chun, B.K. Eosinophilic granuloma of the spine: An unusual radiographic presentation. *Clin. Orthop.*, 99:57-63, 1974.

4. Green, W.T., and Farber, S. "Eosinophilic or solitary granuloma" of bone. *J. Bone Joint Surg.*, 24:499-526, 1942.

5. Jaffe, H.L., and Lichtenstein, L. Eosinophilic granuloma of bone: A condition affecting one, several or many bones, but apparently limited to the skeleton and representing the mildest clinical expression of the peculiar inflammatory histiocytosis also underlying Letterer-Siwe disease and Schuller-Christian disease. *Arch. Pathol.*, 37:99-118, 1944.

6. Kaye, J.J., and Freiberger, R.H. Eosinophilic granuloma of the spine without vertebra plana: A report of two unusual cases. *Radiology,* 92:1188-1191, 1969.

7. Mirra, J.M. *Bone Tumors: Diagnosis and Treatment.* J.B. Lippincott Co., Phila: 1980, pp. 376-382.

§4-30. Gaucher's Disease.

Gaucher's disease is a lipid metabolism disorder associated with the accumulation of ceramide glucoside in histiocytes. The massive accumulation of this lipid in cells of the reticuloendothelial system results in an enlarged spleen, destruction of bone, and abnormalities of the bone marrow. In adults, the axial skeleton is frequently involved. Gaucher's disease causes vertebral body osteolysis, compression fractures and spinal deformities. In the adult it has a protracted course, associated with periods of remission and relapse. The disease was first described by Gaucher in 1882 (entry 1).

§4-30(A). Prevalence and Etiology.

Gaucher's disease is an uncommon illness which can become manifest at any time of life. Ashkenazik Jews are at greatest risk to develop this illness although caucasians, blacks, and orientals can also be afflicted (entry 4). Men and women are equally affected.

In the past, the etiology of Gaucher's disease was thought to be excessive production of a lipid, ceramide glucoside. More recently, the abnormality has been traced to a defective enzyme, beta-glucosidase, which is unable to degrade accumulating lipid (entry 7). The accumulation of this material in cells throughout the body results in the manifestations of Gaucher's disease. The autosomal recessive inheritance of the disease suggests that a single biochemical defect does account for the abnormalities associated with this illness.

315

§ 4-30(B). Clinical Findings.

Clinical symptoms depend on the organ involved and the degree of involvement. Adult patients have abdominal distention secondary to hepatosplenomegaly. Patients with skeletal disease have persistent bone pain, tenderness, difficulty walking, back pain, and loss of height. Neurologic symptoms are usually reserved for children with more severe disease. Constitutional symptoms of generalized fatigue and weakness are also present.

§ 4-30(C). Physical Findings.

Patients with Gaucher's disease have massive splenomegaly, moderate hepatomegaly and minimal lymphadenopathy. Deposits of lipid in the dermis give exposed skin a tan color, while deposits in the sclerae of the eye cause tan pingueculae to appear. Patients may complain of bone tenderness on palpation.

§ 4-30(D). Laboratory Findings.

Anemia, leukopenia, and thrombocytopenia, which is associated with occasional episodes of bleeding, result from replacement of bone marrow cells with lipid-filled histiocytes. Serum acid phosphatase of non-prostatic origin is elevated.

Bone marrow aspirate is usually adequate in obtaining tissue which reveals characteristic Gaucher cells. Gaucher cells have two round-to-oval eccentric nuclei with striated cytoplasm which readily stains with periodic acid-Schiff reagent.

§ 4-30(E). Radiographic Findings.

The infiltration of marrow with Gaucher cells results in characteristic radiographic abnormalities. In the spine, cellular infiltrates result in increased radiolucency of vertebral bodies, accentuation of vertical trabeculae, and

316

compression fractures (entry 2). Fractures may cause complete flattening of a vertebral body (vertebra plana) or the development of depressions in the superior and inferior margins of a vertebral body, referred to as an "H vertebra" (entry 5). Gaucher's disease is also associated with aseptic necrosis of bone, usually at the ends of long bones but it has also been described near the sacroiliac joint. This process results in apparent obliteration of the articulation, simulating changes associated with ankylosing spondylitis (entry 3).

§ 4-30(F). Differential Diagnosis.

The diagnosis of Gaucher's disease is confirmed by the presence of characteristic cells obtained by biopsy of affected tissues. The radiographic findings have considerable overlap with other disorders which produced osteoporosis and aseptic necrosis. The differential diagnosis should include metabolic, hematologic and neoplastic disorders. The possibility of osteomyelitis complicating Gaucher's disease, causing increased bone pain and fever, should also be considered.

§ 4-30(G). Treatment.

Therapy for Gaucher's disease is palliative. There is no effective treatment to reduce accumulations of lipid or increase bone calcium.

§ 4-30(H). Course and Outcome.

Gaucher's disease has a protracted course punctuated with periods of increased symptoms. Compression fractures of the spine, aseptic necrosis of bone and pathologic fractures in long bones can cause severe morbidity. Adult patients usually die of infection, bleeding, anemia or severe weight loss.

317

BIBLIOGRAPHY

1. Gaucher, P. De l'epithelioma primitif de la rate, hypertrophie idiopathique de la rate sans l'eucemie. *These de Paris,* 1882.

2. Greenfield, G.B. Bone changes in chronic adult Gaucher's disease. *Am. J. Roentgenol.,* 110:800-807, 1970.

3. Kulowski, J. Gaucher's disease in bone. *Am. J. Roentgenol.,* 63:840-850, 1950.

4. Novy, S.B., Naletson, E., Stuart, L., and Whittock, G. Gaucher's disease in a black adult. *Am. J. Roentgenol.,* 133:947-949, 1979.

5. Schwartz, A.M., Homer, M.J., and McCauley, R.G.K. "Step off" vertebral body. Gaucher's disease vs. sickle cell hemoglobinopathy. *Am. J. Roentgenol.,* 132:81-85, 1979.

6. Tuchman, L.R., and Swick, M. High acid phosphatase level indicating Gaucher's disease in patients with prostatism. *JAMA,* 164:2034-2035, 1957.

7. Volk, B.W., Adachi, M., and Schneck, L. The pathology of the sphingolipidoses. *Semin. Hematol.,* 9:317-348, 1972.

§ 4-31. Sacroiliac Lipomata.

Sacroiliac lipomata are fatty tumors which are located over the sacroiliac joints. They herniate through weak areas in the overlying fascia and become painful when they are strangulated by the fascia. The pain associated with the herniation of these fatty tumors may be severe, may radiate into the buttock and thigh, and may be associated with limitation of flexion of the lumbosacral spine. Appropriate physical examination will identify the presence of the lipomata over the sacroiliac joints. Injection therapy with a topical anesthetic is adequate to control symptoms in most patients, but surgical removal of the lipomata may be required for patients who do not respond. Ries in 1936 was the first to describe the presence of sacroiliac lipomata and the clinical syndrome associated with them (entry 6).

§ 4-31(A). Prevalence and Etiology.

The prevalence of sacroiliac lipomata in the general population is unknown. Studies which report groups of patients with sacroiliac lipomata suggest that the incidence is relatively high (entries 7, 5). They become symptomatic most frequently when the patient is in the fifth decade. The male to female ratio is 1:4.

The etiology of sacroiliac lipomata is related to weaknesses in the fascia which runs from the cervical to the lumbosacral spine. The lipoma may herniate through a weakened area where there are deficiencies in fascia fibers or through one of the foramen where the lateral branches of the posterior primary division of the first, second, and third lumbar nerves pass (entry 2). The three types of herniations are pedunculated, non-pedunculated and foraminal. With certain motions the fascia strangulates the herniated fatty tumor, compressing its blood supply and nerves and this results in local pain with possible radiation in the distribution of the cutaneous nerves.

§ 4-31(B). Clinical Findings.

The typical patient is a middle aged, obese female with unilateral low back pain radiating to the buttock or anterior thigh. Flexion of the lumbo-sacral spine increases the pain, as does activity. The pain has an aching quality and occasionally may be bilateral. Compressing the area at night while sleeping may cause severe distress. There are no constitutional symptoms such as fatigue, weight loss, fever, or anorexia associated with the lesion.

§ 4-31(C). Physical Findings.

Physical examination demonstrates a tender nodule near the dimples in the sacroiliac area. The sacroiliac lipomata are very tender to palpation. Forward flexion causes pain and that motion can be limited. Neurologic examination, including straight leg-raising tests, is normal.

§ 4-31(D). Laboratory Findings.

Laboratory tests, including hematologic, chemical and immunologic studies, are normal in patients with sacroiliac lipomata. The gross pathological findings demonstrate rounded cylindrical bodies which measure from 1 to 5 cm. in diameter. Microscopically, the lipomata consist of normal adipose tissue with little interposed connective tissue and a fibrous capsule. Nerve fibers are detected in some specimens (entry 4). Signs of edema and hemorrhage may be seen in some portions of the specimens (entry 3).

§ 4-31(E). Radiographic Findings.

Sacroiliac lipomata are unassociated with any radiographic abnormalities.

§ 4-31(F). Differential Diagnosis.

The diagnosis of sacroiliac lipomata is not made unless physicians are aware of the existence of this entity as a cause of low back pain. Frequently, patients undergo extensive examinations before the correct diagnosis is made. Patients who have a history of pain which is increased while rolling over in bed and have a tender nodule which recreates their pain, can have sacroiliac lipomata as the cause of their symptoms.

§ 4-31(G). Treatment.

The initial treatment for sacroiliac lipomata is injection of the nodules with local anesthetic such as xylocaine and these injections frequently produce relief from pain. Patients may notice relief for extended periods of time even from single injections (entry 7). Surgical removal of the painful sacral lipoma is beneficial if injections are unable to improve symptoms (entry 1).

§ 4-31(H). Course and Outcome.

Since patients with sacral lipomata may experience severe pain and limitation of motion of the lumbosacral spine, they frequently undergo extensive evaluations including myelograms and these are consistently normal. When no specific abnormality is discovered, these patients are thought to have psychogenic rheumatism and are referred for psychiatric evaluation. Some have had symptoms for as long as seven years before the correct diagnosis was made (entry 4). The prognosis is excellent when the correct diagnosis is made and appropriate therapy instituted. These patients suffer no functional impairment once they respond to therapy.

BIBLIOGRAPHY

1. Bonner, C.D., and Kasdon, S.C. Herniation of fat through lumbosacral fascia as a cause of low-back pain. *N. Eng. J. Med.,* 251:1102-1104, 1954.
2. Copeman, W.S.C. Fibro-fatty tissue and its relation to certain "rheumatic" syndromes. *Brit. Med. J.,* 2:191-197, 1949.
3. Herz, R. Herniation of fascial fat as a cause of low back pain. *JAMA,* 128:921-925, 1945.
4. Hittner, V.J. Episacroiliac lipomas. *Amer. J. Surg.,* 78:382-383, 1949.
5. Hucherson, D.C., and Gandy, J.R. Herniation of fascial fat: A cause of low back pain. *Amer. J. Surg.,* 76:605-609, 1948.
6. Ries, E. Episacroiliac lipoma. *Amer. J. Obstet. Gynec.,* 34:490-494, 1937.
7. Singewald, M.L. Sacroiliac lipomata: An often unrecognized cause of low back pain. *Bull. Johns Hopkins Hosp.,* 118:492-498, 1966.

§ 4-32. Multiple Myeloma.

Multiple myeloma is a malignant tumor of plasma cells. Plasma cells are the cells that produce immunoglobulins and antibodies, and are located throughout the bone marrow. The multiplication of these cells in the bone marrow is associated with diffuse bone destruction characterized by bone pain, pathological fractures and increases in serum calcium. Back pain is a common symptom of multiple myeloma. Patients with multiple myeloma or solitary plasmacytoma, a localized form of myeloma, can have neurologic symptoms secondary to vertebral body compression fractures or extradural extension of tumor associated with compression of neural elements. Not infrequently, patients will describe an episode of minimal trauma as the initiating factor in the onset of their pain. The course of this disease may be modified by drug and radiation therapy but most patients die within five years of the discovery of the illness. The disease was first described by three physicians who identified the bone lesions, urinary protein and the bone lesions in the 1840's (entries 10, 4, 19). Von Rustizky in 1873 was the first to refer to the disease as multiple myeloma (entry 25). Other names synonymous with the illness include myelomatosis, plasma cell myeloma, and Kohler's disease.

§ 4-32(A). Prevalence and Etiology.

Multiple myeloma is the most common primary malignancy of bone in adults accounting for 27% of biopsied bone tumors (entry 20). The incidence is three cases per 100,000 people in the United States (entry 16) and the patients are usually in an older age group ranging between 50 and 70 (entry 18). Multiple myeloma is rare in a patient below the age of 40 (entry 15). There is a slight increase in the male to female ratio but the ratio may be increased further for solitary plasmacytomas (entry 23).

The etiology of this plasma cell tumor is unknown. Viral infections, chronic inflammation, and chronic myeloproliferative diseases have been suggested as possible initiating factors. Although the symptoms of a fracture associated with trauma are frequently the reason for a patient's initial evaluation by a physician, trauma is not a factor in the etiology of multiple myeloma.

§ 4-32(B). Clinical Findings.

Pain is the most common initial complaint of patients with multiple myeloma and occurs in 75% of patients (entry 16). Low back pain is the presenting symptom in 35% of patients. The pain is mild, aching and intermittent at the onset, and is aggravated by weight-bearing and relieved by bed rest. Some patients have radicular symptoms and are diagnosed as herniated intervertebral discs, sciatica, and arthritis. Approximately 20% of patients give a history of insignificant trauma which results in pathological fractures of vertebral bodies. Paraplegia more often occurs with solitary plasmacytoma than with multiple myeloma (entry 24). Involvement of the ribs, sternum, and thoracic spine may result in kyphosis and loss in height.

Pain is not confined to the back alone in this illness since bone pain may be found in any part of the skeleton and may be secondary to bone marrow expansion or to microfractures (entry 8).

As a consequence of widespread bone destruction, abnormal immunoglobulin production and infiltration of bone marrow, patients with multiple myeloma develop a broad range of clinical symptoms. Hypercalcemia, due to bone destruction, is associated with bone weakness, easy fatiguability, anorexia, nausea, vomiting, mental status changes including coma, and kidney stones. Increased abnormal immunoglobulin concentrations cause progressive renal insufficiency, increased susceptibility to infection, and amyloidosis. Amyloid is a protein which forms from portions of abnormal myeloma immunoglobulins. This protein infiltrates certain structures including bone, muscle, and perivascular connective tissue. Primary amyloidosis in the absence of multiple myeloma may infiltrate bone marrow and simulate radiographic changes in the vertebral column which are similar to those of multiple myeloma (entry 2). Infiltration of bone marrow is associated with anemia, bleeding secondary to a deficiency of platelets, thrombocytopenia, and generalized weakness.

§ 4-32(C). Physical Findings.

In the early stages of the illness, the physical examination may be unremarkable. As the duration of illness increases and bone marrow infiltration progresses, diffuse bone tenderness along with fever, pallor, and purpura becomes a prominent finding on examination. Neurologic examination may demonstrate signs of compression of the spinal cord or nerve roots if vertebral collapse has progressed to a significant degree (entry 11).

§ 4-32(D). Laboratory Findings.

Laboratory examination may reveal many abnormalities including normochromic normocytic anemia, elevated leukocyte count, thrombocytopenia, or positive Coomb's test and an elevated erythrocyte sedimentation rate. Abnormal serum chemistries include hypercalcemia, hyperuricemia, and elevated creatinine.

Serum alkaline phosphatase is normal in most patients with multiple myeloma. Characteristic serum protein abnormalities occur in the vast majority of patients with multiple myeloma. The total serum protein concentrations are increased secondary to an increase in the globulin fraction. The increase in globulins is due to the presence of abnormal immunoglobulins of the G, A, D, E or M classes. Immunoglobulins are composed of light and heavy chains. The light and heavy chains of an antibody combine to form a site which is directed against a specific antigen. In multiple myeloma, instead of having a multitude of antibodies, one single antibody composed of a light and heavy chain, an M-protein, is produced to the exclusion of others. The balance between light and heavy chains may also be disturbed with excess light chain production resulting in Bence-Jones protein in the urine, or excess heavy chain production resulting in heavy chain disease. Serum protein electrophoresis demonstrates the elevation in globulin levels, while quantitative immunoglobulin determination detects the class of immunoglobulin which is present in increased concentration. Urine protein electrophoresis will detect the presence of Bence-Jones proteinuria. Other tests associated with abnormal immunoglobulins include increased serum viscosity which results in the blockage of blood vessels, and a positive test for rheumatoid factor.

Examination of a bone marrow aspirate and biopsy show characteristic changes of multiple myeloma. Bone marrow aspirate demonstrates increased numbers of plasma cells, at levels greater than 30%. Bone marrow biopsy demonstrates diffuse infiltration of bone marrow with plasma cells. Plasma cells may be classified into three histologic grades: well differentiated, moderately differentiated, and poorly differentiated (entry 3). The well differentiated plasma cells look much like normal plasma cells. Anaplastic myeloma cells may mimic undifferentiated carcinomas or small cell sarcomas.

§ 4-32(E). Radiographic Findings.

The predominant radiographic abnormality of multiple myeloma is osteolysis. Diffuse osteolysis of the axial skeleton resembles osteoporosis (entry 6). A characteristic finding is the absence of reactive sclerosis surrounding lytic lesions. In the spine, preferential destruction of vertebral bodies with sparing of posterior elements helps in differentiating multiple myeloma from osteolytic

325

metastasis which affects the vertebral pedicle and body (entry 17). Paraspinous and extradural extension of tumor is seen in patients with multiple myeloma.

Solitary plasmacytomas, when located in the spine, have variable radiographic features. A purely osteolytic area without expansion or a multiple expansile lesion with thickened trabeculae may be observed. An involved vertebral body may fracture and disappear completely or the lesion may extend across the intervertebral disc to invade an adjacent vertebral body, simulating the appearance of an infection (entry 24). Occasionally, osteoblastic lesions have been reported in patients with multiple myeloma and plasmacytoma (entries 5, 22).

While bone scans are used in the early detection of metastatic lesions, they are not helpful in multiple myeloma because the osteolytic lesions which lack bone forming (blastic) activity will not be positive (entry 26). Computed tomography may demonstrate vertebral body involvement before plain radiographs. This is due to the fact that radiographs are able to detect abnormalities in bone calcium only after 30% of it is lost (entry 14).

§ 4-32(F). Differential Diagnosis.

The diagnosis of multiple myeloma requires the inclusion of clinical, radiographic and laboratory data along with the presence of abnormal plasma cells on histologic examination. The presence of characteristic abnormalities, Bence-Jones protein and M-protein on electrophoresis, make the diagnosis an easy one but the diagnosis is more difficult in the patient who presents with diffuse osteoporosis and no detectable myeloma protein (entry 1). Low back pain in the middle aged to elderly patient with osteoporosis on radiographs must be evaluated thoroughly for possible myeloma. Other diseases which must be considered in the differential diagnosis of multiple myeloma include metastatic tumor, osteoporosis, chronic infections, and hyperparathyroidism.

§ 4-32(G). Treatment.

Clinically active multiple myeloma usually requires systemic therapy with melphalan and prednisone over an extended course (entry 9). Approximately 70% of patients respond to this therapy and have a reduction in bone pain and destruction, decreased concentration of abnormal proteins and a normalization of hematocrit, urea nitrogen, creatinine and calcium. Patients with more aggressive disease may benefit by the M-2 drug program, combining vincristine, melphalan, cyclophosphamide, prednisone, and BCNU (entry 7). In patients with cord compression secondary to multiple myeloma, decompressing laminectomy and/or local radiotherapy are indicated (entry 12).

§ 4-32(H). Course and Outcome.

The usual course of multiple myeloma is one of gradual progression. Therapy may have effects on clinical symptoms and amounts of myeloma protein but the average survival remains about five years. Patients with D and G myeloma have poorer prognoses than those A myeloma patients (entries 13, 21). Patients with solitary plasmacytomas have a better prognosis than those patients who initially have multiple lesions (entry 24).

BIBLIOGRAPHY

1. Arend, W.P., and Adamson, J.W. Nonsecretory myeloma: Immunofluorescent demonstration of paraprotein within bone marrow plasma cells. *Cancer,* 33:721-728, 1974.
2. Axelsson, U., Hallen, A., and Rausing, A. Amyloidosis of bone: Report of two cases. *J. Bone Joint Surg.,* 52B:717-723, 1970.
3. Bayrd, E.D. The bone marrow on sternal aspiration in multiple myeloma. *Blood,* 3:987-1018, 1948.

4. Bence-Jones, H. On a new substance occurring in the urine of a patient with mollities ossium. *Philos. Trans. R. Soc. Lond. (Biol.)*, 1:55-62, 1848.

5. Brown, T.S., and Paterson, C.R. Osteosclerosis in myeloma. *J. Bone Joint Surg.*, 55B:621-623, 1973.

6. Carson, C.P., Ackerman, L.V., and Maltby, J.D. Plasma cell myeloma: A clinical, pathologic and roentgenologic review of 90 cases. *Am. J. Clin. Pathol.*, 25:849-888, 1955.

7. Case, D.C., Jr., Lee, B.J., III, and Clarkson, B.D. Improved survival times in multiple myeloma treated with melphalan, prednisone, cyclophosphamide, vincristine and BCNU — M-2 protocol. *Am. J. Med.*, 63:897-903, 1977.

8. Charkes, N.D., Durant, J., and Barry, W.E. Bone pain in multiple myeloma: Studies with radioactive 87m SR. *Arch. Intern. Med.*, 130:53-58, 1972.

9. Costa, G., Engle, R.L., Jr., Schilling, A., Carbone, P., Kochwa, S., Nachman, R.L., and Glidewell, O. Melphalan and prednisone — an effective combination for the treatment of multiple myeloma. *Am. J. Med.*, 54:589-599, 1973.

10. Dalrymple, J. On the microscopical character of mollities ossium. *Dublin Q.J. Med. Sci.*, 2:85-95, 1846.

11. Davison, C., and Balser, B.H. Myeloma and its neural complications. *Arch. Surg.*, 35:913-936, 1937.

12. Gilbert, R.W., Kim, J.H., and Posner, J.B. Epidural spinal cord compression from metastatic tumor: Diagnosis and treatment. *Ann. Neurol.*, 3:40-51, 1978.

13. Gompels, B.M., Votaw, M.L., and Martel, W. Correlation of radiological manifestations of multiple myeloma with immunoglobulin abnormalities and prognosis. *Radiology*, 104:509-514, 1972.

14. Helms, C.A., and Genant, H.K. Computed tomography in the early detection of skeletal involvement with multiple myeloma. *JAMA*, 248:2886-2887, 1982.

15. Hewell, G.M., and Alexanian, R. Multiple myeloma in young persons. *Ann. Intern. Med.*, 84:441-443, 1976.

16. Huvos, A.G. *Bone Tumors: Diagnosis, Treatment and Prognosis,* W.B. Saunders, Phila: 1979, pp. 413-431.

17. Jacobsen, H.G., Poppel, M.H., Shapiro, J.H., and Grossberger, S. The vertebral pedicle sign: A roentgen finding to differentiate metastatic carcinoma from multiple myeloma. *Am. J. Roentgenol.*, 80:817-821, 1958.

18. Kyle, R.A. Multiple myeloma: Review of 869 cases. *Mayo Clin. Proc.*, 50:29-40, 1965.
19. Macintyre, W. Case of mollities and fragilitas ossium accompanied with urine strongly charged with animal matter. *Med. Chir. Soc. Trans.*, 33:211-232, 1850.
20. Mirra, J.M. *Bone Tumors: Diagnosis and Treatment*, J.B. Lippincott Co., Phila: 1980, pp. 398-406.
21. Pruzanski, W., and Rother, I. IgD plasma cell neoplasia: Clinical manifestations and characteristic features. *Can. Med. Assoc. J.*, 102:1061-1065, 1970.
22. Roberts, M., Rinaudo, P.A., Vilinskas, J., and Owens, G. Solitary sclerosing plasma-cell myeloma of the spine: Case report. *J. Neurosurg.*, 40:125-129, 1974.
23. Todd, I.D.H. Treatment of solitary plasmacytoma. *Clin. Radiol.*, 16:395-399, 1965.
24. Valderrama, J.A.F., and Bullough, P.G. Solitary myeloma of the spine. *J. Bone Joint Surg.*, 50B:82-90, 1968.
25. Von Rustizky, J. Multiples myelom. *Dtsch Z. Chir.*, 3:162-172, 1873.
26. Woolfenden, J.M., Pitt, M.J., Durie, B.G.M., and Moon, T.E. Comparison of bone scintography and radiography in multiple myeloma. *Radiology,* 134:723-728, 1980.

§ 4-33. Chondrosarcoma.

Chondrosarcoma is a malignant tumor which forms cartilagenous tissue. It is frequently located in the pelvis, sacrum or lumbar spine. Since the tumor has extremely slow growth and lesions are usually painless, chondrosarcoma of the pelvis and spine may be present for a long period of time before it is discovered. If the lesion is accessible, aggressive surgical treatment may provide a cure; however, if there is a marked delay in diagnosis or if the tumor is inaccessible to excision, high grade chondrosarcoma of the pelvis and spine has a poor prognosis, with less than a 20% ten-year survival (entry 5).

§ 4-33(A). Prevalence and Etiology.

Chondrosarcoma makes up 17% to 22% of biopsied primary tumors of bone (entry 6). Among malignant tumors, chondrosarcoma is the third most common neoplasm, following multiple myeloma and osteogenic sarcoma. The usual age of onset is between 40 and 60 years of age and the ratio of men to women is 1:5.

The etiology of chondrosarcoma is unknown. Primary chondrosarcomas arise *de novo* from previously normal bone, while secondary chondrosarcomas develop from other cartilagenous tumors, such as an osteochondroma or enchondroma. Chondrosarcoma may be induced by irradiation, accounting for 9% of radiation-induced bone sarcoma (entry 4). They also develop in a small number of patients with Paget's disease, fibrous dysplasia, and Maffucci's syndrome (enchondromas with soft tissue hemangiomas) (entries 11, 3, 8).

§ 4-33(B). Clinical Findings.

The pelvis and vertebral column are involved in 34% of patients with chondrosarcoma and they may be symptomless or may present with only mild discomfort and palpable swelling. Tumors in the pelvis are detected when they are palpable through the abdominal wall or cause nerve compression with radicular pain, mimicking symptoms of a herniated disc (entries 6, 10).

§ 4-33(C). Physical Findings.

Physical examination may demonstrate a painless tumor mass or one that is mildly painful on palpation. Rectal examination may be helpful in detecting mass originating in the pelvis or sacrum. Neurologic examination may be abnormal if neural elements are compressed by the tumor.

§ 4-33(D). Laboratory Findings.

Laboratory findings may not be abnormal until late in

330

the course of the tumor. Abnormalities may correlate with tumor size and extent of metastasis. Useful tests to obtain include blood counts, serum chemistries, and sedimentation rates.

The microscopic appearance of chondrosarcoma is graded by the degree of abnormality in the nuclei of the cells. Grade 1 is the most benign form of chondrosarcoma and does not have any associated cellular atypia. Grades 2 and 3 are increasingly malignant, and are associated with higher degrees of cellularity, nuclear size and mitoses (entry 9). In a study of 152 pelvic chondrosarcomas, 56 were Grade 1, 53 were Grade 2 and 43 were Grade 3 (entry 6).

§ 4-33(E). Radiographic Findings.

The characteristic radiographic findings of chondrosarcoma include a well-defined lesion with expansile contours. The interior of the lesion may demonstrate lobular or fluffy calcification with scalloping of the interior cortex of bone. Periosteal and endosteal reactive bone formation leads to a thickened cortex of bone which is typical of a slow-growing tumor (entry 1). Cortical destruction and soft tissue invasion are indicative of more aggressive lesions. In the spine, the vertebral body or posterior elements may be the site of origin.

Plain radiographs are useful in detecting tumors which have calcified. However, soft tissue extension may not be appreciated on these radiographs. Computed tomography and/or arteriography may be useful in determining the extent of extraosseous involvement associated with a chondrosarcoma.

§ 4-33(F). Differential Diagnosis.

The diagnosis of Grade 1 chondrosarcoma must be based upon clinical, radiographic and pathologic findings. The

histologic appearance may be similar to a cellular
enchondroma, a benign lesion. Tumors with similar
histologic appearances may have different aggresive
properties. The findings of pain, rapid growth, cortical
destruction, soft tissue extension, and anaplastic cells on
biopsy are characteristic of higher grade, more malignant
chondrosarcoma. The other tumors which might be differentiated from chondrosarcoma include osteogenic sarcoma,
fibrosarcoma, giant cell tumor, and enchondromas.

§ 4-33(G). Treatment.

Surgery is the treatment of choice for chondrosarcoma.
En-bloc resection of the tumor with a margin of normal
tissue so that malignant cells are not implanted in the surgical wound offers the best chance of long-term survival.
The five-year survival for Grade 1, 2, and 3 pelvic
chondrosarcoma after excisional surgery was 47%, 38% and
15% respectively (entry 10). Tumors which are partially
resected frequently recur with increased cytologic malignancy (entry 2). Chondrosarcomas are radioresistant and
radiotherapy is reserved for tumors which are inaccessible
to excision. Chemotherapy may play an adjunctive role in
these difficult situations.

§ 4-33(H). Course and Outcome.

The patients with chondrosarcoma who present with pain
frequently have more malignant tumors (entry 7). Those
with low grade, well differentiated chondrosarcoma have a
longer survival rate and longer interval between treatment
and recurrence than those with higher grade tumors. These
facts hold true for chondrosarcoma of the pelvis, spine and
sacrum. Chondrosarcoma with low degree of malignancy
grows slowly, recurs locally, and metastasizes late. High
grade tumors grow rapidly and metastasize easily. The
patients with the best outcome are those with low grade

tumors and successful en-bloc excision. Patients with high grade pelvic chondrosarcomas have a poor outcome with a five-year survival of only 20%.

BIBLIOGRAPHY

1. Barnes, R., and Catto, M. Chondrosarcoma of bone. *J. Bone Joint Surg.*, 48B:729-764, 1966.
2. Dahlin, D.C., and Henderson, E.D. Chondrosarcoma: A surgical and pathological problem — review of 212 cases. *J. Bone Joint Surg.*, 38A:1025-1038, 1956.
3. Feintuch, T.A. Chondrosarcoma arising in a cartilagenous area of previously irradiated fibrous dysplasia. *Cancer,* 31:877-881, 1973.
4. Fitzwater, J.E., Caboud, H.E., and Farr, G.H. Irradiation-induced chondrosarcoma: A case report. *J. Bone Joint Surg.*, 58A:1037-1039, 1976.
5. Henderson, E.D., and Dahlin, D.C. Chondrosarcoma of bone: A study of 280 cases. *J. Bone Joint Surg.*, 45A:1450-1458, 1963.
6. Huvos, A.G. *Bone Tumors: Diagnosis, Treatment and Prognosis.* W.B. Saunders, Phila: 1979, pp. 206-232.
7. Kaufman, J.H., Douglass, H.O., Jr., Blake, W., Moore, R., and Rao, U.N.M. The importance of initial presentation and treatment upon the survival of patients with chondrosarcoma. *Surg., Gynecol. Obstet.,* 145:357-363, 1977.
8. Lewis, R.J., and Ketcham, A.S. Maffucci's syndrome: Functional and neoplastic significance — case report and review of the literature. *J. Bone Joint Surg.*, 55A:1465-1479, 1973.
9. Lichtenstein, L., and Jaffe, H.L. Chondrosarcoma of bone. *Am. J. Pathol.,* 19:553-574, 1943.
10. Marcove, R.C., Mike, V., Hutter, R.V.P., Huvos, A.G., Shoji, H., Miller, T.R., and Koslof, F.R. Chondrosarcoma of the pelvis and upper end of the femur: An analysis of factors influencing survival time in 113 cases. *J. Bone Joint Surg.*, 54A:561-572, 1972.
11. Thomson, A.D., and Turner-Warwick, R.T. Skeletal sarcomata and giant cell tumor. *J. Bone Joint Surg.*, 37B:266-303, 1955.

§ 4-34. Chordoma.

Chordoma is a malignant tumor which originates from the remnants of embryonic tissue, the notochord. The notochord is the structure which develops into a portion of the vertebral bodies of the spine in the embryo. Chordomas are located exclusively in the axial skeleton. These tumors are slow-growing and may be present for an extended period of time before symptoms secondary to the compression of vital structures brings the patient to a physician. Virchow in 1857 was the first to suggest the persistence of cells from the notochord in skeletal structures. Horwitz in 1941 proposed that chordomas arose from abnormal chordal remnants in vertebral bones (entry 4).

§ 4-34(A). Prevalence and Etiology.

Chordomas account for 3% of primary bone tumors (entry 7). The tumors usually become evident between the ages of 40 to 70, and the tumor is rarely reported in patients 30 years of age or younger. The ratio of men to women with sacrococcygeal chordomas is 3:1, while the ratio is 1:1 for chordomas in other locations in the spine.

The etiology of the factors which initiates the regrowth of notochordal vestigial cells in the spine is unknown although trauma has been suggested as a possible initiating factor (entry 9). At Memorial Hospital in New York, 15% of patients with sacrococcygeal chordoma gave a history of previous trauma to the low back of significant degree to require medical attention (entry 5). Whether trauma is a causative factor or a chance event in the pathogenesis of the lesion remains conjectural at this time.

§ 4-34(B). Clinical Findings.

Chordomas are located in the sacrum or lumbar spine in 60% to 70% of patients and the symptoms are dependent on the location and extent of the tumor. Patients with sacrococcygeal chordoma present with nondescript lower back pain. The pain may be characterized as dull, sharp, intermittent or constant and is localized in the sacrum. It may be of long duration since the patient may not have thought it to be a significant problem. Some patients present with severe constipation, urinary frequency, hesitancy, or dysuria, incontinence or muscular weakness (entry 1). These symptoms are secondary to direct pressure on pelvic structures or compression of neural elements. Patients with chordomas of the lumbar spine may experience pain in the hip, knee, groin, or sacroiliac region. This pattern of referred pain to the lower extremities may be confusing and delay the discovery of the true location of the spinal tumor.

§ 4-34(C). Physical Findings.

The rectal and neurologic examinations are helpful in detecting the presence of chordomas in the sacrum and spine. Chordomas of the sacrum extend anteriorly into the pelvis and the presacral extension of a chordoma is detected on rectal examination. The tumor is a firm, roundish mass which is palpable through the posterior wall of the rectum. Neurologic examination may demonstrate flaccidity or spasticity on muscle testing. Lesions in the cauda equina (lumbar spine) are associated with muscle flaccidity, while lesions higher in the axial skeleton are associated with muscle spasticity.

§ 4-34(D). Laboratory Findings.

Laboratory findings may be unremarkable early in the course of this tumor. Abnormalities in hematologic and chemical parameters may only appear late once the tumor has grown extensively or has metastasized.

On gross pathologic examination, a chordoma is a soft, lobulated, gelatinous tumor. In the vertebral column, it originates in the vertebral body and spreads either along the posterior longitudinal ligament or through the intervertebral disc. Histologically, chordomas are characterized by cells of notochordal origin, physaliphous cells. These cells contain a large, clear area of cytoplasm, with an eccentric, flattened nucleus; and form columns which are interspersed in fibrous tissue (entry 1). The tumor is also characterized by the production of large amounts of mucin and this histologic appearance bears a close resemblance to that of an adenocarcinoma.

§ 4-34(E). Radiographic Findings.

Sacrococcygeal and vertebral chordomas produce lytic bone destruction with calcific foci and a soft tissue mass (entry 3). Sacral chordomas produce destruction of several sacral segments with a presacral soft tissue mass. Soft tissue extension may be determined best by evaluation with intravenous pyelography, ultrasound, arteriography, venography or computed tomography (entry 11). Vertebral chordoma initially causes destruction of a vertebral body without intervertebral disc involvement. Subsequently, intervertebral discs become narrowed and opposing vertebral end plates are eroded (entry 10). Myelography is helpful in determining extradural extension of the tumor even in the absence of neurologic symptoms.

§ 4-34(F). Differential Diagnosis.

The diagnosis of chordoma is suggested by its location and radiographic features, but the definitive diagnosis is dependent on the examination of a biopsy specimen. Needle biopsy of a vertebral chordoma is usually adequate. Open biopsy is frequently required for sacral chordoma because of the tumor location. Other lesions which must be differentiated from chordoma include liposarcoma, chondrosarcoma, and metastatic adenocarcinoma.

§ 4-34(G). Treatment.

The definitive treatment for chordoma is en-bloc excision.

Unfortunately, because of tumor size at time of diagnosis and the approximation of the tumor to vital structures, partial resection may be the only surgical option. Sacrococcygeal tumors are best treated by surgical excision if the superior portion of the sacrum is uninvolved, followed by irradiation (entry 2). In patients with inaccessible sacral chordomas, radical radiation therapy may slow tumor growth (entry 8). Vertebral chordomas are treated by decompression laminectomy with excision of accessible tumor located in bone, soft tissues and the extradural space. Chemotherapy is usually ineffective (entry 6).

§ 4-34(H). Course and Outcome.

Chordomas are slowly growing tumors which metastasize in 10% of patients late in the course of the illness (entry 12). The five-year survival for sacrococcygeal tumors and vertebral tumors is 66% and 50% respectively. The ten-year survival ranges between 10% to 40% (entry 11). A few patients have survived with a chordoma for twenty years. These data support the impression that chordoma is a tumor with a wide spectrum of behavior, from slow indolent to rapidly progressive destructive growth. The prognosis of each patient must be determined by taking into account the location, pathological characteristics, and invasiveness of the tumor.

BIBLIOGRAPHY

1. Congdon, C.C. Benign and malignant chordomas: A clinico-anatomical study of twenty-two cases. *Am. J. Pathol.,* 28:793-821, 1952.
2. Dahlin, D.C., and MacCarthy, C.S. Chordoma: A study of fifty-nine cases. *Cancer,* 5:1170-1178, 1952.
3. Higinbotham, N.L., Phillips, R.F., Farr, H.W., and Husty, H.O. Chordoma: Thirty-five-year study at Memorial Hospital. *Cancer,* 20:1841-1850, 1967.

4. Horwitz, T. Chordal ectopia and its possible relationship to chordoma. *Arch. Pathol.*, 31:354-362, 1941.
5. Huvos, A.G. *Bone Tumors: Diagnosis, Treatment and Prognosis.* W.B. Saunders, Phila: 1979, pp. 373-391.
6. Kamrin, R.P., Potanos, J.N., and Pool, J.L. An evaluation of the diagnosis and treatment of chordoma. *J. Neurol. Neurosurg. Psychiatry*, 27:157-165, 1964.
7. Mirra, J.M. *Bone Tumors: Diagnosis and Treatment.* J.B. Lippincott Co., Phila: 1980, pp. 243-256.
8. Pearlman, A.W., and Friedman, M. Radical radiation therapy of chordoma. *Am. J. Roentgenol.*, 108:333-341, 1970.
9. Peyron, A., and Mellissinos, J. Chordome, tumeur tramatique. *Ann. Med. Legale*, 15:478-488, 1935.
10. Pinto, R.S., Lin, J.P., Firooznia, H., and LeFleur, R.S. The osseous and angiographic manifestations of vertebral chordomas. *Neuroradiology*, 9:231-241, 1975.
11. Sundaresan, N., Galicich, J.H., Chu, F.C.H., and Huvas, A.G. Spinal chordomas. *J. Neurosurg.*, 50:312-319, 1979.
12. Wang, C.C., and James, A.E., Jr. Chordoma: Brief review of the literature and report of a case with widespread metastases. *Cancer*, 22:162-167, 1968.

§ 4-35. Lymphomas.

Lymphomas are malignant diseases of lymphoreticular origin. They usually arise in lymph nodes, rarely initially in bone, and are classified into two major groups: Hodgkin's and non-Hodgkin's lymphoma. Hodgkin's lymphoma occasionally, and non-Hodgkin's lymphoma rarely, may present as back pain in an adult patient. Axial skeleton involvement may progress to spinal cord compression and associated neurologic symptoms.

§ 4-35(A). Prevalence and Etiology.

The incidence of Hodgkin's and non-Hodgkin's lymphomas is approximately between 40 to 60 cases per million per year. Primary Hodgkin's and non-Hodgkin's disease of bone unassociated with lymph node involvement are rare tumors occurring in less than 1% of biopsied tumors (entry 7). Most bone involvement is secondary to hematogenous spread or direct extension of the tumor (entry 10). A majority of patients who develop lymphomas are between the ages of 20 and 60 and the male to female ratio is approximately 2:1. Although viral infections have been implicated as the etiologic agents which result in lymphomas, the exact pathogenetic factors causing lymphoreticular malignancy remain to be identified.

§ 4-35(B). Clinical Findings.

Patients with primary Hodgkin's or non-Hodgkin's lymphoma of bone develop persistent pain over the affected bone. Bone pain may increase when the patient goes to bed. The pain often precedes the radiographic changes of lymphoma by months. A peculiar clinical finding relates to an increase of bone pain after the consumption of alcohol (entry 1). Radicular pain down the leg may occur with sacroiliac joint diseases or vertebral lesions associated with nerve compression. Occasionally bone lesions may be painless (entry 8).

§ 4-35(C). Physical Findings.

The bone affected by a primary lymphoma is tender to palpation. Soft tissue swelling may be associated with bony tenderness. Patients with axial skeleton disease may demonstrate neurologic deficits.

§ 4-35(D). Laboratory Findings.

The patient with primary disease of bone will not demonstrate hematologic, chemical and immunologic abnormalities associated with disseminated disease. The appearance of anemia, elevated sedimentation rate, and increased serum proteins in a patient with bone disease suggests that either the disease has extended to other tissue or that the bone lesion was secondary to disseminated disease and was late in its development.

339

The histologic picture of Hodgkin's disease includes typical Reed-Sternberg cells, atypical mononuclear cells and an inflammatory component composed of lymphocytes, plasma cells and scattered eosinophils. The reactive histiocytes and eosinophils look superficially like eosinophilic granuloma. In an older adult, a lesion that resembles eosinophilic granuloma may be Hodgkin's disease (entry 6). Non-Hodgkin's lymphomas may exhibit marked histologic variation (entry 9). These lesions lack Reed-Sternberg cells and demonstrate different combinations of abnormal lymphocytes and supporting cells.

§ 4-35(E). Radiographic Findings.

Both primary Hodgkin's and non-Hodgkin's lymphoma have a predilection for the axial skeleton (entry 5). The bone changes of Hodgkin's disease may include lytic (75%), sclerotic (15%), mixed (5%) or periosteal lesions (5%) (entry 4). In the axial skeleton, Hodgkin's disease most frequently involves a vertebral body and involvement of posterior elements of a vertebral body is much less common. Occasionally, an osteoblastic lesion, an "ivory" vertebra, may be seen with Hodgkin's disease (entry 2). Hodgkin's involvement of the axial skeleton may result in compression fractures which spare the vertebral discs. Non-Hodgkin's lymphoma of bone has similar features characterized by lytic and blastic areas, with cortical destruction and little reactive new bone.

§ 4-35(F). Differential Diagnosis.

The diagnosis of the lymphoma is based upon the careful examination of adequate biopsy material. Even with adequate histological material, however, the diagnosis of a specific lymphoma is a difficult one to make because of the pleomorphic forms of the disease. The differential diagnosis of a single osteoblastic vertebral body must include Paget's disease and carcinoma of the breast or prostate. In younger patients with vertebra plana the possibility of eosinophilic granuloma must be investigated.

§ 4-35(G). Treatment.

The treatment of lymphomas is based upon the extent of the illness. All patients must be staged before treatment is initiated. Once the stage of disease is known, the patient should receive appropriate therapy for that degree of involvement. Treatment may include radiation therapy and/or chemotherapy (entries 5, 3).

§ 4-35(H). Course and Outcome.

The therapy for lymphomas has become more effective in the control of the disease and patients have the potential for cure if the disease is not too extensive. In many circumstances, they are able to live productive lives for extended periods of time.

BIBLIOGRAPHY

1. Conn, H.O. Alcohol-induced pain as a manifestation of Hodgkin's disease. *Arch. Intern. Med.*, 100:241-247, 1957.
2. Dennis, J.M. The solitary dense vertebral body. *Radiology*, 77:618-621, 1961.
3. DeVita, V.T., Jr., Serpick, A.A., and Carbone, P.O. Combination chemotherapy in the treatment of advanced Hodgkin's disease. *Ann. Intern. Med.*, 73:881-895, 1970.
4. Granger, W., and Whitaker, R. Hodgkin's disease in bone, with special reference to periosteal reaction. *Br. J. Radiol.*, 40:939-948, 1967.
5. Huvos, A.G. *Bone Tumors: Diagnosis, Treatment and Prognosis.* W.B. Saunders, Phila:1979, pp. 392-412.
6. Jaffe, H.L. *Metabolic, Degenerative and Inflammatory Diseases of Bones and Joints.* Phila: Lea and Fegiber, 1972, p. 887.
7. Mirra, J.M. *Bone Tumors: Diagnosis and Treatment.* J.B. Lippincott Co., Phila: 1980, pp. 406-418.
8. Perttala, Y., and Kijanen, I. Roentgenologic bone lesions in lymphogranulomatosis maligna; Analysis of 453 cases. *Ann. Chir. Gynaecol. Fenn.*, 54:414-424, 1965.

9. Reimer, R.R., Chabner, B.A., Young, R.C., Reddick, R., and Johnson, R.E. Lymphoma presenting in bone: Results of histopathology, staging, and therapy. *Ann. Intern. Med.*, 87:50-55, 1977.
10. Steiner, P.E. Hodgkin's disease: The incidence, distribution, nature and possible significance of lymphogranulomatous lesions in bone marrow; review with original data. *Arch. Pathol.*, 36:627-637, 1943.

§ 4-36. Skeletal Metastases.

A principal characteristic of malignant neoplastic lesions is the growth of tumor cells distant from the primary lesion. These distant lesions are referred to as metastases and are found commonly in the skeletal system. Skeletal lesions result from either dissemination through the blood stream or by direct extension. The axial skeleton and pelvis are common sites of metastatic disease and certain tumors such as prostate, breast, lung, and kidney, have a predilection for skeletal metastasis. Radiographic findings associated with metastatic disease include lytic, blastic, and mixed bony changes. The symptoms and signs of metastatic disease are non-specific and their diagnosis usually requires a biopsy. While a cure of these lesions is extremely unusual, radiation therapy and surgical decompression can be effective in controlling pain, pathologic fractures, and spinal cord compression.

§ 4-36(A). Prevalence and Etiology.

Metastatic lesions in the skeleton are much more common than primary tumors of bone (entry 10) with the overall ratio being 25:1 (entry 16). Tumors which are frequently associated with skeletal metastases include prostate, breast, lung, kidney, thyroid, colon, neuroblastoma, and Ewing's sarcoma (entry 17). Data from autopsy material suggests that 70% of patients with a primary neoplasm will develop pathologic evidence of metastasis to vertebral bodies in the thoracolumbar spine (entry 9).

342

The propensity of bone and the axial skeleton, in particular, to be the site of metastases may be explained in part by the presence of Batson's plexus around the vertebral column and bone marrow inside bone. Batson's plexus is a network of veins located in the epidural space between the bony spinal column and the dura mater covering the spinal cord, and it is connected to the major veins which return blood to the heart, the inferior and superior vena cava. This plexus of veins is unique in that there are no valves to control blood flow and any increased pressure in the vena caval system results in increased flow into Batson's plexus. Metastatic cells may enter this plexus and be deposited in the venous and sinusoidal system of bone which are connected to Batson's plexus (entry 1). Supporting data for the importance of Batson's plexus in the distribution of skeletal metastasis is the frequency of axial skeleton metastases and the predilection of metastasis to the lumbar spine (entries 15, 11). The red bone marrow, located inside of vertebral bodies, long and flat bones, have a rich sinusoidal system. Sinusoidal vessels are usually under low hemodynamic pressure, allowing for pooling of blood. The pooling of blood along with other factors such as fibrin deposits and thrombosis may encourage tumor growth.

§ 4-36(B). Clinical Findings.

A high index of suspicion for the presence of metastasis is important in the evaluation of the patient with a prior history of malignancy or the adult over 50 years of age with back pain unassociated with trauma. Pain is usually the earliest and most significant symptom. The pain has a gradual onset and increases in intensity over time. It tends to be localized initially, but may radiate in a radicular pattern over time. Pain in the lumbar spine is commonly increased with motion, cough or strain. Radicular pain may increase at night as the spine lengthens with recumbency. Patients with spinal cord or nerve root compression secondary to bony or epidural lesions develop neurologic dysfunction which correlates with the location of the lesion. Neurologic symptoms may include numbness, tingling, unsteadiness of gait, weakness, bladder or bowel incontinence, or sexual dysfunction (entries 19, 8).

§ 4-36(C). Physical Findings.

Physical examination demonstrates pain on palpation over the affected bone. Muscle spasm and limitation of motion are associated findings. Careful attention to neurologic deficits (hyperesthesia, segmental muscular weakness, asymmetric reflexes, sphincter dysfunction) may help locate the lesion in the axial skeleton.

§ 4-36(D). Laboratory Findings.

Early in the course of the lesion, laboratory parameters may be unremarkable. However, subsequent evaluation may demonstrate anemia, elevated erythrocyte sedimentation rate, abnormal urinalysis or abnormal chemistries including increased serum alkaline phosphatase concentrations and increased prostatic acid phosphatase in metastatic prostate carcinoma. Therefore, initial negative laboratory data should not dissuade a physician from pursuing further diagnostic evaluation in an older patient with recent onset of back pain.

In patients without a known primary tumor, bone biopsy may provide the first evidence of a malignancy. Occasionally, the histologic features of the biopsy specimen may suggest the source of the lesion such as colloid material with thyroid adenocarcinoma, clear cells with hypernephroma, or melanin with melanoma. On many occasions, the histologic features, squamous cells or mucin cells, may be associated with primary lesions in various organs. Some lesions may be so undifferentiated that the pathologic findings offer no clue to the possible source of the tumor (entry 16).

§ 4-36(E). Radiographic Findings.

Radiographic abnormalities associated with the axial skeleton include osteolytic, osteoblastic or mixed lytic and blastic lesions (entry 22). Osteolytic lesions which affect vertebral body or a posterior element such as a pedicle are associated with carcinomas of the lung, kidney, breast and thyroid. Multiple osteoblastic lesions are associated with prostatic, breast, and colon carcinomas and bronchial carcinoid. A single blastic vertebral body may be seen with prostatic carcinoma, but is more closely associated with Hodgkin's disease or Paget's disease. Vertebral lesions which contain both lytic and blastic metastases are associated with carcinoma of the breast, lung, prostate or bladder. Kidney and thyroid carcinoma, due to their usual slow growth, may cause an expansile lesion from periosteal growth without destruction. Osteolytic lesions are more frequently associated with vertebral body collapse than are osteoblastic lesions. Vertebral body destruction is not associated with changes in the intervertebral disc so the presence of vertebral body destruction and loss of intervertebral disc space suggests infection. However, radiographic and pathologic studies suggest that intervertebral discs may degenerate more rapidly, indent weakened vertebral bone and form a Schmorl's node. Rarely, it may be invaded by tumor, and result in loss of disc integrity (entries 13, 18).

Early in the course of a metastatic lesion, plain roentgenographic examination will be unremarkable since between 30% and 50% of bone must be destroyed before a lesion is evident on plain radiographs (entry 7), however, scintigraphic examination with bone scan makes it possible to detect areas of symptomatic and asymptomatic bone involvement in up to 85% of patients with metastases (entries 8, 5). A bone scan may also suggest the presence of tumor in patients with coincident degenerative disease or osteoporosis. Computed tomography may also be useful in localizing lesions which are difficult to identify on plain radiographs (entry 21).

Myelography is the definitive diagnostic procedure for any patient with a metastatic lesion and spinal cord or nerve root compression. Injections of dye in two locations, lumbar and cisterna magna (cervical spine), may be necessary to locate all potential areas of neural compression since lesions in the lumbar spine may be accompanied with "silent" lesions in the proximal spine (entry 12). Myelography may also identify the location of the lesion to the extradural or intradural space.

§ 4-36(F). Differential Diagnosis.

In the patient with a known primary tumor who develops low back pain, a destructive spinal lesion is associated with the primary neoplasm in the vast majority of cases. These patients may not require a biopsy of the spinal lesion for diagnosis; however, patients with no known primary neoplasm who develop destructive lesions of the spine require a biopsy for tissue diagnosis. Closed needle biopsy of lesions in the lumbar spine can safely yield useful information (entry 6). Other conditions which must be included in the differential diagnosis include primary tumors of bone, infection including Paget's disease, and metabolic diseases including osteoporosis.

§ 4-36(G). Treatment.

Treatment of metastatic disease of the spine is directed toward palliation of pain. A cure is rarely possible since most solitary metastatic lesions are accompanied by a number of "silent" deposits which only become evident over time. The pain of the metastatic lesion of the spine may be secondary to bony destruction or pathologic fracture (entry 2). Therapy directed specifically at vertebral and spinal cord lesions may include radiation therapy, corticosteroids, or decompressive laminectomy. Radiotherapy may be used alone as primary treatment to decrease pain and slow growth, or as adjunctive therapy after surgical decompression (entries 14, 3). Metastatic lesions from the breast, thyroid and lymphomas are most sensitive to radiotherapy. Corticosteroids may help reduce edema and alleviate symptoms in patients with spinal cord compression (entry 4). Decompressive laminectomy is usually of no help in returning function to patients with long standing paraplegia, but it is recommended for those who have recently developed neurological symptoms (entry 20). Patients with breast and prostate carcinoma who undergo laminectomy improve to a greater degree than those with lung or kidney carcinoma (entry 15).

§ 4-36(H). Course and Outcome.

The course of each patient with skeletal metastasis is dependent on a number of factors: the type of tumor, extent of involvement, sensitivity to therapy, and degree of neurologic symptoms but, in general, the prognosis is poor.

BIBLIOGRAPHY

1. Batson, O.V. The function of the vertebral veins and their role in the spread of metastasis. *Ann. Surg.*, 112:138-149, 1940.
2. Bhalla, S.K. Metastatic disease of the spine. *Clin. Orthop.*, 73:52-60, 1970.
3. Bruckman, J.E., and Bloomer, W.D. Management of spinal cord compression. *Semin. Oncol.*, 5:135-140, 1978.
4. Clark, P.R.R., and Saunders, M. Steroid-induced remission in spinal canal reticulum cell sarcoma; Report of two cases. *J. Neurosurg.*, 42:346-348, 1975.

5. Craig, F.S. Metastatic and primary lesions of bone. *Clin. Orthop.*, 73:33-38, 1970.
6. Craig, F.S. Vertebral body biopsy. *J. Bone Joint Surg.*, 38A:93-102, 1956.
7. Edelstyn, G.A., Gillespie, P.G., and Grebbel, F.S. The radiological demonstration of skeletal metastases: Experimental observations. *Clin. Radiol.*, 18:158-162, 1967.
8. Fager, C.A. Management of malignant intraspinal disease. *Surg. Clin. North Am.*, 47:743-750, 1967.
9. Fornasier, V.L., and Horne, J.G. Metastases to the vertebral column. *Cancer*, 36:590-594, 1975.
10. Francis, K.C., and Hutter, R.V.P. Neoplasms of the spine in the aged. *Clin. Orthop.*, 26:54-66, 1963.
11. Galasko, C.S.B., and Doyle, F.H. The detection of skeletal metastases from mammary cancer: A regional comparison between radiology and scintigraphy. *Clin. Radiol.*, 23:295-297, 1972.
12. Gilbert, R.W., Kim, J.H., and Posner, J.B. Epidural spinal cord compression from metastatic tumor: Diagnosis and treatment. *Ann. Neurol.*, 3:40-51, 1978.
13. Hubbard, D.D., and Gunn, D.R. Secondary carcinoma of the spine with destruction of the intervertebral disc. *Clin. Orthop.*, 88:86-88, 1972.
14. Khan, F.R., Glicksman, A.S., Chu, F.C.H., and Nickson, J.J. Treatment by radiotherapy of spinal cord compression due to extradural metastases. *Radiology*, 89:495-500, 1967.
15. Lenz, M., and Fried, J.R. Metastasis to skeleton, brain, and spinal cord from cancer of the breast and effects of radiotherapy. *Ann. Surg.*, 93:278-293, 1931.
16. Mirra, J.M. *Bone Tumors: Diagnosis and Treatment.* J.B. Lippincott Co., Phila: 1980, pp. 448-454.
17. Murray, R.O., and Jacobsen, H.G. Metastatic disease of the skeleton in *The Radiology of Skeletal Disorders*, 2nd ed., Edinburgh: Churchill Livingstone, 1977, p. 586.
18. Resnick, D., and Niwayama, G. Intervertebral disc abnormalities associated with vertebral metastasis: Observations in patients and cadavers with prostatic cancer. *Invest. Radiol.*, 13:182-190, 1978.
19. Rodriguez, M., and Dinapoli, R.P. Spinal cord compression with special reference to metastatic epidural tumors. *Mayo. Clinic. Proc.*, 55:442-448, 1980.
20. Vieth, R.G., and Odom, G.L. Extradural spinal metastases and their neurosurgical treatment. *J. Neurosurg.*, 23:501-508, 1965.

21. Wilson, J.S., Korobkin, M., Genant, H.K., and Bovill, E.G. Computed tomography of musculoskeletal disorders. *Am. J. Roentgenol.*, 131:55-61, 1978.

22. Young, J.M., and Fung, F.J., Jr. Incidence of tumor metastasis to the lumbar spine: a comparative study of roentgenographic changes and gross lesions. *J. Bone Joint Surg.*, 35A:55-64, 1953.

§ 4-37. Intraspinal Neoplasms.

While bone is the tissue in the axial skeleton most frequently affected by primary and metastatic tumors, less commonly, tissues inside the spinal column may also be affected by neoplastic processes. These intraspinal neoplasms may be extradural, between bone and the covering of the spinal cord, the dura; intradural-extramedullary, between the dura and the spinal cord; and intramedullary, in the spinal cord proper. Extradural tumors are most commonly metastatic in origin. Intradural-extramedullary tumors are predominantly meningiomas, neurofibromas, or lipomas. Intramedullary tumors are ependymomas and gliomas. Symptoms of intraspinal neoplasms range from severe pain with rapid neurologic dysfunction to minimal pain associated with slowly progressive sensory loss or motor weakness. Radicular symptoms in these patients are frequently ascribed to a herniated lumbar intervertebral disc (entry 12). Appropriate evaluation of spinal fluid and myelographic studies help differentiate patients with intraspinal neoplasms from those with disc disease. While radiographic features of intraspinal neoplasms help localize the lesion and suggest its origin, definitive diagnosis requires histologic examination of the lesion. Intradural-extramedullary tumors are usually benign and have a good prognosis. Extradural and intramedullary lesions are rapidly growing or locally invasive, and are associated with a poorer prognosis.

§ 4-37(A). Prevalence and Etiology.

Extradural tumors are metastatic lesions which have invaded the intraspinal space from contiguous structures. These are the most common tumors of the spinal canal. Of the intradural lesions, extramedullary neoplasms occur more commonly than intramedullary neoplasms. Intraspinal neoplasms occur in adults over the age of 20, with a predominance in patients between 30 and 50 years.

The extradural, or epidural, space is the predominant site for intraspinal malignant tumors. Metastatic tumors in the spinal canal are extradural in location because the dura is resistant to invasion from lesions which extend from foci in vertebral bone. The extradural space is also the location of Batson's plexus which is a site for hematogenous spread of tumor (entry 2). Major structures in the intradural-extramedullary space are meninges and spinal nerve roots. Meningiomas and neurofibromas arise from these structures. In one study, 25% of intraspinal neoplasms were meningiomas. Intramedullary tumors arise in the spinal cord itself and are composed of cells that make up the support structure of the cord, ependymal and glial cells. Metastatic lesions to the spinal cord are extremely rare (entry 7).

§ 4-37(B). Clinical Features.

Intraspinal tumors may demonstrate a wide variety of clinical symptoms. Patients with extradural metastatic disease have pain as their initial complaint. Pain may localize to the affected area in the spine or may radiate to the lower extremities if neural elements are compressed. The pain characteristically increases in intensity and is unrelenting. The pain is increased at night with recumbency due to the lengthening of the spine. Activity may also exacerbate the discomfort. The pain is unresponsive to mild analgesics and requires narcotics for control. Not uncommonly, neurologic dysfunction rapidly follows axial pain. Neurologic symptoms include weakness, loss of sensation and incontinence.

Intradural-extramedullary tumors grow in proximity to nerve roots and are associated with radicular pain or axial skeletal pain. Meningiomas and neurofibromas are slow growing tumors and this corresponds with slow evolution of symptoms in these patients. Nocturnal symptoms are increased in these patients. Activity during the day may not be associated with symptoms. Neurologic symptoms are slower to develop than in patients with metastatic disease.

Intramedullary tumors are painless. Neurologic dysfunction in the form of sensory deficits, weakness and incontinence is the hallmark of these tumors.

§ 4-37(C). Physical Findings.

Examination of patients with extradural tumors may demonstrate pain on palpation with associated muscle spasm and limitation of motion. Neurologic findings correspond to the level and extent of compression on the spinal roots and cord. Patients with intradural-extramedullary tumors demonstrate slowly changing neurologic abnormalities including gait disturbance, sensory changes and urinary or rectal incontinence. They may also demonstrate lower extremity muscle atrophy. Patients with multiple neurofibromas may demonstrate spinal angulation and kyphosis (entry 18). Patients with intramedullary tumors may have specific sensory changes which correlate with the location of these tumors in the center of the cord. Light touch and position sensation are normal, while pain and temperature sensation are lost. Hyper-reflexia is a result of pressure on the pyramidal tracts. This finding helps differentiate patients with intramedullary tumors from those with herniated disc where hyper-reflexia is a distinctly unusual finding (entry 12). Hyper-reflexia may also be associated with spasticity.

§ 4-37(D). Laboratory Findings.

Abnormal laboratory values are most closely associated with extradural metastatic lesions. The location, degree of spread, and histologic type of tumor will have an effect on the pattern of laboratory abnormalities. Intradural tumors do not metastasize outside of the spinal canal and are not associated with abnormal hematologic or chemical factors. Evaluation of cerebrospinal fluid obtained by lumbar puncture may demonstrate marked elevation in spinal fluid protein. The fluid is usually obtained during myelographic study in a patient with suspected intraspinal tumor.

Histologic findings depend on the cell of origin of the tumor. The most common primary tumor causing extradural metastases in women is carcinoma of the breast while carcinoma of the lung is most common in men (entries 16, 1). Meningiomas are encapsulated, nodular soft tumors with a wide range of histologic patterns including meningothelial, fibroblastic and metaplastic types. Neurofibromas may form fusiform swelling of a nerve root or a pedunculated mass. Intraspinal neurofibromas may take on a dumb-bell form with a central mass inside the vertebral canal connected by a shaft of tumor passing through the intervertebral foramen forming a peripheral mass. Histologic patterns of neurofibromas may vary but usually contain fibrous tissue in an interlacing configuration. Ependymomas arise from cells which line the central ventricular system of the spinal cord. Gliomas arise from glial cells which are the supporting cells for the nerve cells of the nervous system. Gliomas may be peripherally or centrally located in the spinal cord. Ependymomas and gliomas are associated with a variety of histologic forms.

§ 4-37(E). Radiographic Findings.

Radiographic abnormalities of extradural tumors are characterized by destruction of bone in proximity to the growing lesion. Malignant tumors are associated with rapid destruction of bone with loss of posterior elements of vertebral body or vertebral body collapse. Intraspinal lesions are associated with posterior scalloping of the vertebral bodies, a consequence of their location and slow growth (entry 13). Neurofibromas may grow through intervertebral foramina which results in uniform dilation when compared to adjacent foramina.

Myelographic studies are most useful in determining the exact location of an intraspinal tumor (entry 14). Extradural tumors frequently cause a complete block of the myelographic dye at the point of spinal cord compression. The block has irregular edges, has varying radiographic densities, and displaces the spinal contents. Intradural-extramedullary lesions produce a sharp, smooth outline since the tumor is in direct contact with the dye (entry 15). The spinal cord may be displaced to one side and spinal nerve roots may be stretched over the lesion. Arachnoid diverticula, located in the intradural space, will be recognized on myelographic examination (entry 5). Intramedullary tumors arise in the spinal cord. The myelogram demonstrates fusiform enlargement of the spinal cord with tapering of the column of dye superiorly and inferiorly. Not all fusiform swellings of the cord are secondary to intramedullary tumors. Extradural tumors may flatten the contralateral aspect of the cord. Therefore, films must be taken at 90 degree angles so that intramedullary lesions are not confused with extradural lesions.

§ 4-37(F). Differential Diagnosis.

The diagnosis of an intraspinal tumor is suggested by the

presence of back pain which is persistent and increased by recumbency, neurologic dysfunction and myelographic abnormalities. The definitive diagnosis requires histologic confirmation. A high level of suspicion is necessary to make this diagnosis. Patients with intradural tumor may present with lower extremity symptoms of long duration. This association must be kept in mind or the correct diagnosis may be missed (entry 15).

Non-neoplastic lesions of the spinal cord may mimic symptoms and signs of intraspinal tumors. High dose corticosteroid therapy may cause the accumulation of fat in an epidural location. Epidural lipomatosis can cause neurologic symptoms of lower extremity weakness and loss of sensation. With cessation of steroid therapy, however, the epidural fat deposits may disappear with resolution of symptoms (entry 4).

Syringomyelia is a fluid-filled cyst lined with benign glial cells which is located in the central portion of the spinal cord. Syringomyelia, by its growth laterally and longitudinally through the spinal cord, first causes loss of pain sensation, then abnormalities of motor function with weakness in the extremities and scoliosis in the axial skeleton, long tract signs (Babinski reflex) and autonomic dysfunction of the bladder and rectum. This lesion is found most commonly in the cervical spine and only occasionally in the lumbar spine (entry 11). Diagnosis is made by myelography or computerized axial tomography (entry 8). The treatment of syringomyelia is surgical removal of the fluid in the cystic cavity by needle aspiration or myelotomy (entry 17). Without effective therapeutic intervention to halt the progressive growth of the lesion, patients with syringomyelia suffer marked disability with spastic paraplegia, arthropathy, and infectious complications.

§ 4-37(G). Treatment.

Therapy directed at extradural metastatic lesions may include radiation therapy, corticosteroids or decompressive laminectomy (entries 10, 3, 6). The treatment of intradural-extramedullary tumors is complete surgical removal. Intramedullary tumors accessible to surgical excision should be removed (entry 9). Some physicians suggest post-surgical radiation therapy of the spinal cord.

§ 4-37(H). Course and Outcome.

In general, extradural and intramedullary tumors are malignant while intradural-extramedullary tumors are benign. The course and prognosis of these tumors corresponds to their invasiveness, rapidity of growth and location. Extradural and intramedullary tumors have poor prognoses while intradural-extramedullary tumors may be cured with surgical removal.

BIBLIOGRAPHY

1. Barron, K.D., Hirano, A., Araki, S., and Terry, R.D. Experiences with metastatic neoplasms involving the spinal cord. *Neurology*, 9:91-106, 1959.
2. Batson, O.V. The function of vertebral veins and their role in the spread of metastases. *Ann. Surg.*, 112:138-149, 1940.
3. Bruckman, J.E., and Bloomer, W.D. Management of spinal cord compression. *Semin. Oncol.*, 5:135-140, 1978.
4. Butcher, D.L., and Sahn, S.A. Epidural lipomatosis: A complication of corticosteroid therapy. *Ann. Intern. Med.*, 90:60, 1979.
5. Cilluffo, J.M., Gomez, M.R., Reese, D.F., Onofrio, B.M., and Miller, R.H. Idiopathic congenital spinal arachnoid diverticula: clinical-diagnosis and surgical results. *Mayo Clin. Proc.,* 56:93-101, 1981.
6. Clark, P.R.R., and Saunders, M. Steroid-induced remission in spinal canal reticulum cell sarcoma. Report of two cases. *J. Neurosurg.,* 42:346-348, 1975.
7. Edelson, R.N., Deck, M.D.F., and Posner, J.B. Intramedullary spinal cord metastases: Clinical and radiographic findings in nine cases. *Neurology*, 22:1222-1231, 1972.

8. Gonsalves, C.G., Hudson, A.R., Horsey, W.J., and Tucker, W.S. Computed tomography of the cervical spine and spinal cord. *Comput. Tomogr.*, 2:279-293, 1978.

9. Greenwood, J. Surgical removal of intramedullary tumors. *J. Neurosurg.*, 26:275-282, 1967.

10. Khan, F.R., Glicksman, A.S., Chu, F.C.H., and Nickson, J.J. Treatment by radiotherapy of spinal cord compression due to extradural metastases. *Radiology*, 89:495-500, 1967.

11. McIlroy, W.J., and Richardson, J.C. Syringomyelia: A clinical review of 75 cases. *Can. Med. Assoc. J.*, 93:731-734, 1965.

12. McKraig, W., Svien, H.J., Dodge, H.W., Jr., and Camp, J.D. Intraspinal lesions masquerading as protruded lumbar intervertebral discs. *JAMA*, 149:250-253, 1952.

13. Mitchell, G.E., Lourie, H., and Berne, A.S. The various causes of scalloped vertebrae with notes on their pathogenesis. *Radiology*, 89:67-74, 1967.

14. Resnick, D., and Niwayama, G. *Diagnosis of Bone and Joint Disorders.* W.B. Saunders Co., Phila: 1981, pp. 432-445.

15. Robinson, S.C., and Sweeney, J.P. Cauda equina lipoma presenting as acute neuropathic arthropathy of the knee. A case report. *Clin. Orthop.*, 178:210-213, 1983.

16. Vieth, R.G., and Odom, G.L. Extradural spinal metastases and their neurosurgical treatment. *J. Neurosurg.*, 23:501-508, 1965.

17. Wetzel, N., and Davis, L. Surgical treatment of syringomyelia. *Arch. Surg.*, 68:570-573, 1954.

18. Winter, R.B., Moe, J.H., Bradford, D.S., Lonstein, J.E., Pedras, C.V., and Weber, A.H. Spine deformity in neurofibromatosis: A review of 102 patients. *J. Bone Joint Surg.*, 61A:677-694, 1979.

PART V. ENDOCRINOLOGIC AND METABOLIC DISORDERS OF
THE LUMBOSACRAL SPINE

§ 4-38. Introduction.

Endocrinologic and metabolic disorders are systemic illnesses which affect components of the musculoskeletal system throughout the body. Those illnesses which produce symptoms of low back pain and characteristic laboratory and radiologic abnormalities include osteoporosis, osteomalacia, parathyroid disease, ochronosis, acromegaly, and microcrystalline disorders. Patients with these systemic diseases may present to a physician with low back pain as their initial or primary complaint and a minor traumatic event is frequently thought to be the cause. Evaluation by the physician, however, demonstrates compression fractures of the vertebral bodies secondary to inadequate bone stock, extensive degenerative disease of the spine or accumulation of crystals, amino acids, or mucopolysaccharides in bone. A characteristic finding of endocrinologic and metabolic disorders is that, although symptoms may be limited to the axial skeleton, these illnesses cause bone changes in many locations. The bone abnormalities may be subtle, but can be discovered if looked for carefully.

The history of back pain in patients with endocrinologic and metabolic diseases is usually non-specific. Back pain is located predominantly in the lumbosacral spine with occasional radiation into the lower extremities. Pain may be insidious in onset or acute, associated with a vertebral body compression fracture or acute gouty arthritis of the sacroiliac joint. Patients frequently have symptoms of bone or joint pain in other locations. In addition, the systemic manifestations of the endocrinologic or metabolic bone disorder may include symptoms of muscle weakness, renal stones, gastrointestinal malabsorption or change in facial configuration. Physical examination may demonstrate the systemic quality of the underlying disorder. Not only will the patient have back pain with percussion, loss of height, and dorsal kyphosis; they may also demonstrate muscle weakness, peripheral joint inflammation, tetany, pigmentation of cartilage, enlargement of the hands and jaw, or tophaceous deposits. Laboratory evaluation may be very helpful in making a specific diagnosis of the underlying disorder.

Abnormalities of calcium and phosphorous may raise the suspicion of osteomalacia or hyperparathyroidism. Bone biopsy confirms the diagnosis of osteoporosis or osteomalacia. Measurement of parathormone confirm the diagnosis of hyper or hypoparathyroidism. Non suppressibility of growth hormone during a glucose tolerance test is characteristic of acromegaly. Detection of homogenetisic acid in urine is diagnostic of ochronosis. Detection of urate or calcium pyrophosphate dihydrate crystals is essential for the diagnosis of gout or pseudogout, respectively. Radiographic findings may be prominent; decreased bone stock, periosteal resorption, disc calcification, widened joint space, or bone erosions, but they are non-specific. Although radiologic evaluation is not diagnostic, it is important in documenting the global involvement of the skeletal system with these diseases.

Therapy is tailored for each endocrinologic or metabolic illness. Calcium and vitamin D supplements are useful in patients with osteoporosis or osteomalacia. Surgical ablation of parathyroid or pituitary tumors is the therapy of choice for hyperparathyroidism or acromegaly respectively. Episodes of acute microcrystalline arthritis are benefitted by courses of nonsteroidal anti-inflammatory drugs.

Disability associated with endocrinologic or metabolic diseases is caused by progressive loss of bone mass, fracture, angular deformity or extensive degenerative disease of the intervertebral discs and vertebral joints. Progressive disability may be prevented by early diagnosis and treatment, particularly in metabolic bone diseases. Patients who have progressive metabolic bone disease are at risk for multiple fractures, deformity, and incapacitating bone pain. Patients with acromegaly may develop progressive degenerative changes of the spine despite ablation of their pituitary tumor. Patients with acromegaly or ochronosis may develop axial skeletal disease which limits their ability to perform their job.

§ 4-39. Osteoporosis.

Osteoporosis, a metabolic bone disease related to several different disorders, is associated with loss of bone mass per unit volume even though the ratio of bone mineral content and bone matrix remains normal. The reduction of bone mass occurs predominantly in the axial skeleton, femoral necks, and pelvis. Loss of bone mass in the axial skeleton predisposes vertebral bodies to fracture, back pain and deformity. Patients with osteoporosis may sustain multiple vertebral fractures over the years, with persistent mechancial pain and limited work potential. The diagnosis of osteoporosis is usually made on clinical grounds in a patient with a history of back pain and radiographic findings consistent with the disease. Other significant causes of osteoporosis, including tumors, hematologic disease, endocrinologic disorders, and drugs, must be considered in patients whose history, physical examination, or laboratory findings are not consistent with idiopathic osteoporosis. The therapy of osteoporosis is directed at slowing down the loss of bone mass, and calcium and vitamin D supplements, estrogens, sodium fluoride, and physical activity are used as tolerated. The major disability of this disease is derived from the fragility of bone, particularly of the axial skeleton and femoral neck. Fractures may occur with minor stresses such as lifting, jumping, or riding in a vehicle over a bumpy road. Therefore, patients with osteoporosis should be limited to light duty so as to decrease the risk of additional fractures.

§ 4-39(A). Prevalence and Etiology.

The absolute prevalence of osteoporosis is unknown. Patients who are symptomatic or who have radiographic changes of osteoporosis are easily recognized; however, many of them are asymptomatic and have not come to medical attention. By radiographic criteria, 29 per cent of women and 18 per cent of men between 45 to 79 years of age have osteoporosis (entry 4). By a more sensitive method for determining vertebral bone mineral density, however, 50 per cent of women past age 65 may have asymptomatic osteoporosis (entry 15). Bone resorption increases with age, with bone loss normally beginning after age 40. In general, males have more bone mass than females, and blacks have more than whites. These facts correlate with the clinical finding of osteoporosis being more common in Caucasian women, particularly of Northern European extraction. In about 3 per cent of patients with osteoporosis, the disease is progressive, disabling, and an impediment to the activities of daily living (entry 21).

The cause of increased bone resorption on the endosteal (inner) surfaces of bone, with inadequate compensatory bone formation, osteoporosis, is not well understood. Bone is living tissue which is constantly undergoing remodelling. Before adulthood, during growth, bone formation is greater than bone resorption. In the adult skeleton, up to the age of 30, the deposition and resorption of bone are equal and bone mass in both men and women is maximum at this age. Soon after, men start losing 0.3 per cent of their skeletal bone calcium per year. In women, the rate is initially the same as in men but at menopause it increases to 3 per cent per year (entry 8). Patients with osteoporosis are individuals who either mature with a decreased skeletal mass or have an accelerated rate of bone loss after achieving peak bone density.

There are many causes of osteoporosis and they are listed in Table I; (entry 10). Primary osteoporosis includes patients with postmenopausal or senile osteoporosis and those with idiopathic, adult, or juvenile osteoporosis. Senile osteoporosis is found in aging men, while postmenopausal osteoporosis occurs in women after menopause.

Postmenopausal osteoporosis is the most common form of this illness. Juvenile idiopathic osteoporosis occurs in early adolescence and is manifested by fragility in the axial skeleton. The disease may be severe and may cause vertebral fractures, kyphosis and loss of height. Remissions can occur spontaneously, but the patients seldom achieve normal adult bone mass.

Idiopathic adult osteoporosis is found in men under 50 years of age. This illness is often associated with idiopathic hypercalcemia and the axial skeleton is the area most severely affected.

Secondary forms of osteoporosis may be caused by specific abnormalities such as hormonal imbalance which affects bone metabolism (entry 5). Endocrine abnormalities which can cause osteoporosis should be investigated in any young or middle aged person with decreased skeletal mass. Hypogonadism with decreased sex hormones can cause bone loss in both men and women and this also plays a role in postmenopausal osteoporosis, particularly in women who undergo oophorectomy at an early age.

Adrenocortical hormone excess leads to decreased bone mass. Glucocorticoids reduce bone calcium by decreasing intestinal absorption of calcium, by stimulating parathormone so as to increase bone resorption and by exerting a direct inhibitory effect on bone metabolism. The same effect on bone mass occurs whether the source of glucocorticoids is endogenous (Cushing's syndrome) or exogenous (cortisone). Hyperthyroidism, by increasing general metabolic levels, increases bone turnover and remodelling. Parathormone exerts a direct effect on bone causing an increase in resorption. Osteoporosis in hyperparathyroidism may either develop because of parathyroid tumors or from illnesses which result in parathyroid gland hypertrophy. Whether or not diabetes contributes to the loss of bone mass is controversial, but studies suggest that bone density is decreased in patients with this illness (entry 13). Growth hormone is important for skeletal growth and maturation, and deficiencies are associated with decreased bone formation and mass.

The mineralization of bone is a complicated process involving a balance between calcium, phosphate, PH levels, hormones, and other factors. A nutritional deficiency or excess of any one of them may have a profound effect on bone mineralization. Low calcium consumption and decreased bone density are typical in women over 45 years of age (entry 1). The ratio of calcium to phosphorous or phosphate is altered by fluctuations in phosphorus concentration in nutritional sources and this may lead to bone resorption (entry 12). Vitamin D in its activated form, 1, 25-dehydroxyvitamin D_3, helps maintain serum calcium and phosphate levels by increasing both absorption of these substances from the intestine and their resorption from bone. Sources of vitamin D include fortified milk and fish along with the endogenous sources created by exposure of skin to the ultraviolet rays in sunlight. Osteoporosis may also be caused by deficiencies of protein, ascorbic acid (vitamin C), as well as diseases such as coeliac disease which result in intestinal malabsorption.

With extended use, certain drugs can lead to osteoporosis but the mechanisms are unknown. They include the anticoagulant heparin, anticonvulsants like diphenylhydantoin and methotrexate, an antimetabolite used in cancer chemotherapy. Chronic alcoholism is a common cause of bone loss in young men. Osteoporosis in alcoholics is probably related to poor dietary habits and decreased intestinal absorption.

Genetic causes of osteoporosis are rare. Osteogenesis imperfecta is a severely deforming congenital disease occurring in children and to a lesser extent in adults and characterized by skeletal fragility, fractures, blue sclerae, and deafness. Specific disorders of collagen metabolism also result in weakened bones. Nomocystinemia, a very rare heritable disorder related to a deficiency in the enzyme cystathionine B-synthetase, causes mental retardation, tall stature, scoliosis, dislocated lens, and skeletal fragility.

A number of unrelated disorders may also cause osteoporosis. Increased blood flow related to the inflammation in rheumatoid arthritis causes osteoporosis, particularly around joints. Immobilization results in focal (cast) or general (bed rest) osteoporosis. Mechanical stress increases bone mass (entry 19). Prolonged immobilization causes a proportionate loss of bone matrix and mineral and elderly patients who are chronically ill and are placed in bed for extended periods are at great risk. Malignancies, particularly multiple myeloma and other hematologic neoplasms, cause local or general osteoporosis. Acidosis causes increased calcium resorption and if present on a chronic basis, results in osteoporosis.

§ 4-39(B). Clinical Findings.

Patients with vertebral osteoporosis may be asymptomatic, with the diagnosis being made on radiographs taken for other purposes. Symptomatic osteoporosis presents as midline back pain localized over the thoracic or lumbar spine, the most common location for fractures. Most vertebral body fractures occur after some mechanical stress such as slipping on a stair, lifting or jumping. The patient with an acute fracture has severe pain localized over the affected vertebral body.

Occasionally the pain radiates into the flanks, upper portion of the posterior thighs, or abdomen. Spasm of the paraspinous muscles contributes to the back pain. Back motion aggravates the discomfort and many patients attempt to reduce the pain by remaining motionless in bed, frequently lying on their side. Prolonged sitting, standing, and the valsalva maneuver intensify the pain. Severe pain usually lasts three to four months and then resolves; however, some patients are left with persistent, nagging, dull spinal pain after vertebral body fracture secondary to osteoporosis and it may persist even in the absence of new fractures on radiographs. The source of this may be due to microfractures too small to be detected by radiographs or from biomechanical effects of the deformity on the lumbar spine below. Vertebral fractures of the dorsal spine results in anterior compression of the vertebral body. This kyphotic deformity (dowager's or widow's hump) can cause a compensatory flattening of the lumbar lordosis and low back pain. Multiple compression fractures may also cause persistent pain due to mechanical stress on ligaments, muscles and apophyseal joints. They also result in a loss in height and this is one of the parameters used to follow the progression of the disease. Since the disease only affects the anterior components of the vertebral bodies, neural compression is not associated with vertebral osteoporosis. Neurologic or radicular symptoms distant from an area of fracture is unusual and if present, other pathologic conditions must be considered.

§ 4-39(C). Physical Findings.

Physical findings including pain over the spinous process of the fractured vertebral body on percussion and associated spasm of surrounding paraspinous muscles are common. Patients with multiple fractures may demonstrate a loss of height and kyphoscoliosis. Neurologic examination is normal. Patients with severe pain secondary to acute fracture may demonstrate a loss of bowel sounds, ileus, on abdominal examination or bladder distention secondary to acute urinary retention.

§ 4-39(D). Laboratory Findings.

Laboratory parameters in primary osteoporosis, such as serum calcium, phosphate, and alkaline phosphatase, are normal. Urinary concentrations of calcium and hydroxyproline, an index of collagen breakdown, are also normal. The presence of anemia, elevated erythrocyte sedimentation rate, rheumatoid factor, or increased serum calcium, suggests a secondary form of osteoporosis and requires further investigation.

§ 4-39(E). Radiographic Findings.

Radiographic abnormalities of vertebral osteoporosis include changes in radiolucency of bone, trabecular pattern and shape of vertebral bone. Early osteoporosis may not be demonstrated on radiographs until 30 per cent of bone mass has been lost (entry 3). An early finding is the accentuation of the mineral density of the vertebral end plates since osteoporosis causes increasing lucency of the central portion of the vertebral body; however, the outer dimensions of the body remain the same with sharply defined margins (entry 20). Horizontal trabeculae are thinned, accentuating the vertical trabeculae which remain. This pattern is similar to that seen with hemangiomas. Prominent vertical trabeculae are limited to one vertebral body with a hemangioma while multiple vertebrae are affected with osteoporosis. Osteoporosis may progress to the point where vertebral body contours may be affected and "fish" or "codfish" vertebrae occur when intervertebral discs expand into weakened vertebral bone, causing an exaggerated biconcavity. "Fish vertebrae" are particularly common in the lower thoracic and upper lumbar spine.

An anterior wedge compression fracture is manifested by a decrease in anterior height, usually 4 mm. or greater, compared to the vertical height of the posterior body. A transverse compression fracture results in equal loss of both anterior and posterior heights. Many patients will demonstrate radiographic changes of "fish," anterior wedging, and compression vertebrae in the thoracic and lumbar spine. The changes in osteoporotic vertebrae are unevenly distributed along the spine (entry 6), with no two affected exactly alike. "Fish" vertebrae may occur asymptomatically. Anterior wedging and compression in osteoporosis indicate a fracture of the vertebral body and are associated with sudden, severe episodes of incapacitating back pain.

§ 4-39(F). Differential Diagnosis.

In most patients with diffuse osteoporosis of the axial skeleton, vertebral compression fractures and no abnormalities on laboratory evaluation, the diagnosis of osteoporosis is made on clinical grounds alone. Suspicion for the diagnosis is also heightened if the patient has any risk factors which are associated with osteoporosis, such as sedentary lifestyle, low calcium intake, early menopause or oophorectomy, cigarette smoking, excessive consumption of alcohol, protein, or caffeine, and a slender build or family history of osteoporosis (entry 9).

Secondary causes of osteoporosis as listed in Table 1 (at end of this section) should be considered in patients with abnormal laboratory findings such as anemia, elevated erythrocyte sedimentation rate, abnormal serum proteins, or hypovitaminosis D. Osteoporosis may be complicated by osteomalacia in 8% of postmenopausal women. Osteomalacia is a metabolic bone disease characterized by decreased bone mineralization with normal bone matrix usually secondary to inadequate vitamin D. Osteomalacia is usually associated with normal or low concentrations of serum calcium, decreased concentration of phosphorus, and increased levels of alkaline phosphatase. Osteoporosis and osteomalacia may be indistinguishable by clinical and radiologic criteria. Bone biopsy is useful in detecting the presence of osteomalacia complicating osteoporosis and is indicated when the diagnoses are in doubt since specific therapy is available for osteomalacia. Other methods for measuring bone calcium include dual photon absorptiometry, neutron activation analysis, and quantitative computed tomography; however, these methods remain experimental and are not utilized to predict patients at risk for osteoporosis with decreased bone mass in usual clinical settings.

§ 4-39(G). Treatment.

The treatment of osteoporosis should be primarily directed at young adult women to help prevent the disease. Preventative factors include a high calcium intake, at least 1000 mg. per day, regular exercise, and avoidance of excessive protein, alcohol, smoking, and caffeine. In postmenopausal women, preventive measures may also be helpful in slowing bone calcium loss. These measures include increased calcium intake, estrogens, and regular exercise.

Patients who sustain an acute compression fracture experience severe pain and require bed rest for relief of the discomfort associated with physical activity. The period of bed rest, however, should be kept to a minimum since immobilization speeds bone resorption. Bed rest is usually kept to about a week. Analgesics in the form of salicylates or other non-steroidal anti-inflammatory drugs are useful in controlling pain. Pain from paravertebral muscle spasm is responsive to muscle relaxants. Lumbosacral corsets restrict motion and provide comfort but are usually poorly tolerated and tend to weaken abdominal muscles. The back pain usually resolves spontaneously over three to four months and patients are encouraged to participate in non-weight bearing exercises, such as swimming, and resume normal activity as soon as possible.

The long term goal in patients with osteoporosis fractures is to slow the rate of bone resorption, thereby reducing the possibility of additional fractures and progressive deformity. Medical therapies are directed at increasing calcium absorption and improving bone mineralization. No one regimen has been shown to be effective for all patients with osteoporosis. A combination of calcium supplements, estrogens, fluoride, and calcitonin may be used. Calcium supplements alone have been shown to decrease the vertebral fracture rate in osteoporosis (entry 16). The daily requirement of calcium in perimenopausal women is approximately 1.3 to 1.5 gm. per day (entry 14). Vitamin D helps increase calcium absorption from the gut. The improvement in bone mineralization after the institution of vitamin D therapy may be related to the presence of osteomalacia in some elderly patients with osteoporosis (entry 11). The principal effect of calcium supplements and vitamin D therapy may be to decrease bone turnover (entry 17). Doses of 50,000 units of vitamin D once a week are usually well tolerated; however, a complication of vitamin D therapy is nephrolithiases (renal stones). Measurement of 24-hour urinary calcium excretion every four to six months is indicated to detect the presence of hypercalcemia, a condition associated with the formation of calcium containing renal stones.

Estrogens, female sex hormones, tend to decrease bone resorption and enhance bone mass by slowing the rate at which bone loss occurs (entry 7). The lack of a long-term beneficial effect may be due to a compensatory increase in serum cortisone and parathormone which occurs in patients receiving them. The usual daily dose of conjugated estrogens is 0.625 to 1.25 mg. These hormones are given daily for three weeks, and then withheld for three weeks to allow withdrawal bleeding and endometrial shedding. Estrogens are not used routinely in patients with osteoporosis because of side effects which include vaginal bleeding and breast engorgement. Another significant potential complication of estrogen therapy is the increase in risk of endometrial cancer (entry 2). Therefore, estrogen therapy is restricted to menopausal women with severe osteoporosis, or those who have had hysterectomies.

Sodium fluoride increases skeletal mass by boosting osteoblastic activity. Calcium supplements must be given with fluoride to insure adequate bone mineralization. Sodium fluoride is given in daily doses of 40 to 65 mg. in conjunction with calcium (1.5 gm. daily) and vitamin D (50,000 units once or twice weekly). Fluoride therapy is associated with a number of potential toxicities. Although fluoridic bone may appear more dense on radiographic examination, it may be less elastic and more prone to fracture. Preliminary data suggest, however, that patients treated with fluoride for osteoporosis may have a decreased incidence of fracture (entry 18). Other adverse side effects of fluorides include synovitis, gastric irritation, plantar fasciitis of the feet, and anemia. Most patients with vertebral osteoporosis fractures will heal spontaneously and will not require therapy other than the usual calcium and vitamin D. In patients with progressive bone loss, recurrent fractures and deformity, sodium fluoride may be a useful drug to prevent progressive osteoporosis. However, the prospective experimental data to prove its efficacy are not available and, at this date, fluoride must be considered an investigational drug in the therapy of osteoporosis.

Calcitonin, a parathyroid hormone that suppresses bone resorption, is increased when postmenopausal women receive estrogen therapy. Intramuscular calcitonin injections along with calcium supplements may retard the rate of bone loss in osteoporotic women. Calcitonin may also prove beneficial by controlling pain of fractures in osteoporotic patients through the release of endogenous analgesics, B-endorphin. It is still considered an investigational drug for the treatment of osteoporosis and has not as yet been subjected to a controlled experimental study.

§ 4-39(H). Course and Outcome.

Vertebral compression fractures occur episodically and are usually self-limited. Patients will usually have more than one fracture and a recurrence usually occurs within a few years of the first incident. Back pain may resolve within months of a fracture or may persist if increasing deformity causes a mechanical strain in the lumbar spine. The course of osteoporosis is variable and it is impossible to predict the severity and frequency of fractures.

Patients with osteoporosis who continue to work are at risk of repeated fractures and they should not be required to lift or carry heavy objects. They should refrain from work activities which jar their spine. Measures, such as structured exercise and drug therapy, may be helpful at slowing the progression of the disease.

BIBLIOGRAPHY

1. Albanese, A.A., Edelson, A.H., Lorenze, E.J., Jr., Woodhul, M.L, and Wein, E.H. Problems of bone health in elderly. *N.Y. State J.Med.,* 75:326-336, 1975.
2. Antunes, C.M.F., Stolley, P.D., Rosenshein, N.B., Davies, J.L., Tonascia, J.A., Brown, C., Burnette, L., Rutledge, A., Pokempner, M., and Garcia, R. Endometrial cancer and estrogen use: Report of a large case-controlled study. *N. Eng. J. Med.,* 300:9-13, 1979.
3. Ardan, G.M. Bone destruction not demonstrable by radiography. *Br. J. Radiol.,* 24:107-109, 1951.
4. Avioli, L.V. The osteoporosis problem. *Curr. Concepts in Nutr.,* 5:99-103, 1977.
5. Avioli, L.V. Osteoporosis: Pathogenesis and therapy. *Metabolic Bone Disease,* Vol. 1. Avioli, L.V., and Krane, S.M., eds. Academic Press, Inc., New York, 1977, pp. 307-385.
6. Barnett, E., and Nordin, B.E.C. The radiologic diagnosis of osteoporosis. A new approach. *Clin. Radiol.,* 11:166-174, 1960.

7. Christensen, C., Christensen, M.S., and Transbol, I. Bone mass in post-menopausal women after withdrawal of oestrogen/gestagen replacement therapy. *Lancet,* 1:459-461, 1981.
8. Cohn, S.H., Vaswani, A., Zanzi, I., and Ellis, K.J. Effect of aging on bone mass in adult women. *Am. J. Physiol.,* 230:143-148, 1976.
9. Heaney, R.P. Prevention of age-related osteoporosis in women. *The Osteoporotic Syndrome,* ed. Avioli, L.V. Grune and Stratton, New York, 1983, pp. 123-144.
10. Lukert, B.P. Osteoporosis: A review and update. *Arch. Phys. Med. Rehab.,* 63:480-487, 1982.
11. Lund, B., Kjaer, I., Friis, T., Hjorth, L., Reimann, I., Andersen, R.B., and Sorensen, O.H. Treatment of osteoporosis of aging with 1-hydroxycholecalciferol. *Lancet,* 2:1168-1171, 1975.
12. Lutwak, L., Singer, F.R., and Urist, M.R. Current concepts of bone metabolism. *Ann. Intern. Med.,* 80:630-644, 1974.
13. McNair, P., Madsbad, S., Christiansen, C., Faber, O.K., Transbol, I., and Binder, C. Osteopenia in insulin treated diabetes mellitus: Its relation to age at onset, sex, and duration of disease. *Diabetologia,* 15:87-90, 1978.
14. Recker, R.R., Saville, P.D., and Heaney, R.P. Effect of estrogens and calcium carbonate on bone loss in postmenopausal women. *Ann. Intern. Med.,* 87:649-655, 1977.
15. Riggs, B.L., Wahner, H.W., Dunn, W.L., Mazess, R.B., Offord, K.P., and Melton, L.J., III. Differential changes in bone mineral density of appendicular and axial skeleton with aging. *J. Clin. Invest.,* 67:328-335, 1981.
16. Riggs, B.L., Seeman, E., Hodgson, S.F., Taves, D.R., and O'Fallon, W.M. Effect of the fluoride/calcium regimen on vertebral fracture occurrence in postmenopausal osteoporosis: Comparison with conventional therapy. *N. Eng. J. Med.,* 306:446-450, 1982.
17. Riggs, B.L., Jowsey, J., Kelley, P.J., Hoffman, D.L., and Arnaud, C.D. Effects of oral therapy with calcium and vitamin D in primary osteoporosis. *J. Clin. Endocrinol. Metab.,* 42:1139-1144, 1976.
18. Riggs, B.L., Hodgson, S.F., Hoffman, D.L., Kelley, P.J., Johnson, K.A., and Taves, D. Treatment of primary osteoporosis with fluoride and calcium: Clinical tolerance and fracture occurrence. *JAMA,* 243:446-449, 1980.
19. Smith, E.L., and Reddon, W. Physical activity: A modality for bone accretion in the aged. *Am. J. Roentgenol.,* 126:1297, 1976.
20. Thomson, D.L., and Frame, B. Involutional osteopenia: Current concepts. *Ann. Intern. Med.,* 85:789-803, 1976.

371

21. Urist, M.P., Gurvey, M.S., and Fareed, D.O. Long term observations on aged women with pathologic osteoporosis. *Osteoporosis*. Barzel, U.S. (ed). New York, 1970, pp. 3-37.

TABLE I

Causes of Osteoporosis

Primary
Involutional (postmenopausal or senile)
Idiopathic

Endocrine
Hypogonadism
Adrenocortical hormone excess (primary or iatrogenic)
Hyperthyroidism
Hyperparathyroidism
Diabetes mellitus
Growth hormone deficiency

Nutritional
Calcium deficiency
Phosphate deficiency
Phosphate excess
Vitamin D deficiency
Protein deficiency
Ascorbic acid deficiency
Intestinal malabsorption

Drug
Heparin
Anticonvulsants
Ethanol
Methotrexate

Genetic
Osteogenesis imperfecta
Homocystinuria

Miscellaneous
Rheumatoid arthritis

Immobilization

Malignancy

Metabolic acidosis

§ 4-40. Osteomalacia.

Osteomalacia is a metabolic bone disease associated with loss of bone mass per unit volume where the ratio of bone mineral content to bone matrix is decreased. In essence, it is an abnormality in the mineralization of bone. Any disorder which affects calcium or phosphorus concentrations or alters the physiologic conditions needed for the formation of hydroxyapatite crystals, may result in osteomalacia. It may occur in any of the long bones, pelvis, scapula, ribs, or axial skeleton. Loss of bone mineralization weakens the bone which makes it vulnerable to fracture. Osteomalacia in the axial skeleton is associated with back pain, vertebral body weakening, fracture, and progressive kyphoscoliosis. It should be suspected in a patient with bone pain, radiographic evidence of decreased bone mass (osteopenia), and depressed levels of serum calcium and inorganic phosphate. The definitive diagnosis of osteomalacia is made by the presence of a widened osteoid seam, and decreased mineralization found in an undemineralized bone section. Osteomalacia may be caused by vitamin D deficiency, intestinal disorders, drugs, metabolic acidosis usually associated with renal disorders including tubular defects, phosphate deficiencies, and mineralization defects, both primary and secondary.

Evaluation of a patient with osteomalacia is directed at identifying the underlying disease process which results in the decreased bone mineralization. Treatment is directed toward the specific illness causing the osteomalacia and may include vitamin D supplements, pancreatic enzyme supplements, dietary modifications to increase intake of calcium and phosphorus, reduction of metabolic acidosis, and the discontinuation of drugs associated with osteomalacia. The major disability of this illness is bone fragility since osteomalacic bone has a propensity to deform and fracture. Progression of this disease can be prevented if effective therapy is available to treat the underlying cause.

§ 4-40(A). Prevalence and Etiology.

The prevalence of osteomalacia is unknown. The number of illnesses associated with osteomalacia prohibits the computation of prevalence from all sources (Table 1) (entry 9). Osteomalacia in children, rickets, is common outside the United States in areas of the world with malnutrition. Up to 9 per cent of young children may demonstrate radiographic findings of rickets in impoverished urban areas (entry 33).

Osteomalacia includes a group of disorders with similar clinical symptoms and signs but with diverse etiologies. The most frequent abnormality associated with osteomalacia is vitamin D deficiency. Vitamin D in its activated form increases intestinal calcium absorption, renal tubular resorption of calcium and phosphate, and potential bone mineralization (entry 22). The sources of vitamin D are exogenous (fortified dairy products) and endogenous (ultraviolet sun exposure of skin). Dairy products are fortified with vitamin D_2 (ergocalciferol, irradiation product of plant sterols) or vitamin D_3 (cholecalciferol). Adequate gastrointestinal absorption of dietary vitamin D requires an intact and functioning mucosal surface of the small intestine as well as an intact biliary system with adequate concentrations of bile salts. A more important source of vitamin D_3 is produced endogenously, generated by the exposure of 7-dehydrocholesterol in the skin to ultraviolet light.

Vitamin D (either D_2 or D_3) is transported to the liver by a carrier protein. Hepatic enzymes hydroxylate the precursor vitamin at the 25 positions to form 25-(OH) vitamin D. The 25-(OH) vitamin D is then transported to the kidney where the most active form of vitamin D is generated by hydroxylation at the 1 position, 1, 25-(OH) vitamin D. Hydroxylation at the 24 position will form 24, 25-dihydroxyvitamin D. This form of vitamin D may contribute to mineralization of bone and modulation of parathyroid function, but it is less potent than the 1,25 form of vitamin D. Increased activity of the enzyme which controls hydroxylation at the 1 position, 1-hydroxylase, is mediated by decreased dietary intake of calcium, increased parathyroid hormone secretion, and hypophosphatemia (entry 22). Inadequate intake of vitamin D supplemented nutrition and an avoidance of the sun is required before osteomalacia occurs on this deficiency basis.

Normal vitamin D absorption in the gut requires an intact intestinal mucosa, normal hepatobiliary circulation of bile salts, and to a lesser degree, exocrine pancreatic function. Osteomalacia may occur despite adequate production of vitamin D_3 in the skin since the hepatic product, 25-hydroxyvitamin D undergoes enterohepatic circulation like bile salts (entry 2). Most vitamin D is absorbed in the mid-jejunum, while a smaller component is absorbed in the terminal ileum (entry 34). Diseases associated with small bowel malabsorption such as sprue, coeliac disease, Crohn's disease, scleroderma, jejunal diverticula, and bowel bypass surgery may all be complicated by osteomalacia (entry 35). Postgastrectomy patients are also prone to develop osteomalacia. Patients with hepatocellular disorders, cirrhosis, alcoholic hepatitis, chronic active hepatitis and those with biliary system disease, primary biliary cirrhosis, are unable to absorb vitamin D and develop osteomalacia (entries 28, 6). Malabsorption associated with pancreatic insufficiency may cause osteomalacia not only by decreasing levels of 25-hydroxyvitamin D but also by decreasing calcium absorption (entry 20).

Abnormalities, hereditary or acquired, in the metabolism necessary for the formation of 1,25 dihydroxyvitamin D result in rickets or osteomalcia. Patients with an autosomal recessive genetic condition of an absence of the renal 25-hydroxyl-1-hydroxylase are incapable of manufacturing adequate concentrations of 1,25 dihydroxyvitamin D. These patients develop vitamin D dependent rickets (entry 15). Patients with chronic parenchymal liver disease are unable to form 25-hydroxyvitamin D (entry 23). Renal osteodystrophy is the bone disease associated with chronic renal failure. Ostemalacia, osteoporosis, osteosclerosis particularly in the axial skeleton, and secondary hyperparathyroidism, are all potential complications of renal disease. Osteomalacia in renal osteodystrophy occurs secondary to the impaired conversion of 25-hydroxyvitamin D to 1,25 dihydroxyvitamin D due to a decrease in kidney cell mass (entry 3). Hyperphosphatemia and systemic acidosis, complications of chronic renal failure, may also contribute to the deficiency of the renal metabolite of vitamin D (entry 27). Anticonvulsant drugs, phenytoin (Dilantin) and phenobarbital, induce hepatic hydroxylase enzymes which alter 25-hydroxyvitamin D, producing inactive metabolites (entry 19). Phenytoin may also decrease calcium absorption from the gut (entry 39). A rare cause of abnormal vitamin D metabolism is mesenchymal, soft tissue tumors associated with low levels of 1-25 vitamin D (entry 11). Patients with hypophosphatemia and mesenchymal tumors may have clinical and radiologic findings in the sacroiliac joints and this may be confused with ankylosing spondylitis (entry 30). Patients may develop osteomalacia in the face of normal concentrations of 1-25-dihydroxyvitamin D when impaired end organ (gut and bone) responsiveness to the renal metabolite exists. This may occur with chronic renal failure, and anti-convulsant drugs (entry 4). When no other etiology is evident, peripheral resistance to vitamin D associated with hypocalcemia, osteomalacia, and secondary hyperparathyroidism is referred to as vitamin D dependent rickets Type II (entry 5).

Phosphate, as well as calcium, in the appropriate concentration is essential for normal bone mineralization and muscle function. Any disorder which results in phosphate deficiency will result in osteomalacia. Phosphorus is present in most food stuffs. Therefore, phosphate deficiency on the basis of malnutrition is rare. Patients who ingest large quantities of phosphate binding aluminum antacids for ulcer or renal disease are at risk of developing osteomalacia (entry 10). The most common cause of hypophosphatemia is related to hereditary or acquired renal tubular wasting of phosphate. A hereditary form of phosphate depletion is x-linked (male) hypophosphatemic osteomalacia. Phosphate reabsorption in the kidney is modulated by parathyroid hormone, predominantly, and serum calcium concentration (entry 17). Patients with x-linked hypophosphatemic osteomalacia (or familial vitamin D resistant rickets) have complete absence of the parathyroid hormone sensitive component of phosphate reabsorption. The disease is transmitted as a dominant trait in men, has onset in childhood and results in adults who are short and bow-legged (entry 29). The disease may have an onset in adulthood (hypophosphatemic osteomalacia), in patients with no evidence of rachitic deformities (entry 16). Patients with familial vitamin-D resistant rickets develop increases in bone density in the axial skeleton along with ectopic calcification which resembles ankylosing spondylitis (entry 36).

Fanconi syndrome refers to a heterogeneous group of disorders which cause dysfunction in the renal tubules (entry 7). The syndromes have been classified by Mankin into proximal, distal, or combination of proximal and distal tubular diseases (entry 29). Tubular disorders may result in phosphate wasting, glycosuria, aminoaciduria, renal tubular acidosis, hypokalemia, or polyurias. A number of disorders, including cystinosis, Lowe's syndrome (oculocerebrorenal syndrome), Wilson's disease, tyrosinemia, nephrotic syndrome and multiple myeloma may cause tubular dysfunction resulting in osteomalacia. Tumors, both benign and malignant, which inhibit renal phosphate absorption cause osteomalacia (entry 21). Tumors associated with osteomalacia include giant cell tumor, nonossifying fibroma, osteoblastoma, and angiosarcoma (entry 18). Fibrous dysplasia, a disorder of unknown etiology, causing ground-glass appearing lesions in bone, precocious puberty, and skin pigmentation, has been associated with osteomalacia (entry 8).

Alkaline phosphatase is an enzyme required for the normal mineralization of bone. Patients with hypophosphatasia have a deficiency in this enzyme resulting in elevated levels of plasma and urinary inorganic pyrophosphate, phosphorylethanolamine, and osteomalacia (entry 24). Adults with the disease have osteomalacia with axial skeletal ligamentous and tendinous calcification and develop marked limitation of motion in the lumbosacral spine (entry 1). Fluoride in increased concentrations causes inhibition of bone mineralization. Patients in endemic areas of fluorosis or those who received sodium fluoride for osteoporosis are at risk of developing osteomalacia (entry 38). Diphosphonates, synthetic analogues of pyrophosphate, are inhibitors of bone formation and are used in the therapy of patients with Paget's disease. Patients receiving diphosphonates at increased doses for extended periods may develop inadequate mineralization of bone and pathologic fractures (entry 26).

A number of miscellaneous disorders may be associated with osteomalacia. Axial osteomalacia is a disease of adult men who develop typical osteomalacic changes in bone limited to the axial skeleton, including cervical and lumbar spine (entry 13). Some of these patients have features on radiographs which are similar to ankylosing spondylitis (entry 31). Fibrogenesis imperfecta ossium is a rare lesion of men over 50 who develop increased bone density, pseudofractures, and bone pain (entry 14). Osteopetrosis, marble bone or Albers-Schonberg's disease, is a rare inherited disorder associated with increased osteosclerosis (entry 25). The benign form, autosomal dominant in inheritance, is associated with bone pain, fracture, osteomyelitis and cranial nerve palsies due to bony overgrowth. The pathogenesis of the illness is related to the absence of osteoclastic activity in resorbing bone.

§ 4-40(B). Clinical Findings.

Patients with osteomalacia may present with a wide variety of symptoms depending on the underlying cause of their bone disease. The major complaint relating to their bone disease is bone pain. It is maximum in the lower extremities and axial skeleton and is worsened by activity. Lumbar pain increases with standing as opposed to the sudden and severe pain associated with osteoporosis. The pain of osteomalacia tends to be more diffuse and less intense, but of longer duration. Muscle weakness and tenderness, along with episodes of muscle spasm, particularly in the proximal muscles in the lower extremities, may occur with hypocalcemia or hypophosphatemia. Patients with osteomalacia with longstanding disease may also have a history of fractures.

§ 4-40(C). Physical Findings.

Physical findings are sparse in osteomalacia. The affected bones are tender on palpation as may be proximal muscles as well. Muscle weakness, exemplified by inability to climb stairs and rise from a seated position, may also be present. Kyphoscoliosis of the thoracic and lumbar spine is prominent if osteomalacia of the axial skeleton is longstanding. Adult patients with a history of rickets may have short stature, bowing of the lower extremities, and enlargement of the costochondral junctions (rachitic rosary).

§ 4-40(D). Laboratory Findings.

Laboratory abnormalities are frequent in patients with osteomalacia, which is in marked distinction from the normal values of osteoporosis (entry 9). Patients with abnormalities in vitamin D concentration or metabolism have low serum calcium and phosphate concentration, elevated serum alkaline phosphatase, a reduced renal phosphate threshold, generalized aminoaciduria, reduced urinary calcium excretion, increased parathyroid hormone (secondary hyperparathyroidism secondary to hypocalcemia) and decreased serum concentrations of vitamin D. Urinary calcium excretion is less than 75 mg. per day in 95 per cent of patients with osteomalacia and its measurement is a useful screening test (entry 32). Laboratory abnormalities associated with phosphate wasting include low serum phosphate, normal serum calcium, elevated serum alkaline phosphatase concentration, a reduced renal phosphate threshold, normal amino acid excretion, reduced urinary calcium excretion, increased parathyroid hormone concentration with therapy and normal vitamin D concentrations. Additional laboratory abnormalities may be present depending on the underlying disease causing osteomalacia. Examples of these abnormalities might include decreased serum carotene, increased fecal fat in gastrointestinal malabsorption, increased urea nitrogen and creatinine in renal failure, and anemia and elevated erythrocyte sedimentation rate with a tumor.

The definitive diagnosis of osteomalacia must be made by examination of undemineralized sections of a bone biopsy. Tetracycline, which will fix to newly forming mineralized bone, is given to patients before biopsy. The extent of tetracycline fluorescence in bone is used to measure the decrease in the mineralization front. There is also an increase in osteoid thickness (more than 20 u) and an increase in osteoid seams covering cortical and trabecular bone associated with osteomalacia.

§ 4-40(E). Radiographic Findings.

The radiographic findings of osteomalacia are those of osteopenia and loss of bone mass, and they mimic the changes of osteoporosis. The generalized loss of bone density in one is indistinguishable from the other. In osteomalacia, the remaining trabeculae are thickened between radiolucent areas. One of the major radiologic findings in osteomalacia is pseudofractures (Loosers zones, Milkman fractures) (entry 37). They occur in long or flat bones, usually oriented at right angles to the cortex, and they incompletely span the diameter of the bone. Pseudofractures, which usually occur symmetrically, may or may not be associated with pain. Radiographic findings in the spine may include expansion of the intervertebral discs with "cod-fish" vertebrae and scoliosis if osteomalacia occurs during periods of growth. Pseudofractures may occur in the ribs, pelvis, femoral neck, ulna, radius, scapula, clavicles, and phalanges. With longstanding disease and minimal trauma, patients with weakened areas of bone secondary to pseudofractures develop true fractures.

§ 4-40(F). Differential Diagnosis.

The diagnosis of osteomalacia can be suspected in a patient with bone pain and deformity, muscle weakness and tenderness, abnormal laboratory data consistent with disordered bone metabolism and radiographic evidence of inadequate bone mineralization; and it can be confirmed, if necessary, by bone biopsy. Osteomalacia, like acute pain, is a sign of an underlying disease process which requires additional evaluation. The possibility that a patient had one of the diseases associated with osteomalacia listed in Table I (at end of this section) would need to be investigated. In addition to osteoporosis, diffuse carcinomatosis of bone, polymyositis or rhabdomyolysis would also have to be considered in a patient with bone pain or muscle weakness. The acute onset of pain, with pain-free intervals and normal calcium and phosphate concentrations, helps differentiate patients with osteoporosis from those with osteomalacia. Anemia and abnormal serum proteins are associated with diffuse carcinomatosis (multiple myeloma). Elevated muscle enzyme levels identify patients with primary muscle diseases.

§ 4-40(G). Treatment.

The treatment of osteomalacia must be directed at the underlying abnormality which results in abnormal bone mineralization. The daily recommended intake of vitamin D is 400 IU daily for children and 100 IU for adults. This amount of vitamin D is adequate to heal bone lesions secondary to vitamin D nutritional deficiency, although higher doses can be given to speed healing (entry 12). Adequate calcium and phosphate intake are also required to assure normal mineralization of bone. Therapy for patients with osteomalacia secondary to gastrointestinal disorders should be directed at the underlying gut disease. For instance, a

383

gluten-free diet may in itself correct osteomalacia in a patient with coeliac disease. Patients with refractory gastrointestinal disease may benefit from pharmacologic doses of vitamin D (1-10 mg. per day). Oral pancreatic enzymes may improve vitamin D absorption in pancreatic insufficiency. Increased oral intake of vitamin D (1-2 mg. per day), along with correction of acidosis or removal of a tumor, will improve bone mineralization in patients with abnormal vitamin D metabolism. Patients with chronic renal failure may benefit with replacement with physiologic doses of 1,25-dihydroxyvitamin D_3.

Patients with phosphate wasting forms of osteomalacia require phosphate supplementation and vitamin D to reduce the possibility of decreased serum ionized calcium concentrations and secondary hyperparathyroidism. Some patients may also require alkali therapy to control renal tubular acidosis.

Mineralization defects caused by fluoride may be prevented if calcium supplements are administered. The pathologic fractures associated with disodium etidronate used in therapy of Paget's disease may be prevented by using doses of 5 mg. per kilogram along with an alternating monthly schedule of giving and not giving the drug. Medical therapy of adult hypophosphatasia is ineffective.

§ 4-40(H). Course and Outcome.

The prognosis of osteomalacia is based not only on the time of onset of the illness (rickets) and the extent of the bone disease, but also on the underlying disease and its reversibility to therapy. Many forms of osteomalacia are reversible with adequate vitamin D, calcium, and phosphorus supplementation. Pseudofractures heal and bone can restore to normal mineralization and strength. Deformities secondary to bone weakening, lower extremity bowing and kyphoscoliosis remain, but do not progress.

These patients are able to resume normal work activities without increased risk of fracture. Patients with hereditary abnormalities tend to be less responsive to therapy and may continue with areas of bone which are osteomalacic despite therapy. Patients with chronic renal failure and those on chronic dialysis may continue with osteomalacia despite efforts at maintaining calcium and phosphorus concentrations close to normal and correcting acidosis. Secondary hyperparathyroidism, which complicates renal osteodystrophy, may necessitate the removal of hypertrophied parathyroid tissue to prevent additional bone destruction. Patients with tumors and osteomalacia will have a prognosis which corresponds with the characteristics of their neoplasm. With so many diseases associated with osteomalacia, the prognosis for that disease must be evaluated on an individual basis, taking into account all factors which deal with the underlying disease process, extent of bone disease, and potential response to therapy.

BIBLIOGRAPHY

1. Anderton, J.M. Orthopedic problems in adult hypophosphatasia. *J. Bone Joint Surg.*, 61B:82-84, 1979.

2. Arnaud, S.B., Goldsmith, R.S., Lambert, P.N., and Go, V.L.W. 25-hydroxyvitamin D_3: Evidence of an enterohepatic circulation in man. *Proc. Exp. Biol. Med.*, 149:570-572, 1975.

3. Avioli, L.V. Controversies regarding uremia and acquired defects in vitamin D_3 metabolism. *Kidney Int.*, 13: Supp. 8:36-38, 1978.

4. Brickman, A.S., Coburn, J.W., and Massey, S.G. 1, 25-dihydroxyvitamin D3 in normal man and patients with renal failure. *Ann. Intern. Med.*, 80:161-168, 1974.

5. Brooks, M.D., Bell, N.H., Love, L., Stern, P.H., Orfei, E., Queener, S.F., Hamstra, A.J., and DeLuca, H.F. Vitamin D-dependent rickets type II: Resistance of target organs to 1,25 dihydroxyvitamin D. *N. Eng. J. Med.*, 298:996-999, 1978.

6. Compston, J.E., and Thompson, R.P.H. Intestinal absorption of 25-hydroxyvitamin D and osteomalacia in primary biliary cirrhosis. *Lancet*, 1:721-724, 1977.

7. Dent, C.E. Rickets (and osteomalacia), nutritional and metabolic (1919-1969). *Proc. R. Soc. Med.*, 63:401-408, 1970.

8. Dent, C.E., and Gertner, J.M. Hypophosphatemic osteomalacia in fibrous dysplasia. *Quart. J. Med.*, 45:411-420, 1976.

9. Dent, C.E., Stamp, T.C.B. Vitamin D, rickets, and osteomalacia in metabolic bone disease. Avioli, L.V., and Krane, S.M. (eds.). Academic Press. New York, 1977, pp. 237-305.

10. Dent, C.E., and Winter, C.S. Osteomalacia due to phosphate depletion from excessive aluminum hydroxide ingestion. *Br. Med. J.*, 1:551-552, 1974.

11. Drezner, M.K., and Feinglos, M.N. Osteomalacia due to 1, 25-dihydroxycholecalciferol deficiency: Association with a giant cell tumor of bone. *J. Clin. Invest.*, 60:1046-1053, 1977.

12. Frame, B., and Parfitt, A.M. Osteomalacia: Current concepts. *Ann. Intern. Med.*, 89:966-982, 1978.

13. Frame, B., Frost, H.M., Ormond, R.S., and Hunter, R.B. Atypical osteomalacia involving the axial skeleton. *Ann. Intern. Med.*, 55:632-639, 1961.

14. Frame, B., Frost, H.M., Pak, C.Y.C., Reynolds, W., and Argen, R.J. Fibrogenesis imperfecta ossium: A collagen defect causing osteomalacia. *N. Eng. J. Med.*, 285:769-772, 1971.

15. Fraser, D., Kooh, S.W., Kind, H.P., Holick, M.F., Tanaka, Y., and DeLuca, H.F. Pathogenesis of hereditary vitamin D — dependent rickets: An inborn error of vitamin D metabolism involving defective conversion of 25-hydroxyvitamin D to 1, 25 dehydroxyvitamin D. *N. Eng. J. Med.*, 289:817-822, 1973.

16. Frymoyer, J.W., and Hodgkin, W. Adult-onset vitamin D-resistant hypophosphatemic osteomalacia. A possible variant of vitamin D-resistant rickets. *J. Bone Joint Surg.*, 59A:101-106, 1977.

17. Glorieux, F.H., and Scriver, C.R. Loss of a parathyroid hormone-sensitive component of phosphate transport in x-linked hypophosphatemia. *Science*, 175:997-1000, 1972.

18. Goldring, S.R., and Krane, S.M. Disorders of calcification: Osteomalacia and rickets. DeGroot, L. (ed.) *Endocrinology*, Vol. 2. New York, Grune and Stratton, 1979, pp. 853-871.

19. Hahn, T.J., Birge, S.J., Scharp, C.R., and Avioli, L.V. Phenobarbital-induced alterations in vitamin D metabolism. *J. Clin. Invest.*, 51:741-748, 1972.

20. Hahn, T.J., Squires, A.E., Halstead, L.R., and Strominger, D.B. Reduced serum 25-hydroxyvitamin D concentration and disordered mineral metabolism in patients with cystic fibrosis. *J. Pediatr.,* 94:38-42, 1979.

21. Harrison, H.E. Oncogenous rickets. Possible elaboration by a tumor of a humeral substance inhibiting tubular reabsorption of phosphate. *Pediatrics,* 52:432-433, 1973.

22. Haussler, M.R., and McCain, T.A. Basic and clinical concepts related to vitamin D metabolism and action. *N. Eng. J. Med.,* 297:974-983, 1977.

23. Imawari, M., Akanuma, Y., Itakura, H., Muto, Y., Kosaka, K., and Goodman, D.S. The effects of diseases of the liver on serum 25-hydroxyvitamin D and on the serum binding protein for vitamin D and its metabolites. *J. Lab. Clin. Med.,* 93:171-180, 1979.

24. Jardon, O.M., Burney, D.W., and Fink, R.L. Hypophosphatasia in an adult. *J. Bone Joint Surg.,* 52A:1477-1484, 1970.

25. Johnston, C.C., Jr., Lavy, N., Lord, T., Vellios, F., Merritt, D., and Deiss, W.P. Osteopetrosis: A clinical, genetic, metabolic, and morphologic study of the dominantly inherited, benign form. *Medicine,* 47:149-167, 1968.

26. Kantrowitz, F.G., Byrne, M.H., and Krane, S.M. Clinical and metabolic effects of the diphosphorate in Paget's disease of bone. *Clin. Res.,* 23:445A, 1975.

27. Lee, S.W., Russell, J., and Avioli, L.V. 25-dihydroxycholecalciferol to 1, 25-dihydroxycholecalciferol: Conversion impaired by systemic metabolic acidosis. *Science,* 195:994-996, 1977.

28. Long, R.G., Skinner, R.K., Willes, M.R., and Sherlock, S. Serum 25-hydroxy-vitamin D in untreated parenchymal and cholestalic liver disease. *Lancet,* 2:650-652, 1976.

29. Mankin, H.J. Rickets, osteomalacia and renal osteodystrophy: Part II. *J. Bone Joint Surg.,* 56A:352-386, 1974.

30. Moser, C.R., and Fessel, W.J. Rheumatic manifestations of hypophosphatemia. *Arch. Intern. Med.,* 134:674-678, 1974.

31. Nelson, A.M., Riggs, B.L., and Jowsey, J.O. Atypical axial osteomalacia: Report of four cases with two having features of ankylosing spondylitis. *Arthritis Rheum.,* 21:715-722, 1978.

32. Nordin, B.E.C., Hodgkinson, A., and Peacock, M. The measurement and the meaning of urinary calcium. *Clin. Orthop.,* 52:293-322, 1967.

33. Richards, I.D.G., Sweet, E.M., and Arneil, G.C. Infantile rickets persists in Glasgow. *Lancet,* 1:803-805, 1968.

34. Schachter, D., Finkelstein, J.D., and Kowarski, S. Metabolism of vitamin D. Part I. Preparation of radioactive vitamin D and its intestinal absorption in the rat. *J. Clin. Invest.,* 43:787-796, 1964.
35. Sitrin, M., Meredith, S., and Rosenberg, I.H. Vitamin D deficiency and bone disease in gastrointestinal disorders. *Arch. Intern. Med.,* 138:886-888, 1978.
36. Steinbach, H.L., Kolb, F.O., and Crane, J.T. Unusual roentgen manifestations of osteomalacia. *Am. J. Roentgenol.,* 82:875-886, 1959.
37. Steinbach, H.L., and Noetzli, M. Roentgen appearance of the skeleton in osteomalacia and rickets. *Am. J. Roentgenol.,* 91:955-972, 1964.
38. Teotia, S.P.S., and Teotia, M. Secondary hyperparathyroidism in patients with endemic skeletal fluorosis. *Br. Med. J.,* 1:637-640, 1973.
39. Villareale, M.E., Chiroff, R.T., Bergstron, W.H., Gould, L.V., Wasserman, R.H., and Romano, F.A. Bone changes induced by diphenylhydantoin in chicks on a controlled vitamin D intake. *J. Bone Joint Surg.,* 60A:911-916, 1978.

TABLE I

Diseases Associated With Osteomalacia

Vitamin D
  Deficiency
    Dietary
    Ultraviolet light exposure

  Malabsorption
    Small intestine
    Inadequate bile salts
    Pancreatic insufficiency

  Abnormal Metabolism
    Hereditary enzyme deficiency (Vitamin D dependent
      rickets Type I)
    Hepatic failure
    Chronic renal failure
    Systemic acidosis
    Anticonvulsant drugs
    Mesenchymal tumors

  Peripheral Resistence
    Vitamin D dependent rickets (Type II)

Phosphate Depletion
  Dietary
    Malnutrition
    Aluminum hydroxide ingestion

  Renal tubular wasting
    Hereditary
      x-linked hypophosphatemic osteomalacia
    Acquired
      Hypophosphatemic osteomalacia
    Renal disorders
    Fanconi's syndrome

Mesenchymal tumors
Fibrous dysplasia

Mineralization Defects
Hereditary
Hypophosphatasia
Acquired
Sodium fluoride
Disodium etidronate

Miscellaneous
Osteopetrosis
Fibrogenesis imperfecta
Axial osteomalacia
Calcium deficiency

§ 4-41. Parathyroid Disease.

Parathyroid hormone is the dominant factor in the maintenance of serum calcium in a normal range. Hyperparathyroidism results in excess concentrations of parathyroid hormone in the bloodstream and in elevated serum calcium levels. Primary hyperparathyroidism is caused by abnormal growth of the parathyroid glands. Secondary hyperparathyroidism results form the secretion of parathyroid hormone in response to persistently low serum concentrations of calcium. Hyperparathyroidism, regardless of type, leads to bone disease and abnormal physiology in a number of organ systems dependent on calcium for normal function (nervous, genitourinary, and gastrointestinal). The loss of calcium from bone results in pain, weakening and fracture. Untreated disease causes marked osteopenia of the vertebral column with progressive vertebral body fractures and spinal deformity.

Hypoparathyroidism, an illness associated with deficient activity of parathyroid hormone, causes hypocalcemia and associated soft tissue calcification and bony overgrowth. Paravertebral calcification of ligamentous structures in the lumbar spine leads to progressive stiffness and limitation of motion. The diagnosis of parathyroid disease is based on the measurement of serum calcium primarily; and serum phosphate, chloride, and parathyroid hormone secondarily. Patients with hypercalcemia must be investigated for primary hyperparathyroidism as well as for malignancies, other endocrine disorders, renal failure, excess dietary intake of calcium, and granulomatous disorders which may also be associated with increased serum calcium concentrations. Hypocalcemia can be caused by inadequate concentrations of parathyroid hormone following parathyroidectomy or peripheral resistance to the physiologic effects of the hormone (hypomagnesemia). The therapy of primary hyperparathyroidism is surgical removal of all but a small portion of the parathyroid glands. Treatment of secondary hyperparathyroidism is directed toward the underlying illness causing the decreased serum calcium concentrations. Therapy of hypoparathyroidism is directed at maintaining normal serum calcium levels with calcium supplements. Patients with parathyroid disease rarely develop severe skeletal involvement since abnormal calcium levels are usually discovered relatively early in the course of the illness.

§ 4-41(A). Prevalence and Etiology.

The absolute prevalence of parathyroid disease is unknown although an incidence of 1 per 1,000 patients was reported from data obtained at a diagnostic clinic (entry 3). The ratio of men to women is 1:3. Rarely, hyperparathyroidism occurs in two familial syndromes associated with multiple endocrine neoplasms, Type I with pituitary and pancreatic tumors, and Type II with medullary thyroid carcinoma and pheochromocytoma.

Parathyroid hormone (PTH) maintains serum calcium levels by stimulating intestinal calcium absorption, activating osteoclasts for bone resorption, and stimulating renal tubular calcium reabsorption, phosphate excretion, and enzyme synthesis of the active form of vitamin D. The concentration of serum calcium perfusing the four parathyroid glands located posterior to the thyroid gland is the dominant factor in the control of secretion of PTH. PTH is secreted in response to low serum calcium concentrations. Primary hyperparathyroidism is a disease process within the parathyroid glands and in approximately 90 per cent of cases, the abnormality is a neoplasm, usually the overgrowth of one gland forming an adenoma; less often, multiple adenomas, diffuse hyperplasia, or a carcinoma of the parathyroids (entry 16). Secondary hyperparathyroidism is the increased secretion PTH in response to low serum calcium levels caused by abnormalities in other organ systems, such as the kidney, or when there is inadequate vitamin D metabolism.

Hypoparathyroidism also occurs in primary and secondary forms. Primary or idiopathic hypoparathyroidism associated with cessation of function of the four parathyroid glands occurs more commonly in female children and is uncommon. The usual form of hypoparathyroidism is secondary and is caused by damage to or accidental removal of the parathyroid glands during thyroid gland surgery.

§ 4-41(B). Clinical Findings.

The vast array of symptoms associated with hyperparathyroidism is related to direct effects of PTH hypercalcemia. The patient with the florid syndrome of hyperparathyroidsim is unusual since many patients are discovered with hypercalcemia by multiphasic blood screening at an early stage of disease. They complain of bone pain and may present with a history of back pain from vertebral compression fractures (entry 6). Dull back pain also may be related to renal colic secondary to nephrolithiasis. Gastrointestinal symptoms associated with hypercalcemia include anorexia, nausea, vomiting, constipation, and abdominal pain secondary to peptic ulcer disease or pancreatitis. Markedly elevated serum levels of calcium may affect mental status and cause coma and band keratopathy in the eye.

The most prominent symptom of hypoparathyroidism is tetany. Tetany is tonic muscle spasm which occurs secondary to abnormally low concentrations of calcium. Persistent hypocalcemia may cause irritability, depression and decrease in mental activity.

§ 4-41(C). Physical Findings.

Patients with hyperparathyroidism show muscle weakness on examination. Examination of the back may demonstrate percussion tenderness over a recently fractured vertebral body. Kyphosis occurs with wedging of vertebral bodies. Musculoskeletal examination may show joint inflammation (swelling, heat, redness, pain, loss of motion) in patients with acute gout or pseudogout. Both illnesses are commonly found in patients with hyperparathyroidism (entries 18, 14).

Patients with hypoparathyroidism may show signs of tetany when stressed with percussion over the facial nerve (Chvostek's sign) or carpal spasm with reducing blood flow from a blood pressure cuff (Trousseau's sign). They also may have limitation of motion of the lumbosacral spine secondary to soft tissue calcification (entry 12).

§ 4-41(D). Laboratory Findings.

The laboratory parameter of greatest importance in the evaluation of a patient with suspected hyperparathyroidism is the serum calcium. Serum calcium is elevated in over 96 per cent of patients with primary hyperparathyroidism (entry 11). Mild disease can be associated with intermittent elevations so repeated determinations are indicated if suspicion is great and the initial calcium value is normal. Other chemical tests are useful but not diagnostic. These include a low serum phosphorus, elevated serum chloride, elevated serum alkaline phosphatase and elevated urinary calcium excretion. Measurement of parathormone by radioimmunoassay will become the definitive test for hyperparathyroidism once the specificities of the antibody for active PTH is improved and the normal range of values is determined (entry 13). Hypoparathyroidism is associated with a low serum calcium, elevated serum phosphorus and normal alkaline phosphatase. Secondary hyperparathyroidism, usually associated with chronic renal failure, seldom produces hypercalcemia, but is associated with elevated phosphorus concentrations. The electrocardiogram may demonstrate a shortened Q-T interval with hyperparathyroidism and a prolonged interval with hypoparathyroidism.

Patients with hyperparathyroidism have little evidence of clinical bone disease at presentation and bone biopsy is rarely obtained; when it is, it usually demonstrates the effects of PTH on bone. Characteristic findings include an increased number of osteoclasts resorbing bone, osteoblasts repairing bone that is being resorbed and numerous fibroblasts producing dense fibrous tissue. Bone resorption may result in bone cyst formation and bleeding into the cysts results in brown discoloration of the fibrous tissue, and is referred to as a brown tumor. Osteitis fibrosa cystica is the term used in reference to the bone disease of hyperparathyroidism.

§ 4-41(E). Radiologic Findings.

A variety of radiologic lesions are associated with

hyperparathyroidism, including subperiosteal bone resorption, particularly on the radial aspects of the middle phalanges, resorption of the terminal tufts of the phalanges, resorption of the symphysis pubis and sacroiliac joints, a "salt and pepper" appearance to the skull, and cystic lesions of the long bones (entry 7). Axial skeletal changes include osteopenia with wedging of vertebral bodies. Marked kyphosis may also be present. Sclerosis may develop at the superior and inferior margins of vertebral bodies, resulting in a "rugger-jersey" spine. Subchondral resorption at the disco-vertebral junction results in bone weakening and Schmorl's nodes. Other axial skeleton and joint manifestations of hyperparathyroidism include instability of the sacroiliac joints, calcium pyrophosphate dihydrate deposition disease, and gout (entries 18, 14). Hyperparathyroidism secondary to renal failure causes renal osteodystrophy and has radiologic similarities with primary hyperparathyroidism. Osteosclerosis with soft tissue and arterial calcification occur more commonly with secondary hyperparathyroidism than with the primary disease (entry 10). Osteomalacia is also more frequently associated with renal osteodystrophy. Hypoparathyroidism is found with osteosclerosis, subcutaneous calcification and calcification of the longitudinal ligaments of the spine (entry 12).

§ 4-41(F). Differential Diagnosis.

The diagnosis of hyperparathyroidism can be suspected in a patient who presents with hypercalcemia. The symptoms and signs of the more severe classic diseases are seldom observed. Evaluation of corroborating chemical and radiographic parameters help firm up the diagnosis which can be definitively made by the surgical removal of abnormal parathyroid tissue.

A number of illnesses cause hypercalcemia and must be considered in the differential diagnosis of hyperparathyroidism (entry 8). The major categories to consider in the patient with back pain, vertebral fracture, and hypercalcemia include malignancies, secondary hyperparathyroidism and granulomatous disorders such as sarcoidosis. Careful examination for the presence of a malignancy, urinalysis, and chemical testing for renal function and chest radiographs should help differentiate these illnesses from primary hyperparathyroidism.

§ 4-41(G). Treatment.

The treatment of hyperparathyroidism is the surgical removal of the malfunctioning parathyroid tissue in symptomatic patients. There is no effective medical therapy to control the effects of excessive parathyroid hormone. Most patients have a single parathyroid adenoma and its removal brings the hyperparathyroidism under control without recurrence (entry 1). In a smaller group of patients with diffuse hyperplasia of all the glands, total removal of all but a small portion of a single parathyroid gland is required (entry 2). Occasionally reoperation is needed for patients who have recurrence of the tumor or hyperplasia (entry 17). Postoperative therapy for patients with clinical bone disease includes supplemental calcium, phosphorus, vitamin D, and magnesium. This is necessary to mineralize bone which has been chronically resorbed (entry 9). Therapy for patients with secondary hyperparathyroidism is directed toward control of the underlying disease process. For example, renal transplantation may slow down the bone regression of secondary hyperparathyroidism (entry 5).

Patients with back pain from hyperparathyroidism are treated for their underlying disease. Patients with compression fractures receive symptomatic therapy, analgesics, anti-inflammatory drugs, and braces, while their skeletal lesions heal but there may be residual deformity of the spine despite bone healing.

§ 4-41(H). Course and Outcome.

Patients who have mildly elevated calcium concentrations and are asymptomatic may be followed closely without surgical intervention. In one study over four years, 20 per cent of 141 patients with hypercalcemia required surgical intervention (entry 15). Many patients may be followed by testing serum calcium, alkaline phosphatase, renal function, and bone erosion twice a year. Patients who become symptomatic require surgical intervention. Surgery is effective at controlling the disease although an occasional patient may require a reoperation. Bone lesions heal after the source of excess parathyroid hormone is removed. Brown tumors usually heal although large cysts may not and are at risk of pathologic fractures through areas of weakened bone. Lesions in the spine may heal, but areas of fracture and angular deformity are not reversible. Hypoparathyroidism is treated with calcium and vitamin D supplements.

BIBLIOGRAPHY

1. Attie, J.N., Wise, L., Mir, R., and Ackerman, L.V. The rationale against routine subtotal parathyroidectomy for primary hyperparathyroidism. *Am. J. Surg.*, 136:437-444, 1978.

2. Block, M.A., Frame, B., Jackson, C.E., and Horn, R.C., Jr. The extent of operation for primary hyperparathyroidism. *Arch. Surg.*, 109:798-801, 1974.

3. Boonstra, C.E., and Jackson, C.E. Serum calcium survey for hyperparathyroidism. Results in 50,000 clinic patients. *Am. J. Clin. Pathol.*, 55:523-526, 1971.

4. Bywaters, E.G.L. Discussion of simulations of rheumatic disorders by metabolic bone disease. *Ann. Rheum. Dis.*, 18:64-65, 1959.

5. David, D.S., Sakai, S., Brennan, L., Riggio, R.A., Cheigh, J., Stenzel, K.H., Rubin, A.L., and Sherwood, L.M. Hypercalcemia after renal transplantation: Long-term follow-up data. *N. Eng. J. Med.,* 289:298-401, 1973.

6. Dauphine, R.T., Riggs, B.L., and Scholz, D.A. Back pain and vertebral crush fractures: An unemphasized mode of presentation for primary hyperparathyroidism. *Ann. Intern. Med.,* 83:365-367, 1975.

7. Genant, H.K., Heck, L.L., Lanzi, L.H., Rossman, K., Horst, J.V., and Paloyan, E. Primary hyperparathyroidism: A comprehensive study of clinical, biochemical, and radiographic manifestations. *Radiology,* 109:513-524, 1973.

8. Goldsmith, R.S. Differential diagnosis of hypercalcemia. *N. Eng. J. Med.,* 274:674-677, 1966.

9. Gonzalez-Villapando, C., Porath, A., Berelowitz, M., Marshall, L., and Favus, M.J. Vitamin D metabolism during recovery from severe osteitis fibrosa cystica of primary hyperparathyroidism. *J. Clin. Endocrinol. Metab.,* 51:1180-1183, 1980.

10. Greenfield, G.B. Roentgen appearance of bone and soft tissue changes in chronic renal disease. *Am. J. Roentgenol.,* 116:749-757, 1972.

11. Hecht, A., Gershberg, H., and St. Paul, H. Primary hyperparathyroidism: laboratory and clinical data in 73 cases. *JAMA,* 233:519-526, 1975.

12. Jimenea, C.V., Frame, B., Chaykin, L.B., and Sigler, J.W. Spondylitis of hypoparathyroidism. *Clin. Orthop.,* 74:84-89, 1971.

13. Posen, S., Clifton-Bligh, P., and Mason, R.S. Testing for disorders of calcium metabolism. (Editorial) *Pathology,* 12:511-515, 1980.

14. Pritchard, M.H., and Jessop, J.D. Chondrocalcinoisis in primary hyperparathyroidism: Influence of age, metabolic bone disease, and parathyroidectomy. *Ann. Rheum. Dis.,* 36:146-151, 1977.

15. Purnell, D.C., Scholz, D.A., Smith, L.H., Sizemore, G.W., Black, B.M., Goldsmith, R.S., and Arnaud, C.D. Treatment of primary hyperparathyroidism. *Am. J. Med.,* 56:800-809, 1974.

16. Pyrah, L.N., Hodgkinson, A., and Anderson, C.K. Primary hyperparathyroidism. *Br. J. Surg.,* 53:234, 316, 1966.

17. Sayle, A.W., and Brennan, M.F. Strategy and technique of reoperative parathyroid surgery. *Surgery,* 89:417-423, 1981.

18. Scott, J.T., Dixon, A.S.J., and Bywaters, E.G.L. Association of hyperuricemia and gout with hyperparathyroidism. *Br. Med. J.,* 1:1070-1073, 1964.

398

§ 4-42. Ochronosis.

Ochronosis is a rare metabolic disorder associated with the deposition of homogentesic acid in connective tissue throughout the body. The accumulation of homogentesic acid results in darkened pigmentation and progressive degeneration of connective tissue. Ochronotic arthropathy develops in the fourth decade of life and is associated with progressive low back pain, stiffness and obliteration of the normal lumbar lordosis. Peripheral joint disease may also occur in the hips, knees, and shoulders. The diagnosis of ochronosis is suspected by the discovery of blue-black pigmentation of ears, sclerae, and skin and is confirmed by the detection of homogentesic acid in the urine. The radiographic abnormalities of ochronotic spondyloisis are extensive and include intervertebral disc narrowing and calcification, osteophyte formation, and osteoporosis. The therapy of ochronosis is symptomatic, including physical therapy and non-steroidal anti-inflammatory drugs. The course of the illness is progressive, causing marked degenerative changes in the axial skeleton and results in stooped posture, spinal rigidity and limited work capabilities.

§ 4-42(A). Prevalence and Etiology.

Ochronosis is a rare disorder with a prevalence of approximately one in ten million (entry 8). The illness has a wide geographic distribution although most large series have been reported from central European countries (entry 10). Men are slightly more commonly affected than women.

The cause of this illness is the congenital absence of the enzyme, homogentesic acid oxidase (entry 3); without the enzyme there is an accumulation of homogentesic acid. Alkaptonuria is the disease associated with the excretion of homogentesic acid in the urine. Ochronosis, which is caused by the same enzyme deficiency, is the discoloration of connective tissue from the deposition of a black pigment which is thought to be a polymer of homogentesic acid (entry 5). The pigment affects the integrity of cartilage matrix and chondrocytes, and results in cartilage damage and degeneration (entry 7). The inheritance of the disorder is autosomal recessive (entry 10).

§ 4-42(B). Clinical Findings.

Alkaptonuria is usually unrecognized during childhood although occasionally, discoloration of urine is detected before adulthood. Symptoms of ochronosis first appear in the fourth decade of life. Low back pain and stiffness are frequently the initial symptoms of the illness. Herniation of an intervertebral disc, particularly in men, can be the initial symptom of disease in some patients (entry 4). Peripheral joint involvement may cause loss of motion and pain in the hips, knees, and shoulders. Pain in some patients is most intense in the morning. Some also describe episodes of acutely swollen, inflamed joints, particularly the knees. This symptom is related to the association of calcium pyrophosphate dihydrate deposition disease (CPPD) and ochronosis (entry 6). Ochronotic deposition in other organs may cause prostatic enlargement with calculi, renal calculi and failure, and myocardial infarction.

§ 4-42(C). Physical Findings.

Physical examination demonstrates limited motion of the lumbar spine and localized tenderness with percussion. Muscle spasm is usually absent. With advanced disease, there is rigidity of the axial skeleton. Chest wall expansion is also limited. Kyphosis may be prominent and is associated with a loss in height. Dark pigmentation may be discovered in the nose, ears, sclerae, and fingernails.

§ 4-42(D). Laboratory Findings.

The characteristic laboratory finding is the presence of homogentesic acid in urine. Alkalinization of a urine specimen will cause darkening, indicative of the presence of homogentesic acid (entry 7). Synovial fluid analysis may demonstrate a "ground-pepper" appearance of the fluid or pyrophosphate crystals (entries 6, 2).

Pathological examination demonstrates pigment deposition in connective tissue, including articular cartilage, tendons and ligaments. Pathologic changes in the axial skeleton first occur in the lumbar spine. Pigment is located in the nucleus pulposus and annulus fibrosis. The discs become secondarily calcified with hydroxyapatite and become brittle (entry 1).

§ 4-42(E). Radiographic Findings.

Radiographic examination of the lumbar spine in ochronosis demonstrates marked disc space narrowing, osteophyte formation, and disc calcification. "Vacuum" phenomena may be seen in an intervertebral disc and are suggestive of the diagnosis of ochronosis when it occurs at multiple levels. Disease of long duration may be associated with total obliteration of disc spaces and bony fusion, and may be confused with the axial skeletal changes of ankylosing spondylitis (entry 9).

§ 4-42(F). Differential Diagnosis.

A diagnosis of ochronosis is based on the characteristic clinical symptoms and radiographic findings along with the presence of homogentesic acid in urine. Other diseases, such as CPPD, hemochromatosis, hyperparathyroidism, and acromegaly may have intervertebral disc calcification and must be considered in the differential diagnosis (entry 11). Patients with ankylosing spondylitis may have symptoms and signs similar to those of patients with ochronosis but will not have skin pigmentation or disc calcification.

§ 4-42(G). Treatment.

Specific treatment for this disease, replacement of homogentesic acid oxidase, is not available. Treatment is symptomatic and includes rest, exercise, analgesics and anti-inflammatory drugs. Low tyrosine — low phenylalanine diets lower homogentesic acid levels but do not seem to produce major benefits in regard to the course of the illness (entry 7).

§ 4-42(H). Course and Outcome.

The course of ochronosis is progressive and disabling. In some patients the arthritis proves to be disabling for any work. In others, ochronotic arthropathy requires a change in lifestyle which limits work potential (entry 7).

BIBLIOGRAPHY

1. Bywaters, E.G.L., Dorling, J., and Sutor, J. Ochronotic densification. *Ann. Rheum. Dis.,* 29:563, 1970.
2. Hunter, T., Gordon, D.A., and Ogryzlo, M.A. The ground pepper sign of synovial fluid. A new diagnostic feature of ochronosis. *J. Rheumatol.,* 1:45-53, 1974.
3. LaDu, B.N., Zannoni, V.G., Laster, L., and Seegmiller, J.E. The nature of the defect in tyrosine metabolism in alkaptonuria. *J. Biol. Chem.,* 230:251-260, 1958.
4. McCollum, D.E., and Odom, G.L. Alkaptonuria, ochronosis, and low back pain: A case report. *J. Bone Joint Surg.,* 47A:1389-1392, 1965.
5. Milch, R.A. Biochemical studies on the pathogenesis of collagen tissue changes in alkaptonuria. *Clin. Orthop.,* 24:213-229, 1962.
6. Rynes, R.I., Sosman, J.L., and Holdsworth, D.E. Pseudogout in ochronosis. Report of a case. *Arthritis Rheum.,* 18:21-25, 1975.
7. Schumacher, H.R., and Holdsworth, D.E. Ochronotic arthropathy. Part I. Clinicopathologic studies. *Semin. Arthritis Rheum.,* 6:207-246, 1977.
8. Seradge, H., and Anderson, M.G. Alkaptonuria and ochronosis: Historic review and update. *Orthopaedic Review,* 7:41-46, 1978.
9. Simon, G., and Zorab, P.A. The radiological changes in alkaptonuric arthritis: A report of 3 cases (one an Egyptian mummy). *B. J. Radiol.,* 34:384-386, 1961.

10. Srsen, S. Alkaptonuria. *Johns Hopkins Med. J.,* 145:217-226, 1979.

11. Weinberger, A., and Myers, A.R. Intervertebral disc calcification in adults. A review. *Semin. Arthritis Rheum.,* 8:69-75, 1978.

§ 4-43. Pituitary Disease.

Excessive growth hormone (GH) secretion from tumors in the anterior pituitary gland causes gigantism in growing children and acromegaly in adults. A number of morphologic and physiologic abnormalities occur secondary to hypersecretion of GH, including increased growth of bone, cartilage and visceral organs; and hypermetabolism resulting in impaired glucose tolerance and osteoporosis. Low back pain is a prominent symptom of patients with acromegaly although the range of motion of the lumbosacral spine remains normal. The diagnosis of acromegaly is confirmed by the lack of suppression of GH levels in a patient undergoing a glucose tolerance test. Abnormal suppressibility necessitates an evaluation of the sella turcica in the skull for a pituitary tumor. The treatment of acromegaly is directed at decreasing GH concentrations by surgical removal or irradiation of the anterior pituitary tumor. The major disability of acromegaly is the progressive degenerative arthropathy which affects the lower extremities and lumbosacral spine despite normalization of GH levels.

§ 4-43(A). Prevalence and Etiology.

The prevalence and incidence of pituitary tumors are unknown. Patients may have small pituitary tumors which are asymptomatic and are only discovered at autopsy (entry 4). The disease usually begins insidiously in the third to the fifth decade and in men and women equally.

Growth hormone has multiple actions including maintenance of serum glucose levels, increased protein synthesis, calcium absorption from the gut, renal tubular absorption of phosphate, epiphyseal or periosteal bone growth, production of connective tissue, and collagen synthesis. GH secretion is stimulated by hypoglycemia primarily, and to a lesser degree by exercise, sleep and stress. Overproduction of growth hormone while the growth plates are still open causes a marked overgrowth of bone, resulting in extreme height and is referred to as hyperpituitary gigantism. In the adult with closed growth plates, new bone formation is endochondral with periosteal growth, widening bones, and excess cartilage formation. The newly formed cartilage is friable and is easily damaged, causing fissuring and ulceration of the articular surface. Degeneration of the cartilage initiates increased cartilage and bone repair and secondary osteoarthritis. Acromegaly is the disease associated with excessive GH production in the adult. The source of GH is usually an acidophilic or chromophobic adenoma of the anterior lobe of the pituitary gland which is located in the center of the skull in the sella turcica.

§ 4-43(B). Clinical Findings.

The symptoms associated with acromegaly are related to the growth of the tumor intracranially as well as the effect of GH on the musculoskeletal system and other viscera. Headaches and visual disturbances are directly related to the growth of a tumor in the sella turcica. Rheumatic disease symptoms of carpal tunnel syndrome, backache, limb pain, muscle weakness and Raynaud's phenomenon are common (entry 1). Back pain is a symptom in 50 per cent of patients. The pain is localized to the lumbosacral spine in most circumstances but may radiate into the lower extremities when there is secondary cauda equina compression (entry 2). A patient may also notice a gradual enlargement of facial features, deepening of the voice, thickening of the tongue, enlargement of the extremities, particularly the fingers, and diminished libido.

§ 4-43(C). Physical Findings.

Facial characteristics of the patient with acromegaly are usually prominent and alert the physician to the potential diagnosis. The jaw is large, the skin over the face is thickened and coarse, and the forehead bossed. The musculoskeletal examination is characterized by joint swelling due to periarticular thickening and non-inflammatory synovial hypertrophy. Coarse crepitation with motion is common. The hands become "spade-like" and broad, with blunted fingers. Compression of the median nerve in the carpal tunnel elicits paresthesias in the sensory distribution of the nerve and there is a positive Phalen's test(wrist flexion test). Examination of the back may demonstrate percussion tenderness over the spine. The range of motion of the lumbar spine remains normal. This preservation of motion may be related to thickened intervertebral discs which retain their turgor but with progressive disease, painful kyphosis of the axial skeleton may occur. Long tract signs indicative of spinal cord compression may also be found in some patients (entry 3). Secondary osteoarthritic changes may appear in the knees, hips and shoulders. Rarely, an acute episode of joint inflammation may be seen secondary to crystal-induced synovitis from calcium pyrophosphate dehydrate deposition (entry 7).

§ 4-43(D). Laboratory Findings.

Abnormal laboratory parameters include elevated growth hormone levels, elevated serum glucose levels indicative of glucose intolerance, and elevated levels of serum phosphorus and alkaline phosphatase in proportion to skeletal growth. Basal GH levels are elevated and are nonsuppressible during a standard glucose tolerance test. Glucose tolerance is impaired and is relatively resistant to insulin therapy.

§ 4-43(E). Radiographic Findings.

Radiographic findings in the axial skeleton are prominent in patients with acromegaly. Anterior and lateral osteophytes of the lumbar and thoracic vertebral bodies are very prominent and may resemble diffuse idiopathic skeletal hyperostosis (DISH). Posterior osteophytes are less prominent. Posteriorly, the vertebral bodies are scalloped as a result of bone resorption (entry 9). Disc spaces are well maintained and may be increased in size. Anterior intervertebral disc calcification, thought to be secondary to calcium pyrophosphate deposition, may also be seen. Other characteristic radiographic findings of acromegaly include increased heel pad thickness and widening of joint spaces secondary to growth of cartilage (entry 8).

§ 4-43(F). Differential Diagnosis.

The diagnosis of acromegaly can be suspected in a patient with the clinical symptoms and signs previously described and confirmed by measurement of GH in basal and suppressible states. Once abnormalities of GH are determined, evaluation of the sella turcica by CAT scan or cranial tomography is essential.

The differential diagnosis of a patient with symptoms and signs of headaches, head and extremity enlargement and glucose intolerance is essentially limited to acromegaly. The radiographic changes of acromegaly with increased articular space and increased bone surface are easily differentiated from other disease processes. The later stages of joint disease of acromegaly are similar to those of primary osteoarthritis and are difficult to differentiate from these disorders. Scalloped vertebral bodies, while associated with acromegaly, may also be seen as other disease processes associated with increased intraspinal pressure, weakness of the dural sac or genetic abnormalities with tissue accu mulation of mucopolysaccharides (entry 6). Lesions associated with increased intraspinal pressure include intraspinal tumors and cysts, syringomyelia and communicating hydrocephalus. Disorders of connective tissue which result in weakness in the covering of the spinal cord, the dura, predispose vertebral bodies to scalloping. These disorders occasionally cause low back pain and include Marfan's and Ehlers-Danlos syndromes. Tumors of spinal nerves, neurofibromas, may also cause vertebral body indentation. The mucopolysaccharidoses, a heterogeneous group of genetic abnormalities which result in the excessive accumulation of mucopolysaccharides in various organs, may also cause skeletal abnormalities. Hurler's and Morquio's syndrome are most closely associated with posterior skeletal abnormalities (entry 5). The physical appearance of these patients helps differentiate them from those patients with acromegaly.

§ 4-43(G). Treatment.

Treatment of acromegaly is directed at the ablation of the tumor producing excessive amounts of growth hormone. Surgical and radiation therapy which are available include pituitary irradiation, cryosurgery, and surgical ablation through a transfrontal or transphenoidal approach (entry 10). Unfortunately, therapeutic measures which are effective in controlling the pituitary lesion have little effect on the progression of acromegalic arthropathy in the axial skeleton or peripheral joints once the joint disease has developed. These patients develop progressive degenerative joint disease and are treated in a similar fashion with nonsteroidal anti-inflammatory drugs, physical therapy and orthopaedic surgery.

§ 4-43(H). Course and Outcome.

Articular symptoms of acromegaly may range from mild

joint pain to severe disabling arthritis of the peripheral joints and axial skeleton. Degenerative changes may be progressive, causing marked joint destruction which requires joint replacement of hip or knee. Bony or disc enlargement in the spine may cause compressive symptoms requiring decompression procedures. The usual course of the illness in general is one of benign chronicity; however, some patients have a premature demise from congestive heart failure, complications of diabetes, or unrecognized hypopituitarism.

BIBLIOGRAPHY

1. Bluestone, R., Bywaters, E.G.L., Hartog, M., Holt, P.J.L., and Hyde, S. Acromegalic arthropathy. *Ann. Rheum. Dis.,* 30:243-258, 1971.
2. Gelman, M.I. Cauda equina compression in acromegaly. *Radiology,* 112:357-360, 1974.
3. Horenstein, S., Hambrook, G., and Eyerman, E. Spinal cord compression by vertebral acromegaly. *Trans. Am. Neurol. Assoc.,* 96:254-256, 1971.
4. Kovacs, K., Bryan, N., Horvath, E., Singer, W., and Ezri, C. Pituitary adenomas in old age. *J. Gerontol.,* 35:16-22, 1980.
5. McKusick, V.A., Kaplan, D., Wise, D., Hanley, W.B., Suddarth, S.B., Sevick, M.E., and Maumanee, A.E. The genetic mucopolysaccharidoses. *Medicine,* 44:445-483, 1965.
6. Mitchell, G.E., Lourie, H., Berne, A.S. The various causes of scalloped vertebrae with notes on their pathogenesis. *Radiology,* 89:67-74, 1967.
7. Silcox, D.C., and McCarty, D.J. Measurement of inorganic pyrophosphate in biologic fluids: Elevated levels in some patients with osteoarthritis, pseudogout, acromegaly, and uremia. *J. Clin. Invest.,* 52:1836-1870, 1973.
8. Steinbach, H.L., and Russell, W. Measurement of the heel pad as an aid to diagnosis of acromegaly. *Radiology,* 82:418-423, 1964.
9. Stuber, J.L., and Palacios, E. Vertebral scalloping in acromegaly. *Am. J. Roentgenol.,* 112:397-400, 1971.
10. Thomas, J.P. Treatment of acromegaly. *Br. Med. J.,* 286:330-332, 1983.

§ 4-44. Microcrystalline Disease.

Microcrystalline disease, gout and calcium pyrophosphate dihydrate disease (CPPD), are commonly associated with peripheral joint arthritis. Occasionally, patients with gouty axial skeletal disease may develop episodes of acute low back pain secondary to spinal or sacroiliac joint involvement. CPPD is associated with radiographic findings of disc calcification and degenerative changes of the discs and vertebral bodies and symptoms may develop secondary to these degenerative changes. The diagnosis of a microcystalline disease is confirmed by the presence of specific crystals, urate or calcium pyrophosphate, obtained from peripheral joints. The treatment for gout is directed at control of inflammation with anti-inflammatory drugs during acute attacks. Chronic therapy attempts to limit the accumulation of uric acid in the body by increasing urinary excretion or decreasing its production by enzymatic inhibition. Although the progression of gout may be halted through drug therapy, CPPD may progress to secondary degenerative joint changes in spite of it.

§ 4-44(A). Prevalence and Etiology.

The actual prevalence of gout and CPPD is not known. Approximately five per cent of a large adult population had hyperuricemia, while six per cent of an elderly population had CPPD in a joint (entries 4, 9). Men develop gout during the fourth or fifth decade; women develop it after menopause. The etiology is related to the inability of the body to eliminate uric acid. This may occur secondary to underexcretion through the kidney or to overproduction during protein metabolism. Uric acid accumulates in tissues throughout the body. The presence of crystals in joints, soft tissues and other areas may initiate the inflammatory response which results in acute symptoms. Uric acid may accumulate into large collections, tophi, which may be located in superficial structures such as the olecranon bursae as well as in deep areas such as the kidney, heart, and sacroiliac joints.

CPPD causes symptomatic disease in about half the number of patients affected by gouty arthritis (entry 10). Like gout, men are more commonly affected than women. The disease becomes symptomatic in patients between the sixth and seventh decade. Calcium pyrophosphate dihydrate is the crystal in CPPD which initiates the inflammatory response but the factors which facilitate the deposition of these crystals in cartilage and surrounding articular structures are poorly understood. The disease may be associated with a number of metabolic conditions including diabetes mellitus, hyperparathyroidism, hemochromatosis, hypothyroidism, Wilson's disease, and ochronosis.

§ 4-44(B). Clinical Findings.

Patients who present with back pain secondary to gout have a long history of peripheral gouty arthritis and are usually over 50 years of age (entry 7). Most patients have non-radiating low back pain due to chronic gouty arthritis. Occasionally they may have a sudden onset of low back pain associated with an acute gouty attack in the sacroiliac joints.

Patients with CPPD of the spine may also have symptoms of low back pain associated with straightening and stiffening of the spine (entry 11). Rarely do they have neurologic symptoms.

§ 4-44(C). Physical Findings.

Physical findings in the lumbosacral spine may demonstrate spinal stiffness, loss of motion and muscle spasm with pain on motion. Examination of extensor surfaces (elbows,

Achilles tendons) and ears may demonstrate tophaceous deposits.

Patients with CPPD disease may have restricted motion form associated degenerative disease of the spine. Neurologic symptoms with CPPD are rare (entry 3).

§ 4-44(D). Laboratory Findings.

Hyperuricemia is a prerequisite for the diagnosis of gout and many patients will have an elevated level of uric acid during an acute attack; however, a normal level does not eliminate the possibility of gout since uric acid concentrations fluctuate, particularly with anti-inflammatory medications. The diagnosis of gout is established definitively by the demonstration of characteristic crystals of monosodium urate monohydrate in synovial fluid or from aspirates of tophaceous deposits. Other synovial fluid characteristics of gout include fair mucin clot, elevated white blood cell count and increased protein concentration. In a patient with gouty nephropathy, renal function as measured by BUN and creatinine may be impaired, and red blood cells may be present in the patient with uric acid renal stones.

Blood studies in CPPD are of no use except to detect associated diseases such as hemochromatosis or hyperparathyroidism. Synovial fluid aspiration of acute effusions will demonstrate calcium pyrophosphate dihydrate crystals in the vast majority of patients. A careful examination for these crystals is necessary since they are less numerous than uric acid crystals in an inflamed joint and they polarize light weakly in contrast to urate crystals. White blood cell count and protein concentration in synovial fluid will be elevated in a similar fashion to that of gout.

§ 4-44(E). Radiographic Findings.

Radiographic abnormalities in the sacroiliac joint and

axial skeleton are unusual in gout but it may cause joint margin sclerosis with cystic areas of erosion in the ilium and sacrum (entry 1). Gout may also cause erosions of end plates of vertebral bodies, disc space narrowing and vertebral subluxation. Occasionally, extradural deposits of urate may cause nerve compression and can be detected by myelographic examination (entry 6). Pathologic fractures in posterior elements of vertebral bodies are also found in patients with extensive gouty involvement (entry 2).

Radiographic manifestations of CCPD in the spine include intervertebral disc calcifications, primarily in the annulus fibrosis. Disc space narrowing associated with vertebral osteophyte formation may also occur. Vertebral body destruction may become severe enough to cause degenerative spondylolisthesis (entry 12).

§ 4-44(F). Differential Diagnosis.

The diagnosis of microcrystalline disease is confirmed by the detection of the specific crystal in a clinical specimen. Patients with back pain secondary to microcrystalline disease usually have extensive disease in other locations so that aspiration of the facet or sacroiliac joints is not necessary. Careful monitoring of response to therapy will show rapid improvement if the diagnosis is correct.

Infection must always be considered if the patient has extreme pain, fever, and an elevated peripheral white blood count. Patients with septic sacroiliac joints or osteomyelitis will not improve with anti-gout therapy and they will require further evaluation with blood cultures and joint aspiration to rule out infection.

§ 4-44(G). Treatment.

Therapy for gout requires the immediate control of inflammation during the acute attack and the chronic control of hyperuricemia to prevent tophaceous deposits. An acute gouty attack may be controlled with colchicine or non-steroidal anti-inflammatory drugs, particularly indomethacin or phenylbutazone, and the medication is continued for seven to ten days or until the attack is alleviated. Once the acute inflammation has subsided, uric acid concentrations may be controlled by increasing uric acid excretion with probenecid or sulfinpryazone or by inhibiting uric acid production with the xanthine oxidase inhibitor, allopurinol, along with colchicine prophylaxis (entries 13, 5).

Therapy for CPPD is primarily directed towards control of inflammation with non-steroidal anti-inflammatory drugs. Occasionally, aspiration of a joint to remove crystals is helpful in controlling joint symptoms but this is not practical for the axial skeleton. Oral colchicine is not as effective in preventing attacks in CPPD as it is in gout. Controlling diseases associated with CPPD may help to arrest its progression but the calcium pyrophosphate deposits are not resorbed (entry 8).

§ 4-44(H). Course and Outcome.

Acute attacks of gout and CPPD do not occur with any specific intervals between them. Some patients have only one attack while others have frequent, painful bouts of inflammatory arthritis. Both the acute and chronic manifestations of gout can be well controlled with available therapy and if diagnosed early enough, patients should have limited dysfunction from the disease. Those with CPPD also have a variable period between attacks but anti-inflammatory therapy is usually effective at controlling the associated inflammation. There is, however, no effective therapy to either control or reverse the calcification of tissues or the secondary degenerative changes associated with crystal deposition. Patients with

413

severe disease may develop progressive axial skeletal involvement with limited function. Fortunately, these circumstances are rare.

BIBLIOGRAPHY

1. Alarcon-Segovia, D., Cetina, J.A., and Diza-Jouanen, E. Sacroiliac joints in primary gout. Clinical and roentgenographic study of 143 patients. *Am. J. Roentgenol.*, 118:438-443, 1973.
2. Burnham, J., Fraker, J., and Steinbach, H. Pathologic fracture in an unusual case of gout. *Am. J. Roentgenol.*, 129:1116-1119, 1977.
3. Ellman, M.H., Vazquez, T., Ferguson, L., and Mandel, N. Calcium pyrophosphate deposition in ligamentum flavum. *Arthritis Rheum.*, 21:611-613, 1978.
4. Hall, A.P., Barry, P.E., Dawber, T.R., and McNamara, P.M. Epidemiology of gout and hyperuricemia: A long-term population study. *Am. J. Med.*, 42:27-37, 1967.
5. Klinenberg, J.R., Goldfinger, S., and Seegmiller, J.E. The effectiveness of the xanthine oxidase inhibitor allopurinol in the treatment of gout. *An. Intern. Med.*, 62:639-647, 1965.
6. Litvak, J., and Briney, W. Extradural spinal depositions of urates producing paraplegia: Case report. *J. Neurosurg.*, 39:656-658, 1973.
7. Malawista, S.E., Seegmiller, J.E., Hathaway, B.E., and Sokoloff, L. Sacroiliac gout. *JAMA*, 194:954-956, 1965.
8. McCarty, D.J. Calcium pyrophosphate dihydrate crystal deposition disease (pseudogout syndrome) — clinical aspects. *Clin. Rheum. Dis.*, 3:61-89, 1977.
9. McCarty, D.J., Hogan, J.M., Gatter, R.A., and Grossman, M. Studies on pathological calcifications in human cartilage, Part I. Prevalence and types of crystal deposits in the menisci of 215 cadavers. *J. Bone Joint Surg.*, 48A:309-325, 1966.
10. O'Duffy, J.D. Clinical studies of acute pseudogout attacks: Comments on prevalence, predispositions and treatment. *Arthritis Rheum.*, 19:349-352, 1976.
11. Reginato, A., Valenzuela, F., Martinez, V., Passano, G., and Daza, S. Polyarticular and familial chondrocalcinosis. *Arthritis Rheum.*, 13:197-213, 1970.
12. Resnick, D., Niwayama, G., Goergen, T.G., Utsinger, P.D., Shapiro, R.F., Haselwood, D.H., and Weisner, K.B. Clinical, radiographic, and pathologic abnormalities in calcium pyrophosphate dihydrate deposition disease (CPPD): Pseudogout. *Radiology*, 122:1-15, 1977.

13. Yu, T.F., and Gutman, A.B. Principles of current management of primary gout. *Am. J. Med. Sci.*, 254:893-907, 1967.

PART VI. HEMATOLOGIC DISORDERS OF THE LUMBOSACRAL SPINE

§ 4-45. Introduction.

Disorders of the hematopoietic system may involve any area of the body where bone marrow is located. Since the axial skeleton contains a significant proportion of an adult's bone marrow, disorders which cause hyperplasia of bone marrow or the replacement of normal bone marrow cells with abnormal ones may be associated with low back pain. A characteristic of hematopoietic disorders is that although symptoms may be localized to various areas of the skeleton, these illnesses are systemic in origin and cause significant abnormalities in a number of other organ systems. The hematologic disorders which produce symptoms of low back pain include the hemoglobinopathies and myelofibrosis.

The symptoms of back pain in a patient with a hemoglobinopathy (sickle-cell anemia) occur in the height of a vaso-occlusive crisis. These crises occur secondary to the blockage of small vessels and the infarction of tissue by sickled cells. Back pain is acute in onset and has a duration of four to five days. Patients frequently have bone pain in the extremities as well. Patients with myelofibrosis have an insidious onset of low back pain which is secondary to the fibrosis and osteosclerosis which occurs as the bone marrow is replaced with fibrous tissue.

Physical examination of a patient with a hemoglobinopathy will demonstrate a chronically ill individual in acute distress during a crisis. Abnormal findings may include fever, tachycardia, tenderness to palpation over the back and extremities with associated muscle spasm. The patient with myelofibrosis will have pallor, splenomegaly and bone tenderness on palpation. Laboratory evaluation of these patients is very helpful in making a specific diagnosis of the underlying disorder. Patients with hemoglobinopathies have characteristic abnormalities on blood smear, including sickle and target cells. The specific hemoglobin abnormality is identified by hemoglobin electrophoresis. The blood smear in myelofibrosis contains abnormal red blood cell forms, mature and immature white blood cell forms, and variable numbers of platelets. The diagnosis of myelofibrosis is confirmed by bone marrow biopsy which characteristically contains marrow fibrosis, an increased number of megakaryocytes, and osteosclerosis. Radiographic findings associated with hemoglobinopathies include evidence of marrow expansion secondary to hyperplasia which are characterized by loss of trabeculae and cortical thinning, distinctive cup-like depression in vertebral bodies ("H" vertebrae), sclerosis and fractures compatible with aseptic necrosis of bone. Radiographic findings of myelofibrosis include diffuse osteosclerosis in the axial skeleton and proximal long bones.

Therapy for these hematopoietic disorders is essentially symptomatic. Patients are educated to avoid circumstances which may precipitate a painful crisis and they are treated with hydration and analgesics during a vaso-occlusive crisis. Transfusions are reserved for life-threatening complications such as a cerebrovascular accident. There is no effective therapy which alters the course of myelofibrosis.

Disability associated with hemoglobinopathies is caused by the systemic nature of the disease which affects the musculoskeletal, pulmonary, cardiovascular, renal, and nervous systems and it is related to the severity of painful crises. Patients with frequent crises and extensive musculoskeletal disease secondary to sickle cell anemia have a limited work potential. Despite appropriate symptomatic therapy, patients with severe sickle cell anemia die prematurely. Patients with myelofibrosis are limited as their disease progresses. Many patients will die within five years of the diagnosis of their illness. In the interim, chronic anemia, splenomegaly and bone pain hinder their ability to work.

§ 4-46. Hemoglobinopathies.

Hemoglobinopathies are a clinical group of disorders associated with defects in the physical properties or manufacture of the polypeptide chains which are the protein parts of hemoglobin. The presence of abnormal hemoglobin in red blood cells causes continuous premature destruction of these cells and chronic hemolytic anemia. Abnormal hemoglobins also change the shape of red cells, causing sickling and obstruction of the vascular microcirculation. Vascular obstruction leads to deoxygenation, tissue necrosis and pain, and it is referred to as a vaso-occlusive or thrombotic crisis. In adults with hemoglobinopathies, particularly sickle cell anemia, acute back pain along with extremity pain, are the most common symptoms of vascular crises. Persistent bone destruction secondary to vaso-occlusion and hyperplasia of bone marrow in the axial skeleton results in compression fractures, accentuated dorsal kyphosis and lumbar lordosis.

Dysfunction occurs in a number of organs, in addition to the musculoskeletal system: the lungs, heart, kidneys, liver, gall bladder, and spleen. These patients are also susceptible to infections and have an increased propensity to develop osteomyelitis and pneumonia. The diagnosis of a hemoglobinopathy can be suspected in the patient with anemia or an abnormal blood smear and is confirmed by the presence of an abnormal hemoglobin on hemoglobin electrophoresis. Treatment of vaso-occlusive crises includes hydration and analgesia, and narcotics are usually required to relieve pain. Patients with a crisis and fever must be observed for the possibility of an underlying infection. The major disability of this disease relates to the chronic effects of tissue necrosis on many organ systems. Patients with sickle cell anemia may have limitations of their work potential due to recurrent crises, avascular necroses of bone and cardiovascular and pulmonary compromise.

§ 4-46(A). Prevalence and Etiology.

The most common clinically significant hemoglobinopathies include sickle cell anemia (hemoglobin SS), sickle cell hemoglobin C disease (hemoglobin SC), and sickle cell beta thalassemia. Human adult hemoglobin consists of two pairs of coiled polypeptide chains: alpha and beta; and it is referred to as hemoglobin A. Substitution of gamma or delta chains for beta chains results in hemoglobin F (fetal) or hemoglobin A2. Patients with hemoglobin S have normal alpha chains, but have glutamic acid replaced with valine at the sixth amino acid position in the beta chain. Hemoglobin C has lysine substituted in the sixth position of the beta chain. Thalassemia is a defect in the production of an entire polypeptide chain of hemoglobin. Patients with beta thalassemia produce normal alpha chains but no beta chains.

Sickle cell anemia (hemoglobin SS) is a relatively common disorder, present in one of 625 black Americans (entry 8). Hemoglobin SC affects one in 833 black Americans and hemoglobin S-beta thalassemia, one in 1,667. Sickle cell trait (hemoglobin having only one abnormal beta chain with valine) occurs in 8% of black Americans. Sickle cell trait may also be seen infrequently in persons from the eastern Mediterranean, India, or Saudi Arabia.

418

The function of hemoglobin is to carry oxygen in red blood cells to cells throughout the body. When hemoglobin S is oxygenated, it has normal solubility. However, upon deoxygenation, hemoglobin S has decreased solubility and polymerizes into rigid, elongated rods that alter the biconcave shape of red cells into a sickle form.

The change in morphology results in the two major clinical features of sickle cell disease: chronic hemolysis with anemia and acute vaso-occlusive crises associated with pain, organ necrosis, and significant morbidity and mortality. Red blood cells which are irreversibly sickled cells are permanently deformed. They are removed by the reticuloendothelial system at an earlier stage in their life span than normal red blood cells. Marrow hyperplasia is unable to produce an adequate supply to replace those prematurely removed from the bloodstream. Chronic anemia is the result.

Sickle cell crises occur when acute sickling of red blood cells causes a rise in blood viscosity, decreased blood flow and vessel obstructions. Vessel blockage leads to ischemia, increased concentrations of deoxygenated hemoglobin and a progression to sickle crises. Some of the initiating factors which may result in sickle crises include infections, acidosis, fever and dehydration (entry 5). The severity of crises varies from patient to patient and may be related to the concentration of S and other hemoglobins in red blood cells. The frequency of painful crisies cannot be predicted.

Sickle cell crises occur most commonly in patients with hemoglobin SS. Patients with sickle trait usually do not have sufficient hemoglobin S in their red blood cells to cause sickling. Patients with hemoglobin SC have equal amounts of hemoglobin S and C, and very small quantities of hemoglobin F and A2 in each red blood cell. Hemoglobin SC causes milder disease and relatively infrequent crises. Crises may occur when patients are stressed during surgery or medical emergencies (entry 1). In addition, patients with hemoglobin SC have a higher frequency of aseptic necrosis of bone (entry 14).

The severity of sickle cell-beta thalassemia is related to the amount of normal hemoglobin that is produced (entry 13). Patients with sickle cell-B° thalassemia, produce no normal beta chains and have a disease very similar to sickle cell anemia. Patients with B+ thalessemia, produce hemoglobin A, but in reduced amounts. These patients have milder disease.

§ 4-46(B). Clinical Findings.

Patients with sickle cell anemia usually present with a painful vaso-occlusive crisis during childhood. The hand-foot syndrome may be the first manifestation of their disease. Diffuse swelling of the hands and feet occurs along with associated warmth and pain. Infarction of bone marrow, which is present in bones of the hands and feet in children, is the cause of this syndrome (entry 9).

In adults, the most common manifestation of vaso-occlusive crises is back and extremity pain. Back pain may be most severe over the axial skeleton but frequently radiates to the flanks. Muscle spasm may also contribute to the severity of pain in the lumbosacral area. In the extremities, pain is usually asymmetrical and unassociated with soft tissue swelling. The duration of symptoms is four to five days. The patient may be left with no residual pain with resolution of the crisis. Sickle crises may also present as severe abdominal pain which may mimic an acute surgical abdomen. The presence of bowel sounds and the absence of peritoneal signs help differentiate one from the other. Other forms of acute crises include the splenic sequestration, the aplastic, and hyperhemolytic crises (entry 7). Patients with hemoglobin SC and S - B+ have milder disease and as adults may present with occasional bone pain along with a history of abdominal or bone pain as a child.

Tissue infarction secondary to chronic vaso-occlusion is associated with abnormalities in a number of organ systems in patients with sickle cell anemia. In the musculoskeletal system, sickle cell anemia causes bone infarctions, joint effusions, hemarthroses, septic arthritis and osteomyelitis (entry 14). Cholelithiasis, from chronic hemolysis and hepatitis of congestive, viral, or intrahepatic cholestatis are complications in the gastrointestinal system (entry 4). Pneumonia and pulmonary infarction are frequently the cause of hospitalization of sickle cell patients (entry 2). Renal function may be impaired by glomerular sclerosis, papillary necrosis, and a renal concentrating defect (entry 3). Stroke and subarachnoid hemorrhage are potential life threatening complications of the central nervous system in sickle cell anemia (entry 10).

§ 4-46(C). Physical Findings.

Physical findings demonstrate a patient in obvious distress when they are in sickle crisis. The affected areas (back, extremities) are tender to palpation. The patient is febrile, often with a tachycardia and systolic hypertension, and tachypneic. Other physical signs found in painful crises include a tender, rigid abdomen with normal bowel sounds. Abnormal breath sounds and signs of pleural disease are present in the patient with pneumonia or pulmonary infarction.

§ 4-46(D). Laboratory Findings.

Laboratory tests demonstrating abnormalities in the hematologic system are universal in patients with sickle cell anemia. Hematocrit values are in a range of 16 to 36 percent with hemoglobins of 5 to 12 gram percent (entry 7). Leukocyte counts are usually elevated in the 20,000/mm^3

range, with increased reticulocytosis to levels of 33 percent. The inability to concentrate urine is manifested by a low urine specific gravity. Evidence of persistent hemolysis is reflected in increased serum bilirubin and lactic dehydrogenase concentrations.

§ 4-46(E). Radiographic Findings.

The radiographic abnormalities of sickle cell anemia are not unique, but are distinctive and are also diagnostic when detected in multiple sites (entry 12). In the spine, marrow hyperplasia causes loss of bone trabecula and cortical thinning. This results in osteoporosis and coarsening of the remaining trabeculae in the axial skeleton. Vertebral bodies develop a distinctive cup-like depression on the superior and inferior end plates (entry 11). Irregular sclerosis of the sacroiliac joints secondary to bone infarctions may mimic the radiologic abnormalities of ankylosing spondylitis (entry 15). Infarction or osteomyelitis may result in bony bridging in the axial skeleton and hip (entry 6).

§ 4-46(F). Differential Diagnosis.

The diagnosis of sickle cell anemia is suspected in an adult black patient with diffuse, back, bone or abdominal pain with anemia. A blood smear shows sickled cells and red blood cells with Howell-Jolly bodies, which are usually removed by a normally functioning spleen. A "sickle prep" helps confirm the diagnosis. Red blood cells are circulated with two percent sodium metabisulfite for one hour which causes sickling. The exact proportion of normal and abnormal hemoglobins is determined by the separation of the individual hemoglobins on hemoglobin electrophoresis.

The diagnosis of sickle cell anemia is not in doubt when a patient demonstrates the symptoms and signs of crises and has abnormal hemoglobin on electrophoresis. However, patients with sickle cell anemia are susceptible to infection and must be evaluated for that possibility when they present with a crisis.

§ 4-46(G). Treatment.

Medical management of sickle cell crises includes hydration, analgesics for pain and antibiotic therapy in crises when there is an ongoing infection. Standard red cell transfusions and exchange transfusions are reserved for patients with severe complications of sickle cell anemia such as strokes. Transfusions may also be useful preoperatively for patients who are undergoing surgery.

§ 4-46(H). Course and Outcome.

The course of sickle cell anemia is difficult to predict. Some patients with the disease have occasional crises, once every year or two, and have little in the way of organ dysfunction from their disease. On the other end of the spectrum are patients who are in almost continuous crises and are frequently hospitalized. These patients commonly show evidence of generalized disease affecting many organ systems. Patients must avoid situations which predispose to sickle crisis (dehydration) and receive comprehensive medical care. Immunization with pneumococcal vaccine, good nutrition, vitamins and psychologic support have improved their long term outlook. However, until the time comes when the genetic defect of sickle hemoglobin is corrected, patients with sickle cell anemia will be at risk for the complications of their disease and can expect a decreased work potential as well as shortened life span.

BIBLIOGRAPHY

1. Bannerman, R.M., Serjeant, B., Seakins, M., England, J.M., and Serjeant, G.R. Determinants of hemoglobin level in sickle cell-hemoglobin C disease. *Brit. J. Haematol.,* 43:49-56, 1979.

2. Barrett-Connor, E. Pneumonia and pulmonary infarction in sickle cell anemia. *JAMA,* 224:997-1000, 1973.

3. Buckalew, V.M., Jr., and Someren, A. Renal manifestations of sickle cell disease. *Arch. Intern. Med.,* 133:660-669, 1974.

4. Cameron, J.L., Maddrey, W.C., and Zuideman, G. D. Biliary tract disease in sickle cell anemia: Surgical considerations. *Ann. Surg.,* 174:702-710, 1971.

5. Diggs, L.W. Sickle cell crises. *Am. J. Clin. Path.,* 44:1-19, 1965.

6. Diggs, L.W. Bone and joint lesions in sickle cell disease. *Clin. Orthop.,* 52:119-143, 1967.

7. Karayalcin, G., Rosner, F., Kim, K.Y., Chandra, P., and Aballi, A.J. Sickle cell anemia: Clinical manifestations in 100 patients and review of the literature. *Am. J. Med. Sci.,* 269:51-68, 1975.

8. Motulsky, A.G. Frequency of sickling disorders in U. S. blacks. *N. Eng. J. Med.,* 288:31-33, 1973.

9. Pearson, H.A., and Diamond, L.K. The critically ill child: Sickle cell disease crises and their management. *Pediatrics,* 48:629-635, 1971.

10. Powars, D., Wilson, B., Imbus, C., Pegelow, C., and Allen, J. The natural history of stroke in sickle cell disease. *Am. J. Med.,* 65:461-471, 1978.

11. Reynolds, J. A re-evaluation of the "fish vertebra" sign in sickle cell hemoglobinopathy. *Am. J. Roentgenol.,* 97:693-707, 1966.

12. Reynolds, J. Radiologic manifestations of sickle cell hemoglobinopathy. *JAMA,* 238:247-250, 1977.

13. Reynolds, J., Pritchard, J.A., Ludders, D., and Mason, R.A. Roentgenographic and clinical appraisal of sickle cell beta-thalassemia. *Am. J. Roentgenol.,* 118:378-400, 1973.

14. Schumacher, H.R. Rheumatological manifestations of sickle cell disease and other hereditary haemoglobinopathies. *Clin. Rheum. Dis.,* 1:37-52, 1975.

15. Schumacher, H.R., Andrews, R., and McLaughlin, G. Arthropathy in sickle cell disease. *Ann. Intern. Med.,* 78:203-211, 1973.

424

§ 4-47. Myelofibrosis.

Myelofibrosis is a disease of the hematopoietic system characterized by fibrosis of bone marrow and myeloid metaplasia or the production of blood cells in non-marrow containing organs such as the liver and spleen. The disease is characterized by anemia, organomegaly, osteosclerosis, and extramedullary hematopoiesis. Bone and joint pain in the axial and peripheral skeleton is associated with myeloid metaplasia. Patients may develop localized masses of hematopoietic tissue near the spinal cord which can cause neurologic symptoms of weakness, hyper-reflexia and sensory deficit. The diagnosis of myelofibrosis is suspected when bone marrow is unobtainable by needle aspiration from a patient with anemia, and it is confirmed by the presence of fibrosis, islands of hematopoietic cells and osteosclerosis on bone marrow biopsy. Treatment for myelofibrosis is directed at relieving anemia through transfusion and controlling the extent of extramedullary hematopoiesis through chemotherapy or radiotherapy. The course of patients with myelofibrosis is one of steady deterioration. Splenomegaly becomes a major problem associated with abdominal pain and increased sequestration of blood elements. Bone pain increases with disease duration. The most life-threatening complication of myelofibrosis is the progression of abnormal hematopoiesis to acute myelocytic leukemia.

§ 4-47(A). Prevalence and Etiology.

Myelofibrosis is a relatively uncommon disorder which appears in patients during the sixth decade of life (entry 11). Both sexes are equally afflicted. The pathogenesis of this disorder is unknown. While initially thought to be a compensatory mechanism for bone marrow failure, myelofibrosis is part of the spectrum of primary myeloproliferative disorders which affect blood stem cells, including erythrocytes, granulocytes and platelets (entry

1). The marrow fibrosis which is characteristic of the illness is a secondary phenomenon.

§ 4-47(B). Clinical Findings.

Patients with myelofibrosis present with symptoms secondary to anemia — weakness, fatigue, weight loss or splenomegaly-abdominal pain, fullness or heaviness. The onset of symptoms is insidious with a usual delay of one to two years before the diagnosis is made (entry 2). Patients may also experience a gradual progressive weight loss, acute gouty arthritis, nephrolithiasis, jaundice, edema, and lymphadenopathy. Bone pain in the extremities and axial skeleton may be severe. Neurologic symptoms of lower extremity weakness and sensory loss may be present with spinal cord compression (entry 5).

§ 4-47(C). Physical Findings.

Physical examination reveals a patient who appears chronically ill with pallor. Abdominal examination is remarkable with a markedly enlarged spleen. Hepatomegaly and ascites occurs less frequently. Bones which are afflicted by myelofibrosis may be tender to palpation. Examination of the extremities demonstrates edema and purpura. Hyper-reflexia and abnormal Babinski signs are seen in the patients with spinal cord compression.

§ 4-47(D). Laboratory Findings.

Myelofibrosis is associated with a number of hematologic abnormalities. Anemia is found in the vast majority of patients and is initially normochromic, but hypochromic with progression of the illness. Blood smears reveal an abnormal configuration of cells, polychromatic cells (reticulocytes) and increased numbers of white blood cells with both mature and immature forms. This blood smear is characteristic of leukoerythroblastic anemia (entry 14).

Platelets may be present in high, normal, or low numbers. The neutrophil alkaline phosphatase score is high.

Bone marrow in myelofibrosis is unobtainable by needle aspiration because of the fibrosis and hypocellularity. Bone marrow biopsy, which protects the positional integrity of bone and marrow elements, shows fibrosis, increased numbers of megakaryocytes and osteosclerosis (entry 10).

Other laboratory features include an elevated serum or urinary uric acid concentration in most patients with myelofibrosis (entry 6). Secondary gout, with tophi and uric acid stones occurs in an occasional patient (entry 16).

§ 4-47(E). Radiographic Findings.

The radiologic findings of myelofibrosis are those of osteosclerosis in the axial skeleton and proximal long bones (entry 9). In vertebral bodies the sclerosis is increased at the superior and inferior end plates. The sclerosis may be uniformly dense or disrupted by small areas of radiolucency. Paravertebral soft tissue masses may be identified in patients with spinal cord compression and neurologic abnormalities (entry 4).

§ 4-47(F). Differential Diagnosis.

The diagnosis of myelofibrosis may be suspected in the patient with anemia, splenomegaly, and osteosclerosis, and can be confirmed by identifying the characteristic abnormalities on bone marrow biopsy. However, splenomegaly and osteosclerosis are not only found in myelofibrosis.

Systemic mastocytosis is a rare disorder associated with the proliferation of mast cells in skin, bone, liver, spleen and lymph nodes (entry 8). Mast cells contain vaso-active compounds, such as histamine, which cause some of the clinical symptoms of the illness, including hives, flushing, diarrhea, and brownish skin lesions. Patients with mast cell proliferation in the axial skeleton may have back pain.

Physical examination demonstrates hepatosplenomegaly, lymphadenopathy, bone tenderness, and skin pigmentation. Laboratory evaluation demonstrates anemia, leukocytosis with increased numbers of mast cells, and fibrosis and osteosclerosis on bone marrow biopsy. Radiologic abnormalities include diffuse or focal areas of osteopenia or osteosclerosis in the axial skeleton (entry 13). The admixture of lytic and sclerotic bone lesions helps differentiate systemic mastocytosis from myelofibrosis.

Patients with chronic myelogenous leukemia have splenomegaly and abnormal blood smears. These patients also may present with back pain (entry 7). Leukocyte counts are higher with chronic myelogenous leukemia and there is an increased proportion of immature cells. Radiograph abnormalities are uncommon (entry 3).

Other disease entities associated with osteosclerosis include skeletal metastases, Paget's disease, fluorosis and renal osteodystrophy. The association of myelofibrosis and tuberculosis is unconfirmed.

§ 4-47(G). Treatment.

The treatment of myelofibrosis is mostly supportive. Anemia is helped by transfusions, and occasionally, androgen therapy. Patients with bone pain or spinal cord compression may benefit from local radiotherapy. Chemotherapy with low doses of busulfan may decrease spleen size but may cause pancytopenia. Splenectomy is not always helpful in improving blood counts and it may be associated with excessive bleeding postoperatively. Hyperuricemia may be controlled with allopurinol. Patients who develop aggressive disease with increasing organomegaly, peripheral blast cells, anemia, and thrombocytopenia are generally resistant to therapeutic intervention (entry 15).

§ 4-47(H). Course and Outcome.

Occasionally, the course of myelofibrosis is benign with a survival greater than five years after diagnosis (entry 11). In the usual circumstance, the disease is progressive with a survival of only two to three years after diagnosis (entry 2). Rarely patients may have an acute myelofibrosis which is highly aggressive with death occurring within a year. Approximately twenty percent of patients with myelofibrosis develop acute myelogenous leukemia (entry 12) and if they do, they usually are resistant to therapeutic intervention.

BIBLIOGRAPHY

1. Adamson, J.W., and Fialkow, P.J. The pathogenesis of myeloproliferative syndromes. *Br. J. Haematol.,* 38:299-303, 1978.

2. Bouroncle, B.A., and Doan, C.A. Myelofibrosis: Clinical, hematologic and pathologic study of 110 patients. *Amer. J. Med. Sci.,* 243:697-715, 1962.

3. Chabner, B.A., Haskell, C.M., and Canellos, G.P. Destructive bone lesions in chronic granulocytic leukemia. *Medicine,* 48:401-410, 1969.

4. Close, A.S., Taira, Y., and Cleveland, D.A. Spinal cord compression due to extramedullary hematopoiesis. *Ann. Intern. Med.,* 48:421-427, 1958.

5. Cromwell, L.D., and Kerber, C. Spinal cord compression by extramedullary hematopoiesis in agnogenic myeloid metaplasia. *Radiology,* 128:118, 1978.

6. Gilbert, H.S. The spectrum of myeloproliferative disorders. *Med. Clin. North Am.,* 57:355-393, 1973.

7. Klier, I., and Santo, M. Low back pain as a presenting symptom of chronic granulocytic leukemia. *Orthopaedic Rev.,* 11:111-113, 1982.

8. Mutter, R.D., Tannenbaum, M., and Ultmann, J.E. Systemic mast cell disease. *Ann. Intern. Med.,* 59:887-906, 1963.

9. Pettigrew, J.D., and Ward, H.P. Correlation of radiologic, histologic, and clinical findings in agnogenic myeloid metaplasia. *Radiology,* 93:541-548, 1969.

10. Roberts, B.E., Miles, D.W., and Woods, C. G. Polycythaemia vera and myelosclerosis: A bone marrow study. *Br. J. Haematol.*, 16:75-85, 1969.

11. Silverstein, M.N., Gomes, M.R., ReMine, W.H., and Elveback, L.R. Agnogenic myeloid metaplasia. *Arch. Intern. Med.*, 120:546-550, 1967.

12. Silverstein, M.N., and Linman, J.W. Causes of death in agnogenic myeloid metaplasia. *Mayo Clin. Proc.*, 44:36-39, 1969.

13. Tubiana, J.M., Dana, A., Petit-Perrin, D., and Duperray, B. Lymphographic patterns in systemic mastocytosis with diffuse bone involvement and hematological signs. *Radiology*, 131:651-652, 1979.

14. Vaughan, J.M. Leuco-erythroblastic anaemia. *J. Pathol.*, 42:541-564, 1936.

15. Ward, H.P., and Block, M.H. The natural history of agnogenic myeloid metaplasia (AMM) and a critical evaluation of its relationship with the myeloproliferative syndrome. *Medicine* (Baltimore), 50:357-420, 1971.

16. Yu, T.F. Secondary gout associated with myeloproliferative disorders. *Arthritis Rheum.*, 8:765-771, 1965.

Part VII. Referred Pain

Back pain does not only occur with diseases which affect the bone, joints, ligaments, tendons, and other component parts of the lumbosacral spine but may also be a significant symptom of disorders of the vascular, genitourinary and gastrointestinal systems. These visceral organs lie in proximity to the lumbosacral spine. Inflammation, infection, or hemorrhage which originates in the aorta, pancreas, or kidney may spread beyond the confines of these organs, stimulating sensory nerves within the lumbosacral spine. This direct stimulation of sensory nerves not only results in pain which is localized to the damaged area but also may be experienced in a location other than the one being stimulated and occurs in superficial tissues supplied by the same segment of the spinal cord which sends afferent sensory fibers to the diseased area. This is called "referred pain."

Referred pain occurs as a result of the organization of the nervous system and the embryological location of the visceral organs. Sensory impulses of somatic origin (skin and parietal peritoneum for example) travel by somatic afferent neurons to the dorsal root ganglia and then into the posterior horn of the spinal cord. They synapse either with a second neuron which crosses to the opposite side of the cord and ascends to the cerebral cortex through the lateral spinothalamic tract or with motor neurons in the anterior horn of the spinal cord at the same level. Sensory impulses from visceral structures, such as the duodenum or pancreas, travel in visceral afferent nerve fibers which accompany fibers of the sympathetic system through the rami communicates and the posterior horn to join somatic sensory neurons in the posterior horn of the spinal cord. The visceral afferent fibers may travel cranially or caudally in the gray matter of the dorsal horn before synapsing with neurons of the spinothalamic tract.

Sensory impulses of visceral origin travel the same path to the brain as somatic afferent nerves. The radiation of viscral afferents to a number of spinal cord segments may explain the diffuse, poorly localized character of visceral pain. The organization of the spinal segment is further complicated by projections of neurons from higher centers in the brain which may intensify or diminish either visceral or somatic pain. Sensory stimulation which is of a visceral origin may spill over in the dorsal horn to affect somatic sensory nerves and result in pain which is only felt in the corresponding segmental somatic distribution (a dermatome of skin). Sensory input may also stimulate motor fibers in the anterior horn and their stimulation results in muscle contraction and spasm.

The segment of the spinal cord which supplies a visceral structure is not dependent on its location in the fully developed adult, but on its original location in the developing human embryo. Visceral organs migrate to their final location taking along their nerve and vascular supplies and referred pain from them will be sensed in the somatic distribution of their embryologic origin. For example, since the 12th thoracic and 1st lumbar nerves supply the visceral sensory input of the uterus, referred somatic pain originating from the uterus is felt in the groin area which receives its somatic sensory input form the same L1 segment.

Patients with visceral disease in the abdomen may experience three types of pain. True visceral pain is felt at the site of primary stimulation and is dull and aching in character. It has a diffuse and deep location. This is particularly true of visceral structures which originate in the midline (small intestine) and have visceral sensory input from both sides of the spinal cord. Visceral pain from the kidney is more easily localized since the sensory innervation to it is unilateral.

Deep somatic pain in the abdomen is related to stimulation of the parietal peritoneum. These impulses are transmitted by somatic pathways. The pain is localized, sharp, and intense in character. These pains are frequently associated with reflex abdominal wall muscle spasm.

Referred pain to the lumbosacral spine from lesions in the aorta, genitourinary or gastrointestinal tract is characteristically sharp and relatively well localized to the skin. Hyperalgesia may be noted in the area of referred pain and reflex muscle contraction may also be present. Although referred pain usually occurs in combination with visceral and somatic pain, occasionally, it may exist in the absence of visceral pain or symptoms of an underlying disease.

In the setting of low back pain and no associated visceral symptoms, a complete history, physical examination and laboratory evaluation is essential to discover the source of the visceral referred pain. Characteristically, back pain that is referred from visceral structures is not aggravated by activity or relieved by recumbency. Abdominal examination may uncover an asymptomatic pulsatile mass indicative of an abdominal aortic aneurysm or blood in the stool, indicative of a hidden malignancy. Laboratory evaluation may show pyuria indicative of urinary tract infection, or increased amylase reflective of chronic pancreatitis. Referred pain from a visceral structure must be considered as the cause of low back pain when rheumatologic, infectious, metabolic, and neoplastic origins of the pain have been eliminated as possibilities.

§ 4-48. Vascular.

§ 4-48(A). Abdominal Aorta.

Low back pain associated with disease of the abdominal aorta may occur due to aneurysmal dilatation and rupture or obstruction of the vessel. The abdominal aorta is located in the retroperitoneum, just to the left of the midline. The abdominal aorta splits at the 4th lumbar vertebra to form the common iliac arteries which supply the lower extremities. An arterial aneurysm is a localized or diffuse enlargement of an artery. One or all three layers (intima, media, and adventitia) of the aorta make up the wall of the aneurysm. A dissecting aneurysm is caused by the formation of a false channel in the wall of the aorta and it splits apart the layers of the vessel. Saccular aneurysms are bulbous protrusions of all three layers on one side of the vessel. A fusiform aneurysm is a diffuse, circumferential expansion of a segment of the vessel.

Abdominal aneurysms occur most commonly in white men between the ages of 60 to 70 (entry 25); however, they have been reported in patients as early as in the fourth decade of life and are quite common in people over the age of 50 (entry 9). The aneurysms are fusiform in configuration and extend from an area just inferior to the renal arteries to the common iliac artery bifurcation. Only five percent have dilatation of the suprarenal aorta (entry 9). The pathogenesis of the aneurysm is due to atherosclerotic degeneration with resultant structural weakening of the vessel wall. Factors which predispose to the development of atherosclerosis include hypertension, diabetes mellitus, hypercholesterolemia, and smoking (entry 25). Patients with abdominal aneurysms may demonstrate manifestations of generalized atherosclerosis including angina, previous myocardial infarction, stroke or peripheral vascular disease. Trauma, in itself, is not an etiologic factor although it may cause rupture of a preexistent aneurysm (entry 6).

Pain associated with abdominal aortic aneurysm occurs secondary to compression of surrounding structures or expansion or rupture of the aneurysm. Patients with stable aneurysms are asymptomatic and this may be the case in 40 percent of those with an abdominal aneurysm (entry 25). The aneurysm may be noted as an incidental finding on a radiograph of the abdomen, or a pulsatile abdominal mass at the level of the umbilicus which is minimally tender if found on physical examination.

Extension of the aneurysm is associated with increased pain and clinical symptoms. Most frequently, patients experience abdominal pain which is dull, steady and unrelated to activity or eating. Back pain, when it occurs, is usually associated with epigastric discomfort and it may radiate to the hips or thighs. Pressure of the aneurysm on lumbar nerves may give rise to this symptom. With increasing expansion, stretching of the mesenteric root or obstruction of the duodenum is associated with acute episodes of pain and gastrointestinal symptoms of nausea and vomiting. This may simulate gastrointestinal diseases such as pancreatitis or peptic ulcer disease.

Rupture of an abdominal aneurysm is associated with excruciating pain, circulatory shock from blood loss and an expanding mass. The pain of rupture may be the first clinical sign in a previously asymptomatic patient with an aneurysm. An exacerbation of previously milder pain in a patient with an aneurysm is a harbinger of extension or impending rupture. Ruptures of the abdominal aorta are frequently located at the junction of the aortic attachment to the vertebral bodies and the portion which is unattached in the retroperitoneum. Blood from the ruptured aneurysm may be contained in the retroperitoneum or may rupture into the peritoneal cavity or a hollow viscus. Rupture into the retroperitoneum is associated with the sudden onset of severe, tearing or piercing pain which is continuous and increasing in intensity. The pain is present in the deep abdomen and is referred into the back, legs, and groin (entry 2). Physical examination demonstrates hypotension, profuse sweating and a pulsatile abdominal mass which is tender. Palpation of the aneurysm may elicit low back pain which radiates in the thigh or lower abdomen (entry 10). The abdominal wall is usually not rigid. Less frequently, patients have hemorrhagic discoloration of the skin in the back and flanks secondary to a retroperitoneal hematoma, or loss of sensation in the distribution of the femoral nerve along with weakness in the quadriceps musculature as a manifestation of rupture in the psoas region (entries 3, 41).

Plain films of the abdomen are the most helpful laboratory test in detecting abdominal aneurysms. Anteroposterior and lateral projections of the abdomen will demonstrate a curvilinear thin layer of calcification in the wall of the aorta in 70 per cent of patients (entry 30). Vertebral body erosion is generally not found in abdominal aneurysm (entry 58).

Ultrasonography and computerized axial tomography of the abdomen are noninvasive diagnostic methods used to identify the presence and extent of an aneurysm. Aortography is not done routinely in patients with abdominal aneurysm because of potential complications or the lack of time to complete the test in the compromised patient with an acute rupture (entries 57, 39). Aortography may be useful in the patient with an aneurysm associated with occlusive disease and ischemic symptoms involving the lower extremities.

Therapy of abdominal aneurysms is surgical. The presence of an aneurysm 6 cm. or more in external diameter is associated with an increased mortality, and if there is no major contraindication, elective resection and bypass of the aneurysm should be considered (entry 20). Improved survival has been documented for patients who have undergone surgery for intact aneurysms (entry 7). Patients with evidence of a ruptured aneurysm require immediate surgical intervention since up to 80 per cent of patients with ruptured aneurysms live at least six hours from the onset of symptoms. Early operation is the most important factor influencing survival. Recent surgical data suggest that over 70 per cent of patients with ruptured abdominal aneurysms can survive surgery (entry 11). Complications associated with surgery include shock, myocardial infarction, necrosis of viscera and renal failure. Infection of the aneurysm may be a complicating factor both pre and post operatively (entries 11, 4, 43).

Obstruction of the abdominal aorta is associated with pain located in the muscles of the low back and gluteal areas. The pain may be either of acute or gradual onset. Patients who develop acute embolic obstruction of the terminal aorta develop acute claudication to the lower extremities and acute severe low back pain (entry 45). The source of the embolus is most frequently a mural thrombosis from the left side of the heart which has been damaged by a myocardial infarction, cardiomyopathy, valvular disease, atrial fibrillation or atrial myxoma. The embolus lodges at the bifurcation of the aorta, blocking blood flow to the lower half of the body. Patients experience pain in the thighs, low back, buttock and lower abdomen. Neurologic function is impaired as evidenced by weakness, numbness, and paresthesias. Pulses are lost in the lower extremities and the skin turns pale. Removal of the clot is required for the patient to survive. Embolectomy may be accomplished with the use of an intraarterial balloon catheter. Surgical intervention is necessary if non-surgical methods are inadequate and fail to restore blood flow. There is a 20 percent mortality among patients who have had a successful revascularization operation because of underlying cardiac disease.

Low back pain may also be a symptom of the patient who develops gradual obstruction of the abdominal aorta (entry 19). Occasionally they will present with low back pain as their initial symptom and it may occur without associated symptoms of claudication, neurologic dysfunction or pallor. Back pain from arterial obstruction may be limited to repeated activity with the pain subsiding with rest. Plain radiographs of the abdomen are useful in documenting calcification in the aorta, confirming the presence of atherosclerosis as well as the location of the obstruction. By-pass grafting is successful in controlling symptoms in 80 percent of patients undergoing such surgery. Mortality from the operation is greatest with those who have cardiac disease (entry 8).

§ 4-49. Genitourinary Tract.

The organs that comprise the genitourinary tract are located in the retroperitoneum and pelvis and lie close to the lumbosacral spine. Diseases which affect the genitourinary organs may be associated with both local and referred pain which radiates to the lumbosacral area.

§ 4-49(A). Kidney.

The kidneys are located at the level of the 10th through 12th thoracic and first lumbar vertebrae (T10-12, L1). Pain from diseases which affect the kidneys is felt at the costovertebral angle just lateral to the paraspinous muscles at T12-L1. It often radiates around the flank toward the umbilicus and is usually dull and constant. The source of the pain is thought to be due to sudden distention of the capsule of the kidney. Diseases which cause or are associated with acute obstruction, stone, hydronephorosis, or acute pyelonephritis are painful. Disease processes which cause only gradual capsular distention are unassociated with kidney pain.

Acute pyelonephritis is an infection, usually bacterial, of the parenchyma of the kidney (entry 44). Bacteria are able to travel retrograde from the bladder to the kidney when there is an obstruction in the genitourinary system which allows access to the kidney parenchyma. Patients complain of severe pain over one or both costovertebral angles and it is not affected by position or movement. They have frequency, urgency and burning on urination. Patients with acute pyelonephritis are systemically ill with high fever and chills. They exhibit exquisite percussion tenderness over the costovertebral angle. Abdominal signs may include muscle guarding of the abdominal wall, and hypoactive bowel sounds. Laboratory tests confirm the presence of bacteria, and cultures of urine and blood will grow the offending organism. Treatment with appropriate antibiotics results in the control of the infection and resolution of the pain. When the infection is resistant to cure because of structural abnormalities such as polycystic kidneys, the patient may develop chronic pyelonephritis and this may be clinically silent except in acute exacerbations of the chronic infection when localized renal pain will be present (entry 47).

438

Infections which escape the confines of the renal capsule will lodge in the perinephric tissues and form a perinephric abscess. Frequently these patients have a history of chronic renal disease from urinary tract infections, urinary calculus with hydronephrosis, trauma, or hematogenous spread from a distant infected area. The site of pain associated with perinephric abscess is the costovertebral angle and the patient may have mild to severe tenderness on palpation, as well as muscle spasm with a scoliotic posture. A mass may be palpated over the flank. Radiologic findings in the abdomen include ptosis of a kidney, and obliteration of the psoas shadow. Treatment includes appropriate antibiotics and surgical drainage (entry 53).

Renal tumors, including hypernephroma, are frequently painless. Initial manifestations of hypernephroma include painless hematuria, fever, anorexia, anemia of unknown cause or an abdominal mass. However, when there is hemorrhage into the kidney, extension of the tumor beyond the capsule, or obstruction, the patient will complain of localized, dull pain (entry 5). Physical findings include a flank mass, and prominent veins over the abdominal wall if the vena cava has been obstructed. The presence of a tumor is suggested by the presence of a mass on intravenous pyelography and is confirmed by biopsy. Therapy frequently includes nephrectomy along with radio and chemotherapy for metastatic lesions (entry 55).

§ 4-49(B). Ureter.

Nephrolithiasis, urinary stones, may also be associated with back pain. Stones located in the pelvis of the kidney will cause dull flank pain if there is capsular distention, and colic if there is obstruction at the uteropelvic junction. Stones that pass through into the ureter may cause back pain if there is complete obstruction and distention of the ureter and renal pelvis. More commonly, colicky pain is experienced locally and is accompanied by pain in the testicle or vulva if the stone is lodged in the upper ureter or bladder. If the stone is located in the lower ureter (entry 52), the pain is usually excruciating, causing the patient to writhe about to find a comfortable position. This is in contrast to patients with disease in the abdomen with peritonitis who lie quietly because motion increases discomfort. Pain is relieved with passage of the stone through the genitourinary system.

§ 4-49(C). Bladder.

Bladder pain is visceral in origin and is suprapubic or lower abdominal in location. It is not usually located in the sacral area. Patients with severe cystitis (local infection or inflammation of the bladder) may experience mild, diffuse low back pain which resolves with the resolution of the inflammation. Patients with acute cystitis describe burning pain with urination, but little fever and no chills. Those with chronic cystitis who develop signs of obstruction, including cystoceles, may have persistent low grade back pain. The diagnosis of cystitis is made in the patient with pelvic symptoms and a positive bacterial culture. Cystoscopy, which allows direct visualization of the bladder wall, documents the extent of bladder inflammation and the obstruction in the patient with persistent symptoms.

§ 4-49(D). Prostate.

The prostate gland is located in the pelvis at the base of the bladder in males. Diseases of the prostate are not usually associated with visceral pain from the gland itself. Instead, pain is localized to the perineal or rectal area and the sacral portion of the lumbosacral spine. Patients with acute bacterial prostatitis experience fever, chills, malaise and acute pain in the perineum and low back (entry 36). Pain on urination is a very prominent symptom. Rectal examination demonstrates a swollen, tender prostate. Prostatic massage results in a purulent prostatic fluid which, when cultured, grows the infecting organism. They may go on to develop chronic prostatitis after repeated episodes of acute infection. Other patients develop symptoms of chronic prostatitis, including dysuria, nocturia, and pain in the pelvic area, genitalia or low back without any history of prior acute prostatitis (entry 38) and they frequently develop recurrent urinary tract infections. Antibiotics are the treatment of choice, but because of the poor diffusion of antibiotics into the prostate, recurrent and persistent infection is the rule and they may have chronic pain (entry 12).

Neoplasms of the prostate are among the most common malignancies found in males. Benign prostatic hypertrophy, a hyperplastic growth of the glandular tissue of the prostate, develops slowly and is not associated with low back pain. Adenocarcinoma of the prostate does not cause back pain when the tumor is confined to the limits of the gland capsule; however, the cancer may spread through the pelvic lymph nodes and through vertebral veins to bones in the pelvis and lumbosacral spine (entry 59). Symptoms of low back pain with radiation down one or both legs, associated with bladder obstructive symptoms in a male usually over 50 years of age, suggests metastases to the axial skeleton by prostatic cancer. Rectal examination documents a hard, fixed nodular prostate. Anemia, hematuria, decreased renal function and elevated serum acid phosphatase may be present on laboratory examination. Osteoblastic lesions from metastases to bone are demonstrated on plain radiographs of the pelvis. Diagnosis is confirmed by biopsy of the gland (entry 32). Therapy for prostatic cancer include surgical, endocrine and radiation modalities (entry 15).

§ 4-49(E). Female Genital Organs.

Females with disease in the pelvic organs may experience visceral, somatic or referred pain (entry 54). Visceral pain from the uterus, fallopian tubes, or ovary is transmitted through nerves which travel with the sympathetic nervous system to spinal segment T11-T12 and with the parasympathetic system to segments S2-S4. The pain is deep, diffuse and not well localized. Somatic nerves supply supporting tissues of the pelvis including muscles, ligaments (uterosacral, for example), peritoneum and periosteum of bone. Irritation, traction, or pressure on these structures results in more localized, sharp pain which is felt in the suprapubic area or sacrum. Referred pain to the low back develops when sympathetic or parasympathetic nerves are stimulated by a disease process in the pelvic organs and causes sensory fibers in the dorsal horn of the corresponding spinal segment (T12, S2) to be activated. Back pain secondary to a gynecologic disorder is almost invariably associated with symptoms and signs of disease in the pelvic organs. A number of pathologic processes which affect the uterus, tubes or ovaries may be associated with low back pain.

§ 4-49(F). Uterus.

Leiomyomas are the most common form of uterine tumor (entry 48). They are benign tumors composed of smooth muscle which arise in the wall of the uterus. Small myomas usually do not produce symptoms; however, these tumors may become quite large. Large tumors produce symptoms of heaviness in the pelvis and may produce back or lower extremity pain if they place pressure on nerves in the sacral portion of the bony pelvis. Myomas are palpable on physical examination of the pelvis. Surgical removal is indicated for persistent symptoms of pain, uterine bleeding or infertility.

Women who have had multiple births may experience weakening of the uterosacral and cardinal ligaments as well as the pelvic musculature which supports the uterus (entry 51). In addition, patients who have a pelvic or presacral tumor or sacral nerve disorder may develop uterine prolapse. Uterine prolapse is the migration of the uterus down into the vagina. Patients with moderate prolapse will complain of a sensation of heaviness in the pelvis, lower abdominal pulling discomfort, and low back pain. Pelvic examination reveals descent of the cervix into the vagina. Therapy includes a vaginal pessary or surgical removal of the uterus. The malposition of the uterus in the pelvis may be associated with back pain. The body of the uterus is normally directed forward in the pelvis (entry 50). Retroversion of the uterus is the term used to describe the position of the body of the uterus when it is directed posteriorly toward the sacrum. The uterus is moveable and its position may vary; however, its position may remain retroverted if diseases in the pelvic organs (salpingitis) cause fixation from adhesions. Pelvic pain, low back pain, abnormal menstrual bleeding and infertility have been associated with retroversion of the uterus. Pelvic examination allows determination of the position of the uterus. Bimanual replacement of the uterus or vaginal pessary may relieve symptoms.

Pain in women with a retroverted uterus may be increased during menses; however, even those with normal uterine position may experience abdominal and back pain associated with their menses. Painful menstruation and dysmenorrhea is the most common of all gynecologic complaints and is the leading cause of absenteeism of women from work. Dysmenorrhea may be primary or secondary. When primary, it is unassociated with an identifiable gynecologic disease. Secondary dysmenorrhea is caused by pelvic disease such as uterine malposition, cervical stenosis

443

or salpingitis. Painful menses occurs in ovulatory cycles and is associated with nausea, vomiting and diarrhea. Dysmenorrhic pain is pelvic in location, crampy, correlating with uterine contractions, and is frequently referred to the low back. It starts just before the onset of bleeding and lasts one to two days after. Many patients with primary dysmenorrhea obtain relief with the use of nonsteroidal anti-inflammatory drugs or oral contraceptives (entry 61).

Endometriosis is a disease associated with the presence of functioning tissue from the lining of the uterus (endometrium) outside the uterine cavity. Endometrial tissue may be located most commonly on the ovaries or dependent portion of the pelvic peritoneum. However, peritoneal surfaces as well as extraperitoneal sites have been remote locations for this tissue. Endometrial tissue outside of the uterus undergoes the same monthly cycle of growth, shedding and bleeding as the endometrial lining in the uterus, and symptoms of endometriosis are correlated with the site of the abnormal tissue. Implants in the rectovaginal septum, colon, and ureter are associated with low back pain which may radiate to the medial or posterior portions of the thighs. The pain may be intermittent or persistent, but it characteristically increases at the time of menstruation and persists throughout the entire period of bleeding. Pelvic examination may show an asymptomatic pelvic mass or tenderness on palpation of the organ affected by the endometriosis. Diagnosis of the disease is confirmed by biopsy. Potential complications of endometriosis include infertility, intestinal or ureteral obstruction, and infection. Therapy for endometroisis may include medications, including hormones and nonsteroidal anti-inflammatory drugs, pregnancy, and surgery to remove foci of abnormal implants and lyse adhesions (entry 29).

§ 4-49(G). Fallopian Tube.

The fallopian tubes are appendages of the uterus and convey ova from the ovary to the uterus. The most common disorder which affects the fallopian tubes and is associated with low back pain is pelvic inflammatory disease (entry 17). Pelvic inflammatory disease is a term for acute or chronic infection of the tubes and ovary. Bacteria, particularly Neisseria gonorrhoeae gain access to the tube by direct spread from the endometrial lining or by lymphatic dissemination. The infection may localize to the tube or spread to involve the ovary (tubo-ovarian abscess) or the pelvic peritoneum.The clinical symptoms of a patient with acute salpingitis is lower abdominal and pelvic pain which may be unilateral or bilateral. Patients have a feeling of pelvic pressure and may also have low back pain with radiation into the thighs. Nausea may also be present. Physical examination finds an acutely ill patient with or without fever, hypoactive bowel sounds, lower quadrant abdominal tenderness, purulent cervical discharge and exquisite tenderness on movement of the pelvic organs on pelvic examination. Laboratory examination of the discharge may show pathogenic organisms on gram stain and on culture. Therapy for patients who are acutely ill is hospitalization for intravenous antibiotics and bed rest. Repeated infections or inadequate treatment of acute salpingitis can result in recurrent episodes of pelvic infection, tubo-ovarian or pelvic abscess. Complications of these very serious infections include infertility, peritonitis, intra-abdominal abscesses, bowel obstruction or spetic emboli.

§ 4-49(H). Ovary.

Ovarian neoplasms, either benign or malignant, are asymptomatic until they grow large enough to produce symptoms of pelvic pressure. Symptoms associated with ovarian cysts are localized to the pelvis and abdomen. Rarely, back pain may accompany pelvic and abdominal pain when the blood supply to the ovary or cyst is compromised. This may also occur with torsion of an ovarian cyst. Patients with malignant neoplasms of the ovary may develop back pain by direct extention of the tumor or spread through the lymphatics (entry 31).

§ 4-49(I). Pregnancy.

Pregnant women complain of low back pain with increasing size of the gravid uterus. Low back pain may be secondary to increasing tension in the uterosacral ligaments or to a marked increase in lumbar lordosis with concomittant muscle strain. Low back pain in pregnancy may also be related to pelvic girdle relaxation. In the nonpregnant state there is practically no motion in the joints of the pelvis, symphysial and sacroiliac joints. During pregnancy, women produce a hormone, relaxin, which allows increased motion in the pelvic joints and this causes tension in the relaxed capsule and ligaments. Pain develops about the sacroiliac joints, symphysis pubis and the medial part of the thighs. It is increased by active movement such as in climbing stairs. Most patients have resolution of their symptoms postpartum. Rarely do they continue with symptoms of pelvic relaxation and require bracing or operative stabilization of joints in order to resolve their complaints (entry 27).

§ 4-50. Gastrointestinal Tract.

The organs of the gastrointestinal tract associated with back pain are those in direct contact with the retroperitoneum or those with referred pain distribution to the back. Diseases of the pancreas, duodenum, gall bladder and colon may be associated with low back pain of visceral, somatic, or referred origin.

446

§ 4-50(A). Pancreas.

The pancreas is located at the level of the first and second lumbar vertebrae in the retroperitoneum. Pain from diseases which affect the pancreas is felt deep in the mid-epigastrium secondary to somatic irritation and is referred to the back in the region of L1. Acute inflammatory diseases of the pancreas cause severe, persistent pain which may be out of proportion to physical findings. Infiltrative diseases of the pancreas, pancreatic carcinoma, may not cause pain until the lesion has metastasized or caused obstruction of the biliary tree. Disease processes which affect the head of the pancreas cause pain to the right of the spine while lesions of the body and tail are felt to the left of the spine.

Acute pancreatitis is an inflammatory disease of the pancreas in which the digestive enzymes produced by that organ act upon pancreatic tissue (entry 23). Conditions which may precipitate episodes of pancreatitis include gallstones, ethanolism, drugs, hyperlipidemia, trauma, hypercalcemia and obstruction to the outflow of pancreatic enzymes. Most often the presenting and most significant symptoms include steady, boring, severe epigastric pain which radiates through to the upper lumbar spine and is increased in the supine position. Patients with pancreatitis are systemically ill with fever, tachycardia, hypotension, and abdominal tenderness without abdominal muscle guarding. They frequently assume a characteristic position, sitting with the trunk flexed, the knees drawn up, and arms folded across the abdomen so as to get relief. Elevated amylase and lipase concentrations on laboratory evaluation reflect pancreatic inflammation. Radiographic findings in the abdomen may include the presence of a single, dilated loop of small bowel with edematous walls in the left upper quadrant, a "sentinal loop." Treatment for acute pancreatitis includes general supportive measures with fluids, nasogastric suction, respiratory support if respiratory failure intervenes, surgical intervention for local complications (pseudocyst) and removal of precipitating factors which may provoke additional attacks (gallstones). Patients who continue with inflammation of the pancreas, however, may develop chronic pancreatitis, and they have intermittent or persistent pain which is boring, dull or sharp, located in the epigastrium, and radiates through to the back (entry 49). Medical complications of chronic pancreatitis include malabsorption, weight loss, pseudocysts, biliary obstruction and diabetes mellitus.

Pancreatic tumors are frequently asymptomatic until they have metastasized to surrounding abdominal viscera. Patients may present with a nondescript, dull, mid-epigastric pain which radiates through to the back (entry 28). They may also have symptoms of anorexia, weight loss, nausea, vomitting, and jaundice (entry 56). Physical findings include a hard abdominal mass which may be associated with a distended, non-tender gall bladder, Courvoisier's sign. Commonly, abnormal laboratory tests include decreased hematocrit, elevated erythrocyte sedimentation rate, elevated serum alkaline phosphatase, and hyperbilirubinemia. A number of radiographic techniques, including upper gastrointestinal series, sonography, angiography, endoscopic retrograde cholangiopancreatography and computerized axial tomography of the abdomen, may be useful in identifying the presence of a pancreatic tumor (entry 46). Diagnosis of pancreatic tumor is confirmed by intraoperative biopsy or fine needle aspiration biopsy of the pancreas (entry 60). Therapy for pancreatic carcinoma includes surgical removal of the pancreas and duodenum as well as lymph nodes, chemotherapy and radiation therapy. Unfortunately cures of this neoplasm are very rare (entry 14).

§ 4-50(B). Duodenum.

The duodenum is a tubular organ which connects the stomach and the small intestine. It is shaped in the form of the letter C and is located retroperitoneally at the level of the first lumbar vertebrae. The primary disease of the duodenum which is associated with low back pain is peptic ulceration of the posterior wall (entry 42).

A duodenal ulcer is a break in the lining of the duodenum which extends into the muscle layer. Some of the factors which promote ulcers include caffeine, alcohol, aspirin-like drugs and cigarette smoking (entry 22). The pathogenesis of duodenal ulcers seems to be related to the presence of increased concentrations of gastric acid and impaired mucosal defense to protect the mucosa from autodigestion. Patients with duodenal peptic ulcer generally present with burning, epigastric pain which occurs one to three hours after meals, awakens the patient from sleep, and is relieved with food (entry 13). A minority have pain which radiates to the back (entry 24). Back pain is most closely associated with ulceration of the posterior wall of the duodenum. Physical examination may reveal little more than epigastric tenderness. Laboratory evaluation is of little benefit unless a decreased hematocrit is present to suggest hemorrhage, or an elevated serum amylase concentration indicates posterior penetration into the pancreas. Upper gastrointestinal radiography may outline the ulcer with barium while endoscopy allows for direct visualization of the ulcer crater (entry 35). Treatment of peptic ulcers, which includes antacids, histamine H2-receptor antagonists and sulcralfate, is successful in up to 85 percent of patients. However, persistence or a change in symptoms suggests a complication of the disease. An increase in pain, the loss of relief with food, and the onset of radiation to the back suggests a posterior penetration through the wall of the duodenum into the underlying pancreas and acute pancreatitis. Approximately 20 percent of patients with duodenal ulcers may develop this complication (entry 40). Back pain may occur in the absence of anterior abdominal pain and may become persistent. The complications of an untreated posterior penetrating ulcer include pancreatitis, obstruction and giant duodenal ulcers (entry 37).

§ 4-50(C). Gall Bladder.

The gall bladder is a muscular sac which stores the bile produced by the liver. The common bile duct joints the pancreatic duct and enters the duodenum. The gall bladder is an appendage of the common bile duct and is connected by the cystic duct. In certain patients, the bile that is stored in the gall bladder crystallizes to form gall stones (cholelithiasis). The presence of these stones may cause inflammation of the gall bladder, acute cholangitis, which is associated with abdominal pain, tenderness, and fever. Patients may develop pain in the back, on a referred basis, at the tip of the right scapula or occasionally in the dorso-lumbar spine (entry 21). Physical examination demonstrates a positive Murphy's sign (inspiratory arrest with palpation in the right subcostal area), abdominal guarding, rebound tenderness, fever and jaundice. Laboratory tests demonstrate an elevated white blood cell count, hyperbilirubinemia, and increased serum amylase. Ultrasonography is a sensitive test to detect the presence of gall stones and thickening of the gall bladder wall (entry 33). Treatment for acute cholecystitis is frequently surgical removal of the gall bladder. Patients who do not elect cholecystectomy may go on to develop chronic cholecystitis and cholelithiasis. These may experience episodes of biliary colic, steady pain which lasts for hours localized to the epigastrium and right upper quadrant and which radiates to the back in over 50 percent of patients (entry 26). Patients with gallstones are at risk of lodging stones in the biliary or pancreatic ducts. Biliary obstruction may result in gall bladder perforation, gangrene, infection, or pancreatitis.

§ 4-50(D). Colon.

The colon is a long, muscular, tubular organ which resorbs water and electrolytes and stores stool before evacuation. Of the portions of the colon, ascending, transverse, descending, sigmoid and rectum, only the transverse colon is located outside the retroperitoneum. Pain of colonic origin is usually felt in the abdomen in the lower quadrants; however, disease processes which affect the rectum may be associated with midsacral back pain. The sigmoid colon and rectum are located just anterior to the sacrum and coccyx.

Diverticulitis of the colon may be associated with low back pain. Diverticuli are outpouchings of the wall of a portion of the gut. The cleft which is made by the nutrient artery passing into the wall of the gut is their most frequent location. Diverticuli in the gut are usually asymptomatic; however, diverticulitis, a disease associated with infection and inflammation of diverticuli, causes many symptoms (entry 1). The patients develop acute, persistent pain which localizes to the left lower quadrant and then radiates to the low back. They frequently have fever and a change in bowel habits. Dysuria will also be present if the bladder is involved. Physical examination demonstrates tenderness in the left lower quadrant, depressed bowel sounds, tenderness and a mass on rectal examination. Laboratory examination will show an elevation in white blood cells and abnormal urinalysis. Plain radiographs of the abdomen may show evidence of an ileus or intestinal obstruction. A barium enema must be performed with care during the acute stage of the disease for fear of potential perforation. Treatment of acute diverticulitis includes antibiotic therapy to control infection and prevent perforation. Surgical drainage of an abscess is necessary if the diverticular disease has perforated into the peritoneum. The damaged segment of bowel is removed and colostomy performed; this is re-attached to the rectum once the infection is cleared (entry 34).

Rectal adenocarcinoma is another colonic cause for low back pain (entry 18). Patients are asymptomatic in the early stages of the illness except for a change in bowel habits but they may show fatigue from occult bleeding. Those who develop systemic symptoms of anorexia, weight loss, and weakness have disease which has invaded through the wall and has metastasized. Patients with infiltration of the rectum with tumor experience deep pelvic and midsacral back pain which may radiate to the lower extremities if local invasion irritates the sacral nerves. Rectal examination may show a hard mass which is fixed and nontender. Biopsy with a flexible sigmoidoscope can confirm the diagnosis. Curative resection for rectal cancer involves removal of the rectal lymph nodes and surrounding perineal structures (entry 16). Palliative surgery is offered from those patients with metastatic disease who are at risk of obstruction. Chemotherapy for rectal carcinoma has been disappointing.

BIBLIOGRAPHY

1. Asch, M.J., and Markowitz, A.M. Diverticulosis coli: A surgical appraisal. *Surgery,* 62:239-247, 1967.
2. Barratt-Boyes, B.G. Symptomatology and prognosis of abdominal aortic aneurysm. *Lancet,* 2:716-720, 1957.
3. Beebe, R.T., Powers, S.R., Jr., and Ginouves, E. Early diagnosis of ruptured abdominal aneurysm. *Ann. Intern. Med.,* 48:834-838, 1958.
4. Bennett, D.E., and Cherry, J.K. Bacterial infection of aortic aneurysms: A clinicopathologic study. *Am. J. Surg.,* 113:321-326, 1967.

5. Berger, L., and Sinkoff, M.W. Systemic manifestations of hypernephroma: A review of 273 cases. *Am. J. Med.,* 22:791-796, 1957.
6. Cannon, J.A., Van De Water, J., and Barker, W.F. Experience with surgical management of 100 consecutive cases of abdominal aneurysm, *Am. J. Surg.,* 106:128-143, 1963.
7. Crawford, E.S., Saleh, S.A., Babb, J.W., III, Glaeser, D.H., Vaccaro, P.S., and Silvers, A. Infrarenal abdominal aortic aneurysm. *Ann. Surg.,* 193:699-709, 1981.
8. Crawford, E.S., Bomberger, R.A., Glaeser, D.H., Saleh, S.A., and Russell, W.L. Aortoiliac occlusive disease: Factors influencing survival and function following reconstructive operation over a twenty-five-year period. *Surgery,* 90:1055-1067, 1981.
9. DeBakey, M.E., Crawford, E.S., Cooley, D.A., Morris, G.C., Jr., Royster, T.S., and Abbott, W.P. Aneurysm of abdominal aorta. *Ann. Surg.,* 160:622-639, 1964.
10. DeHoff, J.B., and Finney, G.G. Sign of ruptured aneurysm of abdominal aorta. *N. Eng. J. Med.,* 281:47-48, 1969.
11. Diehl, J.T., Cali, R.F., Hertzer, N.R., and Beven, E.G. Complications of abdominal aortic reconstruction: An analysis of perioperative risk factors in 557 patients. *Ann. Surg.,* 197:49-56, 1983.
12. Drach, G.W. Prostatitis and prostatodynia: Their relationship to benign prostatic hypertrophy. *Urol. Clin. North Am.,* 7:79-88, 1980.
13. Earlam, R. A computerized questionnaire analysis of duodenal ulcer symptoms. *Gastroenterology,* 71:314-317, 1976.
14. Edis, A.J., Kiernan, P.D., and Taylor, W.F. Attempted curative resection of ductal carcinoma of the pancreas: Review of Mayo Clinic experience, 1951-1975, *Mayo Clin. Proc.,* 55:531-536, 1980.
15. Elder, J.S., and Catalona, W.J. Management of newly diagnosed metastatic carcinoma of the prostate. *Urol. Clin. North Am.,* 11:283-295, 1984.
16. Enker, W.E., Laffer, U.T., and Block, G.E. Enhanced survival of patients with colon and rectal cancer is based upon wide anatomic resection. *Ann. Surg.,* 190:350-360, 1979.
17. Eschenbach, D.A., and Holmes, K.K. Acute pelvic inflammatory disease: Current concepts of pathogenesis, etiology, and management. *Clin. Obstet. Gynecol.,* 18:35-56, 1975.
18. Falterman, K.W., Hill, C.B., Markey, J.C., Fox, J.W., and Cohn, I., Jr. Cancer of the colon, rectum and anus: A review of 2,313 cases. *Cancer,* 34:951-959, 1974.

19. Filtzer, D.L. and Bahnson, H.T. Low back pain due to arterial obstruction. *J. Bone J. Surg.,* 41B:244-247, 1959.
20. Foster, J.H., Bolasny, B.L., Gobbel, W.G., and Scott, H.W., Jr. Comparative study of elective resection and expectant treatment of abdominal aortic aneurysm. *Surg. Gyn. Obst.,* 129:1-9, 1969.
21. French, E.G., and Robb, W.A.T., Biliary and renal colic. *Br. Med. J.,* 2:135-138, 1963.
22. Friedman, G.D., Siegelaub, A.B., and Seltzer, C.C. Cigarettes, alcohol, coffee and peptic ulcer. *N. Eng. J. Med.,* 290:469-473, 1974.
23. Geokas, M.C., Van Lancker, J.L., Kadell, B.M., and Machleder, H.I. Acute pancreatitis. *Ann. Intern. Med.,* 76:105-117, 1972.
24. Gilson, S.B. Back pain in peptic ulcer. *N.Y. State J. Med.,* 61:625-626, 1961.
25. Gore, I., and Hirst, A.E., Jr. Arteroisclerotic aneurysms of the abdominal aorta: A review. *Prog. Cardiovasc. Dis.,* 16:113-150, 1973.
26. Gunn, A., and Keddie, W. Some clinical observations on patients with gallstones. *Lancet,* 2:239-241, 1972.
27. Hagen, R. Pelvic girdle relaxation from an orthopaedic point of view. *Acta. Orthop. Scand.,* 45:550-563, 1974.
28. Hermann, R.E., and Cooperman, A.M. Current concepts in cancer: Cancer of the pancreas. *N. Eng. J. Med.,* 301:482-485, 1979.
29. Ingram, J.M. *Endometriosis in Current Obstetric and Gynecologic Diagnosis and Treatment.* Ed., Benson, R.C., 4th ed., Lange Medical Publication, Los Altos, Calif., 1982.
30. Janower, M.L. Ruptured arteriosclerotic aneurysms of the abdominal aorta. *N. Eng. J. Med.,* 265:12-15, 1961.
31. Julian, C.G., Goss, J., Blanchard, K., and Woodruff, J.D. Biologic behavior of primary ovarian malignancy. *Obstet. Gynecol.,* 44:873-884, 1974.
32. Kass, L.G., Woyke, S., Schreiber, K., Kohlberg, W., and Freed, S.Z. Thin-needle aspiration biopsy of the prostate. *Urol. Clin. North Am.,* 11:237-251, 1984.
33. Laing, F.C., Federle, M.P., Jeffrey, R.B., and Brown, T.W. Ultrasonic evaluation of patients with acute right upper quadrant pain. *Radiology,* 140:449-455, 1981.
34. Larson, D.M., Masters, S.S., and Spiro, H.M. Medical and surgical therapy in diverticular disease: A comparative study. *Gastroenterology,* 71:734-737, 1976.
35. Laufer, I., Mullens, J.E., and Hamilton, J. The diagnostic accuracy of barium studies of the stomach and duodenum: Correlation with endoscopy. *Radiology,* 115:569-573, 1975.

36. Meares, E.M., Jr. Prostatitis syndromes: New perspectives about old woes. *J. Urol.,* 123:141-147, 1980.

37. Mistilis, S.P., Wiot, J.F., and Nedelman, S.H. Giant duodenal ulcer. *Ann. Intern. Med.,* 59:155-164, 1963.

38. Moller, P., Vinje, O., and Fryjordet, A. HLA antigens and sacroiliitis in chronic prostatitis. *Scand. J. Rheumatology,* 9:138-140, 1980.

39. Nano, I.N., Collins, G.M., Bardin, J.A., and Bernstein, E.F. Should aortography be used routinely in the elective management of abdominal aortic aneurysm? *Am. J. Surg.,* 144:53-57, 1982.

40. Norris, J.R., and Haubrich, W.S. The incidence and clinical features of penetration in peptic ulceration. *JAMA,* 178:386-389, 1961.

41. Owens, M.L. Psoas weakness and femoral neuropathy: Neglected signs of retroperitoneal hemorrhage from ruptured aneurysm. *Surgery,* 91:363-366, 1982.

42. Ross, J.R., and Reaves, L.E., III. Syndrome of posterior penetrating peptic ulcer. *Med. Clin. North Am.,* 50:461-468, 1966.

43. Russinovich, N.A.E., Karem, G.G., and Luna, R.F. Radiology rounds: Persistent lumbar pain and low-grade fever in a 62 year-old man. *Ala. J. Med. Sci.,* 19:67-58, 1982.

44. Sanford, J.P. Urinary tract symptoms and infections. *Ann. Rev. Med.,* 26:485-498, 1975.

45. Schatz, I.J., and Stanley, J.C. Saddle embolus of the aorta. *JAMA,* 235:1262-1263, 1976.

46. Simeone, J.F., Wittenberg, J., and Ferrucci, J.T. Modern concepts of imaging of the pancreas. *Invest. Radiol.,* 15:6-18, 1980.

47. Smith, J.W., Jones, S.R., Reed, W.P., Tice, A.D., Deupree, R.H., and Kaijser, B. Recurrent urinary tract infections in men — characteristics and response to therapy. *Ann. Intern. Med.,* 91:544-548, 1979.

48. Stearns, H.C. Uterine myomas: Clinical and pathologic aspects. *Postgrad. Med.,* 51:165-168, 1972.

49. Strum, W.B., Spiro, H.M. Chronic pancreatitis. *Ann. Intern. Med.,* 74:264-277, 1971.

50. Symmonds, R.E. *Relaxations of pelvic supports in Current Obstetric and Gynecologic Diagnosis and Treatment.* Ed., Benson, R.C., 4th ed., Lange Medical Publication, Los Altos, Calif., 1982, pp. 273-291.

51. Te Linde, R.W. Prolapse of the uterus and allied conditions. *Am. J. Obstet. Gynecol.,* 94:444-463, 1966.

52. Thomas, W.C., Jr. Clinical concepts of renal calculus disease. *J. Urol.,* 113:423-432, 1975.

53. Thorley, J.D., Jones, S.R., and Sanford, J.P. Perinephric abscess. *Medicine,* 53:441-451, 1974.

54. Walde, J. Obstetrical and gynecological back and pelvic pains, especially those contracted during pregnancy. *Acta. Obst. Gynec. Scandinav.,* 41:11-43, 1962.

55. Waters, W.B., and Richie, J.P. Aggressive surgical approach to renal cell carcinoma: Review of 130 cases. *J. Urol.,* 122:306-309, 1979.

56. Weingarten, L., Gelb, A.M., and Fischer, M.G. Dilemma of pancreatic ductal carcinoma. *Am. J. Gastroenterol,* 71:473-476, 1979.

57. Wheeler, W.E., Beachley, M.C., and Ranniger, K. Angiography and ultrasonography: A comparative study of abdominal aortic aneurysms. *Am. J. Roentgenol.,* 126:95-100, 1976.

58. Wheelock, F., and Shaw, R.S. Aneurysm of abdominal aorta and iliac arteries. *N. Eng. J. Med.,* 255:72-76, 1956.

59. Whitmore, W.F., Jr. Natural history and staging of prostate cancer. Urol. Clin. North Am., 11:205-220, 1984.

60. Yamanaka, T., and Kimura, K. Differential diagnosis of pancreatic mass lesion with perculaneous fine-needle aspiration biopsy under ultrasonic guidance. *Dig. Dis. Sci.,* 24:694-699, 1979.

61. Ylikorkala, O., and Dawood, M.Y. New concepts in dysmenorrhea. *Am. J. Obstet. Gynecol.,* 130:833-847, 1978.

PART VIII. MISCELLANEOUS DISORDERS

§ 4-51. Vertebral Sarcoidosis.

Sarcoidosis is a disease of unknown etiology which causes the formation of granulomas, a form of inflammation consisting of epithelioid cells surrounded by a border of mononuclear cells, in any organ in the body. It is most closely associated with granuloma formation in the lung and thoracic lymph nodes. A much smaller proportion of patients develop bone involvement, including vertebral bodies. The clinical history of vertebral sarcoidosis is one of intermittent dull or stabbing, nonradiating low back pain which is increased with activity and relieved with rest. There is tenderness to percussion over the involved vertebrae and this may be associated with fever and

456

generalized lymphadenopathy, or it may be present with no other physical findings. Abnormal laboratory tests include elevated erythrocyte sedimentation rate, serum calcium, alkaline phosphatase and globulins. Radiographic findings of vertebral sarcoidosis include bony destruction and reactive sclerosis of vertebral bodies and pedicles. These also may show intervertebral disc narrowing and a paravertebral soft tissue mass. The diagnosis of sarcoidosis can be suspected in a patient with noncaseating granulomas on biopsy, who has no other identifiable cause for granulomatous inflammation. The therapy of vertebral sarcoidosis includes corticosteroids and surgical decompression and fusion for those patients with axial skeleton disease who have neurologic symptoms of cord compression. The course of vertebral sarcoidosis is benefited by medical and surgical therapy and symptoms are relieved in most patients reported in the literature. The major disability of the disease is related to the degree of irreversible pulmonary dysfunction caused by the granulomatous inflammation and the extent of extrathoracic disease affecting the kidneys, heart and nervous systems. The disease was first described by Hutchinson in 1877 (entry 8).

§ 4-51(A). Prevalence and Etiology.

The exact prevalence of sarcoidosis is unknown but may be as high as one case per 10,000 population. Vertebral sarcoidosis is a rare entity with twelve reported cases (entry 12). Most patients with vertebral sarcoidosis were black males with a mean age of 26 years.

The etiology of sarcoidosis is unknown. The presence of granulomas in the lung and other tissues suggest an abnormality in immune function. T lymphocytes from patients with active sarcoidosis release substances which promote the formation of granulomas. T helper cells as opposed to T suppressor cells are found in increased numbers in the lung while the ratio of helper to supressor cells in the peripheral blood is decreased. In addition, T lymphocytes from patients with sarcoid cause B lymphocytes to produce immunoglobulin. These abnormalities in the number and function of immune cells correlate with the clinical findings of granuloma formation, anergy to delayed hypersensitivity reaction, and hypergammaglobulinemia (entries 7, 4, 6). The granuloma in an organ causes disorganization of normal tissue and the process of healing results in the production of fibrosis in areas of granulomatous inflammation. Granuloma fibrosis in the lungs, heart, kidney, eyes, and musculoskeletal system may be associated with dysfunction in all these organ systems and correlates with the wide range of clinical findings associated with this illness.

§ 4-51(B). Clinical Findings.

The incidence of osseous involvement in systemic sarcoid is up to 9 percent. The most common location for these lesions is in the small bones of the hands and feet. Osseous sarcoidosis almost invariably occurs when there is clinical or radiographic pulmonary involvement. Pulmonary symptoms include cough and shortness of breath. Osseous lesions may be asymptomatic or discovered by chance on radiographs, but this is rarely the case in vertebral sarcoidosis since it is usually painful. Patients complain of a dull or stabbing pain localized at the involved vertebrae. It may radiate from the back to the thighs and is relieved by rest and increased with activity. Patients with spinal cord compression complain of neurologic symptoms, including lower extremity weakness, loss of sensation, and abnormalities of bladder and bowel function (entry 11). Patients with generalized sarcoidosis may also give a history of anorexia, weight loss, and fever.

§ 4-51(C). Physical Findings.

Physical examination of patients with vertebral sarcoidosis demonstrates percussion tenderness over the involved area of the axial skeleton. Limitation of motion of the spine may be an accompanying finding. Those with neurologic involvement may demonstrate impaired sensation, weakness and depressed or absent lower extremity reflexes. General physical examination may demonstrate other organ system involvement with sarcoidosis including skin rash, abnormal breath sounds, generalized lymphadenopathy, and eye inflammation.

§ 4-51(D). Laboratory Findings.

Several biochemical abnormalities have been described in sarcoidosis. These include hypercalcemia, increased serum alkaline phosphatase and hypergammaglobulinemia. Serum angiotensin converting enzyme (ACE) is elevated in patients with sarcoidosis (entry 10). ACE is most active in lung lining cells. Granulomatous inflammation of the lung and lymph nodes may be the source of increased concentrations of ACE. Cutaneous anergy (loss of delayed hypersensitivity) when tested with exposure to four antigens (PPD, mumps, candida and streptokinase/-streptodornase) occurs in a minority of patients (entry 14).

Pathologic specimens from patients with sarcoidosis demonstrate multiple noncaseating granulomas consisting of multinucleated giant cells. They are present in any organ involved with sarcoid, including the lung, skin, bone and muscle. Nonsarcoid causes of granulomas (tuberculosis, berylliosis) must be considered before a diagnosis of sarcoidosis may be entertained.

§ 4-51(E). Radiographic Findings.

Radiographic abnormalities associated with vertebral

sarcoidosis include bone lysis with marginal sclerosis which involves the vertebral body (entry 3). Occasionally the posterior elements of vertebra may also be affected (entry 2). Contiguous vertebrae may be involved; some with narrowing of the intervertebral disc (entry 1). Other patients have noncontiguous vertebral body disease (entry 16). The inflammatory process may progress to cause vetebral body collapse (entry 5). Paravertebral ossification with anterior bony bridges simulating ankylosing spondylitis has been reported (entry 12). Myelography may demonstrate defects compatible with soft tissue compression of spinal cord elements (entry 3). The lower thoracic and upper lumbar vertebrae are the ones most frequently involved in vertebral sarcoidosis.

§ 4-51(F). Differential Diagnosis.

The definitive diagnosis of vertebral sarcoidosis requires a biopsy of the lesion in patients with posterior element involvement or disc space narrowing. Diseases which may mimic sarcoid involvement of the spine include tuberculosis, pyogenic osteomyelitis, Hodgkin's disease and metastatic carcinoma. These diseases may also cause posterior element destruction and/or disc space narrowing. Biopsy of the bone lesion may not be necessary in the patient with biopsy proven pulmonary and skin disease secondary to sarcoid. However, if an anterior lytic lesion of a vertebral body does not respond to therapy, further diagnostic tests, including biopsy, are indicated.

§ 4-51(G). Treatment.

Most patients with vertebral sarcoidosis require cortiocosteroids in the range of 15 mg. to 80 mg. per day to control the symptoms and the granulomatous inflammation (entries 12, 1, 5). Bone lysis and sclerosis may remain the same or improve with therapy. In patients with persistent back pain or neurologic symptoms of cord compression, surgical decompression of the involved vertebrae is required (entries 12, 13).

§ 4-51(H). Course and Outcome.

The prognosis of patients with vertebral sarcoidosis has been generally good. Surgical intervention for biopsies of lesions is important in eliminating other causes of vertebral lysis and sclerosis and confirming the presence of noncaseating granuloma. Decompression and/or fusion of severely affected areas of the axial skeleton has helped prevent the progression of potentially life-threatening neurologic complications. In many circumstances, corticosteroids for sarcoidosis in general, and vertebral sarcoidosis in particular, have been effective in controlling the systemic inflammatory component of this illness. However, it should be remembered that patients with vertebral sarcoidosis have extrathoracic disease and patients with extensive systemic manifestations of sarcoidosis have a less favorable prognosis than those patients with intrathoracic disease (entry 15).

BIBLIOGRAPHY

1. Baldwin, D.M., Roberts, J.G., and Croft, H.E. Vertebral sarcoidosis: A case report. *J. Bone Joint Surg.,* 56A:629-632, 1974.

2. Berk, R.N., and Brower, T.D. Vertebral sarcoidosis. *Radiology,* 82:660-663, 1964.

3. Brodey, P.A., Pripstein, S., Strange, G., and Kohout, N.D. Vertebral sarcoidosis: A case report and review of the literature. *Am. J. Roentgenol.,* 126:900-902, 1976.

4. Daniele, R.P., Dauber, J.H., and Rossman, M.D. Immunologic abnormalities in sarcoidosis. *Ann. Intern. Med.,* 92:406-416, 1980.

5. Goobar, J.E., Gilmer, S., Jr., Carrol, D.S., and Clark, G.M. Vertebral sarcoidosis. *JAMA,* 178:1162-1163, 1961.

461

6. Hunninghake, G.W., and Crystal, R.G. Mechanisms of hypergammaglobulinemia in pulmonary sarcoidosis: Site of increased antibody production and role of T-lymphocytes. *J. Clin. Invest.*, 67:86-92, 1981.

7. Hunninghake, G.W., and Crystal, R.G. Pulmonary sarcoidosis: A disorder mediated by excess helper T-lymphocyte activity at sites of disease activity. *N. Eng. J. Med.*, 305:429-434, 1981.

8. Hutchinson, J. *Illustrations of Clinical Surgery*, Vol. 1, pp. 42. Churchill, London, 1877.

9. James, D.G., Neville, E., and Carstairs, L.S. Bone and joint sarcoidosis. *Semin. Arthritis Rheum.*, 6:53-81, 1976.

10. Lieberman, S. The specificity and nature of serum angiotensin converting enzyme in sarcoidosis. *Ann. N.Y., Acad. Sci.*, 278:488-497, 1976.

11. Moldover, A. Sarcoidosis of the spinal cord: Report of a case with remission associated with cortisone therapy. *Arch. Intern. Med.*, 102:414-417, 1958.

12. Perlman, S.G., Damergis, J., Witorsch, P., Cooney, F.D., Gunther, S.F., and Barth. W.F. Vertebral sarcoidosis with paravertebral ossification. *Arthritis Rheum.*, 21:271-276, 1978.

13. Rodman, T., Funderburk, E.E., Jr., and Myerson, R.M. Sarcoidosis with vertebral involvement. *Ann. Intern. Med.*, 50:213-218, 1959.

14. Tannenbaum, H., Rocklin, R.E., Schur, P.H., and Sheffer, A.L. Immune function in sarcoidosis. *Clin. Exp. Immunol.*, 26:511-519, 1976.

15. Wurm, K., and Rosner, R. Prognosis of chronic sarcoidosis. *Ann. N.Y. Acad. Sci.*, 278:732-735, 1976.

16. Zener, J.C., Alpert, M., and Klainer, L.M. Vertebral sarcoidosis. *Arch. Intern. Med.*, 111:696-702, 1963.

§ 4-52. Retroperitoneal Fibrosis.

Retroperitoneal fibrosis is a disease of unknown etiology which causes fibrosis of the retroperitoneum and renal dysfunction secondary to ureteral obstruction. A grayish plaque of fibrosis envelops the retroperitoneum from the level of the renal arteries to the pelvic brim and laterally to the psoas margins. The structures enveloped in the fibrosis include the aorta, inferior vena cava, ureters and spinal nerves. The clinical history of retroperitoneal fibrosis is one of dull abdominal, back or flank pain of insidious onset. Physical examination may show an abdominal or rectal mass in some patients. Abnormal laboratory tests include a decreased hematocrit, elevated erythrocyte sedimentation rate and increased serum creatinine and a urea nitrogen level indicative of renal dysfunction. Radiographic evaluation with an intravenous pyelogram is useful in documenting ureteral narrowing, dilatation of the urinary collecting system and medial deviation of the ureters. Diagnosis of the retroperitoneal fibrosis is confirmed by biopsy of the retroperitoneum. The therapy includes surgical ureterolysis to reverse urinary obstruction and corticosteroids to relieve systemic symptoms. The course of the disease is improved by treatment. The major disability of the disease is caused by renal failure secondary to chronic obstruction. Retroperitoneal fibrosis became a recognized clinic entity in 1948 as described by Ormond (entry 16).

§ 4-52(A). Prevalence and Etiology.

The incidence of retroperitoneal fibrosis is unknown. Although the literature contains reports which review large numbers of patients with retroperitoneal fibrosis, it is an uncommon disorder (entries 8, 9). Patients develop the disease between the fifth and sixth decades. The male to female ratio for retroperitoneal fibrosis is 3:1.

The etiology of retroperitoneal fibrosis is unknown. The association of vascular inflammation and panniculitis has suggested an immunologic basis similar to that seen with other collagen vascular diseases (entries 17, 5). Another possible etiology is a disorder of uncontrolled fibrous proliferation. Retroperitoneal fibrosis may occur in combination with other disorders, such as Dupuytren's contracture, Reidel's struma, or sclerosing cholangitis, which are associated with excessive fibrous proliferation in a number of organs (entries 6, 3).

§ 4-52(B). Clinical Findings.

The presenting symptom of patients with retroperitoneal fibrosis is pain located in the lower quadrants of the abdomen or the lumbosacral spine. Pain is insidious in onset, dull and non-colicky. It may radiate, in the referred pain pattern of the ureter, from the flank, to peri-umbilical area, to the testes. Patients may also give a history of weight loss, anorexia, fever, and joint pain. Symptoms of urologic compromise, hematuria or oliguria, occur in a later stage of the illness.

§ 4-52(C). Physical Findings.

The most common physical findings are masses found on abdominal and rectal examination. Compression of inferior vena cava is associated with peripheral edema in the lower extremities. Patients may complain of pain with percussion over the lumbosacral spine. Hypertension may be noted in those with renal dysfunction.

§ 4-52(D). Laboratory Findings.

The most commonly abnormal laboratory test associated with retroperitoneal fibrosis is an elevated erythrocyte sedimentation rate. Less often, but still in a majority of patients, a decrease in hematocrit and an elevation in blood urea nitrogen have been noted. Occasionally, autoantibodies (red blood cells, smooth muscle) are positive. Pathologic specimens from biopsies of the retroperitoneum in patients with retroperitoneal fibrosis reveals dense fibrosis with distinct areas of chronic inflammation (entry 24). The chronic inflammation is centered around adipose tissue and blood vessels (entries 21, 12).

§ 4-52(E). Radiographic Findings.

Radiographic techniques, which demonstrate the location of retroperitoneal structures or obstruction of the genitourinary system, help in making the diagnosis. The classic triad of findings on intravenous pyelography is bilateral ureteral narrowing at the level of the 5th lumbar vertebra, medial deviation of the ureters and dilatation of the calices, pelvis, and ureter. Retrograde pyelography may show ureteral obstruction. The sonographic appearance of the retroperitoneum in retroperitoneal fibrosis is one of a smooth-bordered echo-free mass over the sacrum (entry 20). Lymphangiography is useful in distinguishing retroperitoneal fibrosis from lymphoma (entry 1). Computerized axial tomography (CAT) may be the most sensitive technique for evaluation and a CAT scan is capable of detecting the presence and extent of the characteristic soft tissue mass as well as its relation to adjacent abdominal structures (entry 2).

§ 4-52(F). Differential Diagnosis.

The diagnosis of retroperitoneal fibrosis can be suspected in a patient with back pain, constitutional symptoms, evidence of genitourinary obstruction and it is confirmed by biopsy of the retroperitoneum. There are many disease processes, including malignancy, trauma, infection, drugs, and connective tissue diseases, associated with retroperitoneal fibrosis. Metastatic carcinoma from abdominal organs or breast may deposit in the retroperitoneum, causing ureteral obstruction and fibrosis (entry 22). Tumors, such as Hodgkin's disease, non-Hodgkin's lymphoma and sarcomas, may originate in the retroperitoneum and may produce extensive sclerotic reactions (entry 14). A presumptive diagnosis of idiopathic retroperitoneal fibrosis, however, can be made until progressive disease and additional biopsy specimens produce evidence of the malignant process (entry 10).

Trauma to the retroperitoneum may result in subsequent fibrosis. Examples of trauma include bleeding into the

465

retroperitoneum secondary to a clotting factor deficiency or anticoagulant therapy (entry 19). Trauma to the suprapubic area with hematoma formation may produce retroperitoneal fibrosis (entry 25). Inflammatory bowel disease which extends beyond the walls of the gut into the retroperitoneum may initiate fibrosis (entry 4).

Infections in the genitourinary tract may cause fibrosis. Infection may spread through the lymphatics from the bladder to the retroperitoneum and cause fibrosis in periureteral tissues (entry 13).

Drugs are associated with the development of retroperitoneal fibrosis and methysergide (Sansert), a drug used in the treatment of migraine headaches, has been implicated as a cause (entry 23). Amphetamines have been blamed in a much smaller number of patients.

Retroperitoneal fibrosis may also be caused by connective tissue diseases. Weber-Christian disease, a connective tissue disease associated with panniculitis (fat inflammation) causes fibrosis in the retroperitoneum and lymph gland inflammation in the abdomen (entry 11).

§ 4-52(G). Treatment.

Treatment for retroperitoneal fibrosis may include surgery to remove ureteral obstruction and corticosteroids. Ureterolysis, freeing the ureters from the retroperitoneum and placing them laterally or intraperitoneally, is essential to relieve obstruction and preserve renal function (entry 21). This surgical procedure may also relieve back pain and normalize both the hematocrit and the sedimentation rate. Corticosteroids are useful in the early stages of the illness before dense fibrosis is encountered and for relapses of obstruction after initial ureterolysis (entry 15).

§ 4-52(H). Course and Outcome.

The prognosis and course of retroperitoneal fibrosis is

variable. Cases of spontaneous remissions have been reported (entry 18). Other patients have had resolution of the disease after surgical biopsy of the lesion (entry 5). An essential part of therapy is to relieve the obstruction of the ureters. Patients who are anemic at initial presentation have a poorer prognosis (entry 21). Patients with early disease treated with surgery and corticosteroid therapy have the potential to achieve a complete remission of their illness (entry 7).

BIBLIOGRAPHY

1. Bookstein, J.J., Schroeder, K.F., and Batsakis, J.G. Lymphangiography in the diagnosis of retroperitoneal fibrosis: Case report. *J. Urol.*, 95:99-101, 1966.
2. Feinstein, R.S., Gatewood, O.M.B., Goldman, S.M., Copeland, B., Walsh, P.C., and Siegelman, S.S. Computerized tomography in the diagnosis of retroperitoneal fibrosis. *J. Urol.*, 126:255-259, 1981.
3. Gleeson, M.H., Taylor, S., and Dowling, R.H. Multifocal fibrosclerosis. *Proc. Roy. Soc. Med.*, 63:1309-1311, 1970.
4. Harlin, H.C., and Hamm, F.C. Urologic disease resulting from nonspecific inflammatory conditions of the bowel. *J. Urol.*, 68:383-392, 1952.
5. Hellstrom, H.R., and Perez-Stable, E.C. Retroperitoneal fibrosis with disseminated vasculitis and intrahepatic sclerosing cholangitis. *Amer. J. Med.*, 40:184-187, 1966.
6. Hoffman, W.W., and Trippel, O.H. Retroperitoneal fibrosis: Etiologic considerations. *J. Urol.*, 86:222-231, 1961.
7. Jones, J.H., Ross, E.J., Matz, L.R., Edwards, D., and Davies, D.R. Retroperitoneal fibrosis. *Amer. J. Med.*, 48:203-208, 1970.
8. Koep, L., and Zuidema, G.D. The clinical significance of retroperitoneal fibrosis. *Surgery*, 81:250-257, 1977.
9. Lepor, H., and Walsh, P.C. Idiopathic retroperitoneal fibrosis. *J. Urol.*, 122:1-6, 1979.
10. LeVine, M., Schwartz, S., Allen, A., and Narcisco, F.V. Lymphosarcoma and periureteral fibrosis. *Radiology*, 82:90-95, 1964.
11. Milner, R.D.G., and Mitchinson, M.J. Systemic Weber-Christian disease. *J. Clin. Path.*, 18:150-156, 1965.

12. Mitchinson, M.J. The pathology of idiopathic retroperitoneal fibrosis. *J. Clin. Path.,* 23:681-689, 1970.
13. Mulvaney, N.P. Periureteritis obliterans: A retroperitoneal inflammatory disease. *J. Urol.,* 79:410-417, 1958.
14. Nitz, G.L., Hewitt, C.B., Straffon, R.A., Kiser, W.S., and Stewart, B.H. Retroperitoneal malignancy masquerading as benign retroperitoneal fibrosis. *J. Urol.,* 103:46-49, 1970.
15. Ochsner, M.G., Brannan, W., Pond, H.S., and Goodlet, J.S., Jr. Medical therapy in idiopathic retroperitoneal fibrosis. *J. Urol.,* 114:700-704, 1975.
16. Ormond, J.K. Bilateral ureteral obstruction due to envelopment and compression by an inflammatory retroperitoneal process. *J. Urol.,* 59:1072-1079, 1948.
17. Ormond, J.K. Idiopathic retroperitoneal fibrosis: A discussion of the etiology. *J. Urol.,* 94:385-390, 1965.
18. Perlow, S. Obstruction of the iliac artery caused by retroperitoneal fibrosis. *Am. J. Surg.,* 105:285-287, 1963.
19. Popham, B.K., and Stevenson, T.D. Idiopathic retroperitoneal fibrosis associated with a coagulation defect (factor VII deficiency): Report of a case and review of the literature. *Ann. Int. Med.,* 52:894-906, 1960.
20. Sanders, R.C., Duffy, T., McLoughlin, M.G., and Walsh, P.C. Sonography in the diagnosis of retroperitoneal fibrosis. *J. Urol.,* 118:944-946, 1977.
21. Saxton, H.M., Kilpatrick, F.R., Kinder, C.H., Lessof, M.H., McHardy-Young, S., and Wardle, D.F.H. Retroperitoneal fibrosis: A radiological and follow-up study of fourteen cases. *Quart. J. Med.,* 38:159-181, 1969.
22. Usher, S.M., Brendler, H., and Ciavarra, V.A. Retroperitoneal fibrosis secondary to metastatic neoplasm. *Urology,* 9:191-194, 1977.
23. Utz, D.C., Rooke, E.D., Spittell, J.A., Jr., and Bartholomew, I.G. Retroperitoneal fibrosis in patients taking methysergide. *JAMA,* 191:983-985, 1965.
24. Webb, A.J. Cytological studies in retroperitoneal fibrosis. *Brit. J. Surg.,* 54:375-378, 1967.
25. Webb, A.J., and Dawson-Edwards, P. Malignant retroperitoneal fibrosis. *Brit. J. Surg.,* 54:505-518, 1967.

CHAPTER 5

ASSESSMENT AND RELATIVE IMPORTANCE OF PHYSICAL FINDINGS AND DIAGNOSTIC PROCEDURES

Rene Cailliet, M.D.

469

Figure 25. Degenerative Disc Disease.
Figure 26. The Myelogram.
Figure 27. Electromyography (E.M.G.).

Numerous tests are performed daily on patients complaining of low back pain, with or without leg pain. These tests are diligently reported by the doctor and are similarly mentioned in subsequent reports to their referring doctors, the insurance company, and often to an interested attorney. These tests have, over recent decades, become so sophisticated and supposedly "diagnostically precise" that they constitute the basis for diagnosis and frequently are the justification of treatment which is unfortunately very often surgical.

One of the unfortunate inadequacies and inconsistencies in the diagnosis of low back pain is the absence of an accepted standardized terminology. Recently a "Glossary of Spinal Terminology" was prepared by a committee of The International Society for Study of the Lumbar Spine and The American Academy of Orthopaedic Surgeons. The section on pathology unfortunately presents a confusing, composite listing of diagnoses and symptoms without recommendations as to preference for the diagnostic tests which lead to that diagnosis. Although Section XI of the glossary relates to "clinical terms" such as the Lasegue test, the Patrick test, trigger points, etc., the presentation is anything but judgmental; worthless, archaic tests are too often glamourized by their complexity while simple, valuable tests tend to be underestimated. It is the intent of this chapter to discuss the relative value of the history, the physical examination, and the diagnostic tests ordered as they relate to the diagnosis, and ultimately, to the recommended treatment.

The physical examination must be considered in the light of the history and physical findings. The history is the subject of a detailed story as given by the patient and

470

interpreted by the examiner. The examination is the object of the physician's evaluation of the patient, and it serves to confirm the physical damage incurred from the injury which was claimed by the patient in his history. Laboratory tests are usually done to objectively document, justify, and clarify the clinical diagnosis reached by the examining physician based on his/her interpretation of the patient's history, signs, and symptoms. This chapter will also discuss numerous terms used when diagnosing and treating the low back and may be construed to be a glossary. Many of these medical terms will also be evaluated as to their relative accuracy, specificity, selectivity, and logical interpretation.

§ 5-1. The History.

The story given by the patient constitutes the history. Many of the patient's voluntary statements are relevant, causative, and accurate; other information will evolve from questions posed by the examiner. Attacks may be sudden and related to an incident or an activity. This history must initially be accepted as the cause of the patient's symptoms. The activity, movement, or position depicted by the patient must also be clarified as to whether it actually caused the ensuing pain or was responsible for a recurrence of the pain. Low back pain is usually acute or recurrent, but if it is continuous or persistent it is considered chronic. Pain may have resulted from a movement or position initiated by the patient or may have resulted from something done to the patient. The exact movement or position initiated by the patient must be clarified as it can be diagnostic or suggestive, and a specific movement may indicate the mechanism of injury. Pain resulting from bending forward suggests disc trauma ligamentous or muscular sprain. Bending forward while twisting to one side is even worse since rotation or torque is considered more injurious to the tissues of the low back. Reextension, that is, resuming erect posture after having been bent over, more frequently causes low back injury than does forward bending, and damage is more apt to occur if the return to the erect position is done improperly (see Figure 1 at end of this chapter).

The weight of an object lifted is less important than the manner in which it is lifted. Lifting without bending the knees, with the object not held close to the body, and with a failure to return to the upright position by rotating the pelvis first before raising the low back: any or all can cause a low back injury (Figs. 2, 3).

Lifting done without concentration, when distracted, when angry or depressed, when fatigued: any or all can result in injury (Fig. 4).

The location of the pain suggests the site of injury in the lumbosacral spine. Low back pain alone implies local injury to the ligaments, muscles or posterior joints of the spine. These posterior joints are called zygo-apophyseal or facet joints. They bear little or no weight with the spine erect but function to guide the direction in which the spine will move (Fig. 5).

Pain that occurs initially or primarily in the leg, radiating distally toward the foot, ankle, and toes, is usually derived from nerve irritation. Pain that occurs when the patient arches backward, increasing the sway (termed lordosis) of the low back, usually results from inflammation of the facet joints which are now made to bear weight. Pain from an increased arching of the back may result from the fact that this position closes the windows (foramina) through which the nerve roots leave the spinal column to descend down the leg (Fig. 6).

In taking the history it is important to find out if there have been previous similar attacks, how they occurred, how they affected the patient, how they were treated, and how the patient responded to the treatment. These, to the expe-

472

rienced examiner, indicate the severity of the injury, the integrity of the patient, and the possible presence of factors other than the tissue injury sustained.

§ 5-2. Examination.

The patient to be examined is first evaluated by observation and palpation. When palpating, the physician applies his hands to the anatomical structures of the patient's back. The spine is viewed from behind to see if the alignment is good and from the side to observe the posture as well as the range of motion in flexion, extension, lateral bending and rotation. The posterior aspect, as visualized and then palpated by the physician, is used to determine the straightness of the spine; the spine and the head must be in true alignment above the sacrum (Fig. 7).

A scoliosis, when it exists, is a lateral curving of the spine — an abnormality expressed by degrees of deviation. It may be structural, having existed since early childhood and gone on to a fixed bony deformity in adolescence, or it may be functional, that is a transient, temporary curvature that disappears when the patient is no longer erect. A structural scoliosis is due to the anatomical changes in the bony elements of the vertebral column which either have developed gradually over the rapid growth period in childhood or else are a direct result of injury or disease of the spine (Fig. 8).

One particularly confusing element of functional scoliosis relates to the fact that the secondary pelvic obliquity can be perceived as the cause, rather than the result, of the superincumbent spinal curvature. Although the legs are usually equal in length and the pelvis is level, the functional scoliosis may erroneously be attributed to an unlevel pelvis from a short leg.

473

X-rays taken of a patient in the standing position frequently are interpreted as showing subluxation, "slipping," or a rotary torque abnormality of the spine. This "misplaced vertebra" is considered to be the result of injury and the cause of the pain. Its assumed presence and its significance then becomes the basis of "adjustment" to remedy the painful problem. It is an oversimplification of a condition that rarely exists and consequently requires inappropriate treatment to correct a supposed malalignment which either is not abnormal to begin with or else whatever abnormality was perceived had no significance in the diagnosis of the patient's condition.

Spinal motion, in terms of range as well as symmetry and rhythm, is of major diagnostic significance. The statement is frequently made that the patient bends forward and reaches to within six inches of the floor, or twelve inches of the floor, or places his palms to the floor. The significance of the latter statement, however, is not nearly as important as is the quality of spinal flexion in terms of its reversal from a normal sway back status or lordosis as it bends forward. This arch of the low back should reverse itself as the spine changes from a lordotic curve to a round back or kyphotic curve (Fig. 9).

Forward flexion of the spine is a segmental motion with bending occurring at each functional unit: each functional unit comprising two adjacent vertebrae along with their interposed disc. These units also contain the ligaments, nerves, and facet joints of the two adjacent vertebrae. The lumbosacral spine is made up of five functional units and from the sacrum up they are: L5-S1 (the lumbosacral unit), L4-5, L3-4, L2-3, and L1-2 (Fig. 10).

Most angulation and movement occurs at the lumbosacral (L5-S1) and L4-5 levels. As a result, most of the damage and most symptoms relate to these two functional units. In forward bending each unit flexes about eight to ten

474

degrees. This means that the entire lumbar spine only goes through a forty-five degree excursion, and as a person reaches to touch the ground the rest of the motion comes from the pelvis rotating about the hip joints (Fig. 11).

Since a measure of how well, how freely, and how far a person's back flexes is expected for documentation, a medical report should first specify the extent of flexion and go on to discuss why and how it does not flex. Precise measurements are not possible in the every day examination, nor are they necessary. The determined range of motion (usually termed R.O.M.) is essentially a comparison to the expected normal, and the degree of reversal of the lordosis (normal arch or sway) should be specified. The low back cannot flex unless the back muscles (the erector spinae) elongate and permit each functional unit to "open" eight to ten degrees. In order to elongate, however, a muscle must be well conditioned, used properly, and free of protective muscle spasm.

When an injured person with an inflammatory reaction about the functional unit attempts to bend forward, his flexion will frequently be inhibited by protective muscle spasm. The back does not have the normal curve in the erect position, nor is there any reversal of the sway of the back on attempting to bend forward. As the person tries to touch the floor all of the motion occurs at the hip joints; the pelvis rotates without significant flexion or flexibility of the lumbar spine.

Although this inability to flex the lumbar spine can be due to inflammation from injury or damage, it also may be voluntary in that the patient is either afraid or does not wish to bend forward. Consequently, this restriction is not necessarily indicative of damage or disease. An astute examiner, however, can usually differentiate between the two, whether there is actual spasm or just a reflex, protective response. Flexion from an upright position should be compared to similar movement while kneeling because in the kneeling position the pelvis is free to move, the sciatic nerve is not under tension, and the hamstring muscles are relaxed; hence the back can move untethered.

It also should be remembered that flexion is relative and its limitation may only be an indication of poor conditioning. The patient's perceived stiffness may actually represent very little loss of flexibility in respect to the pre-injury state.

It is possible to further analyze the nature of spasm by observing flexion in the sitting position. If the head can be brought down to the knees with the patient seated and the spine bends smoothly, it is unlikely that there is enough pathology in the lumbar spine to prevent flexion; apprehension and protective guarding, rather than pathological spasm, can be assumed. As noted before, however, flexing of the head from the kneeling position will often show normal flexibility even though standing flexion was markedly guarded (Fig. 12).

If, when the patient has flexed as far as he can, pain is felt not only in the back but also in the buttocks and down either leg, there may be what is known as a positive dural sign. This will be described later in relation to the straight leg raising or Lasegue test. The dural sign can also be provoked by forcing the chin down to the chest while the spine is in maximum flexion. This test is known as nuchal flexion and if this movement intensifies the leg pain, it implies more than mere ligamentous or muscular involvement.

If the protective spasm is unilateral (one sided) due to damage or irritation of the tissues on one side of the spine, a scoliosis develops; the spine is tilted to the side because of one sided muscle spasm. Scoliosis is difficult to mimic and may go unnoticed by the patient with back pain; it frequently intensifies with forward flexion (Fig. 8).

Although disc herniation with nerve root irritation, usually at the L4-5 level, can cause a scoliosis by irritating nerves on one side of the spine, this diagnostic finding must be supplemented by other signs and symptoms to justify a diagnosis of sciatic radiculopathy, nerve root entrapment, or disc herniation.

After the patient has fully flexed, he must be requested to assume the erect posture and how this is done reflects past habits as well as the constraints of any tissue injury. Normally a return to the erect position is accomplished by a derotation of the pelvis without altering the forward flexed spine (kyphosis) until the person has raised up to about forty-five degrees. Then, during the remaining forty-five degrees of reextension, the low back resumes its lordosis, along with slight additional derotation of the pelvis (see Fig. 9). Patients with legitimate back pain, however, tend to resume the erect position without any spine movement. It is all done by the pelvis with the help of knee and hip flexion.

The ability to bend sideways in lateral flexion has no major diagnostic significance as long as the patient is not simultaneously flexing or extending. Lateral bending simply causes ligamentous or muscular stretching and can be free or restricted without diagnostic significance.

A test that supposedly uncovers "functional" problems is done with the patient standing erect, arms dependent and hands firmly clasping the hips on both sides. In this test the pelvis is twisted to one side, then the other, by the examiner but the spine cannot move since the arms and hands fix it to the pelvis. Since the rotation of approximately thirty degrees to either side occurs at the hip joints, the claim of low back pain from this maneuver is considered non-anatomic (Fig. 13).

Hyperextension or bending backward changes anatomical relationships and can cause pain. Arching the

477

back and increasing the sway forces the facet joints together, narrows the foramina or windows through which the nerves leave the spine to go into the leg, and compresses the discs posteriorly. A combination of these three factors can create pressure on the nerves as they leave the spine and cause back pain, sciatic leg pain, or both. Interpretation of the symptoms produced by this maneuver, coupled with confirmatory x-ray findings, can be a firm basis for diagnosis.

§ 5-3. The Neurological Examination.

Because of the intimate anatomical relationship that the musculo-skeletal system has to the nervous tissue in and about the spine, the initial biomechanical observations of the examiner must next be supplemented by a thorough neurological assessment, the most important and misunderstood component of which relates to the tension signs. Tension sign is a generic name for any test designed to identify nerve root pressure and/or inflammation, and in turn, dural irritation. The straight leg raising test (SLR) as well as the Lasegue test are two commonly used adaptations of the concept but eponyms of various modifications abound, to the confusion of all. To comprehend the subtilties of the issues involved, however, basic understanding of the anatomy of the area is necessary.

Each nerve root, as it leaves the spinal canal through the intervertebral foramen, is enclosed within a sleeve which contains spinal fluid and very small blood vessels about and within the nerve. This sleeve is known as the dural sac and it provides the circulation to that particular nerve root. Any compression and/or traction on the dura will compress its contents and encroach upon the nerve and its blood supply. Pain can be produced in that nerve along with numbness and even loss of function, and these symptoms can persist if

478

the pressure is severe. The dura is exquisitely sensitive and when inflamed becomes a source of pain, both in the back and down the leg. The tension test or straight leg raising test is meant to determine the sensitivity of this nerve and its inflamed dural sleeve. Additional testing is required to determine the origin of the problem and the basis for the inflammation (Figs. 14, 15, 16).

The purpose of the straight leg raising test is to stretch the dura. When the leg is raised with the knee straight, the sciatic nerve as well as its dural sheath will be placed under tension. If the dura and nerve root are inflamed, the patient will experience pain along its distribution to the lower leg, foot, and ankle. Pain that is not referred below the knee immediately raises a question as to whether there actually is sciatic nerve root involvement. Occasionally a patient with true nerve root compression will define his pain as limited to the buttock or thigh but as a rule, for the test to be positive, it should be felt below the knee (Fig. 17).

The test can then be further refined to rule out discomfort from tight hamstrings, a condition which can confuse the issue, by checking for dural signs in one of two ways: once the leg has been raised as far as permissible, either the foot is passively dorsiflexed (stretched upward toward the head) or else the neck is flexed forward by bringing the chin to the chest. Both maneuvers stretch the dura and confirm the neurological nature of the test.

Straight leg raising inevitably will stretch the hamstring muscles which attach to the pelvis. Thus, if the leg is raised high enough, the pelvis will rotate; the back will bend and become painful if its range is restricted from protective muscle spasm. This type of response should not be construed as a positive tension sign as it does not imply the stretching of an inflamed nerve root or its dura.

In situations where there is a suspicion of malingering, a modified straight leg raising test may be done with the

479

patient seated on the edge of the examining table with legs dangling. Since the thigh is already flexed to ninety degrees in this position, straightening the knees to the horizontal essentially performs a tension test on the whole leg. By simultaneously distracting the patient, one usually can gauge the true state of dural irritability.

In proceeding further with the neurological examination, each nerve root must be examined. The nerve roots are segments of the cauda equina or the extension of the spinal cord as it descends through the lumbosacral area. Each root goes to a specific group of muscles or myotome, or from a specific area of skin or dermatome.

The first sacral (S1) nerve roots are the last ones to emerge from the lumbosacral canal and they contribute to the composition of the sciatic nerve running down the back of the leg. They are also the ones most closely associated with the lumbosacral (L5-S1) intervertebral disc. The nerve fibers in the S1 root go to the gastroc-soleus or calf muscles which permit the person to rise up on his toes. Although the patient is usually just tested for toe walking ability, a more sensitive way to gauge calf muscle strength is to have the subject rise up and down on his toes a dozen or more times. If there is nerve involvement anywhere along its distribution, the muscle on the involved side will fatigue before the other one does, often after only four or five times.

Another muscle innervated by the first sacral root is the gluteus maximus or buttock muscle, and it can be tested with the patient lying supine (face up) with the knee on the involved side flexed at ninety degrees. The strength of the buttock muscle (gluteus maximus) as well as the integrity of the S1 nerve can then be tested unilaterally by having him raise his buttocks off the examining table five or six times (Fig. 18).

The fifth lumbar (L5) nerve root, the one immediately above S1, is closely associated with the L4-5 intervertebral disc, and it also has its own myotome: a specific muscle function. It supplies the anterior tibial muscle, the one which raises the foot at the ankle. It also controls big toe elevation through the extensor hallucis longus muscle. If the L5 nerve root is involved, the patient will have weakness or fatigability in raising the foot at the ankle as well as in holding the big toe up. He will also have difficulty walking on his heels and after a few steps the involved foot will no longer stay up. The L5 root is tested by having the patient sit with legs dangling, heels on the floor and feet elevated. The examiner forces the feet down, testing for strength first, then endurance. When there is L5 root involvement, the foot and/or the big toe cannot be held up against resistance (Fig. 19).

Whereas we have been up to now relating exclusively to the sciatic nerve which is formed by the L5 and S1 nerve roots, it would be well to remember that the L2, L3, and L4 nerve roots up above can also be affected, and when they are, symptoms are referred to the anterior thigh along the course of the femoral nerve down to but no further than the knee (Fig. 20).

Dural irritability at this level is checked by the femoral stretch test. With the patient prone, the femoral nerve is stretched by bending the knee and passively elevating the thigh up from the examining table. The test is positive if pain develops in the front of the thigh. The strength of the muscles innervated by these roots can best be checked by having the patient do a deep knee bend on one leg and then the other, comparing the two for strength and endurance. An alternative method is to have him simulate stair climbing by stepping up and down from a small stool (Fig. 21).

Reflex testing is another important part of the physical examination. The first sacral (S1) root innervates the ankle jerk. There is no reflex relating to the fifth lumbar (L5) root.

481

The knee jerk tests the integrity of the third and fourth lumbar (L3-4) roots. These innervate the quadriceps or thigh muscles and do not contribute to the sciatic nerve.

The ankle jerk is usually tested in the sitting position with the legs dangling but other positions can also be used. The test is done by tapping the Achilles tendon behind the ankle and comparing the response of one leg to the other. Involvement of the first sacral root, from whatever cause, diminishes the response of the ankle reflex. To accentuate the test and make it more precise, however, it can be done with the patient kneeling on a chair and squeezing the back of the chair.

The knee jerk or patellar tendon reflex is usually tested with the patient seated on the edge of the examining table with legs dangling and the tendon just below the knee cap is tapped sharply. It too is evaluated as compared to the response on the opposite side and it can be reinforced by having the patient try to pull his clasped hands apart (Fig. 22).

The Babinski sign is elicited by running a sharp instrument along the sole of the foot. Normally, in a negative reaction, the toes plantarflex or go down. A positive Babinski sign occurs when the great toe extends or goes up, as the other toes spread or splay. It is a pathological reflex which indicates an upper motor lesion, that is, an interruption of impulses from the brain or spinal cord. It is found in demyelinating diseases such as multiple sclerosis as well as in paraplegia; and when found associated with low back pain, possible ominous problems at a higher level should be considered.

The term "functional" is used medically in contradistinction to "organic" and implies that the complaints far exceed the objective findings. There is usually a psychological component as well as secondary gain involved, whether it be monetarily or psychosocially rewarding.

Waddell and his colleagues in Toronto have been more successful than most in standardizing and clinically testing a group of nonorganic physical signs, and for this effort, they received the 1979 Volvo Award for outstanding clinical low back research (entry 2 in Bibliography at end of this chapter). The physical signs they describe provide a simple and rapid screen to help identify the few patients who require more detailed evaluation.

The following group of five types of physical signs was developed and will be briefly described. Any individual sign counts as a positive sign for the type; a finding of three or more of the five types is clinically significant. Isolated positive signs are ignored.

1. *Tenderness* — when related to physical disease is usually localized to a particular skeletal or neuromuscular structure. Nonorganic tenderness is nonspecific and diffuse.

Superficial — the skin is tender to light pinch over a wide area of lumbar skin.

Nonanatomic — deep tenderness is felt over a wide area and is not localized to one structure; it often extends to the thoracic spine, sacrum, or pelvis.

2. *Simulation Tests* — these should not even be uncomfortable so if pain is reported, a nonorganic influence is suggested.

Axial loading — there is a complaint of low back pain with vertical pressure on the standing patient's skull.

Rotation — back pain is reported when the shoulders and pelvis are passively rotated in the same plane (Fig. 13).

3. *Distraction Tests* — a positive finding is demonstrated in the routine manner; this finding is then

checked while the patient's attention is distracted. Findings that are present only on formal examination and disappear at other times may have a nonorganic component.

Straight leg raising — the various ramifications of this test have been described elsewhere in this chapter.

4. *Regional Disturbances* — involve a divergence from accepted neuroanatomy.

Weakness — nonanatomic "voluntary release" or unexplained "giving way" of muscle groups will be explained elsewhere.

Sensory — sensory disturbances fit a "stocking" rather than a dermatomal pattern.

5. *Overreaction* — this may take the form of disproportionate verbalization, jumping, cringing, facial expression, etc. Judgement should, however, be made with caution as it is very easy to introduce observer bias.

The standardized group of nonorganic physical signs described above is easily learned and can be incorporated unobtrusively to add less than one minute to the routine physical examination. The organic component of the history, physical examination, x-rays, and laboratory tests often serves to legitimize the complaint but a grey area obviously exists, in that the physician is the final arbiter of the objectivity of the patient's symptoms. The difficulties are compounded by the undisputed evidence from recent studies that many of our most sacred tests, such as myelograms and CAT scans, have an extremely high false positive rate; that is, a surprisingly high percentage of normal people tested will show up with significant findings. This will be discussed in greater detail as we analyze each of the confusing array of tests used in the examination of the low back.

§ 5-4. Commonly Used Low Back Tests.

Voluntary Release — this test is used to differentiate the patient with true organic nerve damage from the one with an extensive emotional overlay. On attempting a sustained muscle contraction, such as dorsiflexion of the ankle, the patient with a functional problem holds up the ankle with intermittent quivering, breaking, contraction, and release in a "jerky" manner; and it is strongly suggestive of less than organic pathology.

Bragard's Test — has already been mentioned as a dural sign. After straight leg raising to the limit of tolerance, passive stretching of the foot in dorsiflexion causes pain along the course of the sciatic nerve. When positive it confirms the presence of nerve and dural inflammation.

Bow String Sign (Cram Test) — with the patient sitting on a chair with the knee at a little more than a right angle, and the body bent forward so as to lengthen the course of the sciatic nerve at the hip and knee joints, the finger is then pressed into the popliteal space so as to make the nerve a little more tense. A positive test suggests the presence of radiculopathy, and pain is felt along the course of the nerve at the back of the thigh. This test is used infrequently since it is extremely difficult to perform.

Contralateral or Well Leg Straight Leg Raising Test — is positive when pain is felt in the involved leg below the knee as the straight leg raising test is done on the uninvolved side. In other words, sciatic pain will be aggravated when the good leg is raised. This test is a most significant indictment of the seriousness of the lesion since there must be extreme dural sensitivity if the negligible excursion of the opposite root is painful; a surgical outcome is not unusual when it is positive.

Naffziger Test — is done by compressing the jugular veins in the neck, for approximately ten seconds, until the

485

patient's face begins to flush. The patient is then asked to cough to increase spinal fluid (intrathecal) pressure; sciatic pain will be aggravated if there is dural sensitivity.

Valsalva Test — increases intrathecal pressure when the patient is asked to bear down like he is trying to move his bowels; here too, there will be sciatic pain in the presence of dural irritation. Coughing and sneezing also raises intraabdominal pressure and can be used as a modification of this test since leg pain will be produced in the same way.

Kernig Test — is another procedure in which the spinal cord and dural sac are stretched. With the patient lying supine, the head is passively forced down upon the chest. If there is neck pain, the condition is considered unrelated to the low back; leg pain suggests a dural sign.

Milgram Test — is a further modification of the tension test. While in the supine position, the patient is asked to raise up both extended legs about one to two inches from the examining table. This is another way to increase intraabdominal pressure, and thus intrathecal pressure. If this position can be held for 30 seconds without leg pain, then dural irritation can theoretically be ruled out. The test should be considered positive if the maneuver creates leg pain. Just as in the straight leg raising test, however, a complaint of back pain does not make it significant (Fig. 23).

Hoover Test — checks the validity of a patient's inability to raise one leg off of the examining table. Normally as patient attempts to raise one outstretched leg, he has to simultaneously press down with the opposite leg. Thus the downward pressure of the normal leg is a good indication of how hard he is trying, and the examiner can check the sincerity of his effort by keeping a hand under the foot on the good side.

Patrick Test — is a test for hip disease and sacroiliac stability and is performed with the patient supine. The heel

486

of the lower limb being tested is passively placed on the opposite knee as high above the patella as possible. The knee on the side being tested is then pressed laterally and downward by the examiner. The test is considered positive if motion is involuntarily restricted and hip pain is produced. This test dates back to the days when most backache was attributed to sacroiliac dysfunction and it has had little legitimate relevance since the early 1940's when it became accepted that sacroiliac symptoms are almost always referred from the lumbosacral spine. We now know that the sacroiliac joint rarely becomes deranged other than in severe trauma, and it is seldom painful except in ankylosing spondylitis, tumor, infection or pregnancy. The prominent use of this test as the key element in the diagnosis of the average low back syndrome should signal a generous sprinkling of boilerplate.

Kneeling Bench Test — the patient, kneeling on a 12-inch bench, is asked to bend forward and touch the floor. Flexion at the hips is the only requisite to accomplish this. If the hips are normal and the fingers do not reach the floor, nonorganic back pain must be suspected (Fig. 12).

§ 5-5. Sensory Testing.

The dermatomes, areas of numbness or impaired sensation about the leg, are tested by light touch, pinprick, or by a width of cotton or gauze. Anatomical dermatomal distribution is well known so there is little difficulty in recognizing falsification or hysteria when the symptoms do not conform. Numbness of the entire side of the body or a "stocking" anesthesia is not anatomical and is therefore not an objective determination. Although sensory loss is by its nature subjective in its perception, its significance relates to the qualitative aspects of its interpretation (Fig. 24).

A valuable test which is often neglected relates to skin sensation about the anus and anal sphincter tone. The sensation is tested by pin prick and the motor tone by a digital rectal examination. It is a truism that a combination of a positive contralateral straight leg raising test, along with saddle anesthesia and a loss of anal sphincter tone, suggests an acute central disc herniation or a cauda equina compression syndrome: the only real surgical emergency in the low back.

§ 5-6. Confirmatory Laboratory Tests.

One cannot emphasize enough the importance of rigid discipline in the diagnostic work-up of a case of low back pain, and in this respect the low back algorithm discussed in Chapter 3 can be a great help in staying on track. For reasonable assurance of a favorable outcome, the core of information derived from a thorough history and physical examination must be the basis for all subsequent decision making. If one is to stay out of trouble, confirmatory tests must be just that: confirmatory of a clinical impression. The trouble begins when they are used for screening since most of them are overly sensitive and relatively unselective. For this reason, most of the iatrogenic disasters in the management of low back pain can be directly attributed to overreliance on x-ray and laboratory findings which have no positive clinical correlation. The tests most frequently used in the diagnostic work-up of the low back will be described and analyzed with this in mind.

Routine X-Rays — In most situations, three x-ray views are all that are required: two 14 by 17-inch films, one an anteroposterior (AP) view and the other a lateral taken from the side; the third, a small spot lateral, is taken at an angle so as to better visualize the lower two interspaces. Two oblique views are also frequently taken on a routine basis because they occasionally can help identify a subtle spondylolysis; however, the limited information they furnish has not been shown to justify the additional radiation exposure and expense. If, after interpreting the basic three views, there is some question of a pars interarticularis defect, the additional exposures can easily be taken.

Unfortunately, x-rays are too often misread: more specifically "overread." Normal asymptomatic people approaching forty or fifty years of age will frequently have disc narrowing. The disc itself, the intervertebral disc, has an 88% water content. It contains a gelatinous nucleus pulposus confined under great pressure between two cartilaginous end plates by an elastic annulus fibrosus and inevitably it will dehydrate with age. Numerous microscopic injuries occur with everyday wear and tear and permit a certain amount of leakage of nuclear material from the central portion of the disc. As this occurs, insidiously and gradually, it does so without symptomatology, without impairment. Consequently as the years pass, the disc spaces will narrow from repeated microtrauma and usually will remain symptom free until there is superimposed injury. Thus this narrowing per se is not necessarily relevant to the clinical picture unless it can be rationally correlated with the patient's symptoms and signs.

If the intervertebral discs in the front of the functional units narrow, the zygoapophyseal joints behind also approximate. These latter are true diarthrodial joints with cartilage surfaces which can wear down and develop degenerative arthritis or arthrosis. This combination of narrowing in both front and back results in the approximation of the bony elements of the spinal canal; the exits or windows (foramina) through which the nerves emerge from the spine are left with limited reserve space to adapt to the swelling and deformation of injury (Fig. 25).

Computerized Axial Tomography (CAT Scan) — otherwise known as Computed Tomography (C.T. Scan), has given us a new perspective in the radiological analysis of the spine through its ability to create cross sectional imaging at the desired level. In addition to the bony configuration in all planes, it also shows the soft tissue in graded shadings so that the ligaments, nerve roots, free fat, and intervertebral disc protrusions can be seen as they relate to their bony environment.

There is little doubt that the CAT scan can be an extremely valuable diagnostic tool if *used appropriately to confirm the clinical findings* as derived from the history and physical examination. It should be used to confirm, rather than be the basis for the diagnosis. Some recent studies dramatically reveal the potential pitfalls of basing clinical decisions on CAT scan findings in isolation from the patient and his whole clinical picture. CAT scans of the lumbar spines of fifty-three normal subjects with no history of back trouble were submitted to three expert neuroradiologists for interpretation; 34.5% were read as abnormal. The implication is that a neurotic with functional back or leg pain would have one out of three chances of becoming the victim of failed back surgery with all of its dire implications. Given the appropriate history and clinical findings, however, the CAT scan frequently can make a very helpful confirmatory contribution.

Myelogram — Among confirmatory tests, the myelogram has long been the standard that all others are measured up to. In spite of this, it also is far too sensitive and not selective enough. As far back as 1968, we knew that 24% of normal subjects will have bulging discs on Pantopaque myelography, yet innumerable patients have been operated upon under just these circumstances. Conditions have improved somewhat over the past few years, however, since that oil based dye has been replaced by the water soluble

490

Metrizamide, but an incidental and irrelevant L4-5 disc bulge is far from an unusual finding. Note should also be made of the other side of the coin. Whereas the myelogram can be overly sensitive at the L4-5 level and come up with false positive readings, it may miss large disc ruptures and extrusions at the lumbosacral (L5-S1) level if the dural sac is too short to reach down to the level of the pathology. In these situations, venography, and more recently the CAT scan have been helpful in identifying such outlying lesions.

These tests are performed so that the contents of the spinal canal and the dural sheaths of the nerve roots can be visualized. As the dye is injected to admix with the spinal fluid within the dural sac, all of the contents of the spinal canal can be visualized on x-ray; any extradural mass such as a herniated disc will show up as a filling defect from without and an intrathecal tumor will be directly outlined by the dye (Fig. 26).

Above all, it should not be forgotten that the myelogram is an invasive procedure and should not be taken lightly. The dye which is injected is a foreign material which can cause an adverse reaction. Fortunately the water-soluble agent which is usually used in this day and age is only a short lived irritant since it is rapidly excreted. Pantopaque, on the other hand, the dye traditionally used in the past and occasionally still used for pan-myelography, can not always be removed entirely and has been known to cause a crippling arachnoiditis.

The clinical diagnosis of a herniated disc offers no great challenge to the trained surgeon who usually reserves this test for preoperative assurance: to make certain of the location; to check for a congenitally anomalous nerve root, tumor, or double disc. When this confirmatory test is used as an "screening test," in the absence of objective clinical findings, however, exploratory surgery and disaster is a frequent outcome.

Electromyogram (E.M.G.) — In this test needles are placed into muscles to determine if there is an intact nerve supply to that muscle. An abnormal E.M.G. depicts impaired nerve transmission and, more specifically, isolates which nerve root is involved. The particular nerve root to be implicated is dependent upon which muscle is found to be abnormal (Fig. 27).

The E.M.G. is particularly valuable in the localization of a specific root. It should be remembered, however, that it takes at least 21 days for the test to show up as being abnormal. After 21 days of significant pressure on a nerve root, signs of "denervation" with so-called fibrillation can be seen. Before then, the E.M.G. will be negative in spite of nerve entrapment; it will only show that the particular muscle tested is irritable.

There is no quantitative interpretation of the test so it cannot be said that the E.M.G. is 25% normal, 50% normal, or 75% normal to suggest an alteration in nerve pressure and change in nerve function. There can be no validity to the quantitative determinations suggested by some examiners.

What cannot be emphasized too often is that the E.M.G., like all of the other confirmatory tests already discussed, is not a screening tool. When dealing with the average low back problem, it rarely comes up with any information which cannot be derived from a careful physical examination, and with much less trouble and cost. A mystique about this test seems to have developed in medical-legal circles, and far more importance is given to it than it deserves since many well known spinal centers seem to be able to furnish quality care while limiting the use of the test to special circumstances. It certainly can be unequivocally stated that the E.M.G. need not be part of the routine low back work-up. It can even serve to confuse the issue since the tests also may be abnormal from other causes such as diabe-

492

tes, previous peripheral nerve involvement, and trauma, as well as from other types of extraspinal pathology. A well known early electromyographer stated that, in his opinion, 25% of all E.M.G.s performed show some abnormality and when there is careful testing, it is rare to find a patient who does not have some degree of irritability, of demyelization.

Thermography — Unfortunately this test seems to have developed into the latest gimmick in medical-legal circles and is even more misused than the others. The thermogram is a test which records changes in superficial temperature in very colorful, vivid displays. These photographs, when shown to a jury in a courtroom, may be extremely spectacular and can depict whatever the thermographer wishes it to depict, but the interpretation of thermography is as yet to be standardized. The exact meaning is yet to be defined.

The only valid, controlled study available to date was done at the University of Toronto where an unusual set of circumstances made the conditions ideal (entry 1). Back in 1982, before chemonucleolysis was approved in the United States, most cases on this continent were being referred there to Dr. J.A. McCulloch. He and Dr. Leo Mahoney, one of the pioneers in thermography, decided to use this high volume flow of documented, clear-cut herniated disc patients for a controlled study. Thermography of the back and both legs was carried out on the subjects preoperatively by an experienced technician. Subsequently, one of the two doctors interpreted the thermograms blindly, without knowledge of the side or level of root involvement. They had intended to do a larger series, but when the thermograms on the first 23 patients failed to demonstrate any clinically valuable information that could assist in localizing the side or level of involvement in acute sciatica resulting from a herniated lumbar disc, they felt that a failure rate of 100% justified discontinuance of the study. Currently other studies are being conducted to correlate the findings of electromyography, myelography, x-ray, CAT scan, and thermography. The results are months away but the potential value of thermography may be further clarified when they are available.

Venography — Although lumbar epidural venography has had its advocates over the past 20 years and has been used as the primary diagnostic test in a few centers, it has usually been reserved for problem patients in whom myelography did not provide an acceptable answer to a clinical dilemma. A symptomatic lesion occasionally cannot be visualized by Pantopaque in a capacious caudal canal, whereas the epidural venogram can help in making a firm diagnosis. Water soluble dyes, however, have made these ambiguities quite rare and with the additional input from the CAT scan, the procedure has become all but obsolete. The findings with epidural venography are quite accurate when positive, but unfortunately, the false negative rate (sensitivity) is reported to be 18%. A mechanical obstruction can be identified, the origin of which can range from a congenital nerve root anomaly to an intrathecal tumor; this nonspecificity can cause difficulties in the operating room for the unwary surgeon.

Discography — When the technique of discography was originally introduced 30 years ago, it was hoped that the demonstration of posterior extravasation of dye from the disc would constitute an adequate radiological demonstration of a disc rupture. Unfortunately, the findings have not been definitive, since small tears in the annulus commonly associated with degeneration permit extravasation of dye, even in the absence of disc rupture. The test is often misused in symptomatic patients without objective findings — as a last resort for pseudojustification of surgical intervention for psychogenic backache. If it is done with the patient awake, however, symptoms reproduced on disc

494

distension can be localized to that particular level. The test is also of some value in identifying a normal disc at the outer border of a projected spinal fusion, and discography is an integral part of the chemonucleolysis (chymopapain injection) procedure.

Nuclear Magnetic Resonance (NMR) imaging of the spinal cord and canal is a diagnostic tool which is presently receiving a great deal of attention. The image is obtained by the evaluation of the differences in proton density among the tissues studied. It is non-invasive since, in contrast to CAT scans and myelography, it uses no ionizing radiation or contrast agents, and direct multiplanar images are obtainable.

The NMR technology is still under development but at present, long imaging time, in the range of minutes with current NMR scanners as compared with several seconds on the late generation CAT scanners, make NMR impractical as an initial screening device. In addition, as yet there have been no large clinical studies evaluating the specificity (false positives) and sensitivity (false negatives) of the NMR in respect to the lumbar spine. In spite of this, the NMR does hold great promise as an imaging method, with its high theoretical limits in image detail and very low attendant risks. More work on the technology and more clinical experience will have to be accumulated before its true value as a diagnostic method can be established.

BIBLIOGRAPHY

1. McCulloch, J. A., Mahoney, L. Thermography as a diagnostic aid in sciatica. Abstracts of the 1982 Annual Meeting of The International Society for Study of the Lumbar Spine (Toronto).
2. Waddell, G., McCulloch, J.A., Cummel, E., Venner, R.M. Non-organic physical signs in low back pain. *Spine,* 5:117-25, 1980.

BEND
TURN
TWIST

KNEES
NOT
BENT

Figure 1

Improper bending and lifting that can cause low back injury. Bending and *twisting* is probably the greatest cause of low back injury. Bending and lifting *without* bent knees also injures the low back. The history, carefully elicited, will reveal this factor.

Figures 1 to 4, 6 to 12, 14, 15, 17, 19 to 22, 24 to 27 are reproduced with permission from Cailliet, R.: Understand Your Backache. F. A. Davis Company, Philadelphia, 1984.

Figure 2

Improper lifting or straightening can cause low back injury.
Prematurely *arching* the low back can also cause injury.

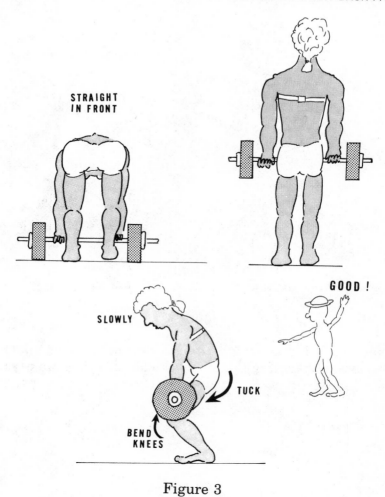

STRAIGHT
IN FRONT

GOOD !

SLOWLY

TUCK

BEND
KNEES

Figure 3

Proper Bending and Lifting Technique

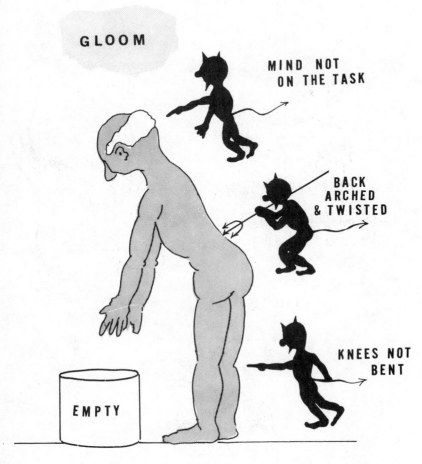

Figure 4

Distraction from anger, fatigue, depression, tension, etc., can cause a person to lift, bend, and reextend improperly — a common factor in low back injuries.

499

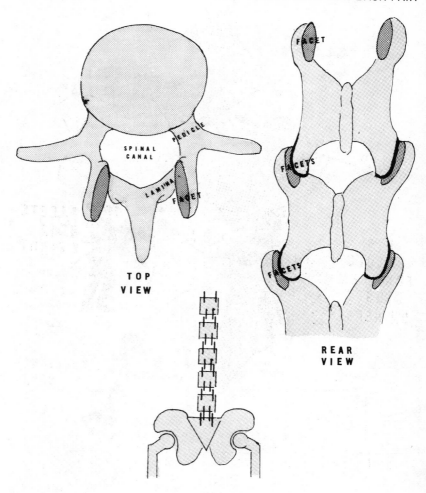

TOP
VIEW

REAR
VIEW

Figure 5

The facet joints (zygo-apophyseal joints) are not primarily
weight bearing. They glide on each other and permit
forward bending and extending, but restrict side bending
and twisting.

500

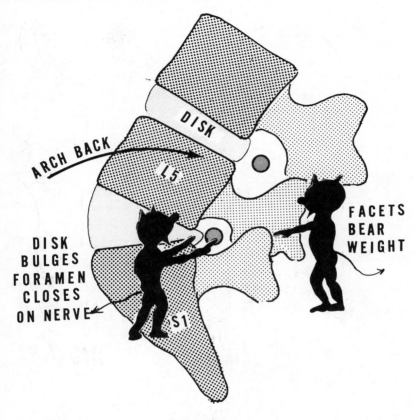

Figure 6

Mechanism of pain from increased sway back (lordosis)

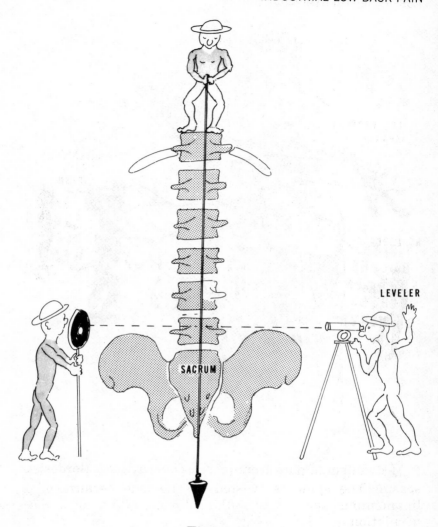

Figure 7

Normal straight spine above level pelvis

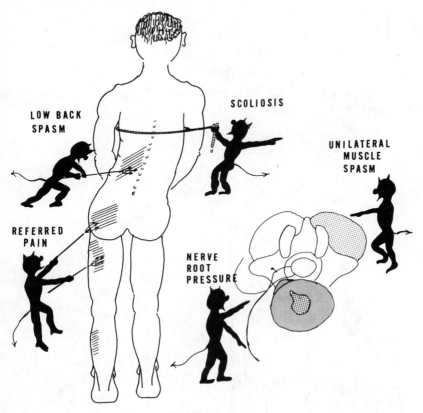

LOW BACK
SPASM

SCOLIOSIS

UNILATERAL
MUSCLE
SPASM

REFERRED
PAIN

NERVE
ROOT
PRESSURE

Figure 8

Functional sciatic scoliosis is due to a one-sided muscle
spasm. The spine is twisted to one side because of a
ligamentous sprain, an inflamed facet joint, or a disc
herniation.

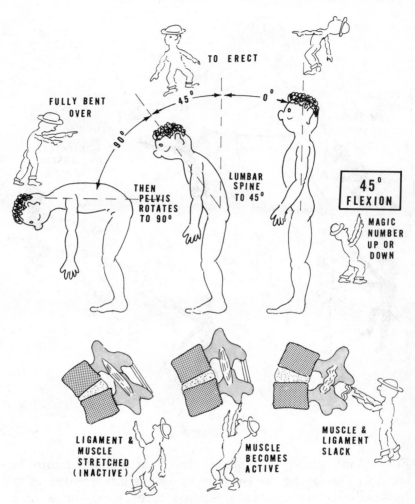

Figure 9

Normal bending and reextending shows the lumbar spine to flex only 45 degrees. The remainder of flexion occurs at the pelvis. This movement is usually smooth and well coordinated in flexion as well as in the return to an erect posture.

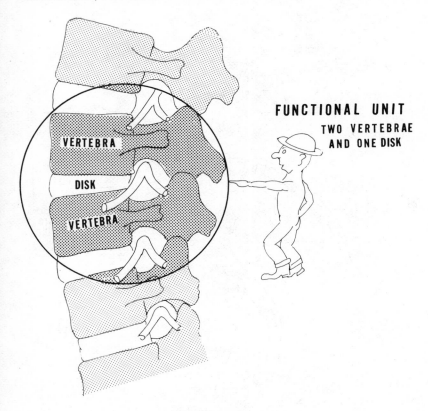

Figure 10

The functional unit is the term used for each mobile segment and comprises: two adjacent vertebrae, the intervening intervertebral discs, nerves, ligaments, and muscles. There are five functional units in the lumbar region: L1-2, L2-3, L3-4, L4-5, L5-S1.

505

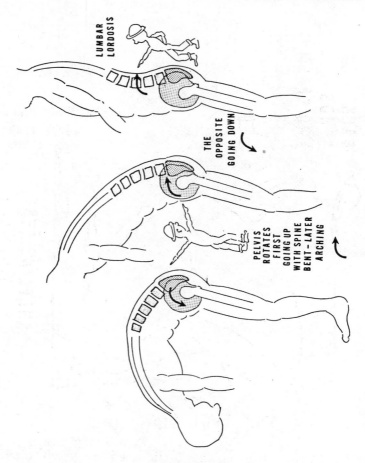

Figure 11

Proper body mechanics of the spine — pelvic movement in bending, reextending, and lifting.

506

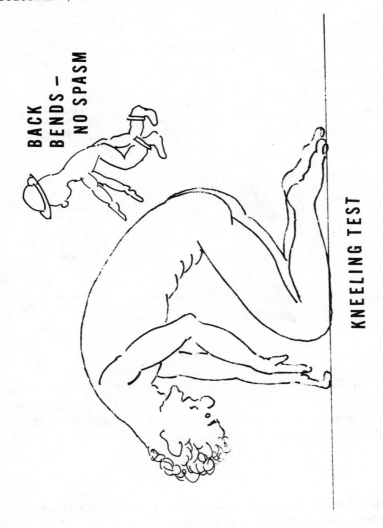

Figure 12

When the patient kneels, all of the hip muscles are relaxed
and only the back bends. If there is a true back injury,
spinal motion will be restricted and painful.

507

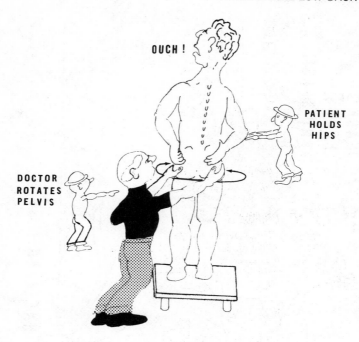

OUCH!

PATIENT
HOLDS
HIPS

DOCTOR
ROTATES
PELVIS

Figure 13

Test for Feigning — Hips are held tightly to prevent any movement of the spine. If the examiner then rotates the pelvis (which does not move the low back) and the patient complains, the pain is *not* organic.

508

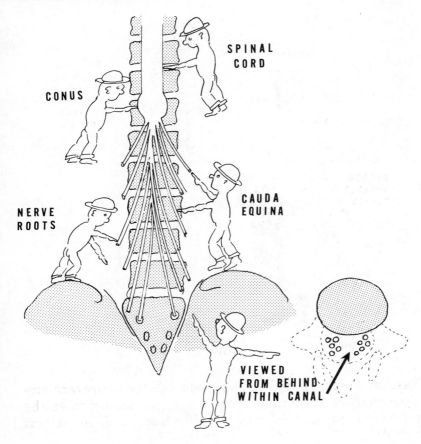

Figure 14

At the level of the first lumbar vertebra, the spinal cord divides into nerve roots and it is then called the cauda equina. Each root descends, on each side, to exit the spine via a foramen and then down to the lower limbs.

509

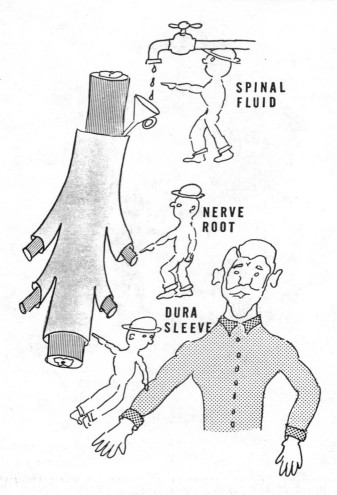

Figure 15

Each nerve is enclosed within a sleeve known as dura. This sleeve contains spinal fluid and the blood supply to that root. It is sensitive and may be the site of origin of sciatic nerve pain.

510

PULL

PRESS

DURA

NERVE

PAIN! in BACK and/or down LEG to CALF

Figure 16

The nerve root leaves the spine via a specific foramen and is enclosed in a dural sleeve. The dural sleeve can be compressed by either traction or pressure with resulting pain over the distribution of that nerve root.

511

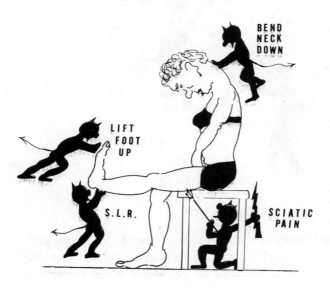

Figure 17

The Straight Leg Raising Test: If a sciatic nerve root is irritated or inflamed, pain occurs when it is stretched as in raising the straight leg. Bending the head down or raising the foot at the ankle further stretches the nerve and increases the discomfort.

512

PICK FANNY
UP & DOWN

Figure 18

Muscle test for integrity of the S1 root. The buttock (gluteus maximus) receives the same nerve supply as does the calf muscle (gastroc-soleus). In the supine position, the leg, bent at the hip and knee, picks up the pelvis by contracting the gluteus muscle. The strength and endurance of that muscle confirms an intact S1 nerve root.

513

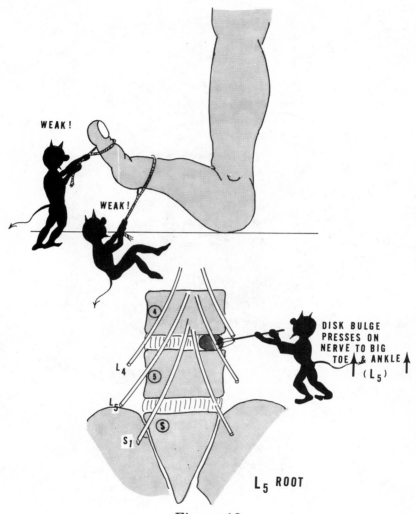

Figure 19

Fifth lumbar (L5) root involvement. If the fifth lumbar nerve root (between L4 and L5 vertebrae) is impaired, the ankle dorsiflexor and big toe extensor are weak. The patient has difficulty picking up his foot and walking on his heels. There are no reflex abnormalities.

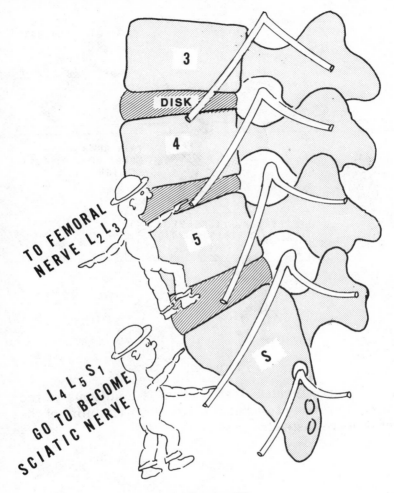

Figure 20

The nerve roots that come out between L4-5 and L5-S1 contribute to the sciatic nerve which goes to the calf, ankle, and foot. The nerves emerging from L2-3 and L3-4 go to the knee muscles in the thigh (quadriceps).

515

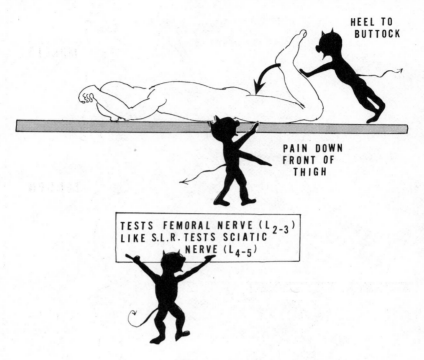

HEEL TO
BUTTOCK

PAIN DOWN
FRONT OF
THIGH

TESTS FEMORAL NERVE (L$_{2-3}$)
LIKE S.L.R. TESTS SCIATIC
NERVE (L$_{4-5}$)

Figure 21

Femoral Nerve Stretch Test — This is analogous to the
straight leg raising test but it checks for irritability of the
femoral nerve instead.

516

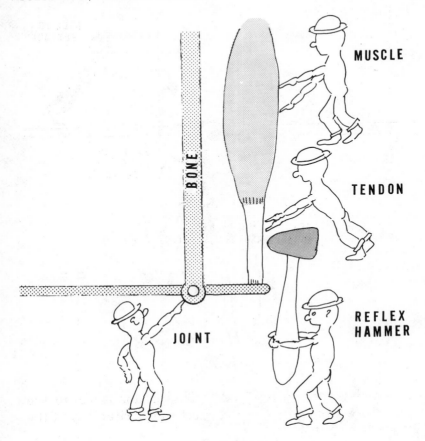

Figure 22

Reflexes: Tapping a tendon normally causes the muscles attached to that tendon to contract. No reflex suggests that the nerve to the muscle may be impaired. The ankle jerk tests the S1 nerve root. The knee jerk tests the L3 and L4 nerve roots. These tests can only be quantitative in the sense that they are present, absent, or as compared to the opposite side, diminished. The fact that they are diminished or absent bilaterally is usually meaningless.

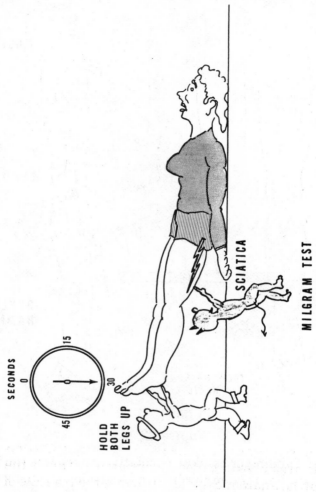

Figure 23

Milgram Test — the legs are held a few inches from the floor by contracting the abdominal muscles. This raises the pressure within the spinal canal and, if there is a bulging disc, further bulging and sciatica will develop.

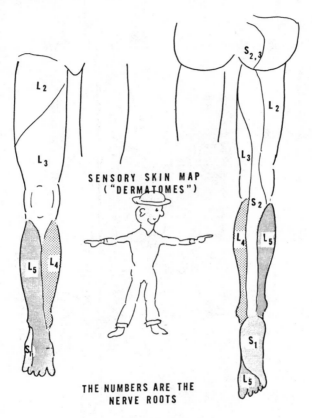

Figure 24

Dermatomes — each nerve root carries skin sensation from a specific area of skin called a dermatome. This is the area which can be anatomically identified as the true site of pain and/or numbness.

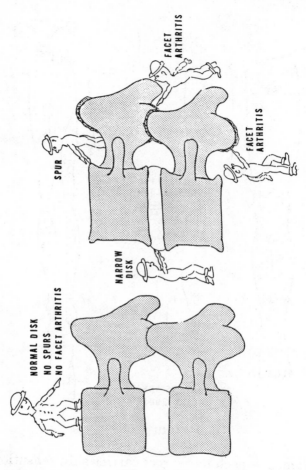

Figure 25

Degenerative Disc Disease — as the intervertebral disc narrows the facet joints behind become closer, their cartilage wears away forming osteophytes, and the foramina close down.

520

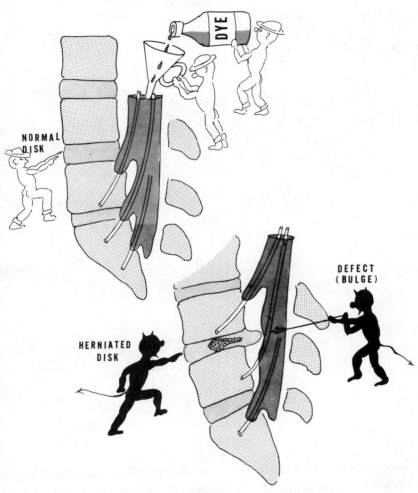

Figure 26

The myelogram — dye injected into the dural sac is visualized by x-ray. A bulging disc, a tumor, or an osteophyte can cause an indentation of the dye column.

521

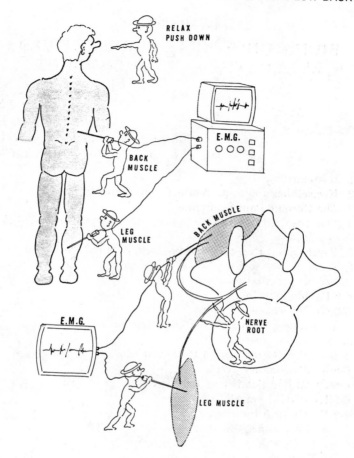

Figure 27

Electromyography (E.M.G.) — wired needles are placed in muscle and the amplified electrical impulses are viewed on a cathode ray tube screen. An impaired nerve root supplying the muscle tested can be identified.

522

CHAPTER 6

BIOPSYCHOSOCIAL PERSPECTIVES EVALUATION AND TREATMENT

Lorenz K.Y. Ng, M.D.
Gregory A. Grinc, Ph.D.
Janice P. Pazar, R.N.
Marianne S. Hallett, M.A., C.R.C.
The Washington Pain Clinic

§ 6-1. Introduction.

The major causes of disability in chronic low back pain may not be in the back! This is not a facetious over-dramatization but serves to emphasize the point that chronic low back pain is a multi-factorial problem and that physical factors/tissue injury are but one aspect to be considered in evaluating and treating the chronic pain-disabled back patient.

Up until the early 1960s, treatment of chronic pain was built mainly on the disease model. According to this model, chronic pain is seen as a symptom of an underlying disease and medical intervention is based mainly on assessment of the organic causes of the pain and their removal (Chapman, 1983; Ng, 1980). Frequently, the avenue of approach is through surgery or, if this does not seem appropriate or proves unsuccessful, the doctor frequently relies on use of medications, bed-rest, and traction. When these fail, patients are told that they have to "live with it" or, failing this, would be referred to a psychiatrist for evaluation or treatment.

Another situation where patients are referred for psychological evaluation is when it is noted that there is a paucity of organic findings in comparison with the extent of the patient's subjective complaints. In these instances, a psychosocial evaluation is requested in an attempt to "explain the discrepancy" between subjective and objective findings, or in an attempt to uncover psychopathology. In either instance, by the time a psychological/psychosocial evaluation is considered, patients are usually at least six months to one year, possibly longer, into the pain and disability process. With the chronic pain syndrome and disability process already far advanced, we end up with a really intractable problem — the Chronic Pain-Disabled Patient. The purpose of this chapter is to review the bio-psycho-social factors that may be involved in the genesis and maintenance of the chronic pain disability syndrome and to discuss the evaluation and treatment strategies that may be utilized.

§ 6-2. Nociception and Pain: Acute Versus Chronic.

Pain is not synonymous with nociception. The latter refers to the noxious stimulation of peripheral tissue receptors ("nociceptors") produced by injury or disease. Stimulation of nociceptors may result in pain which is the subjective sensory experience elicited by the perception of nociception. The sensation of pain frequently activates a negative affective state ("suffering") which, in turn, generates a response which has been described as the "pain behavior." As Fordyce has pointed out (entry 17 in the Bibliography at end of this chapter), as one progresses through each step from nociception to pain to suffering to pain behavior, new factors become involved.

In acute injuries, the linkage between nociception and pain or tissue injury and pain behavior is often direct and unambiguous. Acute pain, or pain of recent onset, usually can be correlated with an underlying organic basis, although social, cultural and psychological factors may affect its expression. Treatment of the underlying pathology or injury often results in resolution of the pain symptom or pain behavior. With the passage of time, however, the connection between the pain behavior and the original tissue injury becomes more complex and farther removed. With chronicity, opportunity is provided for the interplay of psychological, social, environmental and behavioral factors in the reinforcing and shaping of pain behaviors, which ultimately may lose all reference to the original nociceptive stimuli and become controlled mainly by conditioned environmental/psychosocial factors.

§ 6-3. The Chronic Pain Syndrome.

When pain persists beyond the usual course of a disease or normal healing time for an injury, or when it is associated with a progressive disease such as arthritis, the pain may be termed "chronic." Many definitions for "chronic pain" have been attempted, most of which involve some element of time, generally accepted as six months or longer (entry 35). As Bonica has pointed out (entry 6), the most significant factor differentiating "chronic" pain from "acute" pain is that the pain is no longer serving a useful biological or survival function but has become an end unto itself. As the pain becomes chronic, a complex constellation of symptoms evolves which frequently includes affective, vegetative, cognitive and behavioral components. The patient undergoes a progressive physical and emotional deterioration caused by progressive loss of appetite, insomnia, depression, and anxiety, a decrease in physical activity with progressive increase in "down-time," resulting in a state of physical deconditioning, and a progressive state of demoralization with increasing feelings of helplessness, hopelessness and worthlessness.

The above-described constellation of symptoms characterizing the patient with a chronic intractable pain complaint is characteristic enough to constitute a distinct entity that has been labelled as "The Chronic Pain Syndrome" (entry 4). Decreased levels of cerebrospinal fluid endorphins have been found in patients with chronic pain (entry 1). Whether this is the cause or effect, in part, of the chronic pain syndrome remains to be further elucidated. Nevertheless, patients with chronic pain are further exposed to a high risk of iatrogenic complications, often at the hands of well-meaning physicians. The usual drugs or surgical interventions that may be successful with acute problems, when applied repetitively to the patient with persisting pain, may produce disastrous results. Excessive use of tranquilizers, narcotic pain medications and polysurgery are often the iatrogenic end results.

§ 6-4. The Disability Process: Behavioral, Environmental and Social Factors.

There have been numerous studies in the past decade that have contributed to our understanding of factors involved in the development of the "disability process." In his revealing description of this process, Weinstein (entry 40) points out that before injury the typical patient may have had significant physical and emotional needs that were not being met. In some individuals, the injury or illness, frequently relatively trivial, may thus serve to convert a pre-existing or concurrent unpleasant or unacceptable problem into one that is "justifiable" by virtue of the accident or illness. Because many of the individual's needs following injury are subsequently met in an acceptable way due to our social values and customs, the symptoms persist and the illness role of the patient is reinforced by continuing treatment.

In patients with chronic pain, other processes may also take place at the behavioral, environmental and social levels, which further contribute to the disability. The works of Sternbach (entry 36) and Fordyce (entry 15) have emphasized the importance of environmental and psychological factors in the etiology and development of chronic pain behavior in patients. Through the process of learning (i.e., conditioning), previously neutral events in the environment may become conditioned stimuli, capable of eliciting a pain behavior response that may interfere with the outcome of any medical intervention. This learned or conditioned pain-state may further be reinforced by our social and health care system through compensation for injuries or reinforcement for illness behavior, thus adding to the conditioning process. The Honorable Judge H. Grossman of the United States Social Security Administration has noted that chronic pain not only can be an iatrogenic disease but also a "nomogenic complication" (Greek: nomos = law; genic = generated). The existing laws in cases of injury compensation and the adversary legal process are frequently major operant factors in conditioning/reinforcing pain and illness behaviors. Added to this, also, is our existing "health" insurance system which, in practice, rewards illness and reinforces illness behavior rather than health or wellness behavior (entry 30). As Chapman, Brena and Bradford (entry 10) have pointed out, patients who focus on pain as a way to gain an income and/or to win litigation, take on the identity of disabled individuals and frequently find themselves less and less functional and more and more helpless. They become very difficult candidates for medical or vocational rehabilitation.

§ 6-5. Psychosocial Evaluation of the Pain-Disabled Patient.

"It is more important to know what kind
of man has a disease than to know what
kind of disease a man has."

Maimonides

In treating the chronic pain-disabled patient, the above quotation is especially relevant. A chronic pain syndrome places an inordinate amount of stress on an individual and those around him. It is important to know as much as possible about the patient's pre-morbid functioning and his current mental status. Other data related to the patient's social, home, and work environments, as they relate to his medical problems, are also of prominent importance in understanding the dynamics of the patient's reaction to his pain. More specifically, the patient's style of coping with significant stressors, his self-concept as it relates to his work, family and own physical integrity, are very important to understanding the course of the patient's treatment process and in making prognostic statements. Psychological testing can also provide much-needed information on the patient's intellectual functioning and personality structure which may significantly influence the manifestations of the patient's chronic pain syndrome.

A complete psychosocial assessment can provide us with much valuable information about a patient as an individual with certain medical symptoms and with a certain pre-morbid level of adjustment, who has various elements influencing the manifestations of his medical problems. A chronic pain syndrome is as much a behavioral/emotional problem as it is a physical/organic problem. Compliance with prescribed treatment plans is a major issue in dealing with chronic pain patients. The ultimate goal is to maximize effectiveness of the treatment plan through tailoring it to the individual patient and to increase the patient's level of compliance with the treatment plan by uncovering and dealing with resistances and emotional blocks that may cause treatment failure.

§ 6-6. General Issues in Psychosocial Assessment of Chronic Pain.

The initial task in a psychosocial assessment is to clarify the clinical picture in terms of the extent of emotional disturbance created by the chronic pain syndrome and to rule out broad differential diagnostic categories that may be encountered. An initial consideration is to determine if there is the presence of a serious psychotic disorder where the pain may be merely a manifestation in the clinical picture and may, in fact, be delusional. A second consideration of the psychosocial evaluation is to assess the patient's motivation and the likelihood of malingering. The major function of a psychosocial evaluation, however, is to assist in the clarification of the role that psychological and social factors play in the chronic pain syndrome of the patient. For example, such information may be vital in helping the surgeon decide upon the patient's suitability or readiness for surgery where indications exist. Unfortunately, there has been a tendency among health practitioners to diagnose the pain as "psychogenic" in origin when it does not fit into either of the above categories or when it cannot be easily explained in terms of the obvious underlying tissue pathology. Our view is that diagnosis of "psychogenic pain" is of little heuristic or practical value in terms of understanding the disease process or for treatment purposes. A more operational approach to the problem is recommended.

§ 6-6(A). Psychotic Illness and the Pain-Disabled Patient.

There are some patients who appear to have a primary thought disorder related to a basic schizophrenic personality organization that is present with chronic pain. These patients are relatively rare in our experience and rather easy to diagnose. Sometimes, however, it is difficult to determine if (i) the pain is, in fact, delusional or hallucinated; (ii) the patient is basically schizophrenic or psychotically depressed and has some significant underlying pathology causing the pain; or (iii) the patient is experiencing a complete personality decompensation in the form of psychotic illness related directly to the stress of a chronic pain syndrome. In our experience, all three of the above possibilities are relatively unusual. A schizophrenic-like reaction to a chronic pain syndrome is rare. If it does occur, it is usually a paranoid reaction and is frequently limited to obsessive thoughts about health-care providers and the insurance company. It is important to be aware that many chronic pain patients with work-related injuries develop hostile and angry feelings towards employers, health-care providers, and insurance companies (sometimes for valid reasons from the patient's perspective), and this does not constitute paranoia. However, a true full-blown paranoid schizophrenic reaction may occur but is quite uncommon in our experience. Psychotic reactions are most likely to occur in the form of a psychotic depression. Some patients become overwhelmed by the pain and develop extreme feelings of helplessness, hopelessness, and worthlessness. After years of unsuccessful treatment and dozens of medical tests, some patients may develop such a psychotic depressive reaction. In many of the above cases, the basic underlying thought disorder or depressive reaction needs to be aggressively treated before any meaningful treatment of the chronic pain syndrome can occur.

§ 6-6(B). Malingering.

The actual percentage of chronic pain patients who are malingering is unknown. It is rather obvious that the workers' compensation system makes malingering an attractive behavior to some individuals, just as unemployment insurance may play some part in an individual leaving his job. However, it is also clear that few individuals are true malingerers who totally feign injury and pain to collect workers' compensation benefits. However, because a malingerer may have a sociopathic and/or antisocial personality, they are clever at avoiding detection.

Most health-care providers and individuals who deal with chronic pain patients have some internalized standard by which they categorize a patient as a malingerer or faker. Sometimes, psychological testing is utilized with the hope that it will uncover a malingerer. Unfortunately, psychological tests are only slightly better at being lie-detectors than most people because they are somewhat less obvious and may provide another piece of objective information that may reveal a significant medical discrepancy or inconsistency.

There have been at least two sets of criteria developed independently that match our own criteria rather well in evaluating patients suspected of malingering (entry 14; entry 9, in press). The Emory Pain Control Center has labelled their scale an "inconsistency profile" (entry 9, in press). This appears to be an apt description of the type of data that may be used in evaluating this dimension. We list below the patient factors that may be considered in assessing the possibility of malingering:

    (1) Discrepancy between a person's complaint of "terrible pain" and a presentation that reflects health and well-being.

(2) Complete negative work-up for organic disease by more than one physician.

(3) Patient describes his pain in a vague or global manner, for example, "I hurt bad."

(4) Excessive reliance on medical terms that were obviously learned by contact with health-care providers.

(5) Overemphasized gait or posture which develops suddenly, has no organic basis and cannot be substantiated objectively, for example, cane not being used or shoes not being worn unevenly, despite a limp.

(6) Excessive resistance to evaluation procedures, especially when non-invasive procedures are involved.

(7) Resistance when patient is informed that the goal of treatment is a return to gainful employment.

(8) Resistance to treatment procedures, especially in presence of intense complaints of pain.

(9) Inconsistencies in the patient's report of compliance with physical therapy exercises and inability to perform them in the office under observation.

(10) Frequent no-shows or appointment cancellations.

(11) Bizarre response to treatment and diagnostic procedures.

(12) No evidence of emotional disturbances or psychotic disorder.

(13) Inconsistent psychological test profile with clinical presentations, for example, MMPI profile indicative of a psychotic disorder with no clinical signs of psychosis.

(14) Unstable personal and occupational history.

(15) A personal history that reflects a character disorder which might include drug and/or alcohol abuse, criminal behavior, erratic personal relationships, violence, etc.

§ 6-6(C). "Psychogenic" Pain.

Frequently, evaluations are requested in certain types of clinical settings or by individual medical practitioners to determine if the patient's pain is "psychogenic in nature." We feel that the term "psychogenic" is rather misleading and is frequently interpreted differently by different people. There are conceptual and practical problems with the concept of psychogenic pain (entry 22). It is usually a pejorative label and interpreted as such by both the patient and health-care providers. The patient who is referred for treatment with this label almost invariably resists it, either consciously or unconsciously. This often leads to the patient dropping out of the treatment and setting out with renewed determination to seek a new medical opinion to substantiate his complaint of chronic pain as having an organic origin. This undoubtedly raises the probability that the patient will be treated in a more radical and possibly invasive manner and, in fact, may end up with the added problem of iatrogenically induced pain.

A second conceptual factor to consider in the diagnosis of psychogenic pain is that it is usually a diagnosis by exclusion. While investigators have attempted to associate certain personality traits with psychogenic pain, this has been a largely unsuccessful endeavor, and the diagnosis itself is based on negative findings, that is, it is based on the fact that there is no evidence for organic pathology and absence of any psychotic illness. The idea of chronic pain as "psychogenic" thus has little conceptual or heuristic value. Of greater value would be to approach the concept of chronic pain operationally in a way that would provide a basis for developing treatment strategies.

§ 6-7. Respondent Pain and Operant Pain.

As discussed earlier, pain can be conceptualized in

numerous ways and, in terms of chronic pain, the most useful characterization of pain is as a behavior. Sternbach (entry 35) has stated that "in order to describe pain, it is necessary for the patient to do something." The assumption is that often some underlying tissue pathology accompanies the behavior and is the "cause" of the pain behavior. However, Fordyce (entry 15) notes that pain may be viewed as a communication to others that may or may not be directly related to underlying tissue pathology. Szasz (entry 38) has theorized that pain is used to communicate a request for help, feelings of anger or hostility, or as an expiation of guilt. Swanson (entry 37) has labelled pain as a "pathologic emotion" which he compares to the emotions of anxiety and depression. These conceptualizations are similar in that they conceptualize pain more in terms of verbal or non-verbal behavior than in terms of mechanical/tissue dysfunction.

Fordyce (entry 16) has provided a useful conceptualization of pain behavior. Fordyce describes two classes of pain behavior: operant and respondent pain. Respondent pain behavior is described as those pain behaviors influenced by or substantially under the control of some underlying nociceptive stimulus. Respondent pain is a type of reflex reaction. It happens automatically and is typified by a case of a broken leg or burned skin. All acute injuries would fall under this category. Many patients who have chronic pain may have a certain amount of pain that can be labelled as respondent pain. It should also be pointed out that, as time progresses, secondary tissue sources of nociceptive stimuli may set in, resulting from physical deconditioning, weight gain, inappropriate weight-bearing and poor posture, all of which may result in some degree of respondent pain. This may not be insignificant in individuals with chronic pain syndrome who may have low pain threshold and pain tolerance.

Operant pain behavior is behavior that is emitted by the organism and its frequency of occurrence is either increased or decreased based on the nature of the consequences that follow the behavior. It has been demonstrated numerous times with both animals and humans that behavior which is reinforced in a positive way increases its probability of occurrence; behavior which is negatively reinforced decreases its probability of occurrence. The effectiveness of this type of learning paradigm has been demonstrated by Skinner (entry 34) in numerous situations. Operant pain behavior is usually developed as a result of respondent pain behavior being present for some period of time. Given the presence of respondent pain, the opportunity for the pain behaviors to be reinforced exists in numerous situations. The direct reinforcement of pain behavior can occur when medications are used in response to pain behavior or when the patient receives desired attention from significant others in his environment because of the pain, such as his spouse, friends or health-care providers. Operant pain behavior may also result from indirect reinforcement. This occurs when pain behaviors lead to the avoidance of some adversive or unpleasant situation. This is most easily conceptualized in the framework of an individual's day-to-day responsibilities. A patient who displays significant pain behavior often may be excused from work or from social responsibilities to his family or spouse. In essence, the patient learns that he can avoid certain unpleasant or adverse circumstances if he displays pain behavior. This type of avoidance learning is especially powerful if the individual fears or is anxious about responsibilities, is not positively reinforced by social encounters or his relationships with his family members, or has a job that is unrewarding. This type of avoidance learning is what has been frequently labelled as secondary gain.

536

It should be emphasized that operant pain is *real* pain. It would be a mistake to classify respondent pain as real pain and operant pain as faked or "in the patient's head." The concept of operant pain behavior enables us to reframe the previously discussed concept of psychogenic pain in a new light. As stated earlier, the concept of psychogenic pain possesses little heuristic or practical value, whereas the concept of operant pain provides numerous possibilities to both intervene and to further our understanding of chronic pain.

§ 6-8. Assessment Instruments.

The psychological assessment of the pain-disabled patient has posed a new problem since the traditional psychometric instruments available do not quite fit the novel assessment situation. Most psychometric instruments, especially those assessing symptoms and personality, are oriented toward traditional psychiatric settings. Usually, the assessment of the pain-disabled patient is done in the context of a traditional medical setting. While the fact that the psychological tests are largely oriented towards a psychiatric population is somewhat of a problem, it is not as problematic as it appears at first glance, since most patients who develop a chronic pain syndrome do experience significant psychological symptoms. New instruments to assess chronic pain and its sequela have been slow in developing and, unfortunately, many of those developed have been done by clinicians who are unfamiliar with psychometric rigor. However, these newly developed instruments do have face validity and meet an important need in the clinical settings.

On the other side of the coin are psychological assessment instruments that have been extensively researched with a wide range of clinical populations (e.g., the Minnesota

537

Multiphasic Personality Inventory). These more traditional instruments can provide very valuable information on the psychiatric status of our target population. The more traditional projective test instruments most frequently associated with psychological assessment, such as the Rorschach and Thematic Aperception Tests, have some value in assessing the pain-disabled patient and may be used in situations where questions about thought disorder exist. However, routinely they are rarely used because of the time-consuming nature of these tests.

The following will be a brief list and discussion of various assessment instruments that are commonly used in the assessment of chronic pain patients.

§ 6-8(A). Self-Rating Report Instruments.

There are a variety of self-report instruments used to assess chronic pain patients. These types of instruments provide a quick and easy source of data, although they are frequently rather general and the accuracy, validity, and reliability of information obtained is sometimes questionable. However, from a clinical management standpoint, such information can be very useful in providing a basis for development of an individualized treatment plan for the patient.

Self-rating of the intensity of pain and the amount of disruption that pain causes in a patient's life is often requested from the patient. The patient is asked to rate the pain on a scale of zero to ten or zero to one hundred. These rating scales are frequently anchored at various points by descriptions such as "no pain," "mild pain" or "unbearable pain." The self-ratings of the patient's pain intensity may also be done on a visual analogue scale. The visual analogue scale (Fig. 1) involves a straight horizontal line (10cms in length) anchored by descriptors, where the patient simply places a slash mark through the line to indicate where he feels his level of pain.

No pain   Unbearable pain

0   10

Figure 1

The amount of disruption the pain causes to various aspects of the patient's daily life can also be measured. The patient is asked to assign a value of zero to ten in terms of increasing disruption to his life in such categories as: appetite, sleep, mood, sexual functioning, physical activities, and social activities. An example of what we use at the Washington Pain Center is illustrated below (Fig. 2).

Appetite	Mood	Activity/Work/Exercises	Pain	Suffering
Sleep	Sex Life	Sense of Control	Social Life	Goals Accomplished

Figure 2

A daily log of the patient's pain level, activity and medication usage can be used to assess the patient's functional status and to monitor progress. An example of such a daily log is shown in Fig. 3, at end of this chapter.

The key problem in this type of assessment is whether these estimations of the intensity of the pain and the extent of disruption to their functioning in key areas of their life correlate with their actual behavior. A recent study by Fordyce and his colleagues (entry 18) suggests there was little or no relationship between a patient's estimation of the severity of his pain and behavioral measures, such as health-care utilization, medication usage, body position and frequency in engaging in commonplace activities. Fordyce, *et al.,* note that this has particular implications for a concept of operant pain. From the standpoint of assessment, it advises caution in relying too heavily on the patient's self-report. The above problem can be counterbalanced to some extent by having others rate the patient and by having medical staff rate these dimensions based on their information on the patient. Nevertheless, in practice, these self-ratings provide helpful information that can be utilized by treatment staff to measure compliance and monitor progress in treatment.

§ 6-8(B). Personality Tests.

The Minnesota Multiphasic Personality Inventory (MMPI) is undoubtedly the most widely used personality instrument in use today. Dahlstrom, *et al.* (entry 13) describe it as being developed to categorize psychiatric patients and to discriminate between psychiatric patients and normal patients. The 566-item true/false test, which can be given in a short-form, has ten clinical scales and three validity scales. The ten clinical scales that make up the MMPI are: hypochondriasis, depression, hysteria, psychopathic deviance, masculinity/femininity, paranoia, psychasthenia, schizophrenia, hypomania and social introversion.

The results of numerous studies using the MMPI on chronic pain patients indicate a consistent elevation on the so-called "neurotic triad" scales — hypochondriasis, depression, and hysteria. The basic interpretation of this profile is that it reflects a basic preoccupation with physical symptoms, bodily functions, indirect expression of hostility, negativism, depression, and the prominent use of repression and denial to cope with psychological conflicts. When the depression scale is significantly lower than the other two scales in this "neurotic triad," then the much noted "Conversion V" pattern is present, which reflects a tendency to somatize under stress. This "Conversion V" pattern is associated with conversion hysteria where psychological conflicts are expressed as physical symptoms. This Conversion V pattern is a very common profile among chronic pain patients, but it would be a mistake to conclude that most chronic pain patients are suffering from an hysterical conversion reaction. With some frequency, this Conversion V profile has been used to justify the diagnosis of psychogenic pain. As stated earlier, the concept of psychogenic pain is conceptually and operationally inadequate. A number of explanations for the elevation in the neurotic triad are plausible. First, it has been argued that a chronic pain syndrome, in the absence of clear-cut organic findings, is caused when an injury occurs to an individual who has a pre-existing neurotic condition. A second possibility is that the elevation in the neurotic triad is caused by the chronic pain syndrome itself. A third option is that it is unfair to assume that the profile reflects a neurotic orientation because of the objective reality of the situation. That is, it is reasonable for a chronic pain patient to be preoccupied with physical symptoms because of the nature of his injury; it is depressing to have chronic pain and it is necessary to deny and repress emotional conflict to cope with the crisis-oriented situation. Watson (entry 39) notes that there is some evidence to support each of these positions in that all of these options are not mutually exclusive. From our experience, it is quite clear there are individuals who have serious adjustment problems of a neurotic or characterological nature who later develop chronic pain syndrome. It is also obvious that a significant number of patients who develop a chronic pain syndrome were very well adjusted prior to their injury and, since their injury, they have had difficulty coping and present clinically with a picture which looks identical to a patient with long-term neurotic problems. The third option is a recapitulation of the second, with the implication that the individual is not really presenting with a significant clinical picture of emotional disturbance, but simply responding to the test items in an affirmative manner as they reflect the various psychopathologies involved.

Our experience at the Washington Pain Center with the MMPI indicates that there are three basic profiles similar to what other investigators have found. Bradley, Prokop, Margolis and Gentry (entry 7), using a multi-variate analysis of the MMPI, identified three homogenous subgroups of MMPI profiles. The first group is composed of essentially normal MMPI profiles with no significant clinical elevations. These are most frequently found in individuals with significant chronic pain who are functioning relatively well in most areas of their lives. Usually, these patients have continued to work or have carried on the main activities in their lives with minimal disruption. Sometimes, a normal profile is also found in very defensive patients who exhibit behavioral dysfunction. A second group, and by far the most numerous, is a subgroup with elevations on the neurotic triad. These are chronic pain patients with significant behavioral dysfunction and psychiatric symptoms. The third group is labelled as a "psychopathologic profile," where almost all of the clinical scales, with the possible exception of the masculinity/femininity scale, are elevated. This group tends to show very significant psychiatric symptoms and may, in fact, carry additional diagnoses of schizophrenia, psychotic depression or borderline personality. It is of note that, in our experience, patients who are suspected of malingering or greatly exaggerating the severity of their symptoms, will often fall into this last category. Armentrout, Moore, Parker, Hewett and Feltz (entry 2) found similar subgroups as Bradely, *et al.*, and correlated

542

them with various behavioral parameters, such as medication use, activity, sleep, sex and pain behavior. They found that the psychopathological group had the most significant adaptation problem and that the normal profile group the least, with the neurotic triad group falling in the middle. More recently, Heaton, *et al.* (entry 21) have delineated seven types of MMPI profiles which they labelled as the following: hypochondriasis, reactive depression, normal, somatization, somatization with depression, psychotic/borderline, and manipulative reaction group. The further discrimination is very helpful in conceptualizing how to deal with individual patients and how to maximize the treatment regimes. It also has significant implications as a research model for outcome studies.

§ 6-8(C). Millon Behavioral Health Inventory (MBHI).

The MBHI was developed by Millon and his colleagues specifically for patients with physical illness (entry 28). The MBHI rates the patient on eight different coping styles which are derived from Millon's (entry 27) theory of personality. The inventory also provides six psychogenic attitude scales which assess psychosocial stressors found in the research literature to be possible precipitators or exacerbators of physical illness. While the MBHI was not developed specifically for the assessment of chronic pain patients, it has clear potential to be useful in this area and is possibly more relevant to the assessment task. However, at present there is little literature examining its usefulness, although as time goes on, it should become more frequently cited in the literature. A further assessment of its usefulness will require more extensive research and more widespread acceptance on the part of clinicians.

543

§ 6-8(D). McGill Pain Questionnaire.

The McGill Pain Questionnaire (entry 25) was developed to assess how patients conceptualize their pain, and has three major classes of word descriptors that patients are asked to select as it applies to their pain. The three major classes of word descriptors are sensory, affective, and evaluative. The sensory word descriptors enable the patient to describe the pain in terms of temporal, spatial, pressure, dermal, and other sensory properties. The affective word descriptors enable the patient to describe the pain in terms of its emotional and autonomic properties, such as levels of tension and fear associated with the pain. The evaluative descriptors enable the patient to describe the overall intensity of the pain experience and can be characterized by such words as unbearable, intense or annoying. There are a total of seventy-eight pain descriptors in the entire questionnaire. One of the major advantages of the McGill Pain Questionnaire is that it provides data that can be easily quantified along the three dimensions it measures. While the clinical utility of the questionnaire is somewhat limited, it does provide information that can be very helpful in research. In addition, further research may begin to develop its clinical utility, as a recent study by Kremer, Atkinson, and Kremer (entry 23) has demonstrated. Kremer, *et al.,* found that the total number of affective descriptors utilized by the patient was a good predictor of psychiatric disturbance. The authors of this study note that chronic pain patients are usually very sensitive to any implication that there is a psychiatric component to their chronic pain syndrome. The McGill Pain Questionnaire is not an obvious measure of the patient's psychiatric status, and therefore can provide a rather unobtrusive measure of the patient's mental status.

§ 6-8(E). Clinical Interview.

The psychological interview for patients presenting with chronic pain has three main components. The first basic element that needs to be addressed in the interview is to obtain as much information as possible about the patient's pain. This falls into the general category that Fordyce (entry 15) labels the behavioral analysis. The second basic purpose of the psychological interview is to obtain a sense of the patient's current mental status. The third major purpose of the psychological interview is to obtain a complete psychosocial history on the patient, to grasp the significant events in the patient's life and to obtain as accurate a picture as possible of the patient's pre-morbid functioning.

§ 6-8(E)(1). Functional Analysis.

Fordyce (entry 15) has written extensively on the use of behavioral analysis. It is extremely important for the psychological interview to be directed into this area, at least initially, because the patient will be rather suspicious of the whole process of the psychological evaluation, and this is an area that he feels most comfortable discussing. It is important to generate an atmosphere that is positive, open, supportive, and noncritical during this section of the interview, as many patients with chronic pain have had very negative interactions with past health-care providers. It is also important for the interviewer to assure the patient that he is not seeing a psychologist because the physicians believe his pain is "in his head" or "imagined." The goal of the behavioral analysis is to identify correlations between the manifestations of the patient's pain (i.e., the patient's pain behaviors), and various activities and social consequences. In essence, the task is to identify antecedents and consequences around the patient's pain behaviors. This analysis is based on the assumption that a certain amount of pain a chronic pain patient experiences will be "operant" pain. The longer a patient experiences pain, the more likely it is that these pain behaviors will be subject to learning and conditioning. The primary source of this behavioral analysis is the patient's self-report. However, it has been found that the patient's self-report of his pain level intensity, his pain behaviors, and other well activities is not always accurate (entry 18). The interviewer cannot assume that he is obtaining totally accurate information from the patient himself, although obviously heavy emphasis needs to be placed on the patient's self-report. Other sources of information that would be extremely important in conducting behavioral analyses are the observations of the interviewer, observations of other medical staff, and the report of the patient's family and possibly employer and co-workers. In this regard, a home visit by a member of the pain team could be extremely valuable in uncovering important relationships in maintaining the patient's pain behavior.

Table I, at end of this chapter, presents a number of factors that are important in the behavioral analysis, along with some pertinent examples for each area. Where an attribution regarding operant or respondent pain is made, it should be kept in mind that these are only tentative indicators and not pathognomonic signs.

§ 6-8(E)(2). Mental Status Examination.

The mental status section of the clinical interview focuses on the traditional factors examined in this type of evaluation. It is a clinical evaluation of the patient's physical presentation, moods and emotions, intellect and thought processes. The patient's physical presentation during the interview provides a great deal of information to the interviewer in terms of his outward manifestation of pain behavior. Certain patients display a very dramatic set of pain behaviors, while other patients display very little pain behavior. The patient's ease of sitting, standing, and walking is also available to observation over time. It is not uncommon for many pain patients to state that they cannot sit for longer than twenty minutes at a time but, nevertheless, appear able to sit through a clinical interview that will sometimes last close to an hour or more.

§ 6-8(E)(3). Social History.

Obtaining a complete social history is extremely important in being able to assess the patient's pre-morbid level of functioning and to determine any critical events in the patient's family and work history that may significantly affect his psychological functioning. While some patients are rather surprised at being asked to describe their childhood and family background, this also provides a further area to discover past traumatic events in the patient's life that may make them more susceptible to the stresses of their current injury. In addition, family background also provides important insights into the patient's value system and his self-concept. Included in this section should be a detailed history of any previous psychiatric history the patient may have. The patient's past work history, as it relates to a number of jobs and past job-satisfaction, is extremely important. These are often significant factors that relate to the patient's self-image. For example, a patient who has a poor work history may not find his work particularly rewarding and, in fact, may find returning to work to be an aversive state. Therefore, this patient will have a built-in negative reinforcer for maintaining his pain behavior, especially if workers' compensation payments are involved in providing a positive reinforcement for staying off work as well. A detailed history of the patient's social relationships also provides some indication of how the patient has managed his personal life. As in the work history, this also provides some indication as to how rewarding the patient's social relationships have been for him. If a patient's social relationships have been unrewarding or rather chaotic, this may also be a negative reinforcing factor in maintaining his pain behavior, since this also tends to ensure his isolation from others. The patient's level of academic achievement is also notable in that it provides some indication of the patient's motivation and his adaptability.

§ 6-9. Treatment.

§ 6-9(A). A Conceptual Framework.

Until recently, treatment of chronic pain was built mainly on the disease model which views pain as a symptom of underlying disease or tissue pathology. Medical intervention derived from this model focuses on assessment of the "organic" causes of the pain and their removal. This approach is appropriate for acute pain which often is a symptom of underlying tissue pathology or injury. In an acute situation, the pain/pain behavior frequently parallel the underlying tissue pathology. This may be depicted schematically as follows:

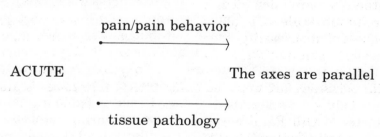

As the condition becomes chronic, the relationship between the reported pain/pain behavior and the tissue pathology may no longer be parallel. As pointed out earlier, environmental, psychological, and behavioral factors can maintain or shape the pain behavior, independent of the underlying tissue pathology or injury. The pain behavior axis, with the passage of time, may lose all direct reference to the pathology axis, and may exist independently of each other in an orthogonal (perpendicular) relationship. This independent and orthogonal relationship in chronic pain between pain behavior and tissue pathology was first proposed by Brena and Koch (entry 8) in the Emory Pain Estimate Model depicted schematically below:

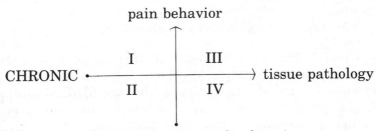

The axes are perpendicular

The construction of this model for chronic pain involves separate ratings of tissue pathology and pain behavior based on objectively obtainable data. Degrees of tissue pathology are depicted graphically on a horizontal zero-to-ten scale. Up to 2.5 points may be assigned for degrees of abnormality on each of the following four factors: physical examination, neurological examination, radiological studies, and pertinent laboratory studies. Measures of pain behaviors are depicted on a vertical zero-to-ten scale. Pain behavior assessment is based on the following four factors: McGill Pain Questionnaire (measuring subjective pain verbalization); Activity Checklist (measuring activities of daily living, ADL); drug use; and the Minnesota Multiphasic Personality Inventory (MMPI) profile. From these ratings, patients can be divided into one of four categories. Each chronic pain classification describes a distinctly different profile, requiring variations in clinical management. In Class I, pain behavior is high and organic pathology scores are low; Class II patients score low on both variables; Class III patients are pain-disabled with high pathology ratings and high pain behavior scores; Class IV patients show a low pain behavior despite high pathology scores.

The fact that chronic pain behavior may exist independently of tissue pathology as a result of learning or conditioning, resulting in operant pain, has important implications for our understanding of chronic pain and its treatment. Failure to recognize this independent relationship can result in serious problems.

While appropriate diagnostic tests need to be performed, the assumption that pain behavior is merely a reflection of tissue pathology frequently leads to a search for tissue pathology to "explain" the pain behavior. This can lead to iatrogenic complications with over-performance of diagnostic tests, inappropriate treatment, and sometimes ineffective or unnecessary surgeries. Furthermore, this focus on tissue pathology can end up producing an unintended side-effect of reinforcing the patient's illness or pain behavior through the "validating" stamp of continued physical treatment. For patients, a self-fulfilling prophecy results: since extensive tests and treatments do not bring improvement, they become reinforced in their belief that they must have a truely serious illness that no one can diagnose or treat. To the claims adjuster, on the other hand, the "discrepancy" that exists between the pain behavior and the tissue pathology is frequently interpreted to mean that there is no medical problem at all. Adversarial positions thus arise all too frequently in these cases, to the detriment of the patient. (Further discussions on the legal issues involved can be found in the article in Chapter 10 by Mr. Herz.)

550

The important point we wish to emphasize here is that the chronic pain syndrome and its attendant chronic pain behavior need to be treated as distinct entities in their own right and not merely as symptoms or manifestations of tissue pathology. This is particularly important to remember when dealing with patients in whom tissue pathology is low in comparison with pain behavior (Class I). These patients, in fact, constitute the predominant population of chronic back pain patients with pending disability claims (entries 9, 31). For these patients, focussing exclusively on tissue pathology may result in over-treatment and iatrogenic problems, while ignoring or rejecting the chronic pain behavior as needing treatment can certainly be counterproductive. Even for patients in Class III, where significant tissue pathology exists in combination with high pain behavior, reversal of pathology alone may not be sufficient to alter the disability (i.e., functional) level, unless pain and illness behaviors are also appropriately modified. Both the structural/anatomic aspects (i.e., tissue pathology) and the function aspect (i.e., pain behavior) need to be evaluated and treated adequately and appropriately if a successful outcome in terms of the well-being of the total person is to be achieved.

§ 6-9(B). Factors Influencing Treatment Outcome.

It is clear that many factors, apart from pathology, influence prognosis. In our experience and in the experience of other centers (see entry 31), the four most important factors determining successful outcome in rehabilitation of chronic pain patients are, in order of importance: motivation, social support system, chronicity, and degree of tissue pathology.

Treatment of tissue pathology relevant to low back pain is discussed at length in several chapters in this book. Adequate and proper treatment of pathology is without doubt necessary and important. However, it bears repetition that correction of pathology alone may not lead to alteration in chronic pain behavior. The factor of chronicity has already been discussed in relation to the multiple biological, psychological, and social conditioning factors that come into play in shaping and maintaining an individual's pain behavior. In patients with chronic low back pain, it is frequently the pain rather than the pathology that prevents them from leading a useful life or from returning to work. By the time pain becomes chronic, patients appear no longer to have any control over their lives which, in fact, appear to be governed by the pain and pain behaviors. They will have developed a condition that has been described as "learned helplessness."

The term "learned helplessness" has been coined to describe the emotions and behaviors of laboratory animals and human beings who have been exposed repeatedly to situations in which they have no control over outcomes. Seligman (entry 33) summarized three major effects of such exposure: reduced motivation to initiate responses; disruption in the ability to learn new responses; and emotional disturbance, particularly depression and anxiety disorders. The same kinds of responses can be seen in many chronic low back pain patients. Many become demoralized and dependent and feel that they have lost control over virtually all areas of their lives. This dependency is further aggravated and reinforced by concurrent stress factors, such as unemployment, financial insecurity, deteriorating home and marital lives and increasing social isolation. In this context, preexisting medical conditions or personality disorders frequently exacerbate, adding to the rapidly worsening situation and producing a cascading effect. The result is an individual who becomes totally dysfunctional and disabled out of proportion to the extent of the underlying pathology or the initial injury, which sometimes may be relatively trivial but, nevertheless, is a crucial event triggering the disability process.

§ 6-9(C). Chronic Pain Treatment Programs: Outpatient Versus Inpatient.

A number of critical factors determine whether the patient may be a suitable candidate for an inpatient or outpatient program. Among these are: the extent of medication abuse; the level of physical and psychological dysfunction; the extent of underlying tissue pathology; and the stability or instability of that pathology. The three major indications for an inpatient treatment program are:

(1) When the patient is heavily addicted, especially to narcotic medications. In this instance, an inpatient program may be necessary to achieve a satisfactory detoxification.

(2) When additional diagnostic work-up is needed, or where the situational factors (such as home environment or transportation issues) are such that participation in an outpatient program is precluded on a regular basis.

(3) When the behavioral or psychological impairment is such as to preclude the patient from actively cooperating on an outpatient basis.

Until recently, most of the successful programs were inpatient, primarily because such programs had the staff and resources necessary to deal adequately with problems of chronic pain patients. However, the major disadvantage of inpatient programs is cost, which usually ranges from $6,000 to over $20,000, depending on the complexity of the case and the duration of hospitalization (entry 31). If the major obstacles for an outpatient program can be overcome, and if the major indications for admissions are not present, then an outpatient treatment and rehabilitation program offers a viable alternative.

Outpatient programs possess several distinct advantages over inpatient programs. One is that of the significantly lower cost of outpatient programs. If logistical problems such as transportation and attendance can be worked out, outpatient programs offer a more effective, longer-term result at lower cost. The major advantage is that the patient does not have to be removed from the home environment into a hospital environment, and the problems that exist at home may be dealt with *in-vivo* during treatment. In addition, from a therapeutic standpoint, the patient is not conditioned to see himself as being sick, and the therapy can focus on assisting him to function or work in his normal home/work environments.

§ 6-9(D). The Washington Pain Center: A Comprehensive Outpatient Program.

§ 6-9(D)(1). Screening Prior to Admission.

Patients may be accepted into the treatment program by a referral from a physician, carrier or an attorney. At this stage, it is imperative that all pertinent medical records be received and reviewed beforehand. Following this, the patient may then be scheduled for an intake evaluation with key members of the treatment team to obtain a more detailed history. The intake evaluation routinely consists of two parts: a medical evaluation to determine the nature and extent of tissue pathology, and a psychological/psychosocial evaluation to determine the degree of affective, vegetative, and functional impairment accompanying the patient's chronic pain syndrome. The results of medical record review, evaluations and clinical impressions are then correlated, and a determination made as to the patient's suitability for participation in the treatment program.

§ 6-9(D)(2). Criteria for Acceptance/Rejection.

(1) Persistence of pain which limits activity and/or results in physical or emotional distress, despite adequate medical work-up or treatment, is the dominant condition triggering referral and acceptance.

(2) In cases where medical work-up is adequate, the requirement is that the underlying pathology be relatively static.

(3) Analgesic or hypnosedative dependency/abuse is an indication for treatment, provided the patient agrees and is motivated to undergo outpatient detoxification. When indicated, inpatient detoxification is arranged prior to starting in the outpatient pain rehabilitation program.

(4) Patients with antisocial personality disorders who show little motivation or who are highly manipulative generally do not benefit from treatment and are accepted only on a trial basis. In these cases, compliance with the treatment plan and the accomplishment of concrete goals are carefully monitored.

(5) Involvement in litigation is not a contra-indication. Some programs do not accept patients whose cases are under litigation, but we have found that this is not a determining factor influencing treatment outcome, provided we have good communication and cooperation among all parties concerned.

§ 6-9(D)(3). Treatment Goals and Principles.

By the time patients are referred for treatment in a pain treatment program, such as ours, they will frequently have been told by their referring physician that they must "learn to live with the pain," or that there is nothing more the referring physician can do for them. Feelings of anger, despair, helplessness, and hopelessness may be pervasive. It is therefore important that treatment goals be established early and that the patients and their significant others, where possible, be included in this process so that they clearly comprehend the goals to be accomplished. Their expectations should be discussed and clarified. Communication also should be established early with the carrier, the patient's attorney, and any other significant party involved, especially the patient's referring physician, so that consistency and uniformity of treatment goals can be maintained.

The primary goals of our treatment program are:

(1) To prevent or reduce iatrogenic complications.

(2) To wean patients from dependence-producing drugs.

(3) To teach use of non-chemical techniques for pain control.

(4) To correct postural and gait abnormalities and increase strength and joint range-of-motion through neuromuscular reconditioning and functional rehabilitation.

(5) To increase activities of daily living by setting gradually increasing minimum standards for functional activity.

(6) To teach greater understanding about the body and improve self-care.

(7) To teach the patient to cope more effectively with pain and with problems of daily living associated with the pain.

(8) To remove reward for pain behavior while encouraging health behaviors.

(9) To prepare and assist patients for re-entry into the workforce.

§ 6-9(D)(4). Components of the Treatment Program.

Recognizing the inadequacies of the disease model for the chronic pain-disabled patient, the focus of treatment shifts from "cure" to "care," and from reliance on "pills" to emphasis on "skills." The point cannot be over-emphasized that the mechanistic, episodic and crisis-oriented approach that characterizes care of acute pain problems may be counter-productive in the management of chronic pain. To be effective, the treatment program should be well-coordinated, with emphasis on a team approach. "Passive" strategies, by which things are done to or for patients, are replaced by "active" strategies, by which patients are taught to assume greater responsibility for their health and rehabilitation.

To accomplish the above goals, the treatment program should comprise medical, rehabilitation and prevention components, employing pharmacological, physical, psychological, behavioral and motivational/educational approaches aimed at restoring the individual to as functional a life as possible.

§ 6-9(D)(5). Pharmacotherapeutic Interventions.

Patients with chronic pain syndrome frequently have a number of significant affective and vegetative symptoms, the most common of which include insomnia, depression, anxiety, emotional lability, increased irritability, difficulty concentrating, and changes in appetite and sexual functioning. It is imperative that these vegetative and affective impairments be adequately treated. Antidepressants have been found to be the most effective (entries 26, 12). Generally, if these functions are improved, the patient's level of pain tolerance and pain threshold are also increased and the patient usually becomes more motivated to participate in the treatment program.

Many patients usually are already taking large doses of prescription analgesic and other depressant drugs on admission to the program. Analgesic medications are gradually withdrawn as the dosage of antidepressant medication is increased. Sometimes, non-steroidal antiinflammatory drugs are substituted and are prescribed on a time-contingent (rather than pain-contingent) schedule in an attempt to extinguish the patient's tendency to reach for a pill whenever discomfort is experienced. Non-chemical techniques for pain relief, including biofeedback/relaxation training, goal-setting, and educational sessions, are started early in combination with a program of physiotherapy and neuromuscular reconditioning exercises directed towards improving the patient's physical functioning.

§ 6-9(D)(6). Physical Modalities and Physical Therapy.

Physical modalities have an important place in the treatment of chronic pain but should be applied with proper behavioral principles. Chemical nerve blocks as well as non-chemical techniques for pain relief, including transcutaneous nerve stimulation, percutaneous nerve stimulation (electrical acupuncture) and other neuro-modulation techniques, may be used to good advantage in a chronic pain management and treatment program. Since it is not the purpose of this chapter to discuss details of these treatment modalities, the reader is referred elsewhere for the specific indications of use (entries 5, 24, 32).

The caveat in the use of physical modalities for chronic pain is that such use should not serve to reinforce pain behavior. Lack of attention to this caveat has led to examples of prolonged physical therapy treatment with little or no beneficial effects and sometimes with counterproductive results. Because of this, some critics have gone to the extreme of stating that physical therapy has no role in the management of chronic pain. This is not the case at all. What is at fault is not what modalities are used, but how they are used. Physical modalities of therapy should neither be used as a reward for pain behavior nor to encourage further passive dependency. Since modalities frequently do provide temporary relief, they could be very useful if used on a contingency management basis to provide a temporary "therapeutic window" during which period exercises may be encouraged and wellness behaviors shaped and reenforced.

The focus should be away from total relief of pain (which is an unrealistic goal) but instead on increasing repetition of exercise, increasing range-of-motion of specific joints, improving tolerance for a particular posture or position and generally increasing the patient's level of activity. These activities should be structured so that the goal is obtainable and measurable and the patient can see subtle progress. Instead of allowing patients to become conditioned to adopt a negative and passive attitude and expect the physical therapist to "take the pain away," the physical therapist should emphasize to the patient the need to be actively involved and to assume a direct responsibility for his own rehabilitation and recovery.

It is important for the physical therapist who works with chronic pain patients to fully understand the unique requirements of working as a member of a treatment team with well-defined common goals, guided by clear behavioral principles.

§ 6-9(D)(7). Psychological and Behavioral Approaches.

Psychological and behavioral interventions used for chronic pain may be characterized in accordance with the basic concepts employed in the definition of the pain problem. The most common techniques we employ include:

(1) Biofeedback, relaxation training and hypnosis.

(2) Psychotherapeutic interventions.

(3) Operant conditioning.

§ 6-9(D)(8). Biofeedback, Relaxation Training and Hypnosis.

The presence of chronic pain creates a situation that may be depicted by the following figure:

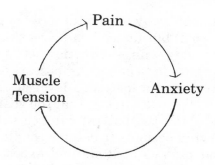

The above figure reflects a basic problem for the patient with chronic pain. Physical pain of long duration creates a state of arousal in the patient, characterized by muscle tension and anxiety. In most situations, this will cause the pain to be more severe and also experienced as more painful. Training the patient through biofeedback techniques or by progressive muscle relaxation to interrupt this cycle, often produces a positive effect on the pain. These techniques also have the effect of teaching the patient a skill that can be relied upon when the pain becomes acute. Aside from the actual biological effects of relaxation training, this technique provides the patient with a sense of some control over his biological functions and his pain. The patient becomes something other than just a helpless victim and can begin to reverse the "learned helplessness."

Hypnosis has been used extensively with chronic pain patients. While it can be a very effective technique, it needs to be presented as something that can be helpful in "controlling" the pain, not as something that will "cure" the pain. With a select group of patients, it can produce the same effects as relaxation training and can go beyond these techniques when used in conjunction with stimulus transformation, time distortion, and dissociation techniques.

§ 6-9(D)(9). Psychotherapeutic Interventions.

Individual and/or group psychotherapy is often an important treatment modality in the treatment of chronic pain patients. For various reasons, it is rejected as a treatment modality by many patients and sometimes by insurance carriers. Often, this is because the patient believes that the recommendation is made because the medical providers believe the pain is "in their head." Insurance carriers may feel that it is a waste of time and money to have the patient "lie on a couch and talk about his childhood," or simply believe that psychotherapy is irrelevant to a physical problem. All of the above assumptions are incorrect. Rarely, if ever, is a patient seen in psychotherapy because we feel his problem is "in his head." As stated earlier, this conceptualization of pain as "psychogenic" leads to significant treatment problems. Psychotherapy for patients with chronic pain, whether it is an individual or group modality, is oriented towards providing the patient with emotional support, crisis intervention, problem solving of situations that the patient sees as unsolvable, and facilitating compliance with other aspects of the treatment program. Insight oriented psychotherapy is not a particularly useful modality, although sometimes it is a part of the overall therapeutic strategy. By far, the most dominant therapeutic strategy is a directive, problem-solving approach.

§ 6-9(D)(10). Individual Psychotherapy.

Individual psychotherapy is often recommended in the initial part of the treatment program to help stabilize the patient, who is often severely depressed, anxious, and possibly suicidal. Later in the treatment program, other therapeutic strategies may be appropriate. Cognitive restructuring is a basic therapeutic technique that involves focussing on a chronic pain patient's dysfunctional thoughts and self-concept (e.g., "I can't stand the pain," or "I am no good for anything or anybody"), and replacing them with more functional thoughts and self-concept (e.g., "I can stand the pain if I relax and follow those plans the doctor and I discuss," and "I can't do everything I used to be able to do but there are still many things I can do that are worthwhile"). The basic process of dysfunctional thought has been linked to anxiety and depression (entry 3), and we have found it very helpful with our patients to help them restructure their self-concepts.

Problem-solving techniques in psychotherapy are employed to help the patients explore possible alternatives to their situations, which they see as unsolvable. Frequently, patients become so obsessed with their situations as hopeless, that they overlook relatively straightforward solutions that a therapist can point out.

Facilitating a patient's compliance with other aspects of the treatment program is another function of psychotherapy. Because of the supportive non-threatening atmosphere of therapy, a patient is often willing to be more honest and straightforward about his situation and feelings about a particular treatment modality. For example, many patients have problems with medications. Patients often refuse to take medication or take too much. Confronting these behaviors and getting the patient to initiate the correct course often prevents a total treatment failure.

§ 6-9(D)(11). Group Psychotherapy.

Group psychotherapy is divided into two separate phases for chronic pain patients. One group, we call a "basic" group, involves those patients active in the program and still trying to cope with their pain. The second type of group is the "re-entry" group and deals with patients who have graduated, in a sense, to dealing with their old responsibilities, most notably employment. The advantages of group treatment involve its economy and the feeling of common purpose, group support and emotional sharing. It is quite surprising that many patients who would normally be considered poor psychotherapy candidates do very well in our group sessions.

Feelings of social isolation are quite prevalent among chronic pain patients. In this context, group therapy has been found to be very helpful by providing a support system where peer pressure can be brought to bear and where shared experiences can provide an empathetic arena for discussion of common feelings, fears, anxieties, and frustrations, as well as hopes and aspirations. Again, these groups should not be open-ended. To be optimally effective, they should focus on specific psychosocial issues of special concern to chronic pain patients.

§ 6-9(D)(12). Operant Conditioning.

Operant conditioning procedures are fully integrated into the treatment program and involve everything from scheduling of a patient's appointments to helping the family learn more adaptive methods of responding to the patient's pain behaviors. Efforts are made not to schedule patients for "emergency" appointments. This can be viewed as a reinforcement of the patient's pain behaviors. Medications are not prescribed on a PRN basis. Staff is careful to reinforce a patient's wellness behaviors and ignore or deal clinically with his pain behaviors. The idea of working to quota, not tolerance, is a prime example of how operant conditioning techniques are utilized. In practical terms, this means that treatment goals must be made very specific and cannot be allowed to become too global or ambitious, as the patient is likely to want. It should be expressed in terms of observable units of activity based on the patient's previous interests, skills and lifestyles. Goals should be established in small bits or units which are both realistic and easy enough to accomplish so that the patient can readily perceive progress.

§ 6-9(D)(13). Pacing and Goal-Setting Is the Key.

This principle may be expressed to the patients as teaching them to "learn to pace yourself," with the aim of gradually increasing the individual's level of physical and emotional endurance and tolerance. Quotas of activities must be set below tolerance levels and the patient should be instructed to work only until the quota is met. The accomplishment of a set quota thus serves as a positive reinforcement for the patient. Through this technique, self-confidence can be re-established to assist patients to overcome the sense of demoralization that so many of them feel by the time they have the chronic pain syndrome. Very often in the life of a chronic pain patient, the pain is the negative reinforcer producing avoidance behavior. Without proper supervision or instruction, patients tend to work at an activity until they reach their pain tolerance level, at which point further effort results in increasing pain, which becomes a negative reinforcer, producing lower and lower levels of activity.

The stipulation of goals and the pacing of activities to achieve specific quotas or units of behavior, constitute the essence of a pain treatment and rehabilitation program.

Similarly, management of psychosocial issues impacting on chronic pain patients should include establishment of specific goals which are measurable and obtainable within a limited time period. Within the workers' compensation system, there is often frustration and impatience with regard to the nature and extent of counseling and rehabilitation services employed as part of the treatment. Unfortunately, it is difficult to predict the pace at which the patient will respond to intervention. However, it is important that treatment be continued if the patient is deriving benefit and moving toward the established goals. This is where motivation and compliance are key elements and, if it appears that the patient is not following through on a consistent basis with much of the treatment regime, then the indications for continuing a treatment program must be reconsidered and alternative strategies employed.

§ 6-10. Summary.

The treatment of the chronic pain patient is clearly a more complicated endeavor than simply the treatment of discrete physical symptoms. It is important to realize that the disease model is not appropriate for most chronic pain patients. Pain/pain behaviors may not be explainable in terms of underlying tissue pathology alone and, while treating the underlying tissue pathology may correct the physical impairment, the patient may still remain symptomatic and dysfunctional. The development of operant pain through learning/conditioning has important implications for the treatment process. Understanding and dealing with operant pain requires the consideration of numerous factors, such as environment, family, self-concept, job, personal responsibilities, as well as legal issues. In terms of the patient, the crucial factors involved are: motivation, social support systems/environment, chronicity, and degree of tissue pathology. The treatment process must take into account the myriad of factors that we have discussed above. We at the Washington Pain Center believe that a well-coordinated, multi-disciplinary team approach utilizing pharmacological, physical, psychological, behavioral, and educational techniques is necessary. Emphasis must shift from "cure" to "care," and from "pills" to "skills." The patient must become more active and less passive if he or she is to reduce the dysfunctional pain behaviors and achieve as optimal a level of functioning as can be obtained.

BIBLIOGRAPHY

1. Almay, B.G.L., Johansson, F., Von Knorring, L., Terenius, L., and Wahlstrom, A. *Pain* 5:153-162, 1978.

2. Armentrout, D.P., Moore, J.E., Parker, J.C., Hewett, J.E., and Feltz, C.J. Pain patient MMPI subgroups: Behavioral and psychological correlates. *J. Behavioral Med.,* in press.

3. Beck, A.T. *Cognitive Therapy and the Emotional Disorders.* New York: International Universities Press, 1976.

4. Black, R.G. The Chronic Pain Syndrome. Surgical Clinics of North America, 55-4:999-1011, 1975.

5. Bonica, J.J. *The Management of Pain,* Lea and Febiger, Philadelphia, PA, 1953.

6. Bonica, J.J. Pain research and therapy: Past and current status and future needs, in *Pain, Discomfort and Humanitarian Care,* L.K.Y. Ng and J.J. Bonica (Eds.), Elsevier/North Holland, New York, 1980.

7. Bradley, L.A., Prokop, C.K., Margolis, R. and Gentry, D.D. Multivariate analysis of the MMPI profiles of low back pain patients. *J. Behavioral Med.,* 1978, 1, 253-272.

8. Brena, S.F. and Koch, D.L. A "pain estimate" model for quantification and classification of chronic pain states. *Anesthesiology Rev.,* 2(2):8-13, 1975.

9. Brena, S.F. and Chapman, S. Pain and litigation in *Textbook of Pain,* P. Wall (Ed.), in press.

10. Chapman, S.L., Brena, S.F., Bradford, L.A. Treatment outcome of a chronic pain rehabilitation program. *Pain,* 11:255-268, 1981.

11. Chapman, S.L. *Behavior Modification in Management of Patients with Chronic Pain,* S.F. Brena and S.L. Chapman (Eds.), S.P. Medical and Scientific Books, New York, 1983.

12. Crue, B.L., Todd, E.M. and Malive, D.B.M., in *Pain: Research and Treatment,* B.L. Crue (Ed.), Academic Press, New York, 1975.

13. Dahlstrom, W.G., Welsh, G.S. and Dahlstrom, L.E. *An MMPI Handbook* (Vols. I and II), Minneapolis: University of Minnesota Press, 1972.

14. Ellard, J. Psychological reactions to compensable injury, *Med. J. Australia,* 8:349-355, 1970.

15. Fordyce, W.E. *Behavioral Methods for Chronic Pain and Illness.* Mosby, St. Louis, Mo., 1976.

16. Fordyce, W.E. Learning Processes in Pain in *The Psychology of Pain,* R.A. Sternbach (Ed.), Raven Press, New York, 1978, pp. 49-72.

17. Fordyce, W.E. A Behavioral Perspective on Chronic Pain in *Pain, Discomfort and Humanitarian Care*, L.K.Y. Ng and J.J. Bonica (Eds.), Elsevier/North Holland, New York, 1980.
18. Fordyce, W.E., Lansky, D., Calsyn, D.A., Shelton, J.L., Stolov, W.C. and Rock, D.L. Pain measurement and pain behavior. *Pain*, 18, 53-69, 1984.
19. Grossman, H.I. Legal aspects of pain and disability in *Management of Patients with Chronic Pain*, S.F. Brena and S.L. Chapman (Eds.), S.P. Medical and Scientific Books, New York, 1983.
20. Haber, L. Disabling effects of chronic disease and impairments, *J. Chronic Disease*, 24(7):469-487, 1971.
21. Heaton, R.K., Getto, C.J., Lehman, R.A.W., Fordyce, W.E., Brauer, E. and Groban, S.E. A standardized evaluation of psychosocial factors in chronic pain. *Pain*, 12, 165-174, 1982.
22. Johnson, S.J. Psychological assessment of chronic pain in *Management of Patients with Chronic Pain*, S.F. Brena and S.L. Chapman (Eds.), S.P. Medical and Scientific Books, New York, 1983, pp. 85-100.
23. Kremer, E.F., Atkinson, J.H. and Kremer, A.M. The language of pain: Affective descriptors of pain are a better predictor of psychological disturbance than pattern of sensory and affective descriptors. *Pain*, 16, 185-192, 1983.
24. Long, D.M. Electrical stimulation for the control of pain. *Arch. Surgery*, 112:884-888, 1977.
25. Melzack, R., The McGill pain questionnaire: Major properties and scoring methods. *Pain*, 1, 277-299, 1975.
26. Merskey, H. *Relief of Intractable Pain*, M. Swerdlow (Ed.), *Excerpta Medica*, Amsterdam, 1974.
27. Millon, T. *Modern Psychopathology*. Philadelphia: Saunders, 1969.
28. Millon, T., Green, C. and Meagher, R. *Millon Behavioral Health Inventory Manual*. Minneapolis: National Computer Systems, Inc., 1982.
29. Ng, L.K.Y. Pain and well-being: A challenge for biomedicine in *Pain, Discomfort and Humanitarian Care*, L.K.Y. Ng and J.J. Bonica (Eds.), Elsevier/North Holland, New York, 1980.
30. Ng, L.K.Y., Davis, D.L., Maderscheid, R.W. and Elkes, J. Toward a conceptual formulation of health and well-being, in *Strategies for Public Health: Promoting Health and Preventing Disease*, L.K.Y. Ng and D.L. Davis (Eds.), Van Nostrand Reinhold Co., New York, 1981.
31. Ng, L.K.Y. (Ed.), New approaches to treatment of chronic pain: A review of multidisciplinary pain clinics and pain centers. U.S. Dept. HEW, NIDA, Monograph Series #36, 1981.

32. Ng, L.K.Y. Acupuncture: A neuromodulation technique for pain control, in *Relief of Chronic Pain,* Aronoff, G. (Ed.), Urban and Schwarzenberg, Baltimore, MD, in press.
33. Seligman, M.E.P. *Helplessness,* Freeman, S.F., Calif., 1975.
34. Skinner, B.F. *Science and Human Behavior.* The Free Press, New York, 1953.
35. Sternbach, R.A. *Pain: A Psychophysiological Analysis.* New York, Academic Press, 1968.
36. Sternbach, R.A. *Pain Patients: Traits and Treatments.* Academic Press, New York, 1974.
37. Swanson, D.W. Chronic pain as a third pathologic emotion. *American J. Psychiatry,* 141, 210-214, 1984.
38. Szasz, T.S. *Pain and Pleasure: A Study of Bodily Feelings.* New York, Basic Books, 1957.
39. Watson, D. Neurotic tendencies among chronic pain patients: An MMPI item analysis. *Pain,* 14, 365-385, 1982.
40. Weinstein, M.R. The concept of the disability process. *Psychosomatics* 19:100-110, 1978.

Figure 3

A Daily Log of Patient's Pain Level

Name _____

Day of Week _____

Time	Pain Intensity		Sitting		Walking		Standing		Medication	
	Rating	(0 - 10)	Activity	Time	Activity	Time	Activity	Time	Type	Amount
am - 1										
1 - 2										
2 - 3										
3 - 4										
4 - 5										
5 - 6										
6 - 7 pm										
7 - 8										
8 - 9										
9 - 10										
10 - 11										
11 - 12										
pm - 1										
1 - 2										
2 - 3										
3 - 4										
4 - 5										
5 - 6 pm										
6 - 7										
7 - 8										
8 - 9										
9 - 10										
10 - 11										
11 - 12										

TABLE I

Behavioral Analyis Outline

Areas of Interest	Examples
A. Temporal factors.	1. Complaints of constant and unremitting pain suggest operant pain. 2. Pain around medication schedule suggests medication habituation and/or operant pain. 3. Respondent pain is suggested by: 　A. Intermittent pain with pain-free periods. 　B. Pain at night or during other periods of inactivity.
B. Changes in pain intensity related to activity.	1. Specifics regarding activity producing an increase or decrease in pain should be noted. For example, pain occurring when lifting objects heavier than a certain weight suggests respondent pain, whereas reports that "all lifting causes me pain" suggest operant pain. 2. Lying down or reclining providing "immediate" relief suggest operant pain. 3. Making adaptive changes but maintaining activity level, e.g., sexual activity, suggest respondent pain.
C. Changes in types of activity.	1. What has changed in the patient's life due to the pain, e.g., physical, social, and occupational? 2. Significant discrepancies in terms of vocational and avocational or optional activities, e.g., continuing to do housework or yardwork but unable to do comparable occupational activity suggests operant pain.

570

Areas of Interest	Examples
	3. Specific decrease in a behavior that was known to be enjoyable suggests respondent pain, e.g., no longer going to the movies or abstaining from sexual contact. Note, however, that many activities, such as sex, may have significant negative or aversive qualities for some patients and so these factors should be closely examined for their reinforcing properties.
D. Environmental factors.	1. Factors that directly or indirectly reinforce the patient's behavior and disability may suggest operant pain, e.g., continuing disability payments, attention of physician, spouse, friends, etc., avoidance of unpleasant job. 2. The more responsibilities avoided due to pain behavior suggests greater likelihood of operant pain.
E. Pain behaviors.	1. Pain behaviors may be verbal or non-verbal. Verbal reports of pain frequency, consistency of description of pain, vague versus detailed description of pain, should be noted. Inconsistency, vague or global descriptions, unchanging nature, are characteristics suggesting operant pain. 2. Non-verbal pain behaviors are more complex and varied. They may range from body movements to illness behaviors. A. Body movements: ability to sit, stand, and walk; posture; grimaces and groans should be

571

Examples

correlated with verbal pain complaints.

B. Medication usage: compliance with directions for prescription; number of physicians prescribing medication; history of use, abuse or tolerance; use of medication on a pain basis or by a time schedule; immediate relief following medication suggests operant pain.

3. Utilization of health care system: number of physicians consulted; attitudes towards physicians and other health-care providers; number of surgeries; number of emergency room visits.

CHAPTER 7

INDICATIONS FOR LOW BACK SURGERY

Richard H. Rothman, M.D., Ph.D.
Ronald J. Wisneski, M.D.

§ 7-1. Introduction.

Modern spinal surgery should produce very predictable and gratifying results. The indications are now clearly delineated, the selection of an appropriate operation is simple, and the surgical technique is fairly easily mastered. If one stays within the accepted guidelines and is judicious in the selection of patients, surgery on the lumbar spine can be most rewarding for both the treating physician and his patient. In this day and age, there is no place for exploratory operations, and this is especially so in the compensation setting. There are objective criteria for exactly when surgery should be considered and when they are not strictly adhered to, poor results can be expected.

The purpose of this chapter is to detail the surgical indications for intervertebral disc herniation, spinal stenosis and spondylolisthesis. The results that can be expected from operative treatment as well as the potential complications will be reviewed, and finally, the indications for chemonucleolysis will be discussed.

§ 7-2. Surgical Indications for a Herniated Disc.

If the various non-surgical treatments discussed in the chapter on conservative management have not been successful, surgical intervention can then be considered. A herniated disc, however, can present clinically with varying degrees of severity, each with its own indications for surgery.

§ 7-2(A). Profound or Complete Neurologic Loss.

The most dramatic presentation of an acute disc herniation is the profound or progressive neurologic deficit referred to as a cauda equina compression (CEC) syndrome. These patients may have loss of all neurologic function below the level of the lesion, including bowel and bladder control, and it is a true surgical emergency. If bowel, bladder, and sexual function are to be preserved, immediate decompression of the cauda equina is imperative. The longer the delay, the less recovery can be expected; and in view of the complex interplay of compression edema and vascular insult to these delicate nerve roots, prediction of surgical results is difficult. Even though these people often will fail to recover their neurological deficit with the most expeditious treatment, once the diagnosis is made, immediate decompression must be performed to keep the patient from losing any further neurologic function.

§ 7-2(B). Progressive Neurologic Deficit.

If one can actually document a progressive neurologic deficit, surgical intervention should be seriously considered. For example, if a patient who is first seen with an absent Achilles reflex is then noted to gradually lose muscle strength in the leg, he must be kept under close observation because if the trend cannot be reversed with non-steroidal anti-inflammatory agents or an epidural steroid injection, then an operation may be the only way to prevent further progression and a permanent neurologic deficit. This syndrome is unusual but must be kept in mind because of the serious consequences of further neurologic loss.

It is more common, however, to find the patient presenting with an acute, stable neurologic deficit and if it is profound, such as when there is a complete paralysis of the quadriceps muscles of the thigh or the dorsal flexors of the foot, surgery is a viable option and in some centers is even considered mandatory. The more prolonged the pressure on the spinal nerves and the more intense the compression, the less likely is function to return. It should be noted, however, that there have been no controlled studies which have prospectively compared the return of nerve function with and without surgery. Anecdotes abound of "miraculous" return to full function after surgery was delayed or refused, so the final decision is often predicated on the quality of the surgery available. Obviously, more caution should be exercised if there is a reasonable concern about potential intraoperative damage.

It is even more of a judgment call when the motor weakness is less severe or when the situation is sub-acute. A stable neurologic deficit is not in and of itself an indication for surgery because an operation does not necessarily lead to a return of function. One should also keep in mind that the deficit may not have any temporal relevance, that is, it could very well be a residual of a prior attack. A well documented past history can be a great help in these situations.

Weber, in an excellent prospective study, took a group of patients with stable neurologic deficits and operated on half of them (entry 33 in Bibliography at end of this chapter). After three years of follow-up, the neurologic residuals were similar in both the surgically and conservatively treated groups. The thing to remember is that in a stable situation the aim of surgery is to relieve pain, not to regain neurologic function; the deficit should not be used as an excuse for inadequate conservative management and premature intervention. Sensory loss and reflex change are helpful in terms of diagnosis, but they are not in themselves indications for surgical intervention. They have no prognostic value in terms of the ultimate outcome (entry 23). After four years, Weber found sensory dysfunction in nearly 46% of his total series of patients. The abnormalities encountered existed either prior to treatment or developed following either conservative or surgical management. No patient was disabled because of the sensory deficit. Similarly, surgical considerations should be delayed when pain is lessening, even though there is some progression of the neurologic deficit.

§ 7-2(C). Unrelenting Sciatica.

Occasionally an acute attack of sciatica will fail to respond to all forms of conservative treatment. The exact time when surgery should be recommended will vary from patient to patient according to their pain tolerance, emotional stability, and the demands of their socio-economic environment. In general, the authors do not recommend surgical consideration in acute sciatica until a period of four to six weeks has elapsed. Since 80% of Weber's non-surgically managed group showed "good" or "fair" results within three months, observation is justified for this period of time. If there has been little or no improvement, however, surgical intervention should be undertaken because further procrastination might adversely affect the end result.

§ 7-2(D). Recurrent Episodes of Sciatica.

Certain individuals, after an initial successful course of conservative treatment, will have recurrent sciatica that becomes incapacitating. Symptoms may be completely absent between the episodes or what begins as low grade pain may become increasingly severe. If the recurrent episodes are not too disabling, and if the intensity of the symptoms is within the patient's tolerance, then continued conservative management is indicated. However, if the frequency and intensity of the attacks are severe enough to interfere with the individual's ability to pursue gainful employment and enjoy the normal activities of daily living, then surgery should be seriously considered. In general, the authors would consider surgery after the third episode, but in this regard there is some difference of opinion.

In most instances, surgery is done to relieve sciatic pain, and the effectiveness of the procedure will depend on the identification and relief of pressure upon neural elements. Ideally, a mechanical nerve root compression will be found whenever an operation is done to relieve sciatica. Unfortunately, this is not always the case and when the pathology cannot be identified intraoperatively, the surgery is rarely successful. One can assume that such failure to discover mechanical compression is due to one of two factors: either there has been an inadequate exploration or else the sciatica was never mechanical in origin. The former factor may be remedied by a more thorough and informed exploration; the latter, that is non-mechanical sciatica, can best be appreciated by a reassessment of the pre-operative evaluation in light of the additional information gained from the surgery. The most thorough study in this regard was made by Hirsch (entry 14) in a review of some 3,000 low back operations. He found that the most predictive and significant preoperative factors in the diagnosis of mechanical paraspinal nerve compression were:

(1) A well defined neurological deficit.
(2) A positive myelogram.
(3) A positive straight leg raising test.

When all of these factors are present, surgery will usually uncover mechanical compression and the result will be a good one. If one or more of these factors are absent, however, additional deliberation is called for before the surgery is undertaken. If, for instance, the patient has a well defined neurological deficit such as an absent Achilles reflex, a positive straight leg raising test, and a positive contralateral straight leg raising test, one can be fairly certain to find a herniated lumbo-sacral disc compromising the first sacral root even though the myelogram was normal. However, if two of the three critical factors are missing, such as a positive myelogram and a neurological deficit, one might well expect a negative exploration with no postoperative relief of the sciatica.

The authors have recently modified these guidelines and feel that in order to predict mechanical root compression, the patient must have:

(1) Either a positive tension sign or a neurological deficit, and

(2) Radiological confirmation by Metrizamide myelography or computerized axial tomography (CAT scan).

With the use of the water soluble myelogram and the CAT scan, both of which are highly sensitive, it would be most unusual to have to undertake surgery without preoperative confirmation of the lesion. Surgery can be most rewarding when these guidelines are strictly adhered to. Certainly, with the advanced technology available in this day and age, there is no excuse for relying on an "exploratory" operation to make the diagnosis.

§ 7-2(E). Personality Factors.

Care must be taken to evaluate both the emotional stability of these patients and their reaction to pain. A person who continues to have minor symptoms with conservative therapy, but who has an overwhelming emotional reaction to this pain, particularly if an element of hostility is present, will usually do poorly even after surgery (entry 21). The authors, however, wish to emphasize that this admonition is not proposed in order to disassociate the "functional" from the "organic" component of back pain. Indeed, it would be naive to overlook the reciprocal interaction between the patient's somatic and emotional state. Rather, effective management of the disabling episode of pain may rest on coincident psychotherapeutic support with or without carefully monitered anti-depressant medication.

If there is any question as to the emotional stability of the patient, psychiatric consultation is called for. In spite of this, one must keep in mind that many patients with chronic low back pain, even though they have functional problems, can be helped with surgery. It has been well demonstrated that longstanding pain can lead to depression even in those who are otherwise emotionally stable, and that depression will sometimes lift after the pain is alleviated. In general, though, it is a good rule of thumb to treat the psychological factors prior to making a surgical decision since intolerable pain just might become tolerable once the depression has lifted.

§ 7-2(F). Selection of the Operation for Acute Disc Herniation.

In most cases of acute intervertebral disc herniation, the primary compelling symptom that leads to surgery is sciatica (leg pain). Although the patient may have had a long history of troublesome but tolerable low back pain, it is the leg symptoms which ultimately will precipitate surgical intervention, and a limited laminectomy with excision of only the herniated disc material is ordinarily the procedure of choice. Since such an operation can usually be done with minimal exposure, it can be relatively atraumatic and little or no bone need be removed when there is a wide interlaminal space. It is essential, however, that the nerve root be completely explored well out into the foramen and freed of all external pressure and tension. There is absolutely no place for any type of fusion in this situation since the problem is not one of instability but rather one of mechanical pressure by the herniated disc on the neural element. The goal of the operation is to relieve this mechanical pressure without creating instability.

§ 7-3. Indications for Surgery in Spinal Stenosis.

Spinal stenosis is the narrowing of the spinal canal because of degeneration of the facet joints as well as the intervertebral disc spaces. If the narrowing is severe, constriction of the neural contents within the spinal canal will develop and cause symptoms. These may vary significantly from patient to patient and can be manifested by back pain, leg pain, or both (entry 30); however, when leg pain constantly interferes with ambulation, surgery should be considered. Unlike the clinical syndrome associated with an acute disc herniation, there may be no objective clinical findings on physical examination. Occasionally, though, the symptoms can be reproduced on walking and positive neurological findings can be observed as they develop. This is referred to as a "stress test" and it is very helpful because of its objectivity. In view of the frequent absence of significant physical findings in canal stenosis, a carefully constructed history is particularly important and the patient's description of his discomfort should be carefully noted.

Routine x-rays are quite helpful in canal stenosis and show disc space narrowing as well as facet joint arthritis and a decreased spinal canal diameter (entry 11). Similar findings can also be observed on a computerized axial tomogram (CAT scan). Unfortunately, however, many asymptomatic patients will show these same radiological findings so a surgical decision cannot be based on them alone in isolation from the complaints; the clinical and radiological data must be compatible if a satisfactory operative result is to be expected. The patient must also be aware that the aim of the surgery is to relieve leg pain; relief of back pain is less predictable.

Although a Metrizamide myelogram is usually not needed in making the decision to operate, it must be done to clear the air in respect to an unsuspected intradural,

581

extramedullary tumor, and it facilitates surgical planning by delineating the extent of the lesion.

§ 7-4. Selection of the Operation in Spinal Stenosis.

The goal of an operation for spinal stenosis is to completely remove all pressure from the neural elements, and the type of pathology found will dictate the extent and nature of the decompression required. If midline ridging is the only abnormality present and the nerve roots are free in the foramen, then a complete laminectomy of the affected levels, with preservation of the facet joints, will suffice. If there is extruded disc material, this obviously should be removed; however, this is not usual in end-stage disc degeneration. The disc space need not be entered in most of these patients since little nuclear material will be present, and in fact one normally tries to avoid the destabilizing effect of such disruption.

Although the symptoms may be unilateral, the authors advise a complete bilateral laminectomy to prevent contralateral symptomatology in the future. If foraminal encroachment is found, a complete foraminotomy is indicated; and if there is a narrow lateral recess, it should be unroofed completely out to the pedicle. In summary, when the operation is completed, there should be no residual mechanical pressure on the neural elements since all impingement should have been removed.

It goes without saying that surgery for patients with spinal stenosis will not return them to heavy work since most of them who become symptomatic are between fifty and sixty years of age. In a compensation setting these employees cannot be expected to return to hard labor and some type of job modification should be under discussion even before the surgery. If this is not available, these people will usually end up with disability retirement. The aim of surgery is only to relieve uncontrollable leg pain and improve ambulation. The often heard statement that "the patient needs an operation so that he can return to (heavy) work" is not realistic.

§ 7-5. Indications for Spinal Fusion.

The exact indication for a spinal fusion is as yet not known. The answers will not be forthcoming until the completion of long-term, prospective studies in which patients within specific diagnostic categories are treated in a controlled, randomized fashion. From what we know today, the authors feel that a spinal fusion should only be considered for the following indications:

(1) The presence of surgical instability created during decompression with bilateral removal of the facet joints.

(2) The presence of neural arch defects (spondylolysis).

(3) The presence of a symptomatic and radiographically demonstrable segmental instability that can be objectively identified by weight bearing lateral flexion and extension x-rays. Instability can be assumed when there is more than a 3.5mm horizontal translation and/or a reversal of the intervertebral angle at a specific interspace.

§ 7-6. Indications for Surgery in Spondylolisthesis.

Spondylolisthesis is a defect or break in the pars interarticularis with forward slipping of one vertebral body on another (horizontal translation). Routine x-rays are all that is necessary to make the diagnosis but this is a relatively common condition which may be unrelated to the back pain in question, and such an incidental radiological finding can on occasion confuse the diagnostic process. Assuming that the low back symptoms are due to related instability, however, the spondylolisthesis can usually be treated successfully by limiting stressful activities, lumbosacral corseting, and spinal and abdominal exercise (entry 20). In the majority of cases, once a patient becomes symptomatic, some form of job modification becomes necessary since, regardless of treatment, heavy work will always be a problem. Surgery should only be considered after an adequate trial of conservative treatment has failed and even then it is a good policy to let the patient ask for it. It is a major operation with extensive morbidity and it requires about six months of recuperation. In spite of the fact that long term results seem quite good, the literature reports no randomized prospective studies of the surgical treatment of spondylolisthesis. In the absence of this essential information, one can only be guided by personal experience as well as by the limited amount of information available from some questionable retrospective surveys (entries 12, 13, 31, 34).

A logical approach to this problem based on symptoms in adults is presented below. The authors have found it to be generally satisfactory and it has been supported by others knowledgeable in the field.

§ 7-6(A). Back Pain Only.

Although spondylolysis and spondylolisthesis have been implicated as causing low back pain (entry 26), the exact mechanism of the pain is uncertain. A recent anatomical investigation of 485 skeletons, however, has confirmed, without exception, the presence of superior facet joint enlargement in the spondylolytic vertebrae, implicating facet joint overload as a pain pathway (entry 8). Most patients with spondylolisthesis complain of back pain; about one third will, in addition, have sciatica. A successful spinal fusion will usually alleviate the low back symptoms in those individuals who have only back pain. It should be kept in mind, however, that even with a successful operation, a return to heavy work should never be a realistic goal since inevitably the intervertebral disc directly above the fusion will be under additional stress and degenerate in time.

§ 7-6(B). Back Pain and Leg Pain.

Those skeletally mature individuals with back and leg pain require both a spinal fusion and a decompression of the neural elements. This combined procedure is the most common operation utilized by the authors for symptomatic spondylolisthesis in the physiologically young and functionally active adult. It has yielded excellent results, with the sciatica relieved by the decompression and the back pain eased by the spinal fusion.

§ 7-7. Results of Operative Treatment.

Even though over four decades have passed since the advent of surgical management of lumbar disc disease, the results of the operation leave much to be desired. Any improvement in the efficacy of the procedure, however, must be predicated on accurate knowledge of the results which can be expected with various diagnoses and techniques. Long term follow-up studies of this type are essential and they are available in the literature.

DePalma and Rothman have reported on over twenty years of experience with fifteen hundred patients who underwent surgery for lumbar disc disease (entry 4). These people were called back for personal interviews, physical examinations, and repeat radiographic studies which included stress films. In order to minimize bias, the evaluations were performed by physicians other than the operating surgeon. The following information is from their studies:

The average age of patients subjected to lumbar disc surgery was forty years and there was a normal frequency distribution above and below this level. This age span is in keeping with the more recent pathologic concepts of disc degeneration. The average age at the follow-up evaluation was forty-eight; the average follow-up period was eight years.

Amelioration of symptoms was the most important criterion for success. Patients were questioned as to the degree and temporal nature of their relief of back and leg pain. Their feelings as to the subjective worth of their surgery were also elicited.

In individuals with surgery on the L5-S1 disc space, fifteen percent had persistent back pain, seven percent had persistent sciatica, and fourteen percent had both back and leg pain. The results were essentially the same when the operation was done at the L4-5 interspace.

When questioned as to overall relief of their pain, approximately sixty percent of both groups of patients stated that they had complete relief from back and leg pain, about thirty percent considered themselves partially relieved, and two to three percent were considered as total failures with no relief whatsoever. Eighty-eight percent of the patients subjectively felt that their surgery was worthwhile.

Physical findings were also evaluated at follow-up and compared with those recorded preoperatively. The nonspecific findings such as muscle spasm, tenderness, limitation of motion, and positive straight leg raising disappeared in ninety percent of the patients who showed these signs preoperatively. Neurologic deficits cleared less often postoperatively. Preoperative motor and sensory deficits were reversed in fifty percent of the patients; however, lost reflexes returned to normal in only twenty-five percent.

Finally, the overall rate of solid fusion for patients who had been stabilized was ninety-two percent, with an incidence of pseudarthrosis (non-solidification of the fusion) of

586

eight percent (entry 5). As far as the subjective evaluation of the worth of their surgery was concerned, eighty-two percent of the patients who developed pseudarthrosis felt that their surgery was worthwhile, as compared to ninety-two percent of the group with solid fusions. It is interesting to note that although there were slightly fewer patients in the pseudarthrosis group who had total relief than in the solid fusion group, three patients in the solid fusion group had no relief, while every patient with a pseudarthrosis had at least partial or temporary relief.

It seems justifiable to draw certain conclusions from the above data on fusions. One of two situations must exist: either the pseudarthrosis represents a fibrous stabilization which is essentially as effective as a bony fusion, or else the fusion component of these procedures was unnecessary in the first place. The former is not unreasonable since the amount of motion which can be demonstrated on flexion-extension films of pseudarthrosis is not significant — usually less than two millimeters. The latter conclusion, however, remains in question. Thus, it would seem prudent to carefully observe patients with pseudarthrosis for rather prolonged periods of time before reoperating to achieve union. There is little rationale in subjecting them to multiple attempts at repair if, as a group, they do just as well as those who have achieved solid fusion.

In summary, if the appropriate indications for low back surgery as presented are followed, one can expect ninety percent overall satisfaction with the operation. Although relief from leg pain is reasonably predictable, only eighty percent of the patients achieve a satisfactory result as far as their back pain is concerned. Although only ninety-two percent of fusion procedures are technically successful, a solid arthrodesis on x-ray does not seem to be the principal determinant of clinical relief since only eighteen percent of those who failed to fuse had significant symptoms. Thus a pseudarthrosis in and of itself is not necessarily an indication for another operation.

§ 7-8. Complications of Lumbar Disc Surgery.

§ 7-8(A). Complications During the Operation.

§ 7-8(A)(1). Vascular and Visceral Injuries.

Trauma to the great vessels, as well as other visceral structures, can occur from penetration of the anterior portion of the annulus by a surgical instrument. Such an injury can have serious implications with a mortality rate of up to 78% if it is arterial, and up to 89% if it is venous (entry 6).

The majority of these injuries occur while trying to clean the anterior portion of the disc space with a curette or a pituitary rongeur. An erroneous estimation of the depth of the interspace can permit the instrument to penetrate the annulus and traumatize one of the large vessels. It is important to remember that the anterior as well as the posterior annulus is involved in the degenerative process and it may very well not be intact enough to act as a protective barrier (entry 19).

The aorta and vena cava lie next to the L4 disc space, while the iliac vessels approximate the L5 interspace. Immediate and massive hemorrhage may occur if the lumen of one of these vessels is violated; if there is an incomplete laceration of the wall of a vessel, however, the hemorrhage may be delayed or else when both the artery and vein are injured, an arteriovenous fistula may develop. This latter diagnosis is one that is easy to miss since in the majority of cases reported, there was no appreciable intraoperative bleeding from the disc space (entry 16). This unique complication more often occurs at the L4-5 interspace because of the close proximity of the iliac artery and vein. Fortunately, in spite of the frequent delay in diagnosis and treatment, a relatively low mortality rate of only about 8% has been reported.

Prevention of these injuries can be facilitated if there is adequate exposure and meticulous hemostasis prior to entering the disc space since uncontrolled epidural bleeding will obscure the field of vision and make it difficult to estimate how deep it is. The depth to which an instrument penetrates the disc space should never exceed one and one-eighth inches, and the surgeon who only performs occasional spine surgery should have his instruments marked in this respect. Even skilled surgeons, however, must constantly keep this dimension in mind while working in the disc space. Pituitary rongeurs should not be closed unless the rasp of metal on bone is felt, nor should the stroke of a curette be continued without this reassuring sensation.

Awareness of the possibility of this complication is important, and prompt surgical intervention is the only effective treatment. Profuse bleeding from the disc space or hypovolemic shock, disproportionate to the observable blood loss, should alert the surgeon to this possibility. The management of such a complication calls for immediate laparotomy and repair of the vascular injury. The spinal wound should be temporarily closed with towel clips and a sterile plastic drape. Blood and fluid replacement must begin at once and the patient turned to the supine position while a vascular surgeon is summoned.

§ 7-8(A)(2). Injuries to the Neural Elements.

Tears of the dura may occur while gaining access to the spinal canal during a laminectomy and decompression although packing cottonoids between the dura and the bony elements of the canal will help prevent this mishap. Reoperations are particularly hazardous in this respect because inevitably one is confronted with extensive scar

589

tissue around the dura and even with meticulous technique, occasional tears of this delicate membrane will occur. If one does occur, the defect should be promptly and adequately repaired so as to prevent the formation of a fistula or spinal extradural cyst.

A nerve root can be injured from excessive retraction, laceration, or thermal burns. Damage from metallic instruments, however, can be avoided by gently packing off the nerve root with cotton pledgets so that it can be safely handled without the necessity of repeated manipulation. Lacerations of a nerve root usually occur when there is inadequate visualization and a flattened nerve root on top of an extruded disc is not recognized. This is less likely to happen when there is good exposure and the bleeding is carefully controlled. There is no particular advantage in attempting to identify a nerve root and disc herniation through a minute opening in the spinal canal, and this aspect of the so-called "microlumbar discectomy" tends to extol doing the operation the hard way. When any difficulty whatsoever is encountered in recognizing pertinent anatomy, such as the shoulder of the nerve root and disc herniations, the laminectomy should be widened so that the nerve root proximal to the herniation can be identified with confidence. No incision into the annulus should be made until the nerve root at that level has been positively identified and retracted. A wide exposure, particularly laterally, will facilitate the procedure.

Thermal burns can be prevented by employing fine-tip, bi-polar electrocautery, but even then it should only be used after the nerve root has been identified and protected. Extreme care also should be taken when setting the current level and the instrument should be checked on muscle tissue before it is applied within the spinal canal.

§ 7-8(B). Complications During the Immediate Postoperative Period.

These complications are not unique to spinal surgery and can occur after any major operation.

§ 7-8(B)(1). Pulmonary Atelectasis from a failure to adequately expand the lungs postoperatively frequently occurs in patients who have had endotracheal anesthesia. This is seen during the first three postoperative days and is a common cause of temperature elevation. Physical findings may be negligible and even x-rays may fail to reveal minimal episodes of atelectasis. The patient will usually respond rapidly to early mobilization along with encouragement towards frequent coughing and deep breathing. The anxious patient, as well as the one who cannot be rapidly mobilized, should be placed on intermittent positive pressure breathing. Blow bottles can also be of some help.

§ 7-8(B)(2). Intestinal Ileus is also an occasional complication of low back surgery. It will produce abdominal distension, nausea, vomiting, and respiratory distress. Bowel sounds are lost to auscultation and once the syndrome is appreciated, nasogastric suction along with fluid and electrolyte replacement is in order until bowel motility has been restored as indicated by peristaltic activity and the passage of flatus. It is our custom to avoid feeding post-laminectomy patients for twenty-four hours, and to only resume meals once we are confident of the restitution of normal bowel function.

§ 7-8(B)(3). Urinary Retention is seen in the immediate postoperative period and is due to a combination of anxiety, pain, and nerve root irritation prior to and during surgery. The use of narcotics can also be somewhat contributory. Effort should be made to manage this problem without catheterization since excessive use of urethral catheters can lead to a retrograde infection. On the other

hand, the bladder can lose tone from prolonged overdistension and then develop an infection from incomplete emptying. In spite of this, reasonable effort should be made to manage the problem without the use of catheterization. The assist of gravity is encouraged by permitting male patients to stand at bedside as early as the night of surgery, and female patients are allowed to use a bedside commode. If this is unsuccessful, parenteral urocholine should be tried. Catheterization is resorted to only when these other measures fail, and the intermittent rather than the indwelling technique is preferred.

§ 7-8(B)(4). **Wound Infection** is a complication dreaded by all surgeons and should be suspected when unexplained pain or temperature elevation develops during the latter part of the first postoperative week. If there is any significant suspicion of sepsis, the wound should be cultured either by needle aspiration or swab and an immediate Gram's stain as well as minimal inhibitory sensitivities obtained. Once an organism is identified, treatment should be started at once. In the presence of a virulent infection, however, the patient should be returned to the operating room and the wound reopened and thoroughly debrided and irrigated. The authors prefer to leave the wound wide open and attend to daily local care in isolation in the patient's room. If the dura is exposed, however, a closed irrigation system is utilized. Parenteral antibiotics appropriate for the specific organism cultured should of course also be used, but it must be emphasized that antibiotics by themselves will not be effective unless there has been adequate surgical drainage. The primary treatment of a surgical infection is surgery.

Spangfort (entry 28), in his computerized review of over 10,000 procedures, reported a post-laminectomy infection rate of approximately 2.9%. Of the more recent cases reported in his monograph, however, the rate had been reduced to under 2% and with the perioperative use of some of the newer bactericidal antibiotics, this complication may prove to be less of a problem in the future.

Occasionally lumbar epidural abscesses will develop and characteristically present with a progression of symptoms from spinal pain, to root pain, to frank paresis and paralysis (entry 1). Under these circumstances, immediate myelography and decompressive laminectomy would be indicated.

§ 7-8(B)(5). **Cauda Equina Syndrome** is, short of death, the most dreaded complication that can occur following disc surgery. The exact mechanism of this rapidly progressive bowel, bladder, and lower extremity paralysis is uncertain although mechanical trauma, hematoma, and vascular injury have been implicated. The artery of Adamkowitz (arteria radicularis anteria magna) is the largest feeder of the lower spinal cord and has been reported to be entering the spinal canal anywhere from T7 down to as low as the L4 level (entry 7). While its predilection is for the left side between T9 and T11, injury to this vessel while retracting or cauterizing about the lumbar nerves may explain this rare and unexpected complication. Spangfort (entry 28) reported five cases of this cauda equina syndrome complication, representing 0.2% of his computerized series of patients undergoing lumbar discectomy. The seriousness of this syndrome is further reinforced by the fact that only 40% of them made a complete recovery.

§ 7-8(B)(6). **Thrombophlebitis** is seen with decreasing frequency since the routine use of early mobilization. Its onset is usually an insidious one although it can begin with a feeling of pain, tightness, and swelling in the affected extremity. Physical examination may reveal tenderness along the course of the vein and swelling of one or both legs. A temperature elevation will frequently occur and a positive

Homan's sign may or may not be present. It would not, however, be unusual for the syndrome to be heralded in by a massive pulmonary embolus in the absence of any prior leg sympthomatology. Treatment should commence at once with intravenous heparin, warm compresses, and elevation of the extremity. The incidence of this complication after spinal surgery, however, is so low that routine postoperative anticoagulation is not indicated unless the patient has had a prior episode of thrombophlebitis or is extremely obese. Spangfort (entry 28) reported an incidence of only 1% thrombophlebitis following laminectomy and fewer than half of these patients showed any evidence of embolic phenomena.

§ 7-8(C). Technical Complications Resulting in Persistent Symptoms.

§ 7-8(C)(1). **Inadequate Nerve Root Decompression** can be the cause of unrelenting postoperative sciatica and upon reexploration, it becomes evident that the true pathology was never uncovered at the time of surgery. This complication can be avoided only if the variety of pathological entities which can cause nerve root compression are understood and appreciated by the surgeon. Uniform success cannot be expected by looking for acute disc herniations through small fenestrations and considering this adequate treatment. The most common conditions in which nerve root compression is unrelieved are unrecognized disc lesions at another level, an unrecognized lateral recess syndrome, an unappreciated migrated fragment of disc material, foraminal encroachment, the tethering of a nerve root about a pedicle, or some anomalous root anatomy. These oversights can be avoided by a generous laminectomy with wide exposure whenever doubt exists as to the true nature of the pathology. Careful exploration of the nerve root should be undertaken from its origin at the dural sac to well out into the foramen. When doubt exists as to the location of the protrusion, two or three disc levels should be carefully explored. An additional half hour spent in this way can save the patient years of misery and disability. The importance of adequate, meticulous exploration cannot be overemphasized since no additional morbidity has been experienced with wide laminectomy (entry 15).

§ 7-8(C)(2). **Fibrosis (Scar Tissue) Around the Nerve Roots and Dura** will occur with great regularity following surgical exploration; however, the reason why this causes symptoms in some patients and not in others is not understood although this scar tissue obviously acts as a tether as well as a constricting force about the nerve roots. Early enthusiasm for the use of a Gelfoam membrane, advocated by LaRocca and MacNab (entry 18), has given way to the concept of the free fat graft as proposed by Lagenskild and Kiviluoto (entry 17). The principle of maintaining a free fat-dural interface was considered important in not only inhibiting perineural fibrosis but also permitting safer dissection in the future, should reoperation become necessary.

§ 7-8(C)(3). **Disc Space Infection** should be considered when there is a rapid, dramatic occurrence of severe back pain one to six weeks after a surgical discectomy. This may be accompanied by the recurrence of sciatic or femoral neuritis, as well as by severe muscle cramps (entry 10). Straight leg raising or reverse straight leg raising will frequently become more restricted than before the operation. The patient may or may not have a temperature elevation and frequently the only laboratory finding is an elevated sedimentation rate. Serial radiograms will show a progressive loss of disc space height, irregular destruction of the end plate, and ultimately, sclerotic reaction.

The diagnosis can usually be made from radiographic and clinical findings, and rarely is a biopsy or needle aspiration needed. To relieve pain, the treatment requires complete immobilization, preferably in a plaster spica from the lower rib cage down to just above the knees. Large doses of parenteral antibiotics effective against staphylococcus organisms are utilized for a period of two to three months. Those patients, however, who do not respond to conservative therapy in a reasonable period of time after the institution of antibiotics should undergo debridement of the disc space either through a retroperitoneal or a direct posterior approach, particularly if a neurological deficit is appreciated. From his computerized study, Spangfort (entry 28) estimated the incidence of this complication to be about 2%.

§ 7-8(C)(4). **Subarachnoid Cysts** may form and occasionally assume dramatic size. Although the symptoms of an extradural pseudocyst are usually the same as those experienced in acute disc herniation, meningeal signs such as headache and neck pain may be present (entry 2). Myelography ordinarily will disclose the true nature of the problem. Treatment of these lesions usually requires operative excision of the cyst with closure of the dural margins either primarily or with the aid of a fascial graft. Mayfield (entry 24) reported on eleven patients who developed this complication from a series of over one thousand operations. Most of them resulted from rents near the dural sleeve, and it was found that free fat plugs will successfully seal such defects.

§ 7-8(C)(5). **Retained Foreign Bodies** are rare in this modern era of accurate sponge counts and radiographic tagging of sponges and cotton pledgets; however, occasionally a pledget will get detached from its identifying string and get lost in the wound. These foreign bodies can cause a local inflammatory reaction, and re-exploration is indicated for their removal. Bone wax as well as starch granules from the glove packaging may on occasion cause a granulomatous reaction and persistent drainage from the wound. Bone wax should therefore be used sparingly, if at all, and the residual starch on the gloves should be wiped off prior to surgery.

§ 7-9. Chemonucleolysis.

Chemonucleolysis is a procedure by which a drug is injected into a herniated nucleus pulposus in order to effect its dissolution (entry 27). If by the absorption of the disc, pressure is relieved on the neural elements (nerve roots), the patient's pain can be expected to disappear.

There are two different types of preparations currently on the market (entry 3). One is chymopapain, which is purified meat tenderizer and it has had most of the publicity; the other is a drug called collagenase. They differ in that they dissolve different components within the nucleus pulposus. Chymopapain reacts with the proteoglycan portion of the disc, while collagenase works on the collagen.

The indications for chemonucleolysis are similar for those used for surgical discectomy. Since the drugs have no influence on bone, chemonucleolysis is not indicated for bone pathology such as spondylolisthesis or osteoarthritis. According to the current state of the art, if a patient has leg pain and a positive straight leg raising test with or without a neurological deficit along with a positive contrast study and has not responded to six weeks of conservative therapy, chemonucleolysis can be considered as an alternative option to surgery.

The side effects of the injection, however, can be quite serious (entries 22, 25). They include major allergic reactions as well as inadvertent neural injury. Occasional deaths have occurred. Because of these serious risks, chemonucleolysis should not be approached as a conservative modality, but rather as an alternative option to surgery. It is in no way a minor procedure.

The reports on the results of the use of chemonucleolysis have been mixed and there is still much controversy as to its exact place in our armamentarium. A satisfactory experience has been reported in about 70% of the patients so treated. When the results of surgery for lumbar disc herniation are compared to those of chemonucleolysis, however, those operated upon have done better (entries 32, 9).

The final place of chemonucleolysis in the treatment of lumbar disc herniations is yet to be defined. More experience with the different agents needs to be accumulated and adequate scientific research is yet to be performed. It undoubtedly will still be several years before the true efficacy of the procedure is known.

§ 7-10. Summary.

This chapter has outlined the indications for successful spinal surgery. As one reviews the overwhelming amount of written material pertaining to low back operations, certain precepts and requirements become obvious:

(1) Accurate knowledge of the variable anatomy and pathology.

(2) Accurate diagnosis of nerve root compression.

(3) Adherence to the proper criteria for surgical intervention.

(4) Selection of the proper operative procedure.

(5) Skillful execution of the procedure by an experienced spinal surgeon.

(6) Prompt recognition and treatment of complications.

(7) Careful postoperative care.

If every patient undergoing a back operation were treated according to these principles, the quality of surgical results would be most gratifying and the grey veil of fear and anxiety that has surrounded spinal surgery for years would be lifted.

BIBLIOGRAPHY

1. Baker, A.S., Ojemann, R.G., Schwartz, M.N. and Richardson, E.P. Spinal epidural abscess. *N. Eng. J. Med.,* 293 #10:463-469, September 4, 1975.

2. Borgesen, S.E. and Vang, P.S. Extradural pseudocysts. A cause of pain after lumbar disc operation. *Acta Orthop. Scand.,* 44:12, 1973.

3. Brown, M.D. *Intradiscal Therapy: Chymopapain or Collagenase.* Year Book Medical Publishers, Chicago, p. 45, 1983.

4. De Palma, A. and Rothman, R. Surgery of the lumbar spine. *Clin. Orthop.,* 63:162-170, 1969.

5. De Palma, A. and Rothman, R. The nature of pseudoarthrosis. *Clin. Orthop.,* 59:113-118, 1968.

6. Desausseure, R.L. Vascular injuries coincident to disc surgery. *J. Neurosurg.,* 16:222-229, 1959.

7. Dommise, G.F. The blood supply of the spinal cord. *J.B.J.S.,* 56B #2:225, May 1974.

8. Einstein, S. *J.B.J.S.,* 60B, 1976.

9. Ejeskar, A., Nachemson, A., Herberts, P., Lysell, E., Andersson, G., Arstam, L. and Peterson, Lars-Erik. Surgery versus chemonucleolysis for herniated lumbar discs. *Clin. Orthop. Rel. Research,* 174:236-242, April 1983.

10. El-Gindi, S., Aref, S., Salama, M., and Andrew, J. Infection of intervertebral discs after operation. *J.B.J.S.,* 58 #1:114-116, February 1976.

11. Epstein, B.S., Epstein, J.A., and Jones, M.D. Lumbar spinal stenosis. *Radiologic Clin. N. Amer.,* 15, #2:227-239, August 1977.

12. Gill, G., Manning, J.G. and White, H.L. Surgical treatment of spondylolisthesis without spine fusion. *J.B.J.S.,* 37A:493-520, 1955.

13. Henderson, E.D. Results of the surgical treatment of spondylolisthesis. *J.B.J.S.,* 48A:619, 1966.

14. Hirsch, C. Efficiency of surgery in low back disorder. *J.B.J.S.,* 47A:991, 1965.

15. Jackson, R.K. The long term effects of wide laminectomy for lumbar disc excision. *J.B.J.S.,* 53B:609-616, 1971.
16. Jarstfer, B.S. and Rich, N.M. The challenge of arteriovenous vistula formation following disc surgery: A collective review. *J. Trauma,* 6 #9:726-733, September 1976.
17. Lagenskild, A. and Kiviluoto, O. Prevention of epidural scar formation after operations on the lumbar spine by means of free fat transplants. *Clin. Orthop.,* 115:92-95, 1976.
18. LaRocca, H. and MacNab, I. The laminectomy membrane. *J.B.J.S.,* 56B:545-550, 1974.
19. Lindblom, K. Intervertebral disc degeneration as a pressure atrophy. *J.B.J.S.,* 39A:933-937, 1957.
20. Magora, A. Conservative treatment in spondylolisthesis. *Clin. Orthop.,* 117:74-79, June 1976.
21. Maruta, T., Swanson, D.V. and Swenson, W.M. Low back pain patients in a psychiatric population. *Mayo Clin. Proc.* Vol. 51:57-61, January 1976.
22. Massaro, T.A. and Jovid, M. Chemonucleolysis, *J. Neurosurg.,* 46:696-697, 1977.
23. Maury, M., Francois, N. and Skoda, A. About the neurological sequelae of herniated intervertebral disc. *Paraplegia,* II:221-227, 1973.
24. Mayfield, F.H. Complications of laminectomy. *Clin. Neurosurg.,* 23:435-439, 1976.
25. McCullouch, J.A. Chemonucleolysis: Experience with 2000 cases, *Clin. Orthop. Rel. Res.,* 146:128, 1980.
26. Nachemson, A. The lumbar spine. An orthopaedic challenge. *Spine,* 1:59, 1976.
27. Smith, L. Enzyme dissolution of the nucleus pulposus. *Nature,* 198 1963.
28. Spangfort, E.V. The lumbar disc herniation — A computerized analysis of 2,504 operations. *Acta. Orthop. Scan. Supp.,* 142:52-78, 1972.
29. Sussman, B.J. and Mann, M. Experimental intervertebral discolysis with collagenase. *J. Neurosurg.,* 31:628, 1969.
30. Verbiest, H. Pathomorphologic aspects of developmental lumbar stenosis. *Orthop. Clin. N. Amer.,* 6, #1:177-196, January 1975.
31. Vestad, E. and Naes, B. Spondylolisthesis. *Acta. Orthop. Scand.,* 48:472-478, 1977.
32. Watts, C., Knighton, R. and Roulhac, G. Chymopapain treatment of intervertebral disc disease. *J. Neurosurg.,* 42:374-383, April 1975.

33. Weber, H. Lumbar disc herniation: A prospective study of prognostic factors including a controlled trial. *J. Oslo City Hospitals,* 28:33-64, 89-120, 1978.
34. Wiltse, L.L. and Hutchison, R. Surgical treatment of spondylolisthesis. *Clin. Orthop.,* 35:116-135, 1964.

CHAPTER 8

VOCATIONAL REHABILITATION, ERGONOMICS, AND JOB DESIGN

James L. Mueller

§ 8-1. Introduction.

This chapter is intended to enable the reader to achieve three major goals:

(1) understanding of the respective roles of the fields of ergonomics and vocational rehabilitation in the management of industrial low back pain;

(2) understanding of the techniques within these fields which can assist in returning the back-injured individual to productive work as soon and in as full a capacity as consistent with resultant physical limitations; and

(3) knowledge of when and how to use the techniques cost-effectively, making optimum use of in-house and external resources in the return-to-work process.

The material presented is based on documented research in the fields of occupational medicine, occupational safety and health, industrial engineering, industrial hygiene, compensation and benefits, and ergonomics, or human factors engineering — the study of the interaction of humans with the built environment. The opinions and viewpoints presented are based on this research and on twelve years of professional experience in the fields of industrial design, ergonomics, rehabilitation research, and disability management. Though the importance of these fields to management of industrial low back pain will be made clear to the reader, their applications to the problem of general industrial disability has begun to develop only recently. The alarmingly disproportionate increases on costs to business such as lost work days, compensation, and medical payments, assure that the importance of these fields will continue to grow.

§ 8-2. The Back-Injured Worker and the Vocational Rehabilitation Process.

In the past, back-injured workers were essentially dumped onto the roles of state or community vocational rehabilitation agencies. These agencies are responsible for providing financial, educational, and technical assistance for returning disabled persons to work. Assistance can include retraining, higher education, innovative work arrangements, and job accommodation. According to the National Safety Council, workers compensation claims for back injuries are over $14 billion in compensation and medical costs (entry 7). The magnitude of the problem has caused many businesses to examine why public rehabilitation systems have been ineffective in dealing with the problem and what steps can be taken to stop this trend of skyrocketing industrial back injuries and resultant costs.

It is a curious application of economic theory that at a time when work disabilities (not only back injuries) cost the United States over 10% of the Gross National Product annually, funding for vocational rehabilitation agencies has been regularly cut. This has placed additional strains on a system already plagued by understaffing and high caseloads, not to mention bureaucratic procedures which emphasize the number of cases processed, rather than the merit of the results. To be fair, some agencies, whether known as DVR's (Division of Vocational Rehabilitation) or BVR's (Bureau of Vocational Rehabilitation) function better than others for a variety of reasons. But the unfortunate result is that few organizations can rely on their local DVR or BVR for help with returning back-injured workers to the job.

A major reason for this lies in the unique characteristics of the back-injured worker within the disabled population. The back-injured worker, for better or worse, rarely has an obvious or demonstrable impairment. In contrast to persons with visual impairments or wheelchair and crutch-users, they appear "less disabled." This handicaps (in the true sense of the word) their access to vocational rehabilitation services. As funds for these agencies have been constricted,

604

counselors have been instructed to concentrate on the most severe disabilities. This tends to leave the essentially mobile, non-blind, non-deaf, back-injured client on the "back burner." The back-injured individual approaching retirement, and in an occupation where a sound back seems nearly a necessity, faces even greater hurdles, since age and occupational experience appear to influence rehabilitation outcome even more than the nature or severity of the injury. In short, case workers will devote considerably more effort to the case of a 25-year-old spinal cord injured computer programmer than to that of a 55-year-old bricklayer with a low back injury.

The lack of demonstrable impairment works against the back-injured individual in other ways. The back-injured individual, often a blue-collar worker, experiences varying degrees of guilt about not being able to "measure up" to the job, not being "tough" in bearing pain, or being equated with malingerers looking for an excuse from work. The largely American "work ethic" phenomenon tends to make the back-injured individual deny the problem until the pain and/or limitation becomes undeniable, and often, irreparable. The elusiveness of the back injury in terms of onset, level of severity, medical treatment protocol, and prognosis all add to the individual's uncertainty about his/her ability to return to work and avoid re-injury.

Unfortunately, many of these fears are shared by the employer, further inhibiting the rehabilitation process. At the root of the problem is the general lack of line-level interest in or responsibility for the injured worker. Line supervisors tend to feel responsible for production first and foremost. Besides, compensation and replacement costs are not line-level costs. But "light-duty," reduced productivity, and costs of job accommodation usually are. With this in mind, it is understandable that a line supervisor may not be very concerned when a back-injured worker seems to be unable to return to work. On the contrary, it may be a welcome opportunity to hire a "whole" person and not worry about how the disabled worker can get back to work as soon as possible. The enormous costs are borne (and raised) elsewhere. Barring personal relationships among the workforce, for the supervisor the back-injured worker is essentially "out of sight, out of mind." And not having to give special consideration to one worker avoids possible problems with other workers, which every supervisor understands, especially under union contract, where light-duty or restricted jobs may be virtually impossible or where senior, job-ready, able-bodied union members are out of work.

The stage is nearly set for loss of the back-injured worker to the never-never land of long-term disability. The worker carries on in guilt and uncertainty with a "patient" identity which is rapidly devouring his/her "worker" identity. The case manager, as a result of, at best, only a vague idea of the costs involved, combined with pressure by the line supervisor to replace the worker, succumbs to inertia. This leaves the worker virtually alone to drift through medical evaluations, therapy, even surgery without motivation or goal. It is a small wonder that some such workers then use this poorly monitored system to their financial gain by returning from leave just long enough to qualify for more leave. By the time someone reviews the case, with alarming results, the "compensation, not rehabilitation" philosophy has taken its toll. The worker has lost the motivation to return to his/her job, the medical prognosis is largely unchanged, and the total care cost has reached an unacceptable level.

§ 8-3. Management and the Vocational Rehabilitation Process.

The problem of vocational rehabilitation of back-injured

workers, and indeed of any disabled worker, is essentially one of the management. The causes for failure of the rehabilitation process as outlined above are essentially caused by lack of management involvement in the rehabilitation process and responsibility for abuse of the system by malingerers, which presents one more economic burden to the organization, must also be shared by management.

The first major step that must be taken is top-level management commitment to a "rehabilitation rather than compensation oriented approach." This apparently simple change may require some not-so-simple reorganization. First, it must be made clear by top-level management all the way to first-level supervisors that the organization supports, and will monitor, return-to-work efforts for back-injured workers. Conversely, it will not support indefinite, unjustified compensation for injured workers. One of the most effective methods for implementing this idea is a modified system of cost accounting. Companies which use this system have removed health care, long-term disability, and workers' compensation costs from the overall corporate realm of fringe benefits and overhead costs. These are reported as line items on the expense reports of each department, division, or facility. The obvious result is that line-level managers are ultimately responsible for the costs of non-rehabilitation of the injured worker. This makes the weighing of the alternatives (job modification, reassignment, assistive devices, light-duty) to compensation a more effective process of cost/benefit analysis. It also offers incentives to line-level managers to demonstrate productivity and cost savings to top-level management.

Some companies maintain a central fund for the costs of assistive devices and other accommodation expenses to further minimize any economic disincentives to the return of injured workers. There is often also line-level concern that a disabled worker may not work to full capacity at first, thus effectively reducing the workforce headcount and decreasing overall productivity figures. To address this, corporate budgets have been arranged to cover the salary of this worker during the initial work-adjustment period, so that the headcount is not affected negatively. In fact, the productivity of the returning worker represents a bonus during this period.

The next major management step in stimulating the vocational rehabilitation process for back-injured workers is to establish clear levels of responsibility for case management. Depending on the organization's size and nature of business, this may involve hiring facility or regional coordinators or assigning this task to an appropriate department. This task often falls to the occupational safety and health staff. Care must be taken that this responsibility is not merely added to those of an already overworked, under-powered staff member, but rather integrated within the department. This may involve realignment of job duties or even the hiring of staff, but the benefits of the program outweigh even these costs.

The problem of low back pain is accepted as one of the most difficult and expensive industrial disabilities. It is obvious by now to the reader that simplistic solutions to the problems of high compensation costs do not exist, any more than do simplistic solutions to medical treatment of the back-injured worker. The multi-faceted nature of the vocational rehabilitation process for back-injured workers is, however, not really more complicated than that for any other disabled worker. It may seem so only because the individual issues involved have historically been so abysmally managed. Poor cost monitoring and accounting, a "compensation rather than rehabilitation" philosophy, and little or no involvement of the injured worker or line-level personnel, have collectively built a problem that can be unraveled only through objective management techniques.

Vocational rehabilitation of the back-injured worker begins with the commitment of top-level management, carried through by motivated line-level personnel and supported and monitored by facility coordinator(s) and clear corporate policies. Next follows the design *and use* of a practical case management system. This includes not only a uniform set of steps in the rehabilitation process, but also practical time intervals and staff task assignments, so that cases do not become stalled indefinitely or fall through the cracks. Especially in the management of back-injured worker cases, time is money. Additionally, and even more importantly, any delay in the rehabilitation process can be fatal. No delay in the rehabilitation process must be allowed to go unjustified or unmonitored. Workers steadily fall into a "patient" identity; case workers and supervisors learn to accept the loss of the worker; and the costs begin to mount, while all concerned seem to be hoping the problem will simply go away. The probability that this will happen is low.

So what is the best procedure for vocational rehabilitation? To answer this essential question, each organization must weigh some alternatives and make some initial decisions. At the beginning is the question of responsibility for monitoring and coordinating the rehabilitation process. As previously mentioned, this individual or group must be identified first, so that all program planning and start-up, as well as subsequent case management, is focused.

Next is the decision about how much of the process can be performed internally and how much is better handled by outside agencies. This requires analysis of internal capabilities as well as various services offered through local agencies or by the organization's own insurance carrier. Some of these services may be available free or at nominal cost.

Awareness of the personnel to be involved in the process of vocational rehabilitation is also important. The inputs of the line supervisor, top-level management (either directly or through some feedback system), compensation and benefits personnel, corporate medical staff, corporate safety and health personnel, the worker's physician, and especially the worker, should be integrated appropriately within the process. Additionally, provision for access to external resources should be included so as to complement internal capabilities.

§ 8-4. The Back Injured Worker — A Sample Case Management Process.

At the end of this chapter is a simplified flowchart illustration of the rehabilitation case management process used by a large chemical and pharmaceutical company. This example will be used as a basis for understanding the components of a timely, well-monitored system.

The process is begun by identification of a reduced ability to perform the job. This may be the result of observation by any of a variety of individuals, including the line supervisor, safety personnel, or the worker himself. The next step is to refer the case for action to the rehabilitation coordinator. If not already done, the employee is immediately contacted and informed of the nature of the problem. At this point, it is determined whether the reduced ability stems from some problem requiring medical evaluation. If not, then the worker's reduced performance is not the result of an impairment and the case is referred for administrative action and documentation. Documentation is stressed throughout this process because of its immense value in program evaluation as well as in case litigation. If medical evaluation is indicated, the information is received from the attending physician, who will determine if the worker is

610

fully capable of work without restrictions. A sample format for medical evaluation reports is shown at the end of this chapter. Note the emphasis on functional assessment. If so, is the employee returning to work? The employee's return to work is monitored and documented. If work restrictions are imposed by the attending physician, the prognosis for the worker's impairment must be defined. If it cannot be defined by the attending physician, a corporate physician evaluates the worker's condition. If the prognosis remains undefined, rehabilitation cannot progress, and the case should be reviewed, either by the insurance carrier, a medical specialist or an external case management specialist or rehabilitation consultant.

Once it has been established that work restrictions must be imposed and a definite prognosis is determined, the return-to-work phase of the rehabilitation process can begin. The first step in this phase is an analysis of the job to describe physical tasks, schedules, materials and equipment used, procedures used, as well as the nature of the work environment. This job analysis is very important, especially in providing to physicians complete and accurate job data which can be correlated with the possible effects on the worker's impairment. It is likely that only through this data will the physician get an objective picture of the job itself. Few physicians are able to gather the data personally. Their insight into the nature of the work is usually based on input from the worker, their patient — a not unbiased viewpoint. The job analysis technique will be thoroughly described later in this chapter.

Application of the job data gathered through the job analysis process is then correlated with the findings of the medical evaluation. At this point, it must be determined whether the back-injured worker is capable of full job performance without job accommodations. If not, the evaluations of accommodation alternatives is begun, integrating the input

611

of line supervisor, worker, and others. The job accommodation process also will be thoroughly described late in this chapter.

Should it be determined that the injured employee can return to work without job accommodations, the rehabilitation coordinator maintains contact with the employee until he/she has returned. In cases where the prognosis is for return to work after a specified recovery period, again the coordinator monitors the employee's progress through recovery and return to work. Should the employee fail to return to work, or if job accommodations are necessary but not feasible, the case should be reviewed by the insurance carrier, an external case management specialist, or rehabilitation consultant. If return to work is not possible, the employee should then be considered for long-term disability status or out-placement. Again, the coordinator continues to monitor and document the case progress and maintain contact at regular intervals with those individuals who return to work.

As mentioned previously, the details of the rehabilitation process, including projected time frames for each step, will be specific to the organization's structure, size, nature of business, and available resources. The preceding process and the accompanying flowchart is intended as an illustration of how the necessary steps might be organized so as to best monitor a cost-effective rehabilitation process.

§ 8-5. Medical Evaluations and Reporting.

Perhaps the most critical phase of the vocational rehabilitation process is the determination of whether the employee can go back to his/her previous job. It is in this phase that far-reaching decisions are based on information which is often inadequate at best. The problem here lies in the interface (or lack of interface) between clinical input concerning the worker's impairment (functional job qualifications) and technical input concerning the job (functional job requirements). In order for management to make the appropriate decision concerning whether the worker can return to his/her previous job, correlation must be possible between what the physician determines the worker *can* do and what the job *requires* him/her to do. Note that this is a management decision, not a medical one. The task of the physician at this point is to determine what, if any, functional limitations the worker has which require work restrictions or accommodations so that he/she can perform safely. The physician's job is not to decide whether the individual should return to the job. Too often, managers prefer to leave the burden of this decision with the physician who is not an expert on the job requirements or possibilities of accommodation. The return to work decision must be a *management* decision based on business factors including employee's work abilities, job experience, functional limitations and/or work restrictions, weighed against the safety, complexity, effectiveness, and cost of accommodation. The physician's report should provide information that the manager can best use in making this decision, but *should not make* the decision, regardless of how severe the limitations may seem. One occupational physician has stated this even more plainly: "Even if the worker is *dead,* the physician does not make that decision."

Most organizations design their own medical examination reporting forms to best suit their specific needs. One such form from the United States Department of Labor's Office of Worker's Compensation Programs is reproduced at the end of this chapter. The best, and most useful, approach tailors the medical report format and the job description format so that the manager can directly assess the impact of the work restrictions upon the performance of the job, as well as what job accommodations might be necessary.

613

Another effective method of soliciting job-related medical input is to provide job data in such a format that the physician can make comments directly on the form, noting job tasks or other components from which the worker must be restricted, unless satisfactory accommodations can be made. Generally, the format can follow the one reproduced at the end of this chapter.

The terms used in the medical reporting format (and the job analysis format) will vary from the example according to the nature of the business. The illustrated format, for example, is used by a number of medium-to-large size manufacturing facilities, hence the inclusion of "VEHICLE OPERATION" and emphasis on WORKLOADS, ENVIRONMENTS, AND RISKS.

The goal of such a format, of course, is to elicit useful functional data concerning the employee without overburdening the reporting tasks of the physician or becoming tangled in medical jargon. The industrial job analysis format is similarly targeted to expressing specific work factors present without extraneous technical data.

§ 8-6. Job Analysis Techniques.

As in the medical evaluation report format, the goal of job analysis is not to compile every bit of job data possible. This is far too expensive and time-consuming and clouds the essential data needed for the determination of whether the worker can perform safely and productively. As previously stated, the aim is not toward voluminous data, but rather toward that data which best describe how the job may affect the worker. Similarly, the medical evaluation format portrays that information which best describes how the worker's impairment may affect job performance.

Job analysis is basically an industrial engineering technique often associated with studies of worker efficiency,

614

time-and-motion studies, and little men with lab coats, stop watches and clipboards. Consequently, the technique of analyzing a job, including observing workers, is often feared and mistrusted by workers and supervisors alike. Gathering accurate and minimally-biased job information, then, requires that everyone understand why the information is important and how it will be used. Experience indicates that the most practical approach is to clearly explain the purpose of the job analysis as an effort to provide physicians with the clearest picture of what their patient actually does on the job. The goal, as in any job placement, is to be as sure as reasonably possible that a given worker can return to the job and that he/she can be productive and safe.

Job analysis is an engineering technique, and various approaches are currently used. These vary from highly sophisticated factor analysis and value assessment to simple narrative. Discussion of the merits and applications of the various approaches has consumed thousands of pages in journals, textbooks, and conference proceedings. Organizations having in-house engineering staffs apply these techniques, tailored to the nature of the business, to generate job descriptions for hiring, compensation, and other uses. None of these techniques, or the resultant data, have been wholly satisfactory in answering the questions involved in returning the disabled worker to the job. Additionally, the time and staff available for gathering job information for this specific purpose is nearly always limited. The author's experience in blue-collar as well as white-collar businesses and organizations has yielded the hybrid approach outlined in this chapter.

Using a JOB FACTORS CHECKLIST, reproduced at the end of this chapter, or other format, nearly a mirror image of the previously illustrated medical evaluation report format, the job analyst (whether a consultant specialist, or

615

internal staff from personnel, safety, or rehabilitation departments) gathers information from a number of sources. This helps to identify obsolete, extraneous, or conflicting job data in order to minimize the bias of any single source. Existing job descriptions, appropriate safety and compensation records, technical literature, organizational charts, and other internal documentation is first gathered to provide a framework for understanding the general nature of the job. Some of the data may help to begin to fill in the JOB FACTORS LIST and formulate an outline of the purpose, nature and output of the job.

Prepared through review of this data, the job analyst is then knowledgeable enough to interview persons with direct experience with the job in question. This may include managers, supervisors, senior employees, the incumbent worker(s), and/or union representatives. The purpose of the interviews is to complete the JOB FACTORS LIST as well as a brief narrative describing the nature of the job, major job duties, expected output, safety considerations, and other general information. At the end of each interview the data is briefly summarized to minimize misunderstanding.

With the line supervisor's previous permission, the analyst visits the worksite to observe the job actually being performed, and correlates this information with data gathered through the interviews. Sufficient time is needed to observe more than one worker, infrequent or emergency procedures, and other variables as appropriate.

It was previously noted that physicians rarely have the benefit of first-hand observation of the job or even the worksite. Therefore, where circumstances allow, the analyst should arrange for photographs of work procedures, body movements, materials and equipment, and the general work environment, to be included in the completed job analysis. These pictures enable the job analyst to save literally thousands of words.

The analyst now must compile the data gathered through the document research, interviews, and on-site observation to complete the JOB FACTOR LIST and the job description narrative, which should be as concise as possible. The remaining specifics are illustrated in the photographs, appropriately reviewed and culled, usually to four or five which show the work environment, any safety risks, specific tasks, and materials and equipment used. This package provides the essential job data most useful, in conjunction with the medical data, to making the *management* decision of whether (and how) the employee may return to work.

§ 8-7. Job Accommodation.

At this point, the *management* decision must be made concerning the suitability of the back-injured worker to perform his/her previous job. The comparison of medical restrictions with job requirements enables the manager to identify potential problems the worker will face (if any). The extent and severity of the problems determine the need for accommodation. Before proceeding with the discussion of accommodation, however, it is important for the reader to know that expense is rarely a factor in making accommodations for injured workers. In fact, national surveys of companies who have done so indicate that 50% of accommodations involve no cost, while another 30% cost less than $500 (entry 1). Compared with the enormous and continuing cost of compensation, these one-time costs seem negligible. As has been stressed earlier, the disabled worker can often be the best information resource to the manager in planning accommodations. Case histories abound illustrating the time and money wasted because subtle problems or simple solutions went unnoticed without the personal insight of the disabled person. Failing to involve the worker has also resulted in their natural resistance to

the rehabilitation process resulting in failure to return to
work and/or even more costly litigation (entry 6). Here
again, cooperation and contact with the worker can speed
and simplify the sometimes complex search for solutions to
unique accommodations problems.

After initial involvement of the disabled worker and
management (including the line supervisor), it may be
helpful to ask help from immediately available in-house
resources. For example, physical plant and engineering
staff may be able to suggest, even produce, changes in the
physical layout of the worksite, assistive devices or equipment, or other hardware modifications. Specifically, this
might include modified posture supports or seating, provision of carts or other transportation devices for carrying
heavy objects or materials, inclined bins or lowered cabinets
for reaching stored items, or power-assisted lifts or
overhead cranes for raising or lowering loads. Physical
changes to the worksite and job redesign will be discussed
more thoroughly later in this chapter. In-house medical,
safety, and therapeutic staff may be able to suggest ways in
which body movements often used in the job might be
altered to reduce specific stresses to the back-injured
worker. Other line supervisors and managers may be able
to suggest ways in which the workflow or production processes may be changed so as to reduce stress on the injured
worker as well as risk of similar injury to other workers.

Should other resources or more innovative solutions be
needed,   in-house   personnel   or   Equal   Opportu-
nity/Affirmative Action Specialists are often aware of other
possibilities. These individuals should also be aware of
available   outside   resources   to   augment   internal
capabilities. Perhaps a special piece of equipment is needed
which is unknown to plant and engineering staff, or a
continuing therapy or fitness program must be set up for the
employee, or perhaps any accommodation solution at all

seems elusive. Access to local expertise outside the organization sometimes makes the difference between failure and successful return to work. This assistance may be free, inexpensive, or expensive, but the cost-effective search for an application of this expertise depends again on the informed management determination of what is "reasonable."

§ 8-8. Reasonable Job Accommodation.

Sometimes, though rarely, the best efforts to accommodate the back-injured worker yield some costly alternatives: expensive or complex modification of the job, retraining and reassignment, or long-term compensation. The decision must again be made by management, weighing the effects of each alternative. Compensation represents the continuing and financially destructive effect of long-term expense, not to mention the previously discussed psychological and financial disincentives for injured workers to return to the job. Nevertheless, it often seems the simplest course. Next, the investment of reassigning and/or retraining the worker in a more suitable job, with the attendant uncertainties, must be compared with its potential for an effective long-term solution. Finally, job accommodation may seem the most complex solution from the perspective of procedure, but it can also be the most cost-effective in solving the case at hand and as a visible example of commitment to "rehabilitate rather than compensate" the back-injured workers wherever possible and reasonable.

When, then, is accommodation a reasonable resolution of vocational rehabilitation of the back-injured worker? The answers include

    (1) when the worker is capable of performing the job safely and productively, with the accommodation(s);

619

(2) when the worker and/or the accommodations do not adversely affect the safety and productivity of others; and

(3) when the cost (time, effort, money, etc.) does not place undue hardship on the organization or its business.

§ 8-9. Implementing the Solution.

The agreement is now reached among management, the worker, physician(s) and others involved that the accommodation(s) needed will be feasible and effective. The accommodation(s) must then be implemented at the line level with any assistance needed, whether in-house or external, provided at the request of the line supervisor and the returning worker. Ultimate success of the vocational rehabilitation process rests largely on these two individuals, and they, more than anyone else, will be responsible for ongoing success. It is essential, then, that all the foregoing steps in the vocational rehabilitation process, as outlined in the flowchart, be focused on the efforts of the line supervisor and the worker in implementing the accommodation(s) and getting back to work. It is also these individuals who should be regularly contacted by the case management coordinator to monitor success and provide any additional assistance needed.

§ 8-10. Ergonomics and Job Design.

In the previous discussion of the vocational rehabilitation process, the many influences on the rehabilitation outcome were presented and integrated within a comprehensive management process. This process brings considerable expertise and resources to bear on the costly problem of returning the back-injured worker to productive employment. Why are all these resources necessary? Why is bringing back an injured worker so different from simply hiring an able-bodied applicant? Don't they both require adjustment to a new situation in order to be productive?

While it is true that any job applicant can represent unknown, even hidden, disabilities, the back-injured worker who returns to the job, any job, is a different kind of "unknown." Some of the specific differences were noted in the early discussion of Vocational Rehabilitation. The worker (and the supervisor) must deal with concerns over re-injury and how to minimize the possibility. As in job accommodation, this requires an understanding of the worker's unique abilities and limitations and how they may actually affect work performance. This is a prime example of practical application of the field of ergonomics or human factors engineering. Maximizing the back-injured worker's opportunity to succeed on the job requires an understanding of the ergonomic effect of back injury. In short, how is this individual functionally different from a "normal" individual? Applying the knowledge of these differences makes job accommodations and reduced risk of re-injury for back-injured workers a reality. This knowledge is also helpful in studying how to redesign at-risk jobs to reduce incidence of injury to currently able-bodied workers. After all, the back injury that does not occur is the least expensive to treat, and job redesign has proven to be one of the most effective methods of back-injury prevention. This section is devoted to applying the physical characteristics (ergonomics) of back-injured persons to job design.

§ 8-11. Ergonomics.

The field of ergonomics is relatively young, having grown primarily out of weaponry design research in World War II. To increase accuracy, safety, and efficiency, the performance capabilities of human beings (in this case soldiers) were studied to determine how to best design the tools of their trade. Though this has resulted in more dramatically effective equipment for destruction, the effects on the design of tools, appliances, vehicles, etc., in the peacetime world have been far less effective. One major reason for this is that the general population is not well-represented by the abilities of the young, fit soldier. Hence older, female, shorter or stouter individuals may have difficulties performing some tasks or operating some equipment. For those persons who experience disabilities, the problem is further compounded. The back-injured worker, for example, faces a similarly hostile work environment. Effective job design for this person requires knowledge of which components of the environment cause and aggravate back injury, and what can be done about them. The ultimate goal of the application of ergonomics is to adapt the environment to the individual in such a way that adaptation on the individual's part is minimized. This is especially true for the back-injured individual in the work environment, since attempts by the individual to adapt to physical stresses often result in re-injury. "A large number (and in some industries the majority) of disabling injuries . . . involve the act of manually handling materials" (entry 5). Though few back injuries can be directly attributed to a single causal event, it seems safe to say that manual materials handling is at least one of the major causes, and 79% of the injuries involving materials handling affect the lower back (entry 8). One might expect, then, that back injuries would be rare among white-collar workers. This is unfortunately untrue. Back problems are common among white-collar workers as well. The increasingly sedentary nature of many of these jobs has caused the white-collar worker to be seriously *understressed* in the work environment, making most workers susceptible to even the slightest strain in performing an occasional light physical task. For this reason, researchers seem to agree that no physical stress at the worksite can be just as detrimental to back health as *overstress* (entry 10). Hence, the first tenet of effective job design is to minimize risk of hazardous physical tasks — *not* to eliminate physical tasks from the job. This is as important in job design for back-injured workers as for others. Since physical fitness is so important in reducing back injuries to both the recovering back-injured worker and to the able-bodied worker, direct medical input concerning optimum design for physical tasks within the job is the best course. Using the JOB FACTORS CHECKLIST described earlier is an excellent vehicle for this input.

During the course of job analysis and design of accommodations for back-injured workers in real job situations, it has become apparent to this author that the inherent nature of most jobs tends to divide workers into two categories: those who have experienced on-the-job back injuries and those who will. Few jobs analyses reveal no potential for back injury based on currently known ergonomic job design guidelines. Keep in mind that these are not guidelines for the injured worker with serious physical restrictions, but rather the "average" worker. The demand for greater productivity, occasional unexpected workloads, the tendency to do a "little more now" so that one can rest later or finish earlier, and a general lack of awareness of work hazards to the back all contribute to making most jobs potential "back risks." Therefore, the following discussion of job design parameters, or ergonomic guidelines, apply to the not yet back-injured worker. As previously stated, medical input concerning specific work restrictions reported on the JOB FACTORS CHECKLIST would further specify job design and accommodation for a given returning back-injured worker.

§ 8-12. Work Position.

The static position(s) from which the worker performs the

623

job has an obvious effect on the relative safety or risk of any individual task. A common job design change on behalf of the back-injured worker is to allow the worker to sit full-time or periodically rather than standing full-time. Union contracts and other restrictions sometimes preclude this simple job change, and it may even impose back stress, rather than relieve it. For example, if the back-injured worker is permitted to perform a standing job from a sitting position, access to materials, equipment controls, visual information and work surface must be adjusted so as not require overhead lifting and reaching, excessive head or body movement to see properly, or unnatural body positions to use the work surface(s) or operate equipment. Changing the work position from standing to sitting effectively lowers the reach envelope, field of view, and comfortable work level approximately 15"—18" (entry 3). In addition, the limited available assistance of the legs for lifting greatly reduces this capability. Lifting from a sitting position can obviously place considerable stress on the back.

Also, the configuration of the seat is of prime importance. The straight back chair, the reclining arm-chair, and the stool with minimal backrest each have their proper place. Too often, the selection of seating is based on prestige rather than on ergonomic need (entry 11). This is not to say that the "harder you work, the softer the chair." Seating selection plays a critical part in effective job design for the returning back-injured worker, especially since excellent choices are commercially available. While such investments are sometimes viewed as extravagant, it should be noted that the ultimate cost to the organization of a single worker on long-term disability could probably cover the one-time cost of a new chair for *every* employee. In selecting the proper chair for a seated worker, one should consider:

(1) stable base (five legs, if casters are used);

(2) height adjustability — to allow various users to set most comfortable height;

(3) seat/back adjustability — to allow worker to change posture or rest when necessary;

(4) rigid, full backrest (unless rearward arm or trunk mobility is required) with lumbar support;

Note: lumbar support is a feature commonly absent in most American chair designs. This is one cause of low back pain in sedentary workers. Newer "ergonomic" chair designs include this feature.

(5) armrests for support and entry/exit, except when working close to the desk, etc.; and

(6) footrest/footbar — to avoid dangling legs if working at elevated position (may be incorporated into worksite or built for worker).

If it is necessary or preferable for the worker to stand at the worksite, provision of a resilient mat over a hard floor and/or leaning board or postural support can help to alleviate stresses and fatigue on the back. Since slips, trips and missteps cause a large percentage of back injuries, floor surface, as well as shoe type, should be considered:

(1) The surface and/or finish should not be slippery, and any change in floor texture should be marked or easily noticed.

(2) Shoe type should suit the surface texture and conditions (wet, dry, oily, etc.).

(3) Changes in floor angle must be clearly marked.

As always, the appropriateness of each of these guidelines is determined with the input of the worker and/or the physician.

§ 8-13. Work Surface.

The proper height for a work surface varies from 29″—30″ for seated workers to 36″—44″ for standing workers. Simple adjustments can be made with standing platforms, blocks under work surfaces, etc. As with floors, the nature of the surface itself is important. Having to stabilize a vibrating or mobile object on a slippery surface induces strains on the back. Likewise, having to drag a bulky object over a rough surface. The nature of the surface should be consistent with the tasks and the materials used. Modifications can be effected with the application of permanent or removable high- or low-friction mats, rollers, fixtures, etc. Where feasible, inclination of the work surface up to 20° may be helpful in minimizing "hunching over" the work surface. For this, solutions range from simple blocks to fully adjustable tables. The effective horizontal dimensions of the work surface should require the worker to reach no further than 18″ forward or 35″ to either side (from center) without changing position.

§ 8-14. Controls.

For the back-injured worker without additional difficulties in manipulation, optimum control design is generally a function of proper placement. The horizontal reach limitations noted for the work surface apply here in addition to vertical limitations. To reach controls which must be regularly used, the standing worker should not have to reach higher than 64″ nor lower than 24″. For seated workers, these dimensions are reduced to 48″ and 18″ respectively.

§ 8-15. Information Display.

In order to prevent poor work posture, display (control feedback, CRT screens, etc.) should be placed whenever possible so as to be easily viewed with the head in a comfortable position. This means that the display should be within a cone generated by a sight line 45° above the normal sight line. The normal sight line is between 10° and 15° below horizontal (entry 4).

§ 8-16. Storage and Materials Handling.

Because of the proportion of back injuries attributable to materials handling, considerable research has been documented on the ergonomics of this range of tasks. Even the simplest materials handling task, say, lifting a carton from the floor to a shelf contains many variables, each representing potential for back risk. For example, in determining whether this task can be performed safely, one might ask:

(1) What is the shape and weight of the carton?

(2) Is the carton's surface slippery? Are handles provided?

(3) Is the load inside the carton evenly distributed? Can it shift inside the carton? Is it fragile?

(4) How many times per minute/hour/day is this task performed?

(5) Is the work environment inside/outside, hot/cold, wet/dry? Is lighting adequate?

(6) Is the floor slippery or rough? Is it level? Are there obstacles to reach or step over?

(7) How high is the shelf?

Obviously the design of materials handling tasks involves many variables, even before one considers whether the worker is back-injured, male or female, tall or short, or physically conditioned for such a task.

It is not surprising then, that so much research has been devoted to the ergonomics of lifting, carrying, reaching, etc., and to the optimal design of the materials handling task.

For the purpose of the back-injured worker, the core issue is, of course, how to determine whether a specific materials handling task represents any risk of reinjury. This is an excellent question yet virtually impossible to answer with certainty given the variables involved. There do exist quantative formulae for redesigning lifting tasks to fall within certain ergonomic limits. These formulae integrate the variables of object weight, horizontal location, vertical location, vertical lift distance, lift frequency, and lift duration. Such formulae are contained in the NIOSH publication, "Work Practices Guide for Manual Lifting" (U.S. Dept. of Commerce, Wash., D.C., March 1981, pp. 121-144) which is a "must" for readers involved in accommodating back-injured workers or seeking to prevent job-related back injuries.

It is this author's experience that the most direct, effective, and practical method of job design to fit the ergonomics of a specific back-injured worker is to intergrate the direct, though essentially subjective, inputs of the worker, the physician, and the supervisor with available guidelines and resources. Reaching an agreeable solution usually means some compromise among the negotiable aspects of the job (available latitude for making accommodations), the worker's motivation and personal insight about his/her capabilities, and the physician's assessment of the worker's limitations vs. the job. Usually quantitative guidelines cannot be followed to the letter. To facilitate this teamwork, illustrated guidelines have been produced to suggest possible methods for job redesign which can be discussed among the individuals in cooperatively reaching an agreeable solution to fitting the job to the back-injured worker. A sample is shown at the end of this chapter (reprinted by permission of Mueller & Zullo, Inc.). The intent of these guidelines is to present some of the technical information available through resources such as the "Work

Practices Guide for Manual Lifting" in simplified form to facilitate application to the vocational rehabilitation/job design process. Through simplified presentations such as this, the individuals involved can more readily identify those variables, among the many affecting safe materials handling, that most directly involve the job and worker in question. From this point, examining alternative solutions can involve these individuals as well as other in-house and/or external technical resources. It is this approach which most often leads to a practical cost-effective solution.

§ 8-17. Prevention.

Though it is somewhat outside the parameters of this chapter, back injury prevention is worth noting in its relationship to job design. Because of the attractiveness of avoiding the entire issue of vocational rehabilitation whenever possible by preventing back injury, various approaches have been studied in depth. Job design, which has been discussed here in some depth, is as important to back-injury prevention as it is to successful vocational rehabilitation. In addition, considerable attention has been given to training in "safe lift" techniques for workers. This is essentially the maintenance of a straight back to minimize torque caused by lifting and carrying objects away from the body. Though these programs have shown some benefit, researchers caution against "cookbooklike" use of these concepts without consideration of risks which may be inherent in the job design or in the physical limitations of the worker. In other words, the job itself, or the worker's own abilities, may make "safe lift" postures impractical or even dangerous. The applicability of such techniques should be evaluated for each job.

Another field of study in the prevention of low back pain has been preplacement screening for at-risk jobs. Since x-rays have proven to be ineffective as a predictor of back injury risk (entry 9), other techniques are being developed, such as standardized strength testing. These techniques face considerable research before being effective as reliable screening tools (entry 2).

Considerable documentation supports physical fitness as the most effective method of back injury prevention (entry 9). Because too little physical activity can be as detrimental as too much, physical fitness programs among white-collar as well as blue-collar workers should continue to be a priority in industrial health care programs.

§ 8-18. Summary.

This chapter has sought to clarify the respective natures and roles of vocational rehabilitation and ergonomics in the management of industrial low back pain. How these disciplines can be applied in returning the injured worker to the job as soon and in as full a capacity as possible has been emphasized.

The back-injured worker represents a unique set of challenges in both vocational rehabilitation and in job design. Though the scope and cost of this occupational disability are vast, available resources are generally quite limited, both within the organization and in the community. The frequent alienation of the back-injured worker from the rehabilitation process predisposes most individuals away from successful return to work. The general business approach of "compensation rather than rehabilitation" compounds the problem, making return to work the overwhelming exception, rather than the rule.

Decisive management control and commitment can and does improve vocational rehabilitation outcome, with dramatic results. Minimized abuse of the compensation system, timely and cost-effective rehabilitation methods, and improved employee morale and productivity are just some of the benefits of a clearly defined and monitored case management process.

The physician's role in the rehabilitation process has been placed in perspective as a *contributor* to the *management* decision of whether and how the back-injured worker can return to work. Other roles in the case management process have been presented in a flowchart illustration through which the status of the rehabilitation process can be monitored to minimize the chance of the worker's being "lost through the cracks."

The direct correlation between medical input and job data input has been illustrated through the JOB FACTORS CHECKLIST. This format presents a common language for relating a worker's abilities and limitations with a given job's requirements. Armed with this information, the manager is capable of making realistic decisions concerning return-to-work and accommodation of the back-injured worker.

Methods of accommodating back-injured workers may range from work rescheduling, to job modification to assistive devices. The most effective way of arriving at a reasonable solution to the accommodation problem(s) has been shown by the integration of key inputs from line supervisor(s), medical staff, in-house and external technical help, and the worker him/herself.

Effectively returning the back-injured worker to productive employment requires understanding of the unique capabilities and limitations of that individual. These capabilities and limitations can be thought of as the human factors, or ergonomic, guidelines. The application of these guidelines is the technique of job design. Job design can involve equipment and materials used in the job as well as work procedures, individual tasks, and the work location itself. Some of the available ergonomic data has been translated into job design guidelines and presented illustratively in this chapter.

The goal of the technical data presented was not to be exhaustive, but rather exemplary. Despite the depth of research supporting certain ergonomic data, especially that of materials handling, none of the data can be applied to effective job design without judgment. That vital judgment must come from those diverse individuals bound together in the vocational rehabilitation process: the manager, the line-level supervisor, medical staff, personnel and safety specialists, technical advisors, and most importantly, the back-injured worker.

The problem of industrial back injury is increasingly costly and complex. The only potential for effective reversal of this trend is decisive management commitment carried through at the line level of the organization — not various desperate "band-aid" solutions. The cost in time and effort may appear high. But the costs are relative, and compared with the scope of the problem, the potential benefits are staggering.

BIBLIOGRAPHY

1. Berkeley Planning Associates. A study of accommodations provided to handicapped employees by federal contractors, U.S. Dept. of Labor, Wash., D.C., 1982, p. ii.
2. Chaffin. Human strength capability and low back pain, *J. Occupational Med.,* Vol. 16, No. 4, April, 1974, p. 254.
3. Diffrient, Niels, Tilley, Alvin R., Bardagjy, Joan C. *Humanscale 1/2/3* MIT Press, Cambridge, MA, 1974, p. la.
4. Diffrient, Niels, Tilley, Alvin R., and Bardagjy, Joan C. *Humanscale 1/2/3,* MIT Press, Cambridge, MA, 1974, p. 7.
5. National Institute for Occupational Safety and Health, Work practices guide for manual lifting, U. S. Dept. of Commerce, Wash., D. C., 1981, p. 113.
6. Office of Federal Contract Compliance Program v. Missouri Pacific R.R., Case #81—OFCCP—8 (Slip Op., filed March 17, 1983).

7. Palisano, P. New Strategies for the Battle Against Back Injuries, *Occupational Hazards,* April, 1963, p. 67.
8. Snook, Stover H. *The Design of Manual Handling Tasks,* The Ergonomics Society, Bedfordshire, England, 1978, p. 1.
9. Snook, Stover H., and Ciriello, Vincent M. Low back pain in industry, *ASSE Journal,* April, 1972, p. 19.
10. Snook, Stover H., and Ciriello, Vincent M. Low back pain in industry *ASSE Journal,* April 1972, p. 21.
11. The big stakes in designing a place to sit, *Business Week,* Apr. 26, 1976, p. 46.

A SAMPLE CASE MANAGEMENT PROCESS

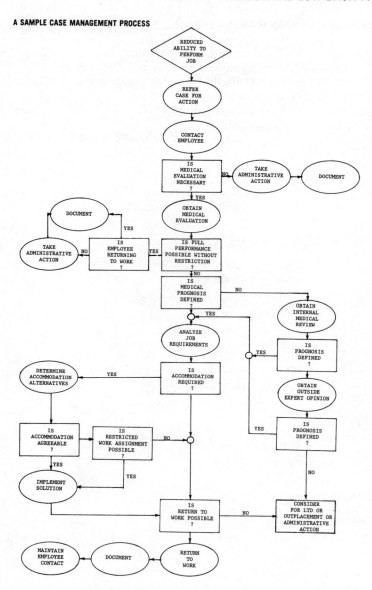

REHABILITATION, ERGONOMICS, AND JOB DESIGN

A SAMPLE FORMAT FOR MEDICAL EVALUATION REPORTS

1. Employee's Name

2. Date of Injury/Illness

3. Nature of Injury/Illness

    a. occupational

    b. non-occupational

Disabled persons are entitled
to privacy of this information.
If this medical report is to be
circulated, #2 and #3 should
be reported on a separate
confidential form.

4. Date of examination

5. Recommendation

    a. defer recommendation pending:

    _____

    b. return to work with no restrictions

    c. return to work with following restrictions:

CHECK ANY JOB FACTORS WHICH EMPLOYEE <u>SHOULD</u> <u>AVOID</u> DUE TO IMPAIRMENT

ENVIRONMENTS TO AVOID

___ indoors ___% (specify)
___ outdoors ___% (specify)
___ extremes of temperature ___ (specify)
___ noise _____ (specify)
___ dry air
___ humidity/dampness
___ extreme temperatures _____ (specify)
___ abrupt temperature change
___ vibration
___ dusts/odors/fumes _____ (specify)
___ solvents/oils _____ (specify)
___ acids/bases _____ (specify)
___ toxins _____ (specify)

RISKS TO AVOID

___ burns	___ working at heights	___ infection
___ explosions	___ silppery surfaces	___ electrical shock
___ moving objects	___ radiation	___ cuts, bruises

SENSORY TASKS TO AVOID

Seeing

    ___ near ___ far ___ wide field ___ color ___ depth

635

Hearing

___ speech   ___ mechanical sounds

Smelling

___ identification of odors

VEHICLE OPERATION TO AVOID

___ automobile
___ forklift
___ hand cart
___ truck
___ other (specify)

WORKLOADS TO AVOID

carrying/lifting	frequent	occasional	rare
less than 10 lbs			
10 - 20 lbs			
20 - 50 lbs			
50 - 100 lbs			
over 100 lbs			

manipulating/handling	frequent	occasional	rare
less than 10 lbs			
10 - 20 lbs			
20 - 50 lbs			
50 - 100 lbs			
over 100 lbs			

pushing/pulling	frequent	occasional	rare
less than 10 lbs			
10 - 20 lbs			
20 - 50 lbs			
50 - 100 lbs			
over 100 lbs			

BODY MOVEMENTS TO AVOID

	frequent	occasional	rare
bending			
climbing stairs			
climbing ladders			
kneeling			
lifting			
pressing (with foot)			
reaching overhead			
running			
sitting			
standing			
twisting			
walking			

	WORK RESTRICTION EVALUATION*	

1. Injured workers' name (First, middle, last)   **2.** OWCP No.

3. Check the frequency and number of hours a day the worker is able to do the following specific types of activities.

ACTIVITY	FREQUENCY		NUMBER OF HOURS A DAY								
	Continuous	Intermittent	0	1	2	3	4	5	6	7	8
a. Sitting											
b. Walking											
c. Lifting											
d. Bending											
e. Squatting											
f. Climbing											
g. Kneeling											
h. Twisting											
i. Standing											

4. Check the lifting restriction.
☐ 0-10 lbs. ☐ 10-20 lbs. ☐ 20-50 lbs. ☐ 50-75 lbs. ☐ 75 & above lbs.

5a. Hand restrictions?   **5b.** Simple grasping?
☐ No ☐ Yes - (check b,c,and d.)   ☐ Yes ☐ No

5c. Pushing and pulling? ☐ Yes ☐ No   **5d.** Fine Manipulation? ☐ Yes ☐ No

6. Can the worker reach or work above the shoulder? ☐ Yes ☐ No

7. Can the worker use his/her feet to operate foot controls or for repetitive movement? ☐ Yes ☐ No

8. Can the worker operate a car, truck, crane, tractor, or other type of motor vehicle? ☐ Yes ☐ No

9. Are there cardiac, visual, or hearing limitations?
☐ No ☐ Yes - (Describe)

10. Are there restrictions concerning heat, cold, dampness, height, temperature changes, high speed ☐ No ☐ Yes - (Describe) working, or exposure to dust, fumes or gases?

11. Are interpersonal relations effected because of a neuropsychiatric condition?
☐ No ☐ Yes - Describe (Ability to give and take supervision, meet deadlines, etc.)

12a. Can the individual work eight hours a day? ☐ Yes ☐ No - (Indicate when)   **12b.** If not eight hours, how many and when?

13. Do you anticipate the worker will need vocational rehabilitation services such as testing, counseling ☐ Yes ☐ No training, or placement to return to work?

14. Has the worker reached maximum improvement?
☐ Yes (Indicate when)   ☐ No (Indicate when)

15. Remarks: (Restrictions from medication or other limitations)

16. Name   **17.** Signature

18. Address   **19.** Telephone No.   **20.** Date

*This chart is modeled after U. S. Department of Labor Work Restriction Evaluation.

637

JOB FACTORS CHECKLIST

CHECK ANY JOB FACTORS WHICH DESCRIBE THE JOB

ENVIRONMENTS

___ indoors ___% (specify)
___ outdoors ___% (specify)
___ extremes of temperature _____(specify)
___ noise _____(specify)
___ dry air
___ humidity/dampness
___ wetness
___ extreme temperatures _____(specify)
___ abrupt temperature change
___ vibration
___ dusts/odors/fumes _____(specify)
___ solvents/oils _____(specify)
___ acids/bases _____(specify)
___ toxins _____(specify)

RISKS

___ burns   ___ working at heights   ___ infection
___ explosions   ___ slippery surfaces   ___ electrical shock
___ moving objects   ___ radiation   ___ cuts, bruises

SENSORY TASKS

Seeing

___ near   ___ far   ___ wide field   ___ color   ___ depth

Hearing

___ speech   ___ mechanical sounds

Smelling

___ identification of odors

REHABILITATION, ERGONOMICS, AND JOB DESIGN

VEHICLE OPERATION

___ automobile
___ forklift
___ hand cart
___ truck
___ other (specify)

WORKLOADS

carrying/lifting	frequent	occasional	rare
less than 10 lbs			
10 - 20 lbs			
20 - 50 lbs			
50 - 100 lbs			
over 100 lbs			

manipulating/handling	frequent	occasional	rare
less than 10 lbs			
10 - 20 lbs			
20 - 50 lbs			
50 - 100 lbs			
over 100 lbs			

pushing/pulling	frequent	occasional	rare
less than 10 lbs			
10 - 20 lbs			
20 - 50 lbs			
50 - 100 lbs			
over 100 lbs			

BODY MOVEMENTS

	frequent	occasional	rare
bending			
climbing stairs			
climbing ladders			
kneeling			
lifting			
pressing (with foot)			
reaching overhead			
running			
sitting			
standing			
twisting			
walking			

639

DEVELOPMENTS IN REASONABLE ACCOMMODATION

A Regular Feature of DISABILITY MANAGEMENT TODAY

MINIMIZING BACK STRESS THROUGH JOB DESIGN

Many of the hazards which can cause excessive
back stress on the job can be minimized through
practical job design. The guidelines on these
illustrations will help to identify and reduce
potential hazards . . .

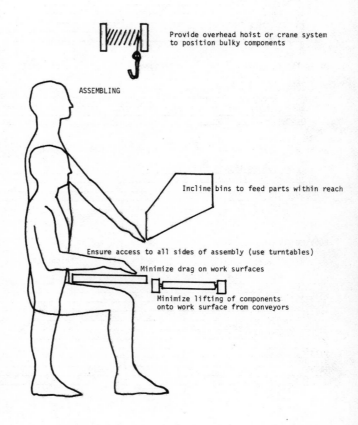

Provide overhead hoist or crane system
to position bulky components

ASSEMBLING

Incline bins to feed parts within reach

Ensure access to all sides of assembly (use turntables)

Minimize drag on work surfaces

Minimize lifting of components
onto work surface from conveyors

640

REHABILITATION, ERGONOMICS, AND JOB DESIGN

WORK PROCEDURE GUIDELINES

1. Assign the job to two or more persons.

2. Change the job from lifting to lowering, from lowering to carrying, from carrying to pulling, from pulling to pushing.

3. Relax the standard time allowed for the job.

4. Reduce the frequency of lifts.

5. Introduce a program of job rotation to limit exposure of employees to stressful lifting demands.

6. Introduce work/rest schedules to maintain acceptable levels of physiological responses (energy expenditure, heart rate, etc.).

7. Introduce appropriate control measures when working in adverse environments, for example, in a hot environment, acclimatize employees, reduce temperatures where possible, assign jobs to only those both physically and psychologically fit.

CONTAINERS

Minimize container weight
Balance load within containers
Distribute heavy loads into two or more containers
Change container shape and/or balance if necessary

Use handles or hooks for easier container handling

Maximum storage height — 52"

Avoid storing items behind others

Use sturdy storage systems
Minimize drag on storage surfaces

CARRYING, LIFTING AND REACHING

When worker must turn around to transfer load, distance should be no step or one full step

Keep load as close to body as possible

Minimum storage height 20" above floor

Use large wheels on carts to minimize effort

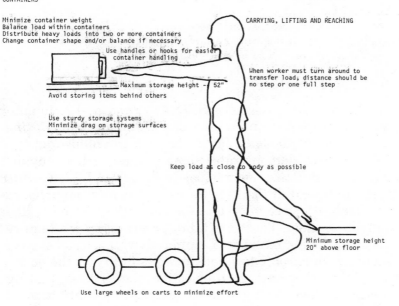

CHAPTER 9

EVALUATION OF THE LOW BACK DIAGNOSIS RELATED IMPAIRMENT RATING

Henry L. Feffer, M.D.

§ 9-1. Introduction.

In the first 1983 issue of SPINE, Brand and Lehmann closed their paper titled *Low-Back Impairment Rating Practices of Orthopaedic Surgeons* with the following:

"This survey demonstrates some of the common impairment rating practices of orthopaedic surgeons. It also points out the practical difficulties in distinguishing impairment from disability as compared to the ease of conceptually distinguishing them. Those statutory programs which require such a distinction make an assumption which, at least with patients with chronic low back pain, is faulty at the onset: mainly that there are good objective means of evaluating patients. In fact, no such means exist to evaluate chronic back pain patients. Any method of rating impairment is arbitrary at best and incorrect and unfair at worst."

Although this statement, unfortunately, is a fair assessment of the current state of the art in low back evaluation, a future based on standardized nomenclature and values is not nearly so bleak; ratings have been found to be replicable, providing conditions are controlled and guide-

642

lines are clear and meaningful to all disciplines involved in the adjudication process. With this in mind, we will first proceed with the generic issues involved and then turn to problems peculiar to the low back. The chapter will conclude with our version of what seem to be the most workable and equitable set of evaluation guidelines available to date.

§ 9-2. Generic Evaluation Issues.

The medical-legal interface is a problem area where physicians, whether they like it or not, work with lawyers to reconcile medicine to the legal world. Although the relationship is difficult at best, misunderstandings in respect to disability evaluation are particularly worrisome. This process of determining whether a worker can do his job can be initiated through requests from individual patients, insurance companies, military services, vocational rehabilitation services, employees, labor unions, attorneys, or state or federal government agencies, each with its own legal environment or set of rules. Thus disability is defined by Social Security regulations as the inability to do any substantial activity; it is all or none. Workers' compensation awards, on the other hand, can be either partial or total, temporary or permanent. In every case, however, when a worker contends that he is unable to do his usual work or compete in the job market and appeals to some outside agency for financial compensation, the developing process of adjudication inevitably requires a medical evaluation.

In any event, although impairment evaluations often begin with a patient's visit to his private physician, certain agencies, like Workers' Compensation Boards or insurance companies, may have physicians and consultants available for that purpose. Since eventually many occupational disease and injury claims are contested, the physician evaluating the disability must be prepared to respond to requests for written testimony as well as depositions and court appearances. It is a difficult area where no official set of guidelines has been established. Medical education generally neglects training in this special type of disability assessment, and the average physician tends to avoid formal education and learn the hard way through his mistakes; he usually has reached senior citizen status before he has become functionally competent.

Any physician involved in disability evaluation soon realizes that physical disability and physical impairment are not synonymous. Physical impairment rating is, hopefully, the objective assessment of body dysfunction. Physical disability, on the other hand, is very complex. It is affected by culture, socioeconomic background, education, experience, and the psychological makeup of the individual. While a physician can objectively rate physical impairment, and can have a valid opinion concerning disability, a full assessment of disability is not strictly a medical matter, and is probably best done by non-medical people.

Until recently, there has been no alternative to the physician's evaluation. Today, however, rehabilitation specialists who are vocational experts have been trained to assess the consequences of impairment; this assessment reflects an individual's ability to procure employment generally and is called *Residual Occupational Access.* It is an evaluation system conceived on the premise that all individuals, based on their age, education, and previous work experience, qualify for a certain percentage of the jobs found in the labor force. Personal injury, when it results in permanent impairment and functional restrictions, reduces that percentage. Therefore, by comparing an individual's pre-injury access to the labor force with the post-injury access to the same labor force, one has a measure of reduced employability or reduced occupational access, and accordingly, a measure of loss of earning power.

644

The Department of Labor in each state publishes statistical information on the existence of specific jobs within the labor force. These standard metropolitan statistical area (SMSA) projections are available for all metropolitan areas with over 50,000 population; and for the total labor force of each state, the United States census information is used as their data base. The *Dictionary of Occupation Titles,* a publication of the United States Department of Labor, provides job descriptions and work requirements for all jobs in the national economy. Comparison of an individual's age, education, physical restrictions, and previous work experience with the requirements of the jobs in the labor force enables one to objectively project the percentage of these jobs for which a worker may be qualified, and calculate it on the following typical work sheet. (See Table I at end of this chapter.)

In order to put the matter of disability evaluation in respect to workers' compensation into some type of logical perspective, it would be helpful to summarize the conclusions of a 1979 task force of the International Association of Industrial Accident Boards and Commissions. They had originally assumed that each state approached compensation in about the same way, with only minor differences; however, it did not take too long to become aware of the existence of fifty different states with markedly different approaches to workers' compensation. They found it impossible to compare North Dakota, with a state operated plan and a large farming population, to Georgia,where the state does not operate the plan and there is a large low-income population. The New York Compensation Board, with designated trained examiners, has difficulty understanding Mississippi, where no trained examiners are present, or even Louisiana, where all cases are decided in Civil Court. The trauma incurred by all workers is similar, and a fracture should heal the same in a black man from West

645

Virginia as it does in a white man from Washington State. A back injury in California should recover as rapidly as one in North Carolina but unfortunately this is not the case since the compensation laws and manner of administration are the ultimate deciding factors. The Canadian system was found to be far superior to ours in this respect.

They noted that it was not too long ago that the state of Florida's workers' compensation system was essentially bankrupt, and industry rebelled at further increases in premiums. The Florida Workers' Compensation Board, the legislature, and even the attorneys, recognized that the underlying cause of this too-near catastrophe was primarily due to confusion regarding payments of permanent partial disability. Attorney's fees were added on top of any award and amounted to approximately forty million dollars in 1978. In an attempt to deal with this problem, Florida has completely abandoned the permanent partial disability approach and now pays only a percentage of the weekly wage loss.

Assuming that the determination of permanent partial disability is a major concern in the majority of states, two solutions would seem practical: (1) Do away with permanent partial disability (as in Florida). (2) Improve and simplify the permanent partial ratings. Since the other forty-nine states are not likely to do away with permanent partial disability (or permanent impairment ratings), it seems that some effort must be made to improve the rating system.

The medical committee report in New York showed evidence that only eighteen of the fifty states gave any real guidelines to physicians in evaluating permanent impairment. Only five states actually assist the physician by giving information referable to specific guidelines. At best, only Nevada has impairment evaluated by physicians who are employed full time by the Workers' Compensation

Board. Unfortunately in the majority of the states, they were able to find little rapport between the Workers' Compensation Board and the practicing physicians; they made a strong recommendation to develop better cooperation between the Workers' Compensation Boards and the medical associations. Finally, they were able to single out Nevada as the state with the most progressive system. It has its own medical department as part of the Workers' Compensation Board and this department has the responsibility of personally interviewing and evaluating any individual who claims permanent partial disability. The physician also has the right to ask for an additional evaluation from an impartial specialist if he feel such is indicated. The state has a rehabilitation center which works closely with the Workers' Compensation Board and the medical department. With this system, the treating physician is relieved of the responsibility of impairment ratings, and the rate of litigation has been materially lowered.

§ 9-3. The Low Back Evaluation Problem.

It is estimated that one out of every three dollars spent on workers' compensation is associated with back pathology and eighty to ninety percent of workers' compensation directors' headaches and time is spent in litigation referable to back problems. In fact the bankruptcy of the Florida system a few years ago was primarily due to the uncontrollable costs of permanent partial back impairments. The guidelines as established by the American Medical Association, the American Academy of Orthopaedic Surgeons, the Veterans Administration, and the Social Security Administration are of little help to the evaluator when it comes to rating any specific low back injury. Information regarding bending and twisting simply does not furnish the answers. The patient bends as far as he

wants to and certainly there is a wide discrepancy among
individuals with normal backs. In evaluating one
extremity, it can be compared to the other but
unfortunately the back stands alone in this respect. Sooner
or later we are going to have to face the fact that physical
impairment ratings of spinal injuries are primarily a
judgement call based on the history and physical findings.
Inevitably, the system is going to have a heavy subjective
element added to it, and this obviously precludes any significant fine tuning. Because of this, using gradations of
less than five percent is just unrealistic.

Although agencies requesting evaluations usually suggest the American Medical Association's *Guides to the
Evaluation of Permanent Impairment* as the ultimate reference, most experienced evaluators have given up on them
for guidance on spinal injuries because these tables consider
motion as the sole criterion of impairment, and do not consider pain in the determination of low back impairment
except when associated with peripheral nerve injury or
when "substantiated by clinical findings." Chronic low back
pain often is associated with little or no objective clinical
signs and these, or the lack thereof, frequently are
unrelated to the injury or disability in question. In addition,
accurate measurement of spinal motion, even in the most
experienced of hands, is often just about impossible. In
effect, the facts that are utilized by most physicians, such as
motivation, age, education, personality, I.Q., and social
environment, strongly suggest that evaluators often rate
disability rather than impairment.

The usefulness of the *AMA Guides* lies primarily in their
"whole man" concept. Each part of the body is considered to
represent only a part of the whole; the percent each part
contributes is based on a notion of function. Since the back
is important to many functions, it contributes a maximum
of sixty percent of the whole man. Thus, once the impair-

ment for the given part is estimated, the whole man impairment can be determined easily. The only other reference with any credibility in this field is the *Manual for Orthopaedic Surgeons in Evaluating Permanent Physical Impairment,* published by the American Academy of Orthopaedic Surgeons. However, even though this is often cited at hearings, it largely relies on subjective symptoms and is of little practical value to the examining physician.

In the final analysis, there is little question that experience is far more important than any reference guide in making these determinations. Pain is the chief limiting factor in spinal disease and there is a great deal of variation between individuals in their response to pain. While there is no foolproof method of assessing pain, one can, through experience, use objective observations which are of assistance in evaluating its effects. In addition to limitation of movement, factors such as muscle spasm, wasting, deformity, tension signs and irrefutable evidence of neurological damage must be taken into consideration.

In order to warrant a rating, a compensable spinal disease must result in some modification of activities. Intermittent symptoms that do not stop the individual from engaging in normal activities will not warrant an award. Intermittent symptoms which are sufficiently frequent or severe enough to cause an individual to avoid the usual duties expected of him, such as heavy lifting, will warrant an impairment rating.

Spinal symptoms during the acute phase of an illness can be totally disabling; however, there are very few cases that can warrant a 100% impairment rating after adequate treatment. Theoretically, the average person earns half of his living by physical activity and half by mental activity. The individual with severe spinal disease who is physically capable of traveling to the work place, then, could conceivably justify a rating of up to 50% total body impairment.

Some states have gotten away from the typical percentage disability rating and California, for instance, has a rather complex system which uses the person's age and occupation modified against an impairment rating given by the physician. This disability rating is derived from the wording of the physician's report and his perception of the impairment of the applicant's ability to perform on the job he was doing at the time of his injury, or to compete for a job on the open labor market. Such disability is based both on subjective factors, such as pain, which impair the ability to work or objective factors, such as limitation of motion, or both. Minimal pain is defined as an annoyance which causes no handicap in the performance of the particular employment activity. Slight pain can be tolerated but would cause some handicap in the performance of the employment activity precipitating the pain. Moderate pain can be tolerated, but would cause marked handicap in the performance of the employment activity precipitating the pain; and severe pain would preclude the employment activity precipitating the pain. In that state, subjective symptoms, even without very much in the way of objective findings, can be the basis for a disability rating. As far as the low back is concerned, the Commission encourages the quantification of the disability into one of the following eight categories:

A. *Disabilities Precluding Heavy Lifting.*
   This rating contemplates that the individual has lost approximately one-quarter of his pre-injury capacity for lifting. (A statement "inability to lift 50 lbs." is not meaningful.) The total lifting effort, including weight, distance, endurance, frequency, body position and similar factors should be considered in reference to the particular individual.

B. *Disability Precluding Very Heavy Work.*
   This rating contemplates that the individual has lost

650

approximately one-quarter of his pre-injury capacity for performing such activities such as bending, stooping, lifting, pushing, pulling and climbing or other activities involving comparable physical effort.

C. *Disability Precluding Heavy Lifting.*

This rating contemplates that the individual has lost approximately one-half of his pre-injury capacity for lifting. (See statement regarding lifting under A, above.)

D. *Disability Precluding Heavy Lifting, Repeated Bending and Stooping.*

This rating contemplates that the individual has lost approximately one-half of his pre-injury capacity for lifting, bending, and stooping.

E. *Disability Precluding Heavy Work.*

This rating contemplates that the individual has lost approximately one-half of his pre-injury capacity for performing such activities as bending, stooping, lifting, pushing, pulling and climbing or other activities involving comparable physical effort.

F. *Disability Resulting in Limitation to Light Work.*

This rating contemplates that the individual can do work in a standing or walking position, with a minimum of demands for physical effort.

G. *Disability Resulting in Limitation to Semi-Sedentary Work.*

This rating contemplates that the individual can do work approximately one-half the time in a sitting position, approximately one-half the time in a standing or walking position, with a minimum of demands for physical effort whether standing, walking, or sitting.

H. *Disability Resulting in Limitations to Sedentary Work.*

This rating contemplates that the individual can do

work predominantly in a sitting position at a bench, desk or table with minimum demands for physical effort and with some degree of walking or standing being permitted.

As one can see, the guidelines above, about as specific as are available in any jurisdiction, can still be mind boggling to interpret and are bound to create inequities. In the final analysis, however, our not inconsiderable frustrations in the past have largely resulted from ambiguities in nomenclature and diagnosis; and no coherent standardized system will be possible until a glossary on spinal terminology, as established in previous chapters, has been recognized and accepted. The Social Security Administration has gone part of the way by classifying jobs as sedentary, light, medium, heavy, and very heavy, and has been very specific in limiting each to ten pounds, twenty pounds, fifty pounds, a hundred pounds, and more than a hundred pounds respectively; but no attempt has been made to establish or standardize the maximum physical exertion requirement of each back condition or combination of back conditions. If a unified system is to evolve, concepts of standardized diagnosis and treatment will have to gain universal acceptance.

Physicians, whether the treating physician or the independent medical evaluator, can be quite glib in the presentation of a dubious finding. For instance, the straight leg raising test is liberally interpreted as being positive if the patient says "ouch" when his leg is elevated 60 degrees. A subsequent diagnosis of herniated disc, predicated on this finding, is meaningless. Most people with spondylolysis go through life without significant symptoms. Thus, the radiological diagnosis of spondylolysis has little significance in the absence of supportive, objective physical findings. Most backache has a functional component to it;

that is, the patient uses the real or imagined condition for some type of personal gain, whether it is to get more attention at home, get an easier job, or make more money from a minor accident. This is exclusive of true neuroses, which often have more deep-seated implications.

A person with severe osteoarthritis of the spine undoubtedly has a certain amount of discomfort every day. This is permanent and probably progressive. Should he sustain a superimposed back sprain, however, even though he may be completely immobilized on a temporary basis, the long-term prognosis can be quite good once he recovers from the acute process; that is, assuming that he has not learned to enjoy the inactivity in the meanwhile. In that case, subjective complaints will not return to their baseline level. In all these situations, it behooves us as physicians to make our reports and evaluations reflect, to the highest possible degree, the true state of impairment in a way that is organized and replicable.

§ 9-4. Standardized Low Back Evaluation Guidelines.

In order to probe for some meaningful consensus, a questionnaire was circulated among 75 American members of the International Society for Study of the Lumbar Spine (ISSLS); they were asked to fill out the ratings they use in 41 specific clinical situations. As would be expected, given the chaotic situation as already presented, the responses from these experts was anything but consistent; however, they could be handled statistically so that in most cases a medium range was obvious. The figures spread from a zero percentage in the completely recovered acute low back sprain to 25% following failed back surgery. Certainly, was this complicated by emotional problems, addiction, and so forth, it was not unusual to see it increased up to 50%, and in practice one often finds even higher ratings in situations where there is obvious bias.

As stated before, the AMA Guides are all but worthless when dealing with the low back. The range of motion of the spine is quite difficult to measure, even for those who are experienced, and the Guides are pretty much dependent on those figures. The emotions, both of the patient and the physician, also play a major role in the evaluation, and under the best of conditions ratings can change widely with changing moods. Most important, it should be remembered that even though we talk about the importance of objectivity of signs and symptoms, probably the objectivity of the evaluator is equally significant.

Although the AAOS Manual gets more specific and recommends ratings for a limited selection of diagnostic categories, the Academy tends to ignore the real sticky problems and reneges in its potential role of leadership in the adjudication process. The raw ISSLS survey data, on the other hand, offer a unique opportunity to once and for all take the initiative and attempt to establish valid linkages among diagnoses, impairment, and physical exertion requirements of various job categories.

The complicated State of California quantification of low back disability has already been presented and it offers a potential classification to work with; however, it is unnecessarily confusing. The physical exertion requirements as defined by the Social Security Administration, on the other hand, are relatively simple and easy to use since these terms have the same meaning as they have in the *Dictionary of Occupational Titles* published by the Department of Labor. Even though the Social Security Administration does not use the permanent partial physical impairment rating system, it is relatively easy to modify their classification to conform to a compensation setting in the following way:

> *Very Heavy Work* is that which involves lifting objects weighing more than 100 pounds at a time, with

frequent lifting or carrying of objects weighing 50
pounds or more.
Heavy Work involves lifting of no more than 100
pounds at a time, with frequent lifting or carrying of
objects weighing up to 50 pounds.

No one with any low back related permanent partial
physical impairment can be expected to perform safely
within either of these categories, and if the applicant cannot
possibly be qualified to do anything lighter, he would have
to be approved for social security.

Medium Work is defined as involving the lifting of no
more than fifty pounds at a time, with frequent lifting
or carrying of objects weighing up to twenty-five
pounds. Workers with five percent or less back related
permanent partial physical impairment can qualify in
this category, but those with higher ratings cannot.
Light Work is described as involving lifting of no more
than twenty pounds at a time with frequent lifting or
carrying of objects weighing up to ten pounds. Applicants with between ten and fifteen percent permanent
partial physical impairment because of a low back
problem should be able to do this type of work.
Sedentary Work is described as that involving no more
than the lifting of ten pounds at a time and occasional
lifting or carrying of articles like docket files, ledgers,
or small tools. Applicants with twenty or twenty-five
percent permanent partial physical impairment should
be capable of this type of work.

Those workers with more than twenty-five percent back
related permanent partial physical impairment will rarely
qualify for any type of productive occupational activity
unless they have special sedentary qualifications which can
be done part time or at home.

Given the state of anarchy in the current low back evaluation scene, it would seem logical to once and for all correlate a broad array of well defined and clearly delineated low back syndromes with the above physical exertion requirements as laid out by the Social Security Administration. Thus the consensus in respect to low back impairment ratings in a standardized set of diagnostic situations as provided by the experienced spine surgeons belonging to the ISSLS, when matched with these physical exertion requirements, could generate the most workable guidelines yet available for evaluating the low back:

A *Contusion* of the back is an injury resulting from a direct blow to the spine. Local soreness can persist for up to two weeks and recovery may be expected without residuals. There should be no difficulty in returning to very hard work.

An *Acute Back Sprain* is defined as a soft tissue lifting or bending injury of an otherwise normal back. X-rays are negative and there is no radiating pain. Complete recovery should be expected in from two to three weeks and there is no reason why the patient cannot return to very heavy work after completing a rehabilitative exercise program. If recovery is delayed, some non-traumatic medical or psycho-social problem should be suspected. Above all, in the absence of true radiating pain, one must maintain a healthy skepticism as to the relevancy of findings generated by invasive and high technology confirmatory tests such as the myelogram, CAT scan, EMG, thermogram, discogram, etc. They are all too sensitive and too little selective. Too often both the physician and the patient are, in this way, led down the garden path to an irreversible and unmitigated surgical disaster.

It should also be noted that one or more myelograms, particularly when the traditional oil based contrast material, Pantopaque, has been used, can on occasion lead to

656

permanent arachnoidal adhesions, that is, irreversible nerve root scarring and chronic pain. The spinal specialists surveyed felt that a tight, painful back with radiological evidence of residual dye in the caudal sac and dural sleeves justifies a 10% permanent partial physical impairment rating. Thus the worker with an acute back sprain who could have been expected to return to very heavy work in two to three weeks, has been restricted to light work for life as a result of a precipitous and unnecessary invasive study — "just to be sure we are not missing anything."

Chronic Back Sprain and *Recurrent Back Sprain* are probably one and the same thing since frequent relapses usually mean that the worker never has really gotten over the original episode. Such a failure of an acute back sprain to respond to simple therapeutic measures usually means a preexisting biomechanical vulnerability from a degenerated disc, facet joint arthrosis, spondylolysis, or from just plain inadequate supportive trunk musculature. In spite of this, but assuming there really was no prior difficulty, compensability on the basis of aggravation is justified and, it was felt that between a 5% and 15% permanent partial physical impairment rating is justified, depending upon the degree of deformity and restriction. Workers in this category should not be expected to do anything heavier than light work, and at that, they occasionally may have to wear a back support. Although these patients usually show chronic radiological changes, they rarely have radiating pain, a neurological deficit, or positive tension signs unless a lumbar canal stenosis has developed.

It is a rare case of *Herniated Nucleus Pulposus* which recovers fully without residual physical impairment, and an initial assumption should be made that none of them will ever get back to Very Heavy Work. That is, of course, assuming that the diagnosis is accurate and has not been made capriciously. The only herniated disc patient who might even have a chance to resume Heavy Work is the occasional one who has made a dramatic recovery under conservative management. Certainly no operation should be expected to do the impossible. Surgery is indicated to relieve an unacceptable level of pain; it does not get anybody back to work. As a rule, an employee who has had a discectomy may be expected to end up with anywhere between a 5% and 25% permanent partial physical impairment. A so-called perfect operative result deserves at least 5%, based on the loss of one "shock absorber" in the spine and the resulting increased vulnerability to reinjury. A 25% rating is not unusual, given a stiff painful back and a substantial neurological deficit. Thus a worker with anything short of the best result will inevitably be restricted to no better than Light Work, and a Sedentary Work limit may be expected in the multiply operated back (failed back syndrome).

It would be possible but certainly very unusual for a *Spondylolysis* to develop from an acute injury on the job. During the last decade we have learned about the epidemiology of this condition. We now know that this break or defect in the posterior arch of a vertebrae is not congenital in origin but rather results from an insidious fracture occurring during the active years between the ages of 5 and 7. This resulting unstable motor unit is vulnerable to acute back sprains either on or off the job and these injuries usually respond to treatment just as any other acute back sprain does. A complete recovery without residual physical impairment should be anticipated. Occasionally, however, this unstable segment can slip and develop into a *Spondylolysthesis,* and in that case, a 5% to 10% permanent partial physical impairment and Medium Work capacity restriction can be expected. If a spinal fusion with or without a decompression laminectomy becomes necessary for stability and there is a good result, a 10% rating and Medium Work restriction is still in order. If chronic or recurrent pain and symptoms persist in spite of the stabilization, the employee deserves a 20% permanent partial physical impairment rating and should no longer be exposed to anything more strenuous than light work.

One of the more difficult situations to evaluate relates to on the job injuries incurred by workers with preexisting osteoarthritis with or without lumbar canal stenosis. Back sprains under these conditions can be slow to respond, and let us face it, the symptoms thereafter tend to be perceived as being more disabling than they were before the compensable injury. In spite of this, the spinal experts surveyed were in favor of awarding a 10% permanent partial physical impairment rating to those who were subjectively worse and a 15% rating to those who were both subjectively and objectively worse. In either case, light work is the most that should be required of them. The ratings were essentially the same if a surgical decompression laminectomy was required.

Although the pathology and resulting deformity associated with fractures of the lumbar spine can be fairly well defined and residual impairment ratings under various specific conditions should be replicable, little has been done in the way of standardization; and related guidelines are as fuzzy here as they are in the more chronic low back situations. The survey, however, was quite helpful in setting standards in the following way:

A *10% Compression* can be expected to end up with a 5% permanent partial physical impairment and a medium work restriction.

A *25% Compression* can be expected to end up with a 10% permanent partial physical impairment and a light work restriction.

659

A *50% Compression* can be expected to end up with a 15% permanent partial physical impairment and a light work restriction.

A *75% Compression* can be expected to end up with a 20% permanent partial physical impairment and a sedentary work restriction.

A *Transverse Process Fracture* should be expected to heal without residual physical impairment, whether displaced or not.

The only remaining difficult evaluation problem relates to the cross over effect between the lower limbs and the spine, and the difficulty in these situations is compounded by the fact that low back pain is such an universal ailment. Since about 80% of industrialized society has low back trouble at one time or another, it is often difficult to either establish or rule out any possible relationship of a short leg to subsequent backache; or on the other hand, the role of a back ailment in the later development of chondromalacia patella from protective squatting. Long experience with such claims, however, strongly suggests that they are, by and large, far fetched and too conjectural for credibility.

BIBLIOGRAPHY

1. Anonymous. Social Security Rulings. Title 20 — Employees' Benefits. 404.1567 — Physical Exertion Requirements.
2. Brand, R. A. and Lehmann, T.R. Low-back impairment rating practices of orthopaedic surgeons. *Spine,* 8:75-78, 1983.
3. Hackley, Susan. Workers' Compensation Administrator — State of California. Personal Communication. July 1, 1980.
4. Ziporyn, Terra. Disability evaluation — a fledgling science? *JAMA,* 250:873-4 — 879-80, 1983.

TABLE I

RESIDUAL OCCUPATIONAL ACCESS

(Model Computation)

Claimant:	John Doe	Disability:	Two laminectomies; 20 percent	
Age	42		body as a whole functional	
Education:	High School		rating	
Work History:	Carpenter	Restrictions:	Lifting 20 pounds	
	21 years			

Occupational Groups	Percent of State Labor Force	Pre-Injury Access	Post-Injury Access	Difference
Professional & Technical	12.19	0.0	0.0	0.0
Managers & Administrators	11.82	0.01	0.01	0.0
Sales Workers	5.60	3.65	3.31	0.34
Clerical Workers	16.09	5.80	5.38	0.42
Crafts & Kindred Workers	12.80	1.32	0.10	1.22
Operatives	12.90	9.56	5.21	4.35
Service Workers	14.70	2.20	1.37	0.83
Laborers, except Farm	4.80	4.70	0.01	4.69
Farmers & Farm Workers	9.10	0.02	0.00	0.02
	100.0	27.26	15.39	11.87
		100.0%	56%	44%

Residual Occupational Access		56%
Loss of Occupational Access	11.87	44%
	27.26	

This individual would have had access to 27.26% of the jobs in his state before the injury. Following the injury, he has access to 15.39% of those jobs. This constitutes a loss of access of 44%.

TABLE II

INDUSTRIAL BACK INJURY
WORK RESTRICTION CLASSIFICATION

(As adapted from Social Security Regulations)

Work Classification	Work Restrictions	PPPI	Relevant Diagnoses
VERY HEAVY WORK	Occasional lifting in excess of 100 pounds Frequent lifting of 50 pounds or more	Zero	Recovered acute back strain Herniated nucleus pulposus treated conservatively with complete recovery
HEAVY WORK	Occasional lifting of 100 pounds Frequent lifting of up to 50 pounds	Zero	Healed acute traumatic spondylolisthesis Healed transverse process fracture
MEDIUM WORK	Occasional lifting of 50 pounds Frequent lifting of 25 pounds	Less than 5%	Chronic back strain Degenerative lumbar intervertebral disc disease under reasonable control Herniated nucleus pulposus treated by surgical discectomy and completely recovered Spondylolysis/spondylolisthesis under reasonable control Healed compression fracture with 10% residual loss of vertebral height

TABLE II (Cont'd)

Work Classification	Work Restrictions	PPPI	Relevant Diagnoses
LIGHT WORK	Occasional lifting of no more than 20 pounds Frequent lifting of up to 10 pounds	10% to 15%	Degenerative lumbar intervertebral disc disease with chronic pain and restriction Herniated nucleus pulposus treated conservatively or operatively, but left with some discomfort, restriction, and neurological deficit Acute traumatic spondylolysis/spondylolisthesis, treated conservatively or operatively, but with residual discomfort and restriction Lumbar canal stenosis Moderately severe osteoarthritis accompanied by instability Healed compression fracture with 25% to 50% residual loss of vertebral height
SEDENTARY WORK	Occasional lifting of 10 pounds Frequent lifting of no more than lightweight articles and dockets	20% to 25%	Multiply operated back (failed back syndrome)

TABLE III

DIAGNOSIS	% PPPI	WORK PERMITTED
Acute Back Sprain — complete recovery	0	(very) Heavy Work
Chronic Back Sprain	5	Medium Work
Degenerated Disc — superimposed sprain		
Complete recovery	0	Heavy Work
Acceptable level of discomfort and restriction	5	Medium Work
Chronic pain and restriction	10-15	Light Work
Herniated Nucleus Pulposus — conservative care		
Complete recovery	0	Heavy Work
Acceptable level of discomfort and restriction	10	Light Work
Herniated Nucleus Pulposus — surgical discectomy		
Complete recovery	5	Medium Work
Acceptable level of discomfort	10	Light Work
Pain and restriction without neurological deficit	15	Light Work
Pain and restriction with neurological deficit (failed back syndrome)	25	Sedentary Work
Acute Traumatic Spondylolysis/Spondylolisthesis		
Conservative care		
Complete recovery	0	Heavy Work
Residual discomfort	10	Light Work
Preexisting Asymptomatic Spondylolysis/ Spondylolisthesis		
Superimposed sprain		
Conservative care		
Complete recovery	0	Heavy Work
Chronic pain	10	Light Work
Recurrent pain	10	Light Work
Preexisting Asymptomatic Spondylolysis/ Spondylolisthesis		
Superimposed sprain		
Laminectomy and/or fusion		
Complete recovery	10	Light Work
Chronic or recurrent pain	20	Sedentary Work

TABLE III (Cont'd)

DIAGNOSIS	% PPPI	WORK PERMITTED
Preexisting Lumbar Canal Stenosis		
Superimposed sprain		
Conservative care		
Status quo	0	Heavy Work
Subjectively worse	10	Light Work
Subjectively and objectively worse	15	Light Work
Preexisting Lumbar Canal Stenosis		
Superimposed sprain		
Decompression with or without fusion		
Status quo	5	Medium Work
Subjectively worse	10	Light Work
Subjectively and objectively worse	20	Sedentary Work
Preexisting Osteoarthritis		
Acute back sprain		
Conservative care		
Status quo	0	Heavy Work
Subjectively worse	10	Light Work
Subjectively and objectively worse	20	Sedentary Work
Acute Back Sprain or Herniated Nucleus Pulposus		
Conservative or surgical care		
No objective residuals		
Confirmed neurosis	5	Medium Work
Compression Fracture — healed with:		
10% compression	5	Medium Work
25%	10	Light Work
50%	20	Sedentary Work
75%	20	Sedentary Work
Transverse Process Fracture — healed with:		
No displacement	0	Heavy Work
Malunited	0	Heavy Work

CHAPTER 10

PERSPECTIVES

§ 10-1. The Federal Government's Perspective.

James L. DeMarce

Since there is no Federal Bad Backs Benefits Act nor any Office of Musculoskeletal Disabilities (Occupational) to administer it, it is tempting to say that there is no comprehensive single federal perspective on the topic of work related back injuries. Although such an approach would be technically correct, such a statement would be both misleading and essentially inaccurate. In fact, a remarkable number of federal agencies play a variety of roles in the four major facets of the problem: prevention, treatment, compensation, and rehabilitation.

The essence of the federal position on work related back injuries and the loss of human resources that flow from them is rather like Calvin Coolidge's summary of the preacher's sermon on sin: "He's against it." That opposition is expressed through programs designed to reduce the incidence of such injuries, secure the prompt availability of appropriate medical care when they occur, the provision of income maintenance during any ensuing period of disability, and the minimization of such disability through the encouragement of rehabilitation and reemployment of the injured worker.

Injury prevention through the development and enforcement of health and safety standards is the responsibility of the Occupational Safety and Health Administration, and for those industries over which it has jurisdiction (primarily mineral and coal mining and preparation facilities), the Mine Safety and Health Administration. Since this volume is concerned with those cases where injury prevention was unsuccessful, the role of those agencies will not be further pursued here. However, it must be borne in mind that minimization of the risk of re-injury is an integral part of the process of treatment, compensation, and rehabilitation of injured workers and it will be discussed in relationship to those stages in the process.

The role of the federal government in the treatment, compensation, and rehabilitation of work-related back injuries, as indeed with regard to all such disabling conditions, is divided into essentially two types. The first type includes those classes of persons for whom the federal government has some form of special responsibility and the second applies to the population of the United States generally.

The largest group of persons for which the federal government has such a special responsibility is its own civilian employees. This group (together with various volunteers and other related individuals) currently includes approximately 2.8 million persons. They are covered by the Federal Employees' Compensation Act (FECA)[1] which defines the obligations of the federal government to them in the event of on-the-job injury or the contraction of an occupational disease while in the course of federal employment. Although the FECA is often perceived by federal employees as being a part of the package of fringe benefits applicable to federal employment, it is in fact the functional equivalent of a state workers' compensation statute.

The FECA has existed since 1916 and was most recently the subject of major amendments in 1974. It includes the full range of treatment, compensation and rehabilitation services provided by most state workers' compensation laws. The benefits provided, however, are in general more generous than those provided by most state workers' compensation programs. The procedures for the determination

---

1. 5 U.S.C. 8101 *et seq.*

of eligibility for benefits are administrative and non-adversarial. The basic philosophy of the Act is summarized in section 8103(a) which defines the injured employee's right to receive medical treatment. It states:

> The United States shall furnish to an employee who is injured while in the performance of duty, the services, appliances, and supplies prescribed or recommended by a qualified physician, which the Secretary of Labor considers likely to cure, give relief, reduce the degree or the period of disability, or aid in lessening the amount of the monthly compensation.

In the typical case, the physician will first encounter the injured worker clutching a form known as the CA-16 in his or her hand.[2] That form serves the multiple purposes of authorizing initial treatment, guaranteeing federal payment for the treatment, and requiring the submission of a report. This reporting requirement is extremely important in the development of a claim for benefits under the FECA.

Although the FECA is non-adversarial in nature, to establish entitlement to benefits beyond the initial treatment provided for by the CA-16, the injured worker's claim file must contain documentation of the fact of injury,

---

2. The example in the text involves a situation in which the worker's back problem is being handled as a traumatic injury. In fact, under the FECA, a back problem may be handled as either a traumatic injury or an occupational disease condition. The distinction between the two approaches is made upon whether the problem is reported as the result of a single traumatic episode or in the course of the gradual worsening of a more generalized condition without the presence of a single identifiable triggering event. The label chosen affects the forms and procedures used to process the claim. However, both conditions are compensable if they involve employment caused disability. The general admonitions concerning the need for prompt, comprehensive and clearly explained medical reporting are equally or even more applicable in the case of occupational disease claims.

its relationship to the federal employment, and the extent and duration of any disability attributed to that injury. Thus, to properly serve such a patient, the physician's obligation is not limited to providing the best available medical care designed to maximize recovery. The physician must also provide clearly stated and explained information concerning the cause of the injury and any resulting disability.[3] Causation here is to be understood not only in the direct sense but also includes whether a specific traumatic event or ongoing factors in the work environment have aggravated, accelerated or exacerbated a pre-existing condition and thus created a greater impairment of physical function than would otherwise have been present. The physician must also assess the expected length of the expected recovery process and describe any limitations on the worker's activities which will be applicable during it.

Basic to the FECA and most injury disability compensation programs are two interrelated concepts: impairment and disability. Impairment refers to a loss of a physical

---

3. The narrative describes the typical situation in which the treating physician's reports will be the primary source of information on the injured worker's condition available to the personnel responsible for adjudicating the worker's claim for benefits under the FECA. However, the treating physician should be aware that those reports are reviewed not only by claims examiners. They may also be reviewed and analyzed by district office medical personnel or consultants. In appropriate cases, the claimant's file and/or the claimant may be referred to a board certified specialist for examination to determine the efficacy of the treatment being received, to resolve conflicting reports received when the claimant is being treated by more than one physician, or to provide information not available from the treating physician's reports. The practice of early referral of injured workers to such specialists for diagnosis and development of courses of treatment designed to minimize long term disabilities was previously pioneered under the FECA as part of a Low Back Project which is now being expanded to include other types of musculoskeletal injuries.

function or capability, for instance, a partial loss in the range of motion in a joint or a situation in which repeated heavy lifting would be contraindicated. Disability refers to a loss of earning capacity which temporarily results from an injury or occupational disease as well as the permanant residual impairments which remain after maximal medical improvement or recovery has been achieved. Impairment is essentially a medical evaluation of a physical function which is usually calculated as a percentage or class of loss of function according to procedures spelled out in the AMA *Guides to the Evaluation of Physical Impairments* or some similar formula. Disability, however, is essentially a legal conclusion about the economic impact of an impairment upon the injured person's ability to earn in the pre-injury job or some other form of employment.

Although a workers' compensation act, such as the FECA, pays for the costs of all necessary injury or occupational disease related medical care, it does not pay for any related pain or suffering borne by the injured worker. The only case in which payments are made to the worker (other than reimbursement for self-procured medical care) are disability payments. These function as income replacement for wages lost due to the injury or occupational disease. They may be either total or partial, temporary or permanent. The interplay of the concepts of impairment and disability may sometimes seem to produce anomalous results. For instance, the loss (or even the loss of use) of a single finger may be totally and permanently disabling to a worker whose earnings are totally dependent upon a high level of manual dexterity, such as a concert pianist, who has no other saleable skills; whereas the loss of an arm or leg may have no continuing economic (that is to say, disabling) impact upon an administrator or an attorney once the initial recovery period has been completed.

The FECA provides relatively generous disability benefits. Normal pay is continued for up to 45 days after the occurrence of a disabling injury. If disability continues beyond that period, tax free disability benefits equal to 66-2/3% of pre-injury earnings are provided (75% of pre-injury earnings, if the injured worker has one or more dependents). In the event of long-term disability, benefits are subject to periodic cost-of-living adjustments.

Since the continuation of disability payments is dependent upon the continuation of the employment-caused disability, periodic reporting by the treating physician on the course of the patient's recovery is crucial to accurate benefit payment. It is important to report not only the point in time when recovery is sufficiently complete that the worker can be released to perform the full range of pre-injury duties, but also the stages at which the worker is able to return to work on a parttime or limited duty basis. A premature return to work may result in prolonging the period required for full recovery or increase the risk of re-injury. Conversely, delayed return to work may give rise to psychological complications as well as increase the difficulty of reintegrating the worker into the workplace.

The FECA imposes a duty upon the injured worker to return to work or otherwise exercise such earning capacity as remains after the recovery process has been completed. In those situations where only a partial recovery is possible and the worker will never be able to return to even a modified version of the pre-injury employment, the treating physician's inventory of the continuing limitations upon the worker's future activities as well as his or her remaining capabilities constitute the starting point for more formal efforts at rehabilitation in the form of retraining and/or alternate work placement or the development of remunerative self-employment skills.

Although most other aspects of the FECA are administered by the Division of Federal Employees' Compensation

672

(DFEC) of the Office of Workers' Compensation Programs (OWCP),[4] the evaluation of a case for the appropriateness of formal rehabilitative services is provided by personnel in the Division of Vocational Rehabilitation (DVR), another component of OWCP. Where such services are found to be appropriate and necessary to facilitate the worker's return to productive activity, the needed services are procured and their performance supervised by DVR personnel at federal expense. The actual testing, training, and placement services, however, are normally performed by state vocational rehabilitation services or increasingly by private vocational rehabilitation firms.

Although the adjudication of claims for benefits under the FECA is non-adversarial in nature, that does not mean that all claims are automatically approved. Any claim which is denied in whole or in part can be further pursued either by requesting reconsideration of the denial or requesting a hearing on the disputed issues or making an appeal to the Employees' Compensation Appeals Board (ECAB). The first two alternatives are ruled upon by DFEC employees; however, the ECAB is an administrative appeals board whose members are directly appointed by the Secretary of Labor and which is organizationally independent of the adminis-

---

4. In addition to the national office and two claims adjudication offices in Washington, D.C., the Division of Federal Employees' Compensation maintains district offices in Boston, New York, Philadelphia, Jacksonville, Chicago, Cleveland, Dallas, New Orleans, Kansas City, Denver, San Francisco, Honolulu and Seattle. Any physician or attorney desiring further information concerning the program can contact the nearest office or submit a request to the Technical Assistance Section, Division of Federal Employees' Compensation, Office of Workers' Compensation Programs, Washington, D.C. 20210. The Technical Assistance Section will provide a package of informational materials answering questions most frequently asked about the program, copies of frequently utilized forms, and the text of the statute and its implementing regulations.

tering agency. The decisions of the ECAB are published and are not subject to judicial review. A claimant may obtain the assistance of counsel in any stage of the proceedings concerning a claim under the FECA; however, less than 1% of all claims involve assistance of counsel. The claimant is liable for the cost of such legal services and no request for an attorney's fee for services related to a claim under the FECA is valid without administrative approval.

The second major group of injured workers for whom the federal government has a special responsibility are those persons whom the federal constitution precludes from coverage under state workers' compensation acts. A series of decisions by the United States Supreme Court that state workers' compensation laws could not be applied to injuries occurring upon the navigable waters of the United States because the admiralty power was exclusively reserved to the federal government led to the passage of the Longshoremens' and Harbor Workers' Compensation Act in 1927.[5] It incorporated most of the substantive provisions of the New York state workers' compensation law, which was regarded as the model state act at that time. It created a federally administered system for the treatment and compensation of workers injured in the course of private employment while working at longshoring and shipbuilding activities over navigable waters.

In 1928, Congress enacted a "temporary" extension of the Longshore Act to the District of Columbia. This temporary extension of a federally administered workers' compensation act served as the functional equivalent of a state workers' compensation act until the District of Columbia assumed responsibility for the administration of private on-the-job injuries under "home rule" in the early 1980s. Subsequent extensions of the Longshore Act extended its

---

5. 33 U.S.C. 901 *et seq.*

674

benefits to other workers who were largely outside the coverage of state workers' compensation acts, such as civilian employees of contractors operating at foreign defense bases, employees at PXs and NCO clubs, and workers employed in petroleum extraction on the outer Continental Shelf. The most recent major amendments to the Longshore Act were enacted in 1972. They increased the benefits available and extended its coverage to include shoreside areas "customarily used by an employer in loading, unloading, repairing, or building a vessel." [6]

In its current form, the Longshore Act provides treatment, disability and rehabilitation benefits for several hundred thousand workers scattered throughout the United States and around the world.[7] Although the Longshore Act generally provides benefits roughly comparable to those available under the FECA, there are a number of significant differences between the two federally administered workers' compensation programs.

Unlike the FECA, most medical and income replacement benefits under the Longshore Act are provided by private

6. As of this writing (early May 1984) the Senate and House of Representatives have each passed major amendments to the Longshore Act. These bills address virtually every major aspect of the law and contain significantly different solutions to the perceived need for its reform. If the two branches of Congress are able to compromise these differences and enact a single bill, the result may significantly alter the current version of the Longshore Act described in the narrative.

7. Of course, the major concentrations of workers covered by the Longshore Act are in the major American port cities. In addition to its national office in Washington, D.C., the Division of Longshore and Harbor Workers' Compensation, which is responsible for the administration of the statute, maintains district offices in Boston, New York, Philadelphia, Baltimore, Norfolk, Jacksonville, Chicago, New Orleans, Houston, Galveston, San Francisco, Long Beach, Honolulu and Seattle. Detailed information concerning the basic act and its extensions is available from any of these offices.

675

resources (either by self-insured employers or commercial workers' compensation insurance companies). As under the FECA, the injured worker has an initial choice of treating physician; however, the injured worker must also submit to examinations by physicians of the employer's choice. Comparable reporting requirements are applicable to the physician's treatment and analysis of the injured worker's progress toward recovery. The treating physician should be aware that most disputed cases under the Longshore Act revolve around conflicting assessments of the rate and extent of the worker's medical recovery as viewed by the treating physician on the one side and the employer's examining physician on the other. The administrators of the law may make use of impartial examining physicians in an effort to resolve such disputes; however, claims involving extended total disability frequently must be referred to an administrative law judge for resolution.

Although the overwhelming majority (approximately 95%) of all injuries and related claims for benefits under the Longshore Act are resolved amicably either directly by the parties or through the mediating efforts of administrative personnel, the basic structure of the statute is an adversarial one. Whether the parties are preparing for a formal contest of a disputed case or merely maneuvering for advantage in settlement negotiations, the tendency of each side is to seek medical and other expert testimony most favorable to its position.

As under the FECA, there are no arbitrary limits upon either the medical or income replacement (disability) benefits available under the Longshore Act. Total disability entitles the injured worker to 66-2/3% of pre-injury earnings tax free up to a maximum of 200% of the national average weekly wage. Permanent disability benefit levels are adjusted annually to reflect changes in the national average weekly wage. In the case of permanent disability, the Longshore Act also provides for survivor's benefits regardless of the cause of the worker's death.

As indicated, when efforts at the informal resolution of a disputed claim are unsuccessful, the case is referred to the Office of Administrative Law Judges for a hearing. The administrative law judge's decision is appealable to the Benefits Review Board. Judicial review of disputed decisions is available in the United States Courts of Appeals and ultimately in the Supreme Court of the United States. A claimant for benefits under the Longshore Act may retain the assistance of counsel at any stage in the process of the resolution of a disputed claim. If the attorney's assistance results in the employer's agreeing or being ordered to provide benefits previously denied, the attorney's fees and related costs must also be paid by the employer. Any fee must be approved by the administrative agency or court before which the attorney practiced. If the attorney is unsuccessful, no fee is payable.

Although the Longshore Act is adversarial, the burden of proof imposed upon the claimant is frequently less than under the FECA. Thus, in difficult cases, such as occupational diseases, stress, heart attack or reported disabling pain not corroborated by objective findings, a claimant under the Longshore Act is more likely to prevail than under the FECA. Similarly, under the Longshore Act once the claimant has established an inability to return to the pre-injury employment, the burden of proof shifts to the employer to show that other employment exists which the worker is capable of performing in his impaired condition. Failure to make such a showing results in the entry of a total disability award against the employer.

Vocation rehabilitation services are also available under the Longshore Act. They may be provided either by the employer in an effort to reduce its liability for long-term disability benefits or, as under the FECA, by personnel in

677

OWCP's Division of Vocational Rehabilitation. However, unlike the FECA, there is no obligation upon the injured worker under the Longshore Act to participate in vocational rehabilitation and benefits may not be withheld for failure to do so.

Not all claims under the FECA or the Longshore Act result in the award of benefits or the claimant may only be found entitled after extended evidentiary development or protracted appeals. The physician or attorney representing such a denied claimant may need to consider other means to meet the injured worker's needs for medical care [8] and/or income. A federal employee with at least five years of civilian service who is unable to perform his or her pre-injury work duties and is not *currently* receiving FECA benefits may be eligible for disability benefits administered by the Office of Personnel Management. Similarly, if the injured worker is a veteran and the disabling condition is related to a pre-existing "service connected" disability, increased benefits and medical care may be available from the Veterans Administration.

The preceding discussion summarizes the benefits available to disabled workers for whom the federal government recognizes some form of special responsibility. The disabled worker who is unable to qualify for any of those benefits may nonetheless be eligible for the income maintenance, medical care, and rehabilitation benefits which the United States provides for all eligible persons in its population.

For this second tier of benefits, administered by the Department of Health and Human Services, the eligibility

---

8. Unfortunately, once the worker has claimed that the cause of the disabling condition is work related, general health care insurers, such as Blue Cross/Blue Shield, are unlikely to accept responsibility for any related medical expenses until any claim for workers' compensation benefits has been definitively denied or abandoned.

criteria are more rigorous and the available benefits less generous than under the FECA and the Longshore Act. Depending upon the length of time the disabled worker has been covered by the Social Security system and other factors based on need, the disabled worker may be eligible for either Title II Social Security Disability Income benefits or Title XVI Social Security Supplemental Security Income benefits. The test of disability is the same for both programs. The claimant must be unable to engage in substantial gainful activity due to an objectively medically determinable physical or mental impairment, or combination of impairments, which has lasted, or is expected to last, at least 12 months or may result in death. A determination of eligibility benefits under either of these disability programs normally brings with it access to medical care under either the Medicare or Medicaid systems. Any disabled person may seek retraining and job placement services through the use of the local state vocational rehabilitation services. Such local services are partially federally funded. Lastly, the disabled person's job search efforts enjoy the protections of section 504 of the Rehabilitation Act of 1973 which forbids discrimination on the basis of a physical or mental handicap.

The future of this rather complex two-tiered system, containing on the one hand relatively generous benefits available to those disabled workers for whom the federal government recognizes some special obligation and on the other hand minimal benefits for all other disabled workers who are devoid of any other state or private sources of support, is unclear. Two conflicting approaches have been applied to the situation. The first contends that the federal government should be a leader in both areas. It contends that programs for federal beneficiaries should be model programs which comparable state or private programs should seek to emulate. This approach led to the liberalizing amendments to the Longshore Act in 1972 and the FECA in 1974. It also inclines toward the belief that the second tier or "safety net" level of benefits should be steadily expanded to cover more classes of beneficiaries and benefits. In general, this approach tends toward a resolution of doubtful cases in favor of the claimant in the interest of avoiding hardship.

The second approach begins with the assumption that the federal government's responsibilities are essentially those of any governmental employer. It contends that federal benefits should be adequate and equitable as defined in terms of comparability with what other large scale employers, such as state governments or major corporations, provide. With regard to the second tier of benefits, it tends to focus upon the Social Security system as a payer of last resort only after all alternate sources of income maintenance and/or medical care have been exhausted or at least offset against any remaining federal liability. It also focuses upon a general tightening up of the administration of all benefits programs. This philosophy largely accounts for recent proposed changes in the Longshore Act, FECA, and Social Security Act benefit systems. Which approach ultimately prevails is essentially determined by a political response to the perceived social and financial costs associated with the two alternatives.

§ 10-2. Managing Back Injury Claims: The Adjuster's Perspective.

Kevin M. Quinley

Back injury claims pose a major challenge to the insurance industry and produce substantial headaches to the average claims adjuster. Good reasons exist for this state of affairs. To a degree, the claims industry mirrors society, and back pain is a societal problem. Eight million people in the United States suffer from some back impairment. Half a million of these cases are caused by on-the-job injury (entry 3 in Bibliography at end of § 10-2(C)). Almost 700,000 such injuries result from a one-time lifting or exertion. Focusing on *low* back pain only, one finds that this problem disables about five million Americans. Sufferers of low back pain pay 1.5 billion dollars per year in hospital and doctor bills, including 200,000 surgeries and almost 19 million doctor visits. Small wonder, therefore, is the conclusion that, "Pain compensation is a national disaster" (entry 5).

Pain compensation has also grown into something of a mega-business. Statistics alone fail to convey the enormity of the problem. Among those handling claims — and particularly workers compensation cases — it is an open secret that back injury cases are the "toughest nuts to crack." Shoptalk among claims adjusters is typically replete with horror stories of back injury claims that just refused all attempts at resolution. For the claimsman, back injury cases are, compared to other types of claims, harder to disprove, more expensive, more prolonged, and more impervious to final closing. One large workers' compensation carrier estimated that long-term low back claims made up only about 10% of its case load, but comprised 90% of the money paid out. As one claims supervisor added, "They also make up 90% of my headaches" (entry 1).

Notwithstanding medical advances, epidemic back pain and injury show no signs of abating. Those insurance carriers and employers which implement rational and coordinated management of back injury cases will be at a competitive advantage in terms of lessening their overall claims expenditures. An emotional reaction of "the system is out to get me" on the part of the claims industry is both understandable and unwise. Avenues are available to the claims adjuster in the quest to manage back injury cases.

Meaningful discussion of back claims from an adjuster's

perspective must include the following questions: What is meant by "managing" back injury claims? What are the goals with regard to managing these claims? What specific strategies may carriers pursue in trying to attain the goals? What are the strengths and weaknesses of these strategies? These are the issues under discussion. The term "carrier" is used as shorthand to include insurance companies, self-insureds and employers. Our focus is confined to *workers compensation* back injury claims for two reasons. First, most back injury claims are incurred while on the job. Second, this writer feels there is very little the adjuster can do nowadays to manage back injury claims in a third-party or tort liability context.

If all is not lost, therefore, what can we say to the harried adjuster who, from his bunker of claim files, cries for help?

§ 10-2(A). Aims of "Back Claim" Management.

What do we seek to do when we try to "manage" back injury claims? What is meant by managing such claims? For the claimsman, the goal of managing back injury claims is primarily to minimize the carrier's exposure. Expressed in its most mundane fashion, this means minimizing the amount of money paid out on back injury claims, and being able to close files. Adjusters are evaluated in part according to how vigilant they are in resisting groundless and exaggerating claims, and how fast they "turn files over."

Wise management of back injury claims should ultimately reduce the amount of money which carriers spend on such cases. To achieve this state of affairs by means reasonable and lawful is the aim of claims management. This can be stated forcefully and without any trace of defensiveness or apology. Policyholders' premiums should not be spent on unmeritorious or built-up claims. Despite their humanitarian aims, workers compensation statutes are not aimed at providing benefits for all injuries, all risks, in all circumstances. Someone must ensure that workers compensation benefits flow to those claimants that such statutes are intended to cover. In this sense, the claimsman represents the public's (and, particularly, industry's) first line of defense against fraudulent or exaggerated claims.

The claims industry must not lose sight of the goal of minimizing claims pay-out on back injury cases. This absolutely legitimate goal should not be confused with a host of related goals which are more appropriately within the bailiwick of other professions. There exist many goals related to the aim of minimizing the carrier's exposure. If attained, these other goals will certainly assist the claims adjuster in his quest, and will be welcome by-products. What are some of these related aims?

Curing the claimant of his back problem is something the medical industry may be able to accomplish. Whether the cure results from surgery or conservative therapies matters less to the claimsman, so long as a cure is effected. Eradication of a claimant's back pain means, for the claimsman, that the claimant is no longer disabled and may no longer be receiving disability benefits. Curing the claimant is a medical goal, not one within the adjuster's power to attain. It may also be an elusive goal, since back problems tend to be recurring conditions of life, rarely amenable to a final solution.

Short of a complete cure, a claimant may receive a release from his doctor to return to work. One may be physically capable of returning to work without being cured in a full and final sense. The distinction makes less difference to the claimsman, however, since a doctor's release to return to duty typically heralds the cessation of disability payments. It may not hearken any end to continued medical treatment, however, and this is why a cure of a back problem is preferable (and rarer) in the claimsman's eyes than a return-to-duty slip. Still, the timing of a release to return to duty remains essentially a medical question, not within the adjuster's realm of control.

What of those cases where it seems unlikely that the treating physician will ever release the claimant to return to work, much less provide a cure? Many back injury cases fall into this category, hence the dread that such cases often hold for adjusters. Worse still, there may be direct medical opinion from the treating doctor that the claimant will *never* be able to return to regular duty. If the adjuster is reluctant to accept this conclusion or has reason to question the doctor's opinion on this point, it behooves the claimsman to seek alternate medical opinion that the claimant is (or soon will be) physically capable of returning to work. At best, however, this presents conflicting medical testimony and the ultimate weighing of the opinions is not within the adjuster's province. Even in seeking a second medical opinion regarding the claimant's disability, the adjuster has no control over what type of conclusion is reached.

Close management of back injury claims is also enhanced if the claimsman can gather *non-medical* evidence of a claimant's ability to return to work. Non-medical means something other than a doctor's opinion or a medical report. Armed with such evidence, the claimsman may be able to successfully deny a new claim, or bring an old one to a decisive conclusion. Common examples of such non-medical evidence would include: evidence that the claimant is working at a different but equally strenuous job while collecting disability benefits; proof that the claimant's physical exertions in a non-job setting undermine subjective complaints made in the doctor's office or before the hearing examiner. This type of substantiation often depends on pure luck — being in the right place at the right time — as much as the adjuster's tenacity in conducting a good *sub rosa* investigation.

At some point, though, the adjuster must concede the unlikelihood of being able to place the claimant back to his old job. Here is where the concept of limited/selective employment/light work comes into play. Adjusters hope that the insured/employer will be able to offer the claimant work within medical restrictions. If so, the adjuster may be able to cease disability payments altogether, or at least drop them down to some smaller proportion. If not, the adjuster must look elsewhere for other strategies and in the interim continue indemnity benefits to the claimant. Some employers recognize the economic virtue of reducing their compensation exposure and virtually invent a job to conform with a claimant's medical restrictions. The maintenance man who is given clerical work is one example. At the other extreme, some employers resist taking back an injured worker so long as there are *any* restrictions. This presents a challenge to the adjuster in educating employers on how they may ultimately save themselves money by restructuring jobs for employees who are not "100%" recovered.

Resistance runs high in many quarters, however, due to various factors such as union rules and fears that claimants will re-injure themselves. If an employer cannot be persuaded to provide modified duty, then the adjuster must look outside of the organization where the claimant worked when the accident occurred for job placement. The goal is to find the claimant alternate employment, using transferable skills, elsewhere in the marketplace. An alternative would be retraining the claimant for a job requiring new and different skills. Both areas are appropriately the venue of the rehabilitation specialist and not of the claims adjuster. To the extent the adjuster can provide suitable alternate employment, either using transferable skills or enhanced

685

skills through "classic" rehab, the adjuster may succeed in
eliminating or at least reducing a carrier's workers compensation exposure.

Persistent investigation may reveal that the claimant's
back injury is not job-related. This affords the adjuster a
legal defense in denying the back injury claim *in toto* — lost
time and medical benefits. To the extent the adjuster with
investigative skills amasses such documentation —
through statements of witnesses, medical records, etc. —
the employer's workers compensation exposure is thereby
reduced. Claimsmen in the field know, however, that a successful denial of a back injury claim is a rarity. In part, this
is due to statutory presumptions built into compensation
laws, liberal interpretation of such laws in the claimant's
favor, and the difficulty in proving that someone did *not*
sustain a back injury. Given the subjectivity of the ailment,
it is well-nigh impossible to prove that someone did not
injure his back on the job.

Settlement of the back injury claim is a related aim which
may aid in the overriding goal of managing back injury
cases. Unlike most tort liability claims, workers compensation claims are not usually "settled" in a full and final
sense. Compensation boards closely scrutinize proposed
settlement agreements and will not hesitate to reject them
if they feel the claimant's interests have not been sufficiently protected. Many compensation statutes do not allow
percentage permanency ratings on the back to be
automatically translated into some type of permanency
award. Frequently, injuries involving so-called "scheduled"
parts of the body (arms, legs, and fingers for instance) can
be resolved in a formulaic manner once everyone agrees on
the percentage of permanent disability. Such leeway is
often denied the claimsman on back claims, making more
attractive the option of full and final settlement versus a
"permanency stipulation." In some states, so-called Special

or Second Injury Funds are inapplicable to back claims, the very cases where the Funds are (from the adjuster's viewpoint at least) most needed. Small wonder, then, that adjusters feel that the proverbial light at the end of the tunnel is merely an on-rushing train.

Limiting the carrier's exposure remains the "bottom line" goal in managing back injury claims. To the extent that certain medical and humanitarian goals are also achieved as a result of these claims efforts, so much the better. But these other goals are merely means to an end, insofar as the adjuster's role is concerned. These other goals are often outside the adjuster's realm of influence. They point to the use of certain strategies to insure that benefits flow to those claimants deserving of such benefits, within the language and intent of the applicable workers compensation law. From a public policy standpoint, no useful purpose is served when unmeritorious or exaggerated back injury claims are paid. What, then, are the chief strategies employed by adjusters in managing back injury claims?

§ 10-2(B). Strategies for Managing Back Injury Claims.

§ 10-2(B)(1). The Investigative Phase.

Effective management of back claims begins at the stage of initial investigation. Through persistent and meticulous investigation, the claimsman may be able to show that the back injury alleged by the claimant was not job-related, was not accident-related, was pre-existing, or is not of sufficient severity to warrant disability. Back pain is for thousands of individuals a condition of life, and workers compensation provides a tempting avenue for an expense-paid trip through a rather liberal benefit system. A good foundation laid at the investigative stage may avoid many problems and headaches later. Though still difficult, successfully denying a back injury claim at its outset is much easier than trying to contain a claim whose costs have later snowballed. Certain investigative avenues, if used wisely, may enhance the adjuster's ability to manage back injury claims: (1) The claimant's statement; (2) Scrutiny of medical records; (3) Researching the claimant's medical history; (4) Researching the claimant's work history.

Ironically, it may be the claimant's own statement which ultimately affords the grounds for successfully defending a back injury claim. Several thorough checklists have been developed for use in workers compensation cases, and will not be duplicated here (entry 4). To determine whether a lifting incident for example qualifies as an accidental injury, the adjuster should broach questions pertaining to the mechanics of the lift, weight of objects lifted, how far objects were lifted or carried, whether the lifting occurrence was out of the ordinary or was performed every day, etc. Some jurisdictions require that compensable lifting incidents be accompanied by a slip, twist, fall, or unusual exertion. Point-blank questions to this effect should be posed. Covering activities — athletic or household chores, for instance — away from the job may also implicate non-occupational factors as the source of the back problem.

"Creep-itis" is adjuster slang for a situation in which the claimant's injury starts as a "simple" knee or shoulder strain, but then a few days (months or, in some cases, years) later works its way to the back. Adjusters are often frustrated when the injured limb appears to be on the verge of healing and the claimant announces that "my back has now started to hurt me too." Even acknowledging a dormant period between injury and symptoms, skepticism may be well-founded. Attempts may be made to "graft" an unrelated back injury upon what began as a shoulder strain case. To prevent "creep-itis" from rearing its ugly head, adjusters are encouraged to incorporate in the claimant's statement language to the effect that: "In this accident, I injured only the following parts of my body . . . ."

688

Much can be gleaned from close review of the medical records regarding the recent injury. Claimants may be more open and candid with a doctor or hospital personnel than they are with the claimsman. What kind of history did the claimant give the emergency room staff during the first treatment? Is there clear indication from the hospital chart that the back problem is job-related? Did the claimant make clear to all the treating doctors and facilities that the back pain was work-related? If not, this may raise a red flag and suggest closer investigation of the claim. Credibility questions regarding the claimant's testimony may also be raised.

Beyond the medical care received by the claimant around the time of the injury, the claimsman should also research the medical history. This includes but is not limited to incorporating in the claimant's statement information pertaining to prior accidents, injuries, hospitalizations, surgeries, names of family physicians, etc. If an adjuster goes to the trouble of meeting with the claimant for a statement, the adjuster should also at least obtain a signed medical authorization. This enables the adjuster later to research the claimant's medical history and to determine whether the back trouble is a pre-existing condition. If so, many compensation statutes allow apportionment of permanent partial disability awards based on pre-existing conditions. Further, many jurisdictions allow Special/Second Injury Fund relief if a combination of a prior disability with a recent injury produces a greater disability overall. The net effect of both types of provisions lies in reducing the ultimate expenditure made by the employer/carrier. This is part and parcel of managing back injury claims from the adjuster's viewpoint.

Work history should also not escape the adjuster's scrutiny. Names and addresses of prior employers should be collected. Job descriptions at each position should be delineated. Reasons for job changes, frequent job changes, and gaps in employment may hint at prior injuries, claims, or medical problems. Employment records can reveal that which claimants can conceal: group health claims, hospitalizations, prior on-the-job injuries, etc. These records may be compared with the claimant's own testimony to indicate a problem with credibility or truthfulness. Work histories also reflect a claimant's background and skills. These are relevant in at least two ways. First, they point toward concurrent/free-lance employment that a claimant might be performing while receiving workers compensation benefits. Second, the prior jobs may later be useful to the rehab counselor if a job is sought in light of a claimant's transferable skills. A thorough summary of work history incorporated into the claimant's statement is one of the building blocks for managing the back injury claim at the investigative stage.

§ 10-2(B)(2). Medical Management.

"Medical management" is a phrase first encountered in rehabilitation literature. It has now become well imbedded in the adjuster's lexicon. More often referred to than precisely defined, medical management involves the direction and coordination of the efforts of health care providers so that the interests of the patient and the carrier are met. Depending upon the provisions of the particular compensation statutes, medical management may aim to avoid undertreatment and overtreatment of the injured claimant. Undeniably, medical management is often a euphemism for making sure that health care providers do not use insurance carriers as "fair game" for all sorts of billing rates and practices. Adjusters frequently tell stories of doctors who openly admit that they send one bill to the patient and a separate/higher bill to the insurance company *for the same procedure.* Claimsmen can be forgiven if they suspect that the back pain industry and mega-business sometimes seems to declare open season on insurance company coffers. To combat the basis for this perception is one aim of medical management.

Part and parcel of medical management on back claims involves providing employers with lists of recommended physicians and facilities. These may be health care providers to whom employers refer their injured workers. In every locale there are known amongst the claims industry certain physicians who seem extremely claimant-oriented. "Liberal" in this context refers not to political affiliation but the apparent eagerness on the part of some physicians to indefinitely certify disability based solely on subjective complaints. Adjusters working "in the trenches" know who these physicians are: frequently the same doctor or group of doctors treat most of the patients who happen to be represented by counsel. Sometimes the injured claimant sees his attorney before he sees the doctor.

Similarly, any adjuster knows of local physicians who are objectively oriented and less inclined to carry along forever the claimant who comes in every other week and says, "Yup, my back still hurts." Managing back injury claims will be easier if the claimsman compiles a medical panel and provides this to employers. Communication with employers is the key. The panel should recommend more than one physician within each specialty. For back injury cases, the most frequently used specialties will be orthopaedics, neurology, and neurosurgery. Claimsmen should ensure that physicians on the panel are impartial, dispassionate and thoroughly qualified. The doctors should be board-certified, if possible. They should not be doctors known as "yes men" for the insurance industry, lest every opinion from such physicians be dismissed as the view of the "company doctor." Employers allowed such leeway should prominently post the names of recommended specialists for injured employees. Some jurisdictions allow employers to limit the claimant's choice of physicians to practitioners on the panel, if the panel is provided to the claimant at an early stage. Adjusters must take the initiative in educating employers as to the usefulness of this tool. What happens too often in the field, however, is that a panel is not given at an early stage to the claimant, who has already seen "Dr. Whiplash." The adjuster, upon spotting the report with Dr. Whiplash's letterhead, attempts to correct matters by belatedly imposing a panel and disclaiming responsibility for the disfavored doctor's bill. Belated attempts to retroactively impose a panel are futile, and underscore the role of the adjuster communicating with the employer at this pre-loss stage.

In medically managing the back injury claim, the claimsman should also watch for new methods of treatment applicable to the back. Holistic pain clinics have become the vogue of late, with an interdisciplinary approach drawing upon many different medical specialties (entry 2). Some parts of the claims industry view the pain center concept as a fad. Such reservations should not blind adjusters to the possibilities of seeking a creative medical approach to a stubborn low back problem. A standardized diagnostic protocol should be considered. If a disc problem is involved, the claimsman should be open to the possibility of an epidural injection as an alternative to the surgical laminectomy. The adjuster should know which practitioners and facilities in the vicinity use rational management and should be prepared to make recommendations if called upon to do so.

§ 10-2(B)(3). Independent Medical Examinations.

If wisely used, the independent medical examination remains one of the most effective strategies for the adjuster in managing the back injury claim. Not every case is a candidate for an independent medical examination. Costs of the exams must be weighed against the exposure involved. From an economic standpoint, it makes little sense to pay for a $350 IME if all that is at issue is a week's worth of temporary total disability. Medical specialties most frequently used for IME's are orthopaedics, neurosurgery, and neurology. Timing is very important. An IME scheduled either too early or too late in the life of a claimant may be worthless.

What does the adjuster seek from an IME in trying to manage the back injury claim? The following types of issues are frequently addressed by the IME physician: (1) Whether the claimant can, from an objective medical standpoint, return to his old job; (2) Whether the claimant could perform some type of limited/modified/light work, i.e., what are the claimant's physical restrictions and capabilities?; (3) Does the claimant have any permanent partial disability and, if so, what percentage?; (4) Whether any pre-existing or unrelated medical problems contribute to the claimant's back trouble; (5) Whether in all medical probability the claimant's back condition is causally related to a specific on-the-job injury; (6) Whether the frequency and type of medical treatment rendered heretofore has been reasonable and appropriate; (7) Whether prospective medical treatment — surgery and hospitalizations, for example — is medically necessary and appropriate. Further, should the claimant refuse without good cause to attend an IME, some jurisdictions will allow termination of disability payments.

Convincing arguments can be made that the IME is an over-rated tool; however, in too many cases, the IME produces the medical equivalent of a swearing contest — the claimant's doctor versus the "company doctor." Compensation boards, aware of the humanitarian aims behind compensation laws, are quick to weigh more heavily the opinion of the treating physician than the judgement of a "one shot" IME doctor. It is reasoned that the treating physician is more familiar with the patient, and therefore is best qualified to speak about medical restrictions. Questions of fact are frequently resolved in the claimant's favor. Some compensation statutes include presumptions favoring the injured employee, presumptions difficult but not impossible to overcome.

Despite these well-founded concerns, the IME can be a very effective cost-cutting weapon in the adjuster's arsenal. In reality, the IME is often used in a slapdash and unthinking manner. Common pitfalls undermining the effectiveness of the IME include:

Choosing the same doctor/doctors all the time. Pretty soon compensation boards give zero credibility to doctors whose opinions are paraded out time and time again by the insurance representative.

Choosing an "insurance company doctor." This is the carrier's counterpart to the "Dr. Whiplash" utilized by the plaintiff's bar. Physicians who do virtually all their work for insurance carriers may provide claimsmen with the "Johnny-One-Note" opinion sought, but these practitioners generally have no more credibility in court than the claimant-oriented physicians. As one claimsman said of an oft-used "conservative physician": "He'd give a dead man a zero PPD rating."

Selecting a doctor who spends all or most of his time performing IME's. Ideally, the doctors chosen for IME's should derive only a fraction of their sustenance from such consults, and have a health practice from the treatment of their own patients.

Using unqualified physicians. If the treating physician or the claimant's expert is board-certified, for instance, best be sure that the IME doctor has comparable credentials. Some doctors have sub-specialties, and one should not send a back injury claimant to an eminent hand specialist just because the latter is an orthopaedist.

Failing to provide the IME doctor with adequate background information. Copies of all prior medical records should be sent to the IME doctor well in advance of the appointment. This provides a keener, more intelligent perspective on the case, and will lend more credence to his opinions. Few tactics will puncture the IME's advantages quicker than opposing counsel revealing that the consulting doctor did not have the benefit of the complete medical history and picture. If the issues of credibility or causation are involved, then the background material need not be limited to medical records, but may include items such as statements of the claimant, supervisors, and witnesses.

Not clearly communicating with the IME doctor exactly what you are after. Form letters have the virtue of efficiency when scheduling IME's. The cover letter to the consulting doctor is extremely important, but must be adapted and worded to the needs of the particular case. What is the purpose of the IME? What is the adjuster's concern? What major issues would the claimsman want the doctor to address in the medical report? These should be spelled out clearly to the consulting physician ahead of time.

Chances of success from an IME are directly proportional to the amount of intelligent forethought and groundwork provided by the adjuster at the pre-IME stage. Used selectively and with appropriate timing, the IME remains perhaps the adjuster's strongest tool in managing back injury claims.

§ 10-2(B)(4). Limited/Modified Duty.

Just because a back injury claimant may not yet be capable of returning to his old job does not foreclose the possibility of the worker performing *some* type of work. Short of an unqualified release for full duty, the physician may provide the claimant with medical clearance for returning to some less strenuous activity. Euphemisms such as modified duty, limited duty and selective employment all boil down to the same thing: returning the back injury claimant to work sooner than would be the case if the adjuster simply waited around for an unconditional release for regular duties. The phrase "light duty" should be avoided because of negative connotations.

To the extent the adjuster can persuade the employer to restructure a job so that it will be within a claimant's physical limitations, the claimsman will abbreviate the workers compensation pay-out and return the claimant to some productive role. In many cases, employers will modify the old job since they are aware that their compensation rates are tied to the level of workmen's compensation payments. Some corporations almost go so far as to invent a job so as to conform with the doctor's limitations. Returning the claimant to limited duty makes good economic sense. If the limited duty involves no pay reduction, then the offer of such work may foreclose any further liability for temporary total disability benefits. Even if the limited duty job pays less than the regular job, the claimant may often be switched from temporary total to temporary partial disability, normally a much smaller weekly amount.

Adjusters must seize the initiative in bringing before the physician the issue of modified duty, and cannot assume that the doctor will let them know when the claimant has reached such a plateau. Often the doctor is totally unaware that limited duty is available: the claimant certainly does not volunteer the information, or the adjuster assumes the doctor knows, or the adjuster is just too bogged down by a caseload to make the appropriate follow-up. Common restrictions for back injury claimants involve: (1) Lifting restrictions by pounds (e.g., "No lifting over 25 pounds"); (2) Length of time permitted to work — two hours, four hours, etc.; (3) Number of rest breaks allowed; (4) Restrictions on bending and stooping; (5) Restrictions on status positions, sitting or standing for prolonged periods; (6) Restrictions on driving or riding in a car; (7) Restrictions on pushing, pulling, and carrying activities.

Once the adjuster has in hand the medical report outlining what types of modified duty are within the back claimant's ability, a first-class selling job is often needed. Despite its apparent economic good sense, resistance to modified duty runs high in some quarters for various reasons. The claimsman must be an effective communicator and salesman to persuade a reluctant employer to accept a worker back at modified duty. Resistance to the practice can be found among employers and employees.

Employers may resist modified duty for any of the following reasons: (1) A fear the claimant will re-injure his back and then they will be facing a more serious claim; (2) They fear morale problems from outsiders: "He's getting paid for doing half the job that I do"; (3) Fears that the employer will not be getting its money's worth from someone they feel is performing "half a job"; (4) Economic factors seem to preclude the luxury of paying people to do jobs for which there is no great demand. If the back injury claim is to be effectively managed, the claimsman must allay these concerns.

Resistance from the employee's side arises for many reasons. Some union contracts forbid employees in a certain work category to be placed outside of the original job description. Fear of re-injury or skepticism regarding the employer's motives is another impediment. Perceptions of limited duty can hinder the search for this option. Take a man, particularly one in an industrial setting (where back injuries are frequent) and who is also a breadwinner, husband, and father. To this man, "light work" may hold all the attractiveness and excitement of a Tupperware party. "Modified work" as an idea may make such a person feel emasculated, dependent, and awkward. Adjusters handling many back injury claims realize that some claimants will fight the concept of light duty, with all of its panty-waisted connotations.

At some point the adjuster must cut losses when it becomes clear that there is employer resistance, worker apprehension, or just plain lack of realistic demand for modified duty jobs. At this stage, the claimsman might look beyond the original employer for a place for the back injury claimant. In cases where return to regular work appears unlikely, and where return to modified duty is unavailable, the adjuster turns to the realm of rehabilitation to help manage the back injury claim.

§ 10-2(B)(5). The Rehabilitation Process.

Rehabilitation in the context of back injury claims is a term loosely used to encompass all those activities aimed at finding a claimant a suitable job in the labor market. Included here is "classic" rehabilitation in the sense of retraining the claimant with new skills. Also involved is a focus on existing and transferable skills which render a claimant employable in some capacity outside of his old job. Rehabilitation focuses on (1) Using present skills to find the back injury claimant a job without the need for retraining; (2) Adding skills to a claimant's repertoire to enhance employability.

From the claimsman's perspective, the goals of the rehabilitation process should be: (1) To succeed in placing the claimant in another job; (2) Document the availability of medically-approved employment; (3) To document non-cooperation if such is present and if cooperation is required by the workers compensation statute; (4) To aid the adjuster in coordinating medical treatment of the back injury claimant and to insure that the latter is receiving the best possible care; (5) To aid cost containment through medical bill audits, identifying duplicative or counter-acting therapies. One observer from the rehabilitation industry estimates that fully 40% of counselors' case loads are comprised of "bad back cases" (entry 7). Assuming that management of the back injury claim can benefit from wise intervention on the part of a rehab counsel or, the question arises of what criteria are used in evaluating candidate cases. Certainly not all back injury claims — not even all those involving lost time — warrant rehab referral. One excellent set of guidelines for evaluating back injury cases for such referral is as follows: (1) Repeat laminectomies; (2) Spinal fusion; (3) Non-surgically treated back injury continuing greater than ninety days; (4) Certain physicians managing the case; (5) Evidence of multiple back injuries (entry 8).

Wisely and selectively used, rehabilitation referral is an investment which may repay itself many times over, *if* it succeeds. Within the claims industry, however, there are some common complaints (with varying degrees of merit) and frustrations with the rehabilitation industry:

Lack of goal-orientation. The "bottom line" is to get the claimant back to work. Adjusters complain sometimes that rehab vendors are more interested in "file building" than in concrete results. Typical objects of this ridicule are the six-page rehab reports which describe on and on the different positions considered and which basically conclude: "We can't find the client a job."

Lack of realism. Every adjuster handling back injury claims has heard, at the rehab vendor's expense, of efforts to retrain the garage mechanic to be a brain surgeon; efforts to send the grocery stock clerk to a two-year computer program; suggestions to send the ditch-digger back to college to get his B.A. degree at the carrier's expense. Even conceding exaggeration in these examples, they reflect a residue of skepticism on the claimsman's part toward the "rehab gurus."

699

Usurping the adjuster's role. The adjuster's role is to manage the overall claim. The rehab vendor's role is to assist in finding alternate suitable employment. Sometimes the dividing line blurs, and this provokes adjusters' annoyance. A real-life example would be the rehab vendor who receives a much sought-after referral, pursues job placement for a year and a half, and then admits defeat with the following suggestion: "We are unable to find this claimant a job. I would suggest that you try to settle this case. Attached is our monthly invoice in the amount of $1,309.17 . . . ."

Burdensome communications. While frequent communication is needed, there can be too much of a good thing. Adjuster case loads are generally many times the size of those of a typical rehab counselor. Adjusters consider their time precious, and they are constantly trying to "tame the telephone." Adjusters do not need daily telephone reports or to be told each time the Ace bandage is changed.

In defense of the rehab vendor, it must be stated that all too often adjusters merely dump the case in the lap of the counselor and seem to want to have nothing to do with the file thereafter. Abandoning the file to rehab allows the adjuster to console himself that he has "done something" on the file. Many referrals made by adjusters amount to "Okay, here's a new case. Do something." Goals must be spelled out and continuously refined and clarified. The adjuster must be clear on precisely what is desired from the rehab process: job placement, labor market survey, job retraining, medical coordination, reassurance of the claimant, cost containment? The adjuster must provide frequent input and guidance to the efforts of the rehab vendor. To the extent that rehab serves merely as a dumping ground for the tough cases, to the extent rehab referral serves merely as a means to assuage the adjuster's need to do something, then the process invites failure and all of the complaints listed heretofore.

§ 10-2(B)(6). Special Injury Funds.

Most state workers compensation laws include provisions pertaining to Subsequent/Second/Special Funds (entry 6). These devices aim to minimize employers' disincentives toward hiring handicapped or previously disabled workers. Though provisions vary, these Funds try to limit the liability of the employer whose employee sustains a subsequent injury or disability. A recent injury combined with a pre-existing disability may produce an overall disability much greater than would normally be the case. To hold employers responsible for these greater disabilities is thought from a public policy view to be unfair to employers, and harsh on disabled individuals seeking work.

In such cases, Second Injury Funds may hold the employer responsible only for that pro-rated portion of disability due solely to the recent injury. Frequently, statutes limit employers' liability in such cases to a finite number of weeks, in many cases 104 weeks. Beyond this stage, the Special Fund may pick up payments to the claimant and even may reimburse the employer for any payments made in excess of 104 weeks. Space constraints do not allow review here of the varieties of Second Injury Funds. Common denominators to these Funds include the following requirements from employers who wish to avail themselves of the funds' relief: (1) That the claimant have a prior disability; (2) That the employer have knowledge — either actual or constructive — of the pre-existing disability; and (3) That the combination of the pre-existing disability produces a greater disability overall.

Though some Special Funds exclude back injury cases, jurisdictions allowing Fund applicability to back injuries

701

provide the adjuster with a powerful tool in managing the claim and limiting the carrier's exposure. The adjuster may be able to lay the groundwork for the carrier tapping into the Second Injury Fund, and thereby succeed in saving the client thousands of dollars — with no hardship on the claimant. Limiting an employer's responsibility to "only" 104 weeks of compensation may sound odd, but those persons handling back injury claims know that for some serious back cases, two years of disability payments is a relative drop in the bucket. Investing time in building a case for Second Injury Fund relief is an investment that may pay incalculable dividends down the road.

Proving that there is a pre-existing disability will show how well the adjuster can research a claimant's medical history. Since for many people back pain is a common condition of life, a painstaking examination of a claimant's medical history may pay off. In the initial statement from the claimant, the adjuster should cover prior back injuries, treatments, accidents, and hospitalizations. A signed Medical Authorization should be obtained. Family doctors should be identified and a complete copy of their charts requested. Prior hospital and medical records should be reviewed. Union health records, personnel files, and group health/medical files may contain information pertaining to pre-existing back problems or disabilities. Prior employers may be a source of information. Checks with the Index Bureau and with local compensation commissions may produce evidence of prior claims or disabilities. All these investigative avenues can lead the adjuster to proof that the claimant had some pre-existing back problem or disability, whether or not the claimant was fully aware of the condition. Examples of common pre-existing back problems that are frequently discovered through research include: degenerative changes in the spine, spondylolisthesis, scoliosis, arachnoiditis, osteophyte formations.

Once the adjuster reaches the point where a pre-existing disability is documented, the next step is proving that the employer had knowledge of this condition. This test can be met in a number of ways. Pre-employment physicals or pre-employment medical questionnaires are useful sources. Testimony from co-workers or supervisors as to a claimant's limitations may suffice. Some jurisdictions have held that the employer has constructive knowledge of the pre-existing condition so long as the information is "manifest" to the employer. In such cases the adjuster need only show that medical evidence reflecting the prior disability existed in material which was available to the current employer.

Demonstrating that a combination of the prior condition and the recent injury has produced a greater disability is usually the easiest of the three steps. This is a medical question which the adjuster must pose to the treating physician or, particularly, to an IME physician. If an independent medical consultation is used, the adjuster must take care to ensure that all of the relevant prior medical records are provided. Further, the adjuster must spell out and communicate clearly to the IME physician precisely what type of medical issue needs to be addressed. Many physicians are very experienced in conducting these types of evaluations, but the adjuster's zeal for using these doctors must be tempered by considerations noted earlier regarding IME's in general.

Neophyte adjusters are often uneasy about seeking Second Fund relief. They are so used to proving that a claimant has no disability that it feels strange to gather evidence showing that the claimant has a greater disability or, as required in some jurisdictions, that the claimant is even permanently and totally disabled. The research required to build a good Special Fund case often appears tedious and boring to some adjusters. Others simply have no grasp of the

Special Fund ramifications. For the adjuster who succeeds in breaking out of this mindset, however, the savings will be substantial. Counsel can set into motion the process aimed at tapping into the Subsequent Injury Fund, but the foundation for successful efforts here can be cemented through claims work at the investigative level. Subsequent Injury Funds provide a very potent tool for adjusters seeking to manage back injury claims.

§ 10-2(C). Conclusion.

This section does not pretend to be an exhaustive discussion of all techniques for managing back injury claims. Only those this writer considers most common and effective are discussed. Settlements as a technique for concluding back claims is not addressed, since it is more an aim of claims management than a technique. If preceding approaches are intelligently employed by the adjuster, then chances of negotiating a favorable settlement should be greatly enhanced.

Nor are the approaches mutually exclusive. For example, IME's can be used in conjunction with rehab efforts or Second Injury Fund pursuit in order to minimize the carrier's exposure in a way consistent with the claimant's interests. Used in combination, these approaches can yield powerful results. Managing back injury claims is not so simple, however, as marking off a checklist so that one can tell one's Claims Manager, "See, I'm doing something"! A flurry of activity should not be confused with purposeful claim management. The approaches described herein must be used thoughtfully, creatively, and not in a knee jerk fashion.

Because the compensation system seems, to the adjuster, to be so heavily stacked against the employer, it is tempting for the claimsman to take a defeatist and caretaker approach. Back injury claims have probably caused more than their share of cases of adjuster burnout and frustration. To manage the back injury claim, the adjuster must not succumb to such pitfalls, and must be more than a "tender of the files." On the other hand, aggressiveness for the sake of pure aggressiveness is no virtue in handling claims. The point is that, far from being a helpless role-player in an employee-oriented compensation system, the adjuster has at his disposal a powerful repertoire of tools to aim the back injury claim toward clear and positive goals. Back injury claims pose a considerable challenge to the claims industry, which must focus its efforts less on hand-wringing and more on constructive approaches to claim management.

BIBLIOGRAPHY

1. Careers: Getting back to work, *Washington Post,* 13 May 1981, p. B5.
2. Hendler, Nelson H. and Fenton, Judith A. *Coping With Chronic Pain* (New York, N.Y.: Potter Clarkson N., 1979), pp. 97-104.
3. Lippitt, Alan B. Acute lumbosacral derangements and their management, *Rehabilitation Forum* Vol. 6, (July—August 1979): 1.
4. Magarick, Pat. *Successful Handling of Casualty Claims* (Brooklyn, N.Y.: Central Book Company, 1974), pp. 648-655.
5. Smoller, Bruce and Schulman Brian. *Pain Control: The Bethesda Program* (Garden City, N.Y.: Doubleday & Company, 1982), p. 5.
6. United States Chamber of Commerce, *Analysis of Workers' Compensation Laws 1984* (Washington, D.C.: U.S. Chamber of Commerce, 1984), pp. 34-37.
7. Willenbrink, Marilyn L. "The Rehabilitation Process," *Medical Management Course* (Atlanta: Crawford & Company, 1984), p. 117.
8. *Ibid.,* pp. 117-118.

§ 10-3. From the Private Lawyer's Perspective.

*Gerald Herz**

§ 10-3(A). Introduction.

My practice is primarily devoted to the legal handling of workers' compensation cases. The majority of cases I handle requires defending against claims made by injured workers, with the balance involving the assertion of claims on behalf of workers. Some of the most difficult and challenging cases, from a legal standpoint, involve injuries to the back. In these cases, there is a particular need for cooperative interaction among doctors, attorneys, insurers, and self-insureds. I have attempted to highlight problem areas in managing industrial low back problems. Proper management can minimize the exposure for compensation, medical benefits, and rehabilitation expenses while providing adequate care to the claimant. If such interaction is successful, it will lead the claimant to an early return to work and reduce or avoid the problems of the claimant's resignation to injury and consequent quest for secondary gain.

In a back case where we have a true objectivity, such as disc injury, the disability is often clear. The complaints of pain and limitation in such cases can be attributed to positive neurological findings which substantiate the complaints of pain and limitation made by a claimant. Under those circumstances, the physician can prescribe a regimen of treatment with an objective in mind of returning the patient to the job market at the earliest possible date so that the claimant does not become preoccupied with his/her disability. The soft tissue injury case, on the other hand, is an enigma to the treating physician and to counsel for the

---

* I would like to give a substantial amount of credit for assistance in helping me prepare this chapter to my associate, Resa J. Toplansky, and to my legal assistant, Christine Gosselin, who helped me with the research and thoughts necessary to prepare this section.

706

employer or carrier; there are often non-existent or limited objective findings followed by long periods of treatment which often reinforce the claimant's disability. These cases warrant additional attention to detail and to the special needs of both the claimant and the carrier.

In preparing this article I obtained comments from other attorneys who predominently practice in the area of workers' compensation and who are respected for their legal abilities in this area. Their comments help me present a broader view of how the lawyer regards the soft tissue type injury case. One of the attorneys,[1] when queried regarding the management of low back injury cases, stated: "To my knowledge, no one has successfully consistently managed industrial low back injury cases. This is particularly true where the original injury is not severe and only involves soft tissue." My experience indicates his comment to be accurate.

The initial management of the back injury case has to start with the treating physician, who often feels compelled to treat the patient based upon his subjective complaints because the physician feels he must give the patient the benefit of the doubt. The physician should understand the basics of how the workers' compensation system works so that he can make medical judgments that are fair to both his patient (who may be entitled to payment of benefits) and the employer or carrier (who is required to pay the cost of the medical care and the workers' compensation payments to the claimant). Protracted treatment leads to protracted disability which also may impose an unnecessary, direct financial burden on many employers, in the form of higher premiums and/or decreased coverage availablility.[2] Accord-

1. Richard W. Turner, Esq., of the law firm of Hamilton & Hamilton, Washington, D.C., is engaged in the defense of workers' compensation cases and represents a substantial number of corporate clients.

2. Many employers are on a retrospective premium basis and their premiums are adjusted annually to reflect the amount of payments made by the insurance company on their behalf.

ingly, the cost of workers' compensation benefits can represent a substantial portion of the overhead and may require an employer to withdraw or limit the performance of work in a given jurisdiction because they are prohibitive. The result is a reduction of jobs with a rippling effect to the claimant and other employees in a local community. In other words, you cannot isolate one compensation case from its economic impact to an employer, industry or the job market in the particular locality.

Low back injury cases represent a substantial portion of the litigation involved in workers' compensation owing to the fact that the objectivity is elusive while the actual or subjective disability is continuing.[3] There are many low back injury cases where the claimant has extensive subjective complaints of debilitating pain with marginal or even non-existent medical findings, yet the physician continues his patient on disability from his usual employment.[4] The soft tissue injury is characteristically unresponsive to medical treatment[5] and often entails permanent or recurring disability wherein the physician is unable to give a definitive medical reason or a finite course of successful treatment.

Iatrogenic disability exists in its most virulent form in low back cases. The treating physician finds slight muscle spasm coupled with complaints of pain and limited motion but cannot find objective signs. In many, if not most, of these cases, the physician prescribes unlimited time off from work and substantial amounts of physical therapy and thereby reinforces the claimant's feeling that he has sustained a severe injury; this situation makes it all the more difficult for the claimant to return to work.[6]

---

3. L. Airola, *Workmen's Compensation—Back Injuries,* 10 Am. Jur. *Trials* 589, 592.

4. Walker v. Rothchild International, 526 F.2d 1137, 1140 (9th Cir. 1975) (dissenting J. Hufstedler).

5. Airola, *supra,* note 3, at 593.

6. Richard W. Turner, Esq., *supra,* note 1.

In addition, we find that in the back injury case a claimant whose treatment is protracted becomes persuaded that he is very disabled and in many cases will develop a conversion reaction, which in and of itself is compensable. Normally this requires a psychologist or psychiatrist to become involved, giving the case a new dimension with a spiraling increase in medical and compensation benefits. The result is that the claimant usually will never return to employment, even though the great majority of his complaints are subjective in nature.

A physician treating a claimant for a low back injury has a responsibility not to reinforce feelings of disability or to impose unrealistic limitations on a claimant but should persuade him to return to work as soon as it is medically advisable. If the circumstances warrant it, the physician should immediately involve the claimant in a regimen of therapy that will reinforce his feelings of ability to work. In those cases where the physician feels that rehabilitation is going to be necessary, then he should urge that the rehabilitative efforts commence during the active treating period so that the claimant can make the transition into an employment environment as quickly as possible.

The back injury case very often creates a sociological explosion and brings into play much more than the claimant/physician relationship. The claimant, who normally has led a rather vigorous life in an arduous occupation, now feels emasculated, as he is forced to spend a great deal of time at home. This causes a disruption of the family life, creating tensions with the spouse and children. Economic hardship can ensue as it becomes necessary to spend savings to maintain the expenses of the household.

The claimant's family necessarily gets drawn in by the claimant's disability. The longer the duration of disability, the more support is needed from the other members of the household. Rehabilitation experts have identified certain stages that claimants and their families experience during the period of disability. For example, V. Christopherson [7] sets forth four such stages and concludes that selected resources, such as coping mechanisms, involvement from the extended family and support from community agencies, must be utilized in order to alleviate the tensions created by the claimant's disability. The treating physician should therefore be aware of these sociological implications when designing a treatment program.

§ 10-3(B). Statutory Analysis.

In order to fully understand the impact of the interrelationship between treatment and disability, a background on the concept of workers' compensation and, specifically, compensation for low back injuries under existing statutory schemes is important.

For purposes of workers' compensation payments, disability is not based merely on the severity of the physical injury sustained but also upon how that injury impacts on the individual's earning capacity. [8] Under the workers' com-

---

7. Christopherson, V., "The Patient and Family," *Rehabilitation Literature,* 1962, February, 34-41, cited in *Role of the Family in the Rehabilitation of the Physically Disabled,* Edited by Paul W. Power and Arthur E. Dell Orto, University Park Press, Baltimore, MD (1980), Chap. 4, p. 148. "Following a period of initial anxiety (acute stage), and then a time when the patient attempts to regain as much residual strength as possible (reconstruction stage), the family members begin to perceive the patient's condition realistically. They become aware that the patient's rehabilitation has reached a point of diminishing returns. This is the plateau stage, and possibly the most difficult time for the family members. They realize that hope for improvement has diminished and perhaps a long period of care faces both the patient and family. It is a time when the family needs realistic objectives. Following this stage, the ill family member may suffer a setback. Grief and anger among the family are aggravated, and constant anxiety may pervade the family (deteriorative stage) unless selected resources alleviate this tension.

8. Hotaling v. General Electric Co., 364 N.Y.S.2d 243 (App. Div. 1975);

pensation statutes, an employee does not receive compensation for pain and suffering but is paid based on the industrial impact of the injury on his present and future (in) capacity to work.[9] In cases involving an injury to a limb, the compensation acts determine the disability according to a fixed schedule of awards, which represents a legislative determination of the extent of wage earning capacity the claimant has lost as a result of the limb injury. For example, the loss of an arm entitles a claimant to 312 weeks of compensation in the District of Columbia and 300 weeks in the State of Maryland.[10] If the individual has sustained only a 10 percent disability of the arm, he receives 31.2 weeks of compensation in the District of Columbia and 30 weeks of compensation in Maryland. The scheduled award would be the final resolution of his compensation claim for that injury unless he could later prove, under the provisions of the law, that his injury had substantially worsened and he was thus entitled to further benefits.

Low back injuries fall under a separate and distinct category, commonly referred to as "other cases." Compensation is awarded based upon a medical evaluation of the limitations and disability, stated by a percentage, in addition to considerations of the claimant's age, education and industrial experience; these factors are then translated into loss of industrial capacity or loss of wage-earning capacity. In short, what the claimant is able to do in a competitive labor market.

The various states have developed different approaches in determining how compensation for a back injury should be awarded. In a minority of jurisdictions,[11] the compensation

---

Employers Liability Assurance Corp. v. Hughes, 188 F. Supp. 623 (S.D.N.Y. 1959).

9. Veley v. Borden Co., 13 A.D.2d 883, 215 N.Y.S.2d 311 (1981).

10. 36 D.C. Code Section 308; Article 101, Section 36, Workers' Compensation Law of Maryland.

11. Virginia, District of Columbia and some other states.

for a low back injury is based on a narrow assessment of the actual loss of wages suffered as a result of the industrial injury.[12] The actual loss of wage theory totally ignores the special circumstances of the claimant's age, education, and experience. The theory also fails to account for those special situations where, for example, a sympathetic employer takes the claimant back at the same wage rate but does not advance him economically as he would have if the injury had not occurred, solely because of his physical limitations. Further, no consideration is given to the fact that the claimant's opportunities in a competitive labor market may be limited, or that he has lost some career development potential as a direct result of the injury.

The majority of jurisdictions utilize a loss of capacity concept and consideration is given to the total impact of the injury on the claimant after considering age, education, etc. Once the limitations are considered, the claimant must make reasonable efforts to secure some work despite the incapacity.[13] But if the injured worker is unable to work, the extent of his disability may be adjudged to be total. Nonetheless, in some jurisdictions there are limitations as to time or cumulative amount of disability payments which a claimant may receive even where total incapacity is proven. For instance, in the states of Virginia and Maryland, compensation benefits are limited to 500 weeks but in cases of permanent total disability in the State of Maryland, the claimant can re-apply for additional benefits after the initial 500 weeks of compensation have been paid.

In the majority of jurisdictions utilizing the loss of earning capacity concept, including New York, Wisconsin and under the federal Longshoremen's and Harbor Workers' Compensation Act (LHWCA), disability resulting

---

12. Mess v. Frust Coal Co., 195 Va. 762, 80 S.E.2d 533 (1954).

13. Fletcher v. Island Creek Coal Co., 201 Va. 645, 112 S.E.2d 833 (1960).

from a low back injury is determined by comparison of pre-injury wages with post-injury wage-earning potential as opposed to actual wages earned. In these jurisdictions, the claimant's condition is evaluated in its totality; all medical and economic factors are relevant and must be considered by the adjudicating body. Post-injury wage-earning capacity may differ from actual wages earned due to the general wage increases, extraordinary efforts on the part of the injured worker, or other similar factors.[14] As previously indicated, the award will be based on the translation of the medical determination of the disability, giving additional weight to the claimant's particular occupation, experience, training, age at the time of injury, and other relevant factors.[15] When the claimant enters a new, alternative occupation after the injury, the true loss of earning capacity of the claimant is determined by assessing what the claimant's earnings would have been in the new occupation on the date of accident, deducted from his wage-earning capacity on the date of accident.[16]

In sum, it can fairly be stated that the claimant's actual disability is more fairly determined and has less of an onerous impact on him under the loss of capacity statutes than it does under the actual loss of wages statutes.

In many jurisdictions,[17] once the medical evidence indicates that the work-related low back injury prevented the claimant using reasonable effort from performing his usual occupation, a presumption of permanent total disabil-

----

14. Lawrence v. Norfolk Dredging Co., 319 F.2d 805 (4th Cir. 1963), *cert. denied,* 84 S. Ct. 443 (1963).

15. Welch v. Leavey, 397 F.2d 189 (5th Cir. 1968), *cert. denied,* 89 S. Ct. 685 (1969); Randall v. Comfort Control, Inc., No. 83-1123, ____ F.2d ____ (1984).

16. Bethard v. Sun Shipbuilding & Dry Dock Co., 12 BRBS 691, BRB No. 78-548 (1980).

17. Wisconsin, New York, under the Longshoremen's and Harbor Workers' Compensation Act, and others.

713

ity arises.[18] The burden of proof then shifts to the employer
who must rebut this presumption by establishing: (1) that
suitable alternative employment is available within the
claimant's restrictions, (2) that the claimant is capable of
re-entering the employment market, and (3) that the claimant has the capacity to earn the same or at least some wages
even though he has the stated medical limitations.[19] If the
employer cannot rebut the presumption and cannot demonstrate that there is work available to the claimant within
his restrictions, a claimant will be awarded permanent total
disability.

In order to prove the availability of jobs that the claimant
can perform, the employer must, at a minimum, demonstrate the general availability of suitable jobs within the
particular claimant's capacity to perform by virtue of his
age, education, skills, physical condition, employment history, and rehabilitation potential. The employer does not
have to actually find the claimant a job [20] or guarantee him
employment,[21] but the employer does have to demonstrate
that employment opportunities exist within the geographic
area of the claimant's home.[22] If the claimant is working
away from home for a prolonged period, then the employer
must demonstrate that there are jobs available to the claim-

---

18. American Stevedores, Inc. v. Salzano, 538 F.2d 933 (2d Cir. 1976);
Balczewsk v. Department of Industry, Labor & Human Relations, 251
N.W.2d 794, 76 Wis. 2d 558 (1977).

19. Newport News Shipbuilding v. Director, 592 F.2d 761 (4th Cir.
1979); American Stevedores, Inc. v. Salzano, 538 F.2d 933 (2d Cir. 1976).

20. Miller v. Central Dispatch, 673 F.2d 773 (5th Cir. 1982).

21. New Orleans Stevedores v. Turner, 661 F.2d 1031 (5th Cir. 1981);
American Mutual Insurance Co. v. Jones, 426 F.2d 1263 (D.C. Cir. 1970).

22. Bumble Bee Seafood v. Director, 629 F.2d 1327 (9th Cir. 1980); Pilk-
ington v. Sun Shipbuilding & Dry Dock Co., 9 BRBS 473, BRB No. 78, 283
(1975).

ant in the area where he was working at the time of injury.[23]

Workers' compensation laws are written for the benefit of the worker and based upon humanitarian principles, thus they favor the claimant with certain presumptions. The fact of physical or psychological harm is *not* presumed and the claimant has the burden of proving the nature and extent of his disability.[24] Even though the claimant does not enjoy a presumption of the nature and extent of his disability, the adjudicator is mandated to resolve all doubtful questions of fact in his favor.[25] Accordingly, even though there are two medical reports, one finding the claimant disabled and another finding him capable of returning to employment without disability, the doubt will normally be resolved in favor of the claimant, assuming that both reports carry equal weight. In most cases the treating physician's report is usually given greater weight than the report of an independent medical examiner who has only seen the claimant on one occasion.[26] As long as the adjudicator has taken all testimony and medical reports into consideration, he is not bound to accept any particular medical theory or conclusory opinion and may allow or disallow benefits based upon all the facts before him.[27]

---

23. Newport News Shipbuilding & Dry Dock v. Director, Office of Workers' Compensation Programs, 492 F.2d 762 (4th Cir. 1979).

24. Young & Co. v. Shea, 397 F.2d 185 (5th Cir. 1968), *cert. denied,* 395 U.S. 920 (1969); Murphy v. SCA/Shayne Brothers, 7 BRBS 309 (1977). *But see contra,* Murphy v. SCA/Shayne, 78-1143 (D.C. Cir. June 25, 1979) (basing its affirmance on the applicability of the presumption to the fact of physical harm and the presumptions adequate rebuttal).

25. Stracham Shipping Co. v. Shea, 406 F.2d 521 (5th Cir. 1969); Army & Air Force Exchange Service v. Greenwood, 585 F.2d 791 (5th Cir. 1978).

26. Harding v. Mother Goose, Inc., 57 O.I.C. 159 (January 24, 1977).

27. Walker v. Rothchild International, 526 F.2d 1137 (9th Cir. 1975) (dissenting J. Hufstedler).

§ 10-3(C). Role of Physicians.

I feel that physicians may not fully understand the power of their reports and opinions and how expensive the consequences can be to industry and to the claimant. The claimant suffers economically and emotionally if the physician is casual or unrealistic in the handling of the case, and he can cause his patient problems that far exceed the bounds of the actual physical injury and limitations imposed.

The physician, during his initial evaluation and treatment of the injured worker, should consider the following:

§ 10-3(C)(1). Causal Relationship.

The physician should, first and foremost, consider whether the claimant truly suffered the injury because of a distinct or extraordinary activity associated with the job [28] or whether the back strain was an isolated event while on the job which has no causal connection to the employment.[29] The physician should not feel compelled to reinforce a connection between the work and the back strain if it is in fact an ordinary ailment to which any of us may be subject. In other words, if it is not truly a problem related to some extraordinary work activity, and the physician determines that it *is* an ordinary ailment of everyday life, then the condition should not become the responsibility of the employer. Medical care should be handled under the health insurance plan which is in effect for the worker.

§ 10-3(C)(2). Pre-Existing Conditions.

It is extremely important that the physician obtain a detailed history of pre-existing injuries or problems that the

---

28. Fellons v. Syracuse Supply Co., 263 N.Y.S.2d 785 (1965); Lewellyn v. Industrial Commission, 38 Wis. 2d 43, 155 N.W.2d 678 (1969).

29. Buchanan v. Bethlehem Steel Co., 302 N.Y. 848, 100 N.E.2d 45 (1951); Nicoletti v. Ett Pomeroy Co., 283 A.D. 1129, 131 N.Y.S.2d 699 (1954).

patient may have had. Aggravation of a pre-existing condition *is* compensable as an industrial injury; however, under many workers' compensation statutes,[30] the employer and carrier are provided with some relief in that they are only obligated to pay for that portion of the disability caused by the present injury. In these jurisdictions, if the disability is serious in nature, the employer generally pays for that portion of the disability attributable to the injury without regard to the pre-existing condition [31] and a "Subsequent Injury Fund" pays the balance of the disability that is attributable to the pre-existing conditions.[32] Under this approach the claimant is paid for the entire disability but there is an apportionment between the employer and a Fund so that each pays its fair share of the claim. Accordingly, the claimant's medical history is critical in back injury cases as the issue of pre-existing conditions commonly arises.[33]

§ 10-3(C)(3). Cumulative Back Problems as an Occupational Disease.

There is a recent trend to handle claims based on continuous or cumulative back injury as occupational diseases.[34] This trend alleviates the necessity of pinpointing a precise time of accident and avoids the possibility that the statute of limitations will bar the claim. A compensable injury based on continuous exposure to excessive stress would be considered to have occurred as of the date of last

---

30. Maryland, District of Columbia, and Virginia, among other jurisdictions, incorporate this feature.

31. Dent v. Cahill, 18 Md. App. 117, 305 A.2d 233 (1973).

32. SIF v. Thomas, 275 Md. 628, 342 A.2d 671 (1975).

33. Tsirigotis v. Samuel Scholsberg, Inc., 3 A.D.2d 1024, 294 N.Y.S.2d 448 (1968).

34. Airola, *supra,* note 1, at 618; Nugent v. Pro Football, Inc., 82-DCWC-270.

717

exposure.[35] In other jurisdictions, however, the cumulative trauma theory is not always recognized, and these claims are thus not compensable, on the theory that the worker has merely suffered a gradual diminution in health over a period of time which is common to workers and non-workers alike.[36]

§ 10-3(C)(4). Subjective Evidence of Ability to Work.

The treating physician normally establishes a relationship with his patient and gives the claimant the benefit of the doubt when the claimant complains of pain or an inability to perform certain activities, even though the physician cannot find any true objectivity to the complaints. In other words, the physician treats the patient based on complaints rather than actual findings and often, if the patient complains long enough, the physician will continue to treat and ultimately give the claimant a substantial disability rating with severe restrictions as to the type of employment for which he is fit. Many employers and carriers have now resorted to surveillance film in order to prove that the claimant has a greater ability to function than he has led the physician to believe, and these films are shown to the physician in order to provide him with some objective evidence as to what his patient can and cannot do. Often, a physician will alter his opinion on the nature and extent of the disability after viewing the films. In other words, a physician may be "fooled" by his patient.

---

35. Beveridge v. Industrial Accident Commission, 176 Cal. App. 2d 592, 346 P.2d 545 (1959).

36. Bruzdowski v. Coleco Industries, Inc., 30 A.D.2d 886, 291 N.Y.S.2d 447 (1968); Land v. Dudley Lumber Co., 32 A.D.2d 977, 301 N.Y.S.2d 682 (1969); Eastern Shore Public Service Co. of Maryland v. Young, 218 Md. 338, 146 A.2d 884 (1958).

§ 10-3(C)(5). The Psychological Implications.

If a physician feels that his patient has a pre-existing weakness in the form of a neurotic tendency and that the patient will develop a disability neurosis precipitated by a back injury, then he should recommend treatment by an appropriate psychologist or psychiatrist in order to ward off a more serious trauma, the neurosis or conversion reaction. The neurosis is considered part of the overall injury and is fully compensable.[37] Accordingly, the physician can perform a valuable service to his patient by identifying and dealing with the psychological aspect of the case early on so that the major effort is directed towards the treatment of the low back injury without the complications caused by involvement of a neurosis.

§ 10-3(C)(6). Evidentiary Considerations.

Disagreement amongst physicians with regard to the nature and extent of the disability and limitations is common in low back injury cases. Opposing sides in a workers' compensation case will typically refer the claimant to physicians of their choice. Physicians often fall into categories of those favorable to claimants and those favorable to insurance companies; the former are those physicians who are liberal in ratings, and the latter are more conservative. Shopping for ratings by the respective parties ultimately overlooks the best interest of the injured worker. Where the claimant's attorney obtains an overstated opinion of the disability, the claimant may, in turn, be persuaded that he is more substantially disabled than he actually is. This determination may prevent the claimant from returning to his usual employment, resulting in an overall reduction in earnings over the period of his work-life

---

37. Pokorny v. Chadbourne, Wallace, Parkes & Whiteside, 14 A.D.2d 662, 219 N.Y.S.2d 130 (1961), Sky Chefs, Inc. v. Rogers, 220 Va. 800, 284 S.E.2d 605 (1981).

expectancy. The claimant has exchanged his livelihood for a compensation award which may provide him with money but which deprives him of his self-esteem and which may impact on him sociologically in the adverse manner discussed above. On the other hand, the physician who examines for the insurance company may understate the disability to the effect that the claimant is deprived of his rightful compensation and is encouraged to go back to the type of work that in reality he cannot and should not perform without consideration of alternative employment or necessary rehabilitation.

§ 10-3(C)(7). Pain Clinics.

It has been my experience that pain clinics are expensive and often nonproductive and are utilized by insurance companies in a desperate attempt to mitigate their exposure. It is felt by many physicians, adjudicators, and attorneys, that this type of clinic reinforces disability and normally does not have any positive effect on getting the claimant back to work; and for that reason, the prospect of having a claimant go to a pain clinic causes great trepidation on the part of the insurance company. That is not to say that some pain clinics are not better than others, but from a general point of observation, they are not normally thought of as being productive or successful. Alternatively, a psychologist or psychiatrist specializing in depression and/or pain management may be brought in during the initial part of the treatment to assist in a better and faster adjustment.*

---

* *Editors' note:* These remarks obviously reflect the frustrations of a plaintiff attorney who has had to deal with innumerable claimants who were misdirected to pain clinics. We agree with him that many pain clinic referrals are inappropriate: the malingerer who is abusing the system or the recently injured worker who is depressed and, as the author noted, should be sent directly to a qualified psychiatrist. In our experience, however, the failed back surgery case with arachnoiditis and chronic drug dependency often requires the team approach as described in Chapter 6.

§ 10-3(C)(8). Vocational Rehabilitation.

In many states, including Virginia, Maryland and the District of Columbia, vocational rehabilitation is prescribed in the compensation act itself and is a right to which the claimant is entitled. In other states, it is prescribed collaterally under a vocational rehabilitation act.[38] The vocational rehabilitation specialist is extremely important to both the claimant and the employer, as they may be the claimant's only opportunity of being re-trained or placed in alternate employment. The provisions for vocational rehabilitation has a concomitant benefit to the employer because its exposure for compensation is reduced or eliminated so that both parties are benefited if the outcome is successful. The physician should therefore impose realistic restrictions upon the claimant (with explanations) so that a proper vocational rehabilitation analysis can be accomplished.

§ 10-3(D). Techniques for the Practitioner.

I queried another well-respected attorney [39] to discuss the technical aspect of handling the low back case and he was kind enough to offer a list of suggestions that I feel are equally important to the physician and attorney handling a back injury case, and are as follows:

§ 10-3(D)(1). "Inadequacy of Medical Information.

"(a) Hospital records and doctors' entries therein, plus physicians' handwritten office notes in many instances are illegible and not decipherable. To the extent possible the individual making the entry should be admonished to write clearly, legibly, and completely. The problem becomes par-

---

38. 29 U.S.C. §§ 331-342 (1977).

39. John C. Duncan, III, Esq., of the law firm of McChesney, Pyne & Duncan, Washington, D.C.

ticularly acute where the records are eventually microfilmed and where the practitioner uses symbols and abbreviations which are not uniformly recognized in the medical community.

"(b) In many instances the past medical data, particularly office records, become critical in workmen's compensation cases. In situations where a doctor has retired or died, information should be left with the local medical society indicating the location of patients' records.

"(c) Physicians' reports often are wholly inadequate to address the various aspects of a workmen's compensation claim which, in many instances, impedes proper management of the claimant's back problems and entitlement to workmen's compensation benefits. At the outset of any new patient evaluation, the physician should specifically record how the incident in question took place and all pertinent past medical history whether related to the back injury or not. It would also be useful to identify by approximate date, hospital, and/or physician, the particulars of the past medical history secured.

"(d) By adequately detailing the foregoing information, the physician will potentially remove any cloud as to the compensability of the injury, thereby insuring adequate medical treatment for the patient as well as payment of the doctor's bill. It will also serve to minimize legal issues down the road where concepts of new injury versus recurrence are involved plus temporary exacerbation. In addition, because of the legal connotation attached to the term *recurrence,* physicians should avoid the use of that term and simply describe the circumstances pertaining to the new incident, the location of the problem, physician examination findings, diagnosis, etc. The legal implications of the diagnosis should be left to the courts.

"(e) Physicians should also be made acutely aware of the fact that in most jurisdictions the legal issues involved will rise or fall, in the main, on the medical reports they submit.

Rarely will the doctor be required to appear in person to testify; (s)he will usually be required to do so, or submit to depositions only where the report is lacking in content or clarity. Where a physician is rendering a medical opinion, whether it be on the issue of medical causality, new injury or recurrence, or any other aspect bearing on the proper medical management, the claimant's entitlement to workmen's compensation benefits, or the patient's capabilities of returning to work, the doctor should include in the report that he is rendering such an opinion based upon a reasonable degree of medical certainty; the supportive reasoning as to why he has come to a particular conclusion should also be stated.

"(f) During the course of treatment, the doctor should indicate when the patient has reached maximum medical improvement and he should also specifically spell out, when indicated, what work restrictions he deems appropriate for the injured worker. A simple statement that the individual is now a candidate for vocational rehabilitation or is permitted to return to work in a light duty status is wholly inadequate and is simply an invitation for future difficulties inuring to the detriment of all concerned.

"In order to facilitate certainty regarding return to or re-employment, the physician should, in his initial report, include a recitation of the patient's occupation and a brief description of what those duties consist of, as related by the claimant. Thus, when the doctor indicates that the injured worker is able to return to his regular work duties, it would be apparent to all concerned that the practitioner was cognizant of what those work duties were. In all of his reports, the doctor should recite whether or not the patient is still totally disabled from his regular work duties, thus addressing the question of whether or not the worker is entitled to continuing total disability benefits.

"(g) Finally, the doctor should *promptly* report the

results of his initial evaluation and ongoing office visits so as to insure proper ongoing medical care plus continuation or cessation of benefits. If the doctor, for example, feels that additional testing is warranted, including hospitalization, the carrier may be forced to decline same unless it has a report justifying the additional measures recommended by the treating physician. Doctors should also be readily accessible to the carrier, the patient, the attorneys, or any vocational rehabilitation specialist, to the extent feasible, so as not to impede proper medical management and continuation of benefits. In addition, forms sent in pertaining to allowable work activities, collateral insurance, Social Security, etc., should be promptly completed and returned, with a copy being kept in the doctor's file.

"(h) At some point along the way, the carrier may well elect to have an independent medical evaluation and the treating physician should be receptive to discussing his continued treatment and other conclusions with such examining physician. Even though they may professionally disagree and come to opposite conclusions, the channels of communication should remain open, not only for the benefit of the patient, but also for discussion of available treatment options.

§ 10-3(D)(2). Pointers for Carrier's Proper Management of Industrial Low Back Injury Cases.

"(a) Upon receipt of the notification of injury, medical authorizations should be promptly secured from the claimant. Carrier investigation should include the taking of a signed or recorded statement detailing specifically how the injury occurred, to whom it was reported, etc., and those sources should be checked for confirmation. A thorough exploration should be undertaken touching upon all prior medical problems, serious illnesses, workmen's compensation claims, personal injury claims, medical malpractice claims, and submissions for non-work related problems to group health carriers, union welfare funds, etc. The policy numbers and identities plus addresses of union trustees, etc., should be ascertained. The identity of claimant's current and past family physician should be secured as well. Appropriate follow-up should be done where warranted, particularly in potentially serious back injuries.

"(b) Where there is some question, from a medical standpoint, as to whether or not the injury is work related, there should be an immediate evaluation by a selected specialist in that field. That doctor should then be furnished with ongoing progress reports from the treating physician and asked to comment on the reasonableness and necessity of the treatment being rendered or suggested. Follow-up evaluations should be scheduled to see if the carrier's doctor concurs in the treatment or the claimant's ability to return to gainful employment."

§ 10-3(E). Summary.

The successful treatment of industrial low back injuries requires the cooperation of the claimant, the treating physician, the attorneys and the insurance carrier. Even these efforts, however, may not always result in a positive outcome for all parties involved. If each party is made more aware of the important considerations enumerated herein, at least some cases may be more effectively handled, providing a better result in terms of superior treatment for the claimant and more reasonable exposure of the carrier and self-insured.

MEDICAL REPORT OUTLINE

Based on experience in my cases, I would suggest the following points be covered in every medical report:

1. Claimant's full name, age, occupation, and the physical requirements of his particular employment.

725

2. Prior medical history. (This should include the details of prior injuries or diseases which produced a residual deficit, with an explanation of the specific complaints and limitations that were imposed by virtue of the prior episodes.)

3. Description of the accident in question with some detail as to exactly how the accident occurred and the nature of the trauma suffered with specifics as to the parts of the body.

4. Present complaints.

5. Clinical examination, explained in lay terms for the benefit of nontrained medical reviewers.

6. X-ray examinations, again in lay terms, with interpretations as to the extent of any prior conditions found on x-rays and how these conditions impact on the extent of the disability sustained by the current trauma.

7. Diagnosis related in lay terms so that the full nature of the diagnosis is readily understood, together with the specific limitations imposed.

8. Recommendations for treatment and a reasonable prognostication as to how long the claimant will be under active medical care.

9. Apportionment, if appropriate during examination, as to the relative contributions of the prior conditions and the present trauma to the overall incapacity of the claimant.

10. Rehabilitation. (Suggestions as to what any rehabilitative actions should be considered, as well as referral to allied medical specialists to assist in preventing the claimant from becoming depressed or becoming involved in secondary gain as a result of the accident.)

If these points are carefully considered by the examining physician, the need for clarification and supplemental reports may be alleviated and the contents of the report would provide appropriate information for development of alternative employment or consideration for rehabilitation.

§ 10-4. Industry's Perspective.

Leslie D. Michelson

The rapid escalation in workers' compensation costs during the past decade and increasing concerns about the quality of health care received by injured employees have moved workers' compensation issues up in the priorities of corporate executives. Traditionally, workers' compensation has been regarded as a cost of doing business over which there was little control. A worker was injured; he received medical care and the employer accepted the quality of the care and costs without accurately monitoring either one. Without controls, inappropriate care was many times rendered and the costs have escalated. Benefits today exceed $15 billion per year and American industry is spending $23 billion per year on workers' compensation.

Due to these increasing economic losses, top executives are presently becoming interested in trying to control and manage their health care problems. The purpose of this section is to first present the workers' compensation issue from the employer's perspective. Next is to establish the goals for controlling the workers' compensation problem by the employer. Finally, various solutions to attain these goals will be discussed.

§ 10-4(A). Problems.

The overall problem facing industry is the existence of a large population of patients at multiple locations under treatment by a broad spectrum of physicians. Most employers are not equipped to monitor the health care delivered to their employees. The *major* problem is that there is no standardized protocol for diagnosis and treatment of low back pain in industry. Diagnostic criteria and treatment regimens are haphazard and may vary from patient to patient in the same doctor's office. Simply stated, different doctors treat similar cases very differently. This makes it difficult, if not impossible, for an employer or insurer to manage health care problems.

Perhaps the best illustration of this difficult feature of the health care profession in presented in the work of John E. Wennberg, which formed the basis for an issue of *Health Affairs.* Wennberg has documented systematic and persistent regional differences in the rates of use for many common surgical procedures and other medical services in the United States. As Wennberg describes it:

> Most people view the medical care they receive as a necessity provided by doctors who adhere to scientific norms based on previously tested and proven treatments. When the contents of the medical care "black box" are examined more closely, however, the type of medical service provided is often found to be as strongly influenced by subjective factors related to the attitudes of individual physicians as by science. These subjective considerations, which I call collectively the "practice style factor," can play a decisive role in determining what specific services are provided a given patient as well as whether treatment occurs in the ambulatory or the inpatient setting. As a consequence, this style factor has profound implications for the patient and the payer of care (entry 2 in Bibliography at end of this chapter).

The examples are dramatic. According to Wennberg, depending on the hospital market in which one is being treated, 20% to 70% of the women under 70 have had a hysterectomy, 15% to 60% of the men under 85 have had a prostatectomy, and 8% to 70% of the children have undergone a tonsillectomy. A study of tonsillectomies in New York provides a vivid illustration:

> A survey of 1,000 eleven-year-old school children in New York City found that 65% had undergone tonsillectomy. The remaining children were sent for examinations to a group of physicians and 45% were selected for tonsillectomy. Those rejected were examined by another group of physicians and 46% were selected for surgery. When the remaining children were examined again by another group of physicians, a similar percent were recommended for tonsillectomy, leaving only sixty-five students. At that point, the study was halted for lack of physicians (entry 1).

728

In addition, the employer has several other obstacles in obtaining control of his health care system. First, most of the people who would be charged with this responsibility do not have the medical training necessary to review and make decisions about the quality and appropriateness of the health care, particularly in complex cases. They typically are responsible for large numbers of cases and are encouraged to handle even more, making it difficult to find the time needed to follow a case carefully and continuously.

Second, the information typically provided by treating physicians is insufficient to allow for meaningful review. The common report: "Patient continues to have spasms and intermittent pain. Off work for two more weeks," is not very helpful to anyone seriously interested in the patient's case. If an employer is to monitor a case effectively, he must have an organized data base that is routinely completed each time a patient is evaluated.

Finally, even if accurate data are sent to the employer, he usually lacks the means to store the information and monitor it on an ongoing basis. Although most large industries have computerized information systems in respect to their billing procedures and inventories, the vast majority keep handwritten medical charts. Most can extract the information by reading each record, and in most cases without consistency. Thus a computerized system is indicated. This would help each industry to easily and accurately identify its major medical problem areas and would give employers a solid basis for comparison when instituting new therapy. To control medical problems, managers must know the problems' dimensions, and accurate and timely data are needed to provide these measurements.

§ 10-4(B). Goals.

Once management has become interested in improving its workers' compensation program and is knowledgeable about the problems, the next question is to determine just what the goals of such a program should be. Ambiguity in the goals, inappropriate goals, or even worse, the failure to set goals, will inevitably result in an ineffective program.

In light of the trend of rapid cost escalation, it is tempting for management to set cost containment as its single goal. Although that is an entirely appropriate focus, it is not a sufficiently detailed goal to provide guidance in managing a worker's compensation program. It leaves many questions unanswered. What costs are to be contained? Benefit costs, medical costs, indirect costs, morale costs, human costs, or all of the above? How are they to be contained? By trying to minimize injuries? By identifying malingerers? By reducing the amount of health care claimants received? By increasing the amount of health care received? Each of these questions is answered, at least implicitly, in every workers compensation program and each answer has different consequences. The four primary goals presented can be applied universally to all size industries and are fair to both the employer and employee:

1. *Quality Health Care* — The first objective for any successful workers' compensation program should be to make certain that all injured employees receive quality health care. This means that the injury should be diagnosed promptly and accurately and that the claimant should receive the indicated treatment for the injury. In addition, the progress of the case should be followed continuously and carefully to see that the selected treatment modalities are having the expected results and to make sure the claimant is referred to a specialist when necessary.

2. *Appropriate Health Care* — Consistent with the first objective, a company should make sure that injured workers receive appropriate health care. A company should try to avoid the wasted expense of ineffective

730

treatment, such as passive physical therapy; of premature or unnecessary treatment, such as a myelogram, CAT scan, or surgery within the first six weeks for a patient with a herniated disc; or of insufficient treatment, such as bedrest without other diagnostic tests or treatments for a patient suffering from an undiagnosed spinal tumor. In each of these instances, as well as in many others, the inappropriate care — whether it be too much, too little, too early, or too late — results in unnecessary risk or exposure for the claimant and unnecessary expense for the employer.

3. *Rapid Return to Maximum Function* — The ultimate goal for the management of each case should be returning the claimant to maximum function at the earliest opportunity. The key to achieving this goal is understanding from the outset what the claimant's maximum improvement is likely to be, predicting when that will occur, and planning to make sure that the patient is prepared to re-enter the work force at that time. Most patients with herniated discs, for example, will never be able to return to heavy duty without unnecessary pain and risk of reinjury; but many such patients will be able to return to restricted duty in a reasonable amount of time. Similarly, a patient with a mild acute back strain should be ready for full duty within a few weeks. Establishing these guidelines and communicating them to the patients and physicians from the outset is the most effective technique for making the results conform to the expectations.

4. *Cost Containment* — As may be apparent, designing a workers' compensation program to achieve the goals of quality care, appropriate care, and rapid return to maximum function is the surest way to achieve cost

containment. When each claimant's health care is carefully monitored to ensure that the patient is being diagnosed and treated properly, cared for in a cost effective way, and returned to duty as rapidly as possible, all of the cost components of workers' compensation — direct and indirect, quantifiable and unquantifiable — will reflect savings.

§ 10-4(C). Solution.

There are a number of potential methods which can be employed to obtain an effective compensation low back program and these include safety, ergonomics and rehabilitation as discussed in Chapter 8. The emphasis in this section, however, is on how the acute injury is managed from the first day that it occurs until its conclusion. Other strategies that seek to reduce the likelihood of injury (preventive and rehabilitative programs) have their place, but regardless of the success of these efforts, there inevitably will be a finite number of injuries.

The basis for a solution is a consistent clinical approach that can be used to guide both physicians and employers in the management of low back care. The approach is reflected in a protocol, or algorithm (Chapter 4), that sets forth the sequence of diagnostic evaluations, treatment modalities, and patient responses that should be expected. This standardized protocol has been computerized and can be used in monitoring the medical and compensation aspects of an individual low back case. Using the computer is attractive because it allows one to track unlimited numbers of cases in a standardized fashion.

The algorithm is not intended to supplant physician judgment about an individual case or to suggest that all physicians agree on all aspects of the management of a low back patient. Rather, it is intended to incorporate generally accepted outer boundaries of treatment. The algorithm recognizes that there are areas about which reasonable physicians may disagree and it accepts the resulting variation. It is not so flexible, however, that it permits evaluations or treatments that are clearly inappropriate in light of existing medical knowledge.

This computerized approach has been applied to industry (entry 3). Two groups were studied in prospective fashion: 5,300 employees at the Potomac Electric Power Company for two years and 14,000 United States Postal Service employees for one year. All work related back injuries were monitored based on the algorithm. If the patient's diagnosis, treatment or clinical progress differed from that predicated by the algorithm, an independent medical examination was obtained in a timely fashion. The results were dramatic and encouraging:

> Both groups demonstrated significant and continuous reductions in number of incidents, in days lost from work, in low-back surgery, and in financial costs. The number of low back pain patients at PEPCO decreased 29% the first year and 44% the second; days lost from work decreased 51% the first year and 89% the second; low-back surgery dropped 88% the first year and 76% the second. Results for the U. S. Postal Service demonstrated a decrease in the number of low back pain patients (41%), in days lost from work (60%) and in financial costs (55%) (entry 3).

Based on these results, this program has been enhanced for use by large employers throughout the country. It is a very attractive system because the program is automated and provides sophisticated medical guidance from the patient's first injury day. This greatly expands the productivity of claims examiners and company doctors and nurses, and supplements their medical training. Also the program uses carefully designed forms to facilitate complete, accurate, and rapid data collection. Thus, employers can obtain the data necessary to review cases in a meaningful way. Finally, the program includes standards of care to ensure quality. By using the program, an employer will no longer have to mediate disputes among expert physicians or endorse a course of treatment just because it is accepted by some physicians in the area. The program continually compares the recommended course of treatment with accepted standards of care and identifies non-conforming cases early in the patient's course.

This type of program achieves the four goals set forth above. First, it ensures that all claimants receive quality health care. The program builds a record on every case in a methodical and organized fashion from the outset. Cases that may become problems can be identified early and referred for expert evaluation within days or weeks when there are still viable options, rather than in months or years, after there may have been surgery or else psychological problems or unnecessary complications have developed. Similarly, those patients who may have been misdiagnosed can be identified and examined by experts.

Second, the program assures appropriate care. Because it builds a profile of each claimant and compares it to the algorithm, it can provide guidance as to when a procedure is being used inappropriately, i.e., too early, too often, or ineffectively under the circumstances. In addition, when it appears that the worker is more seriously ill than his diagnosis would indicate — the patient failed to respond to treatment or to return to work in the time period expected — the patient can be identified for the appropriate expert evaluation.

Third, the program achieves the objective of rapid return to maximum function by setting a goal for resumption of work, continuously monitoring the progress to that goal, and intervening if the patient fails to achieve the goal.

Finally, the program achieves cost savings by ensuring that all patients receive quality health care, appropriate health care and return to maximum function at the earliest opportunity. As the pilot programs at PEPCO and USPS demonstrate, continuous monitoring of health care to achieve these goals can result in dramatic savings in benefits, efficiency, morale, and human costs.

This type of program is currently being expanded and developed for other areas of the body. The cervical spine (neck) has been tested (entry 4). Total body programs of this type will probably be the standard for medical practice in the 1990's.

§ 10-4(D). Conclusion.

The rapid escalation in workers' compensation costs has quite properly resulted in workers' compensation programs moving up in the hierarchy of corporate concerns. To implement a successful workers' compensation program one should look beyond the simplistic notion of cost containment to the real goals of quality care, appropriate care, and rapid return to maximum function. Programs designed to achieve these goals inevitably will also achieve cost containment. The key obstacles to the achievement of these goals, however, are the absence of accurate data from treating physicians, and the wide variations in practice.

Sophisticated, computerized, monitoring programs are available to overcome these obstacles and achieve these goals. The key to the program's success is based on the constant monitoring of the patient's medical problem so that intervention, when necessary, can be instituted early rather than late in the patient's clinical course.

BIBLIOGRAPHY

1. Eddy, D.M. Variation in physician practice: The role of uncertainty. *Health Affairs,* Vol. 3, 2:76-89, 1984.
2. Wennberg, J.E. Dealing with medical practice variation: A proposal for action. *Health Affairs,* Vol. 3, 2:6-32, 1984.

735

3. Wiesel, S.W., Feffer, H.L., and Rothman, R.H. Industrial low back pain: A prospective evaluation of a standardized diagnostic and treatment protocol. *Spine,* 9:199-203, 1984.
4. Wiesel, S.W., Feffer, H.L., and Rothman, R.H. The development and prospective application of a cervical spine algorithm to industrial patients. Accepted for publication to *Journal of Occupational Medicine,* September 1984.

Index

740

743

748